T0202437

Universitext

Universitext

Series Editors

Sheldon Axler
San Francisco State University

Carles Casacuberta
Universitat de Barcelona

Angus MacIntyre
Queen Mary University of London

Kenneth Ribet
University of California, Berkeley

Claude Sabbah
École Polytechnique, CNRS, Université Paris-Saclay, Palaiseau

Endre Süli
University of Oxford

Wojbor A. Woyczyński
Case Western Reserve University

Universitext is a series of textbooks that presents material from a wide variety of mathematical disciplines at master's level and beyond. The books, often well class-tested by their author, may have an informal, personal even experimental approach to their subject matter. Some of the most successful and established books in the series have evolved through several editions, always following the evolution of teaching curricula, to very polished texts.

Thus as research topics trickle down into graduate-level teaching, first textbooks written for new, cutting-edge courses may make their way into *Universitext*.

More information about this series at http://www.springer.com/series/223

Arturo Kohatsu-Higa · Atsushi Takeuchi

Jump SDEs and the Study of Their Densities

A Self-Study Book

Springer

Arturo Kohatsu-Higa
Department of Mathematical Sciences
Ritsumeikan University
Kusatsu, Japan

Atsushi Takeuchi
Department of Mathematics
Tokyo Woman's Christian University
Tokyo, Japan

ISSN 0172-5939 ISSN 2191-6675 (electronic)
Universitext
ISBN 978-981-32-9740-1 ISBN 978-981-32-9741-8 (eBook)
https://doi.org/10.1007/978-981-32-9741-8

© Springer Nature Singapore Pte Ltd. 2019
This work is subject to copyright. All rights are reserved by the Publisher, whether the whole or part of the material is concerned, specifically the rights of translation, reprinting, reuse of illustrations, recitation, broadcasting, reproduction on microfilms or in any other physical way, and transmission or information storage and retrieval, electronic adaptation, computer software, or by similar or dissimilar methodology now known or hereafter developed.
The use of general descriptive names, registered names, trademarks, service marks, etc. in this publication does not imply, even in the absence of a specific statement, that such names are exempt from the relevant protective laws and regulations and therefore free for general use.
The publisher, the authors and the editors are safe to assume that the advice and information in this book are believed to be true and accurate at the date of publication. Neither the publisher nor the authors or the editors give a warranty, expressed or implied, with respect to the material contained herein or for any errors or omissions that may have been made. The publisher remains neutral with regard to jurisdictional claims in published maps and institutional affiliations.

This Springer imprint is published by the registered company Springer Nature Singapore Pte Ltd.
The registered company address is: 152 Beach Road, #21-01/04 Gateway East, Singapore 189721, Singapore

Preface

The subject of jump-type stochastic processes has seen exponential growth in applications during the past 50 years. Before that, it was mostly restricted to certain areas of operations research and insurance mathematics. Nowadays it is commonly applied to particle systems, neurology, geology and finance, among many other disciplines. Therefore, it appears naturally in many applications problems.

On the other hand, it is a subject that may be easier to start studying, in comparison with Brownian motion, as the dynamics involved are much simpler at the beginning.

For this reason, during the past few years, we have introduced this subject to advanced undergraduate students and graduate students. The starting point of this book was a seminar series directed by the first author at the Department of Mathematical Science, Ritsumeikan University, given to advanced undergraduate students. This book consists of two parts: the first part focuses on the stochastic calculus for Lévy processes, while the second part studies the densities of stochastic differential equations with jumps. The joining theme is the study first of densities of Lévy processes and then of the densities of solutions of equations driven by Lévy processes.

Our goal in the first part of the book is to give a simple and at the same time, a somewhat broad overview of different types of jump processes starting from basic principles. First, we start introducing simple Poisson processes with some basic properties related to their jump times. Then, one quickly proceeds to the construction of compound Poisson processes, and there we study their dynamics and introduce the Itô formula in a simple way. That allows a quick introduction to stochastic calculus which deals with the time evolution of these processes. We also discuss stochastic equations driven by compound Poisson processes which can be simply understood as random difference equations. Also important is the introduction of Poisson random measures and the definition of Lévy processes so that a more general structure to be introduced later is laid down gradually.

Through these two chapters, a first goal is achieved. That is, to grasp gradually a first idea of the interest of studying Itô's formula and stochastic equations and the tools that will allow the study of more general Lévy-type processes. In these two chapters, a number of important exercises are proposed with some hints of solutions, which appear at the end of the book. This is done on purpose so that readers try to enhance their knowledge through self-questioning and simple exercises. In this sense, our primary goal was not to design a book to be used as a textbook (which can be done too) but a self-study book.

Therefore, an important warning for readers: Do not skip the exercises! With this in mind, we propose hints in a separate chapter to promote interaction of the reader with the text. We have also tried to give some explanations on the meaning of various important results. This is done because we have in mind a reader, who does not have a lecturer guiding the reading. If at some point you do not understand completely such an explanation, do not despair! It is probably our fault.

As in other textbooks, we also propose complementary exercises which may be used to deepen your understanding of the subject. These are named "Problem" throughout the text.

After two introductory chapters, we start describing the arguments through a limit procedure in order to construct processes with the infinite frequency of jumps. This is done in two steps: First, for processes of finite variation and then processes of infinite variation. Again, our focus is not on generality but on simple principles that will allow the young readers to understand the subject. Again, the topics of Poisson random measures, Itô's formula and stochastic equations appear. In particular, we clearly explain the need for compensation in order to define jump processes with infinite variation. It is important to note here that the experience of the results and exercises in Chaps. 5 and 6 are strongly used towards the end of Chap. 6 and many results concerning the Itô formula and stochastic equations have to be developed by the reader as exercises.

Up to this point, the processes introduced so far are one dimensional. At the end of Chap. 6, we also describe other types of jump processes such as non-homogeneous Poisson processes, multi-dimensional Lévy processes and subordinated Brownian motions.

As a preparation for the second part of the book we have introduced in Chap. 7, the study of flows associated with stochastic equations. The style here is also modern, and aligns with current techniques striving for a simple proof which uses approximations.

This could be also used as the entrance door to more advanced courses or advanced textbooks (e.g., [2, 8, 41, 51, 52]). A reader interested in learning the deeper and exciting theory of jump processes and its associated calculus should proceed this way.

This chapter also serves as an introduction to the second part of the book, where we will be interested in the infinite-dimensional integration by parts formula for stochastic equations driven by jump processes. This topic is not available in the previous given references.

In Chap. 8, we briefly give a general overview of the second part of the book.

The second part of the book is slightly more advanced, although the first three chapters use only very basic principles. It tries to take the reader into learning an infinite-dimensional integration by parts formula which can be used at research level. Chapter 9 introduces the analytical techniques through Fourier analysis that one may use in order to study densities. This also serves to understand why the integration by parts formula is important in order to understand the behavior of densities of random variables associated with solutions of stochastic equations. Chapter 12 is written in a simple style trying to introduce the basic ideas of the integration by parts formula in infinite-dimensional analysis. It definitely differs from other books on the subject, because we strive to explain various approaches using simple ideas and examples and avoid the use of advanced mathematics or giving a general theory to deal with the most general situations.

We proceed in Chap. 11 to apply these ideas to some simple examples that appear in the calculation of the Greeks in finance. This may be also interpreted and used in other jump models and gives the ideas that will be fully developed in subsequent chapters.

Chapter 12 is the entrance door to a different level of complexity. Here, we start describing the Norris method (cf. [45]) for the integration by parts formula. Towards the end of this chapter we compare this technique with the Bismut method, which is based on Girsanov's change of measure theorem in order to achieve an integration by parts formula.

Then we proceed to apply this technique to solutions of stochastic equations. In a final challenging chapter for the reader (as it was for the authors) we use as a research sample a problem for which the application of all the introduced concepts is not easy. We took the model studied by Bally-Fournier [7] related to the Boltzmann equation in physics.

Through this streamlined path, we hope that the book will help young students to get a glimpse of what jump processes are, how they are studied and to achieve a level which may be close to a research-level problem. Clearly, the present book does not discuss many other interesting topics related to jump processes. Again, our goal was not to give a broad view of the subject with details but just draw a path that may allow the inexperienced reader to have a glimpse of one research topic.

For other detailed properties of Lévy processes, we refer the reader to other books (e.g., [8, 41, 51]). For books dealing with other related techniques in the area of infinite-dimensional calculus for jump-type processes, we refer to e.g., [9, 15, 31]. Notably, we do not touch on the topic of continuous-type Lévy processes such as

Brownian motion. The reason for this is that we believe that the path characteristics of Brownian motion may cloud the explanation about jump processes. Also, we believe that there are many other textbooks that the young reader may use in order to obtain some basic information on Brownian motion, its basic properties and associated stochastic calculus and stochastic equations. See [30, 34, 43, 44, 46, 49, 53].

The style of the text is adapted to self-study and therefore an instructor using this book may have to streamline the choice of topics. In this sense, we have modeled the book so that readers who may be studying the topic on their own can realize the importance of arguments that otherwise one may only see once in an advanced book.

For instructors using this book:

1. The text has repetitive ideas so that the students will find how they are used in each setting and understand their importance.
2. The first part of the book is basic, while the second is more advanced and specialized, although it uses basic principles.
3. We assume knowledge of basic probability theory. No knowledge of stochastic processes is assumed, although it is reccomended.
4. Some of the deeper results about measurability or stochastic analysis are given but not proven. Such is the case of the Burkholder–Davis–Gundy inequality for jump processes or Lebesgue–Stieltjes integration.
5. There are certain issues that are avoided in the text in order not to distract the reader from the main text. These are, for example, the BDG inequalities, modifications of stochastic processes, etc. They are briefly mentioned in the text referring to other textbooks.
6. As a way to guide the reader through the book in the best order some possibilities are proposed in the following flowchart. The dotted lines mean that they are not essential to design a course but they could be used as topics to conclude the instruction. The topics at Level 1 are easily accessible for a student with basic knowledge of probability theory. In Chap. 5, one may exclude Sects. 5.4, 5.5 and 5.6 in a first reading.

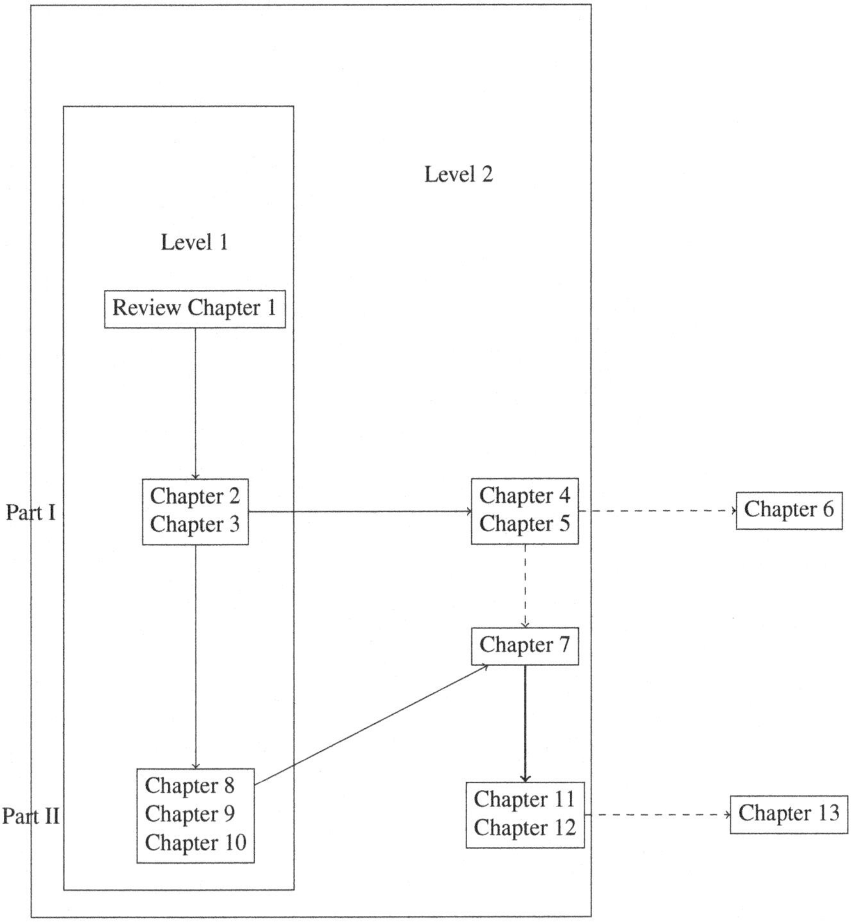

7. In Sect. 5.3 and Chap. 7, we hint at how the proofs can be carried out using the experience built in previous chapters. Still, one may use [2] in order to complement the proofs which are not given in detail.

Kusatsu, Japan
Tokyo, Japan
April 2019

Arturo Kohatsu-Higa
Atsushi Takeuchi

Acknowledgements

AK-H would like to thank all the colleagues and students, who have shared their views with him about this project, in particular, all the students who actively participated in the seminar series at Ritsumeikan University. Their comments and active participation have shaped the current form of the book.

AT used these notes as a basis for lectures at Osaka City University. That helped the book to be definitive. He thanks the students who participated in these lectures.

Contents

Notations

Throughout the text, we use the following notation. Other notation which is not basic to this text and which is defined through the developed theory can be found at the end of the index.

A^*	The transpose of the matrix A
$I_d \equiv I$	The identity matrix on $\mathbb{R}^d \otimes \mathbb{R}^d$ is the identity matrix
e_k	$e_k = (0, \ldots, 0, 1, 0, \ldots, 0)^*$. The k-th unit vector in $\mathbb{R}^d$ for $k = 1, \ldots, d$
$\mathbb{R}_0^m$	$:= \mathbb{R}^m \backslash \{0\}$
$\mathbb{R}_+$	$:= [0, \infty)$
$\mathbb{N}^*$	$:= \mathbb{N} \cup \{0\}$
r.v.	Random variable or random vector
i.i.d.r.v.'s	Independently and identically distributed random variables (vectors)
a.s.	Almost surely
$\mathcal{M}(\Sigma)$	Space of Σ-measurable (real-valued) functions
$\mathcal{M}_b(\Sigma)$	Space of bounded Σ-measurable functions
(*LHS*)	When dealing with long equalities. Left-hand side. Similarly (*RHS*)
càdlàg	Functions whose domain is included in $\mathbb{R}$, which are right-continuous with left-hand limits
$\sharp(A)$	The cardinality operator, which is the number of elements in the set A. Sometimes may also be denoted by $\|A\|$ if no confusion arises
$\langle x, y \rangle \equiv x \cdot y$	Inner product between two vectors x and $y \in \mathbb{R}^d$
x^*	The transpose of the vector x. All vectors are column vectors unless stated otherwise
$\|x\|$	The norm of the vector $x \in \mathbb{R}^d$
$\|\|A\|\|$	The norm of the matrix $A := \sup\{\|Ax\|; \|x\| = 1\} = \max\{\lambda; \exists x \neq 0, Ax = \lambda x\}$

$\mathscr{B}(\mathscr{S})$	The Borel σ-algebra (σ-field) on S. In the case that $\mathscr{S} = \mathbb{R}$ we use $\mathscr{B} := \mathscr{B}(\mathbb{R})$
$\mathscr{B}_1 \otimes \mathscr{B}_2$	The product σ-field completed with the subsets of sets of measure zero
$\sigma(\mathscr{A})$	The σ-field generated by the collection of random variables $\mathscr{A}$ or collection of sets $\mathscr{A}$. This σ-field is always assumed to be completed with the null sets of the probability space
$Leb(A)$	The Lebesgue measure of the Lebesgue measurable set A
$\mathbb{E}[X], \mathbb{E}(X)$	Expectation of the random variable X
$\mathbb{E}[X\|\mathscr{F}]$	Conditional expectation of X conditioned to $\mathscr{F}$
$X \stackrel{\mathscr{L}}{=} Y$	The random variables or processes X and Y have the same law
$Exp(\lambda)$	Exponential distribution with parameter λ
$Gamma\,(\theta, r)$	Gamma distribution with scale parameter θ and shape parameter r
$X \sim D$	X is a random variable with distribution D
$\delta_a(A)$	The Dirac point mass measure evaluated at the set A
$\mathbf{1}_A(x)$	Indicator function. In the case that $A \subset \Omega$ we may simplify the notation to $\mathbf{1}_A$. Sometimes we may also use, e.g., $\mathbf{1}\,(x > 0)$
$\operatorname{sgn}(x)$	$\operatorname{sgn}(x) := \mathbf{1}_{(x>0)} - \mathbf{1}_{(x<0)}$. The sign (or signum) function
$X_{t-} = \lim_{s\uparrow t} X_s$	Limit on the left of X at t
$X^+ := X \vee 0$	The positive part of X
$X^- := (-X) \vee 0$	The negative part of X
$\lfloor x \rfloor$	The floor function (also called greatest integer function)
$\mu * \nu$	The convolution of measures μ and ν
μ^{n*}	$\mu^{n*} = \mu * \cdots * \mu$. The n-th convolution power of the measure μ. Similarly, we use f^{n*} as the n-th convolution power of the function f
$\mu \otimes \nu$	The product of the measures μ and ν
$\mu^{\otimes n}$	Similarly, defines the n-th product of the measure μ with itself
$\\|f\\|_\infty$	$\\|f\\|_\infty := \sup_x \|f(x)\|$. The uniform norm
$\int_{\mathbb{R}^d} f(x)dx$	Multi-dimensional integral for $f : \mathbb{R}^d \to \mathbb{R}$
$\partial_\alpha f$	Given the multi-index $\alpha = (\alpha_1, \ldots, \alpha_\ell)$ with length $\|\alpha\| = \ell$, this symbol denotes the multiple partial derivative of f with respect to the variables indicated in α. Sometimes, we may use the notation $\partial_z f$ to indicate the gradient derivative with respect to the variable z. In the one-dimensional case, we use f' for $\ell = 1$
$(\partial_z)_a f$	This is a short notation for $\partial_z f(a)$ used when f depends on many variables
∇f	The gradient of f. In the one-dimensional case may be used to indicate derivative with respect to space variables
$\operatorname{div} f$	The divergence of a function f with respect to all its variables. We may use $\operatorname{div}_z f$ if f depends on other variables and we want to take the divergence with respect to the vector variable z

$C^k(A;B)$	The space of $[k]$ times continuous differentiable functions with $[k]$-th order derivative which is a $k-[k]$-Hölder continuous function
$C_0^k(A;B)$	The subspace of functions in $C^k(A;B)$ which tend to zero at infinity
$C_c^k(A;B)$	The subspace of functions in $C^k(A;B)$ with compact support
$C_b^k(A;B)$	The subspace of functions in $C^k(A;B)$ with bounded derivatives. Note that the functions themselves do not need to be bounded
$C_p^k(A;B)$	The subspace of functions in $C^k(A;B)$ with derivatives which may grow at most with a polynomial order $p \in \mathbb{N}$
$B_r(x)$	The ball in $\mathbb{R}^d$ of center $x \in \mathbb{R}^d$ with radius $r \geq 0$. Sometimes, we simplify the notation with $B_r = B_r(0)$
$L^p(\mathbb{R}^d)$	$L^p(\mathbb{R}^d) \equiv L^p(\mathbb{R}^d;\mathbb{R}^k)$. The space of $\mathbb{R}^k$-valued measurable functions defined on $\mathbb{R}^d$ with respect to the Lebesgue measure such that its p moment is finite
$L^p(\Omega;\mathbb{R}^k)$	The space of random vector of dimension k under a measure $\mathbb{P}$ which is understood from the context. Sometimes we may use the simplified notation $L^p(\Omega)$
$\phi(x)$	$\phi(x) := \frac{1}{(2\pi)^{d/2}}\exp(-\frac{\lvert x\rvert^2}{2})$. The density for the standard Gaussian distribution with mean zero and identity covariance matrix
$\Phi(x)$	The distribution function associated with the standard Gaussian distribution with mean zero and identity covariance matrix

We use i to denote the imaginary unit. Sometimes the same symbol is used as an index $i \in \mathbb{N}$. The meaning should be clear from the context. In the space of complex numbers, $Re(z)$ and $Im(z)$ denote the real and imaginary parts of the complex number $z \in \mathbb{C}$.

When describing collections of items such as functions, stopping times or random variables which are understood as processes, we may use one of the following various notations:

$$X \equiv \{X_t; t \in A\} \equiv \{X_t\}_{t\in A}$$

Domains of integration are sometimes explicitly written. Otherwise, we may omit it in cases where it is clear what the domain of integration (usually the whole space) should be from the context. A similar statement is valid for the notation of function spaces and their range of values.

Chapter 1
Review of Some Basic Concepts of Probability Theory

In this chapter many mathematical details or proofs are not given so we refer the reader to the appropriate references in basic probability theory. See for example [10, 60].

We will always work on a complete probability space $(\Omega, \mathscr{F}, \mathbb{P})$ which we assume contains all the r.v.s that are required for the arguments. In particular, one may have to consider products of probability spaces in order to construct a probability space containing certain sequences of i.i.d.r.vs.

Also we remark for the inexperienced reader that the monotone class theorem has different versions and therefore you should pick up on the one that is used in each proof. For a list of some versions of this theorem, see Chapter 0 in [49].

Definition 1.0.1 Let $(\Omega, \mathscr{F}, \mathbb{P})$ be a probability space. A mapping $X : \Omega \to \mathbb{R}^d$ is an $\mathbb{R}^d$-valued random vector if it is $\mathscr{M}(\mathscr{F})$-measurable. The measure $P_X(B) := \mathbb{P}(X \in B)$ is called the law or distribution of X. It may be further denoted by F_X when using it as a distribution function $F_X(x) := \mathbb{P}(X_i \leq x_i, i = 1, ..., d)$.

Lemma 1.0.2 *If $\sigma(X_n; n \in \mathbb{N})$ and $\sigma(Y_n; n \in \mathbb{N})$ are independent σ-fields*[1] *and* $\lim_{n\to\infty}(X_n, Y_n) = (X, Y)$ *a.s., then X and Y are independent.*[2]

Exercise 1.0.3 Prove Lemma 1.0.2.

1.1 Characteristic Function

The concept of the characteristic function replaces the moment generating function when one wants to study the fine details of the distribution function of the random variable.[3] If fact, it is not true that there is a one-to-one correspondence between

[1] We assume as usual that these σ-fields are complete.

[2] The notation lim denotes limits for functions or in the case of random variables this denotes limits in the $(a.s.)$ sense. This may also be denoted using the symbol $\to$.

[3] This is somewhat equivalent to replacing Laplace transforms by Fourier transforms.

© Springer Nature Singapore Pte Ltd. 2019

A. Kohatsu-Higa and A. Takeuchi, *Jump SDEs and the Study of Their Densities*, Universitext, https://doi.org/10.1007/978-981-32-9741-8_1

moment functions and distribution functions. This fact, becomes true only when one studies characteristic functions as they are always well defined. The concept of characteristic functions is usually not fully covered in standard undergraduate level courses and for this reason we slightly enlarge the contents with some exercises.

The intention of this section is to give some information so that the reader may grasp a first idea of the concepts. A more detailed discussion is presented in Chap. 9.

Definition 1.1.1 (*Characteristic function*) The characteristic function $\varphi(=\varphi_X)$ of a random variable X is defined as the map $\varphi : \mathbb{R} \to \mathbb{C}$ such that for all $\theta \in \mathbb{R}$,

$$\varphi(\theta) := \mathbb{E}[e^{i\theta X}] = \mathbb{E}[\cos(\theta X)] + i\mathbb{E}[\sin(\theta X)].$$

Note that the above expectations are always well defined. This is the main reason for using characteristic functions in comparison with the moment generating function which most students learn in basic probability courses. See also the next exercise.[4] We may also say sometimes that φ is the characteristic function of the measure P_X. In particular, P_X is the image law $\mathbb{P} \circ X^{-1}$. That is, $P_X(A) = \mathbb{P}(X \in A)$ for $A \in \mathscr{B}$.

Exercise 1.1.2 Let $X = e^Z$ where $Z \sim N(0,1)$ is the standard Gaussian (normal) distribution. In general, $N(a,b)$ denotes the normal distribution with mean a and variance $b > 0$. Prove that $\mathbb{E}[\theta^X] = \infty$ for all $\theta > 0$.

Still the moment function is not that bad. That is, if it exists in a neighborhood of 0 then the corresponding distribution is uniquely characterized by the moment generating function.[5] Note that another version of the moment generating function is the Laplace transform of X which is defined as $\mathbb{E}[e^{\theta X}]$ and is very useful in many situations when it is finite. The Laplace transform can be defined for $\theta < 0$ and X a positive random variable. It is also a very useful tool to solve ordinary and partial differential equations. Still, it is a good idea that one starts to get acquainted with the characteristic function which is always a well-defined function.

Corollary 1.1.3 *Let $\varphi = \varphi_X$ be the characteristic function of a random variable X. Then the following properties are satisfied:*

1. $\varphi(0) = 1$,
2. $|\varphi(\theta)| \leq 1, \forall \theta \in \mathbb{R}$,
3. $\theta \mapsto \varphi(\theta)$ *is uniformly continuous on* $\mathbb{R}$,
4. $\varphi_{(-X)}(\theta) = \overline{\varphi_X(\theta)}, \ \forall \theta \in \mathbb{R}$,
5. $\varphi_{aX+b}(\theta) = e^{ib\theta}\varphi_X(a\theta)$.

Theorem 1.1.4 *If X, Y are independent random variables, then*

$$\varphi_{X+Y}(\theta) = \varphi_X(\theta)\varphi_Y(\theta), \ \forall \theta \in \mathbb{R}.$$

[4] If you have more experience in analysis, maybe you have learned this concept as the so called Fourier transform. Although there is a slight variation here as in most books, one starts with periodic functions.

[5] See e.g. Section 30 in [10].

Exercise 1.1.5 Prove that if $\varphi(\theta)$ is the characteristic function of a random variable X then $|\varphi(\theta)|^2$ and $Re(\varphi(\theta))$ are also characteristic functions. Find random variables associated with these characteristic functions.

Theorem 1.1.6 (Lévy's continuity theorem) *Let X, X_n, $n \in \mathbb{N}$ be a sequence of random variables with characteristic functions φ_X, φ_{X_n}, $n \in \mathbb{N}$. Then $\varphi_{X_n}(\theta) \to \varphi_X(\theta)$ for all $\theta \in \mathbb{R}$ if and only if X_n converges in law to X.*

Theorem 1.1.7 (Bochner theorem) *Let $\varphi : \mathbb{R} \to \mathbb{C}$ be a positive definite function,*[6] *continuous at* 0 *with* $\varphi(0) = 1$. *Then there exists some probability measure so that its characteristic function is given by φ.*

Definition 1.1.8 For a given n-dimensional random vector X, we say that $f : \mathbb{R}^n \to \mathbb{R}_+$ is its density function if for any bounded measurable function with compact support g, we have the following equality:

$$\mathbb{E}[g(X)] = \int g(x) f(x) dx.$$

The above integral is understood as a Lebesgue integral.

Exercise 1.1.9 Suppose that the distribution of X given by $F(x) := \mathbb{P}(X_i \leq x_i; i = 1, ..., n)$ in the above definition is differentiable. Then prove that its derivative $\partial_{(1,...,n)} F(x)$ is the density function of X.

Note that the uniqueness of the density function is only true in the a.s. sense.

Theorem 1.1.10 (Lévy's inversion theorem) *Suppose that φ is the characteristic function of some random variable. Furthermore, assume that it is integrable in the sense that*

$$\int_{\mathbb{R}} |\varphi(\theta)| d\theta < \infty.$$

Then there exists a positive function f such that

$$\varphi(\theta) = \int_{\mathbb{R}} e^{i\theta x} f(x) dx.$$

A sharper statement which is always true is that if F denotes the distribution function of the random variable for which φ is its characteristic function, then

$$F(b) - F(a) = \frac{1}{2\pi} \lim_{T \to \infty} \int_{-T}^{T} \frac{e^{-i\theta a} - e^{-i\theta b}}{i\theta} \varphi(\theta) d\theta. \tag{1.1}$$

[6]That is, $\sum_{i,j=0}^{n} \varphi(\theta_i - \theta_j) z_i \bar{z}_j \geq 0$ for any sequence of real numbers $\theta_i \in \mathbb{R}$ and $z_i \in \mathbb{C}$, $i = 1, ..., n$.

At any rate, this result implies that for every characteristic function only one distribution function can be associated with it. In other words, two different distribution function can not have the same characteristic function.

Another interesting way of looking at this result is that the linear space generated by complex exponential functions contain indicator functions. In fact, on the (RHS) of the above equation we see an integral combination of exponential functions in the form of $\int_{-T}^{T} \frac{e^{-i\theta a}-e^{-i\theta b}}{i\theta} e^{i\theta x} d\theta$ and on the (LHS) we have the $\mathbf{1}_{(a,b]}(x)$. Therefore again, complex exponential functions generate all the space of measurable functions and therefore this is the reason as to why the one-to-one correspondence between characteristic functions and probability measures. We will see more about this in Chap. 9.

Exercise 1.1.11 Deduce from (1.1) that if $\int_{\mathbb{R}} |\varphi(\theta)|(1+|\theta|)^p d\theta < \infty$ for any $p > 0$ then the density function associated with φ exists and is infinitely differentiable with bounded derivatives. A similar result is also valid in the multi-dimensional case.

Theorem 1.1.12 (Characteristic functions and independence) *The sequence of random variables* $X_1, \cdots, X_N$ *are independent if and only if for any* $\xi = (\xi_1, ..., \xi_N) \in \mathbb{R}^N$,

$$\mathbb{E}\Big[\exp\Big(i\sum_{j=1}^{N}\xi_j X_j\Big)\Big] = \prod_{j=1}^{N}\mathbb{E}[\exp(i\xi_j X_j)].$$

Proof We just prove the "only if" part. Let $X = (X_1, \cdots, X_N)$. Then

$$(LHS) = \mathbb{E}\Big[\exp\Big(i\xi^* X\Big)\Big] = \int_{\mathbb{R}^N} e^{i\xi^* x} P_X(dx),$$

$$(RHS) = \prod_{j=1}^{N}\int_{\mathbb{R}} e^{i\xi_j x_j} P_{X_j}(dx_j) = \int_{\mathbb{R}}\cdots\int_{\mathbb{R}}\prod_{j=1}^{N} e^{i\xi_j x_j} P_{X_1}(dx_1)\cdots P_{X_N}(dx_N)$$

$$= \int_{\mathbb{R}^N} e^{i\xi^* x} P_{X_1}\otimes\cdots\otimes P_{X_N}(dx).$$

From Theorem 1.1.10 (Lévy's inversion theorem), we have

$$P_X = P_{X_1}\otimes\cdots\otimes P_{X_N}.$$

Hence,

$$\begin{aligned}
\mathbb{P}(X_1 \in E_1, \cdots, X_N \in E_N) &= \mathbb{P}(X \in E_1\times\cdots\times E_N) = P_X(E_1\times\cdots\times E_N)\\
&= P_{X_1}\otimes\cdots\otimes P_{X_N}(E_1\times\cdots\times E_N) = P_{X_1}(E_1)\cdots P_{X_N}(E_N)\\
&= \prod_{j=1}^{N}\mathbb{P}(X_j \in E_j).
\end{aligned}$$

Therefore $X_1, \cdots, X_N$ are independent.

Exercise 1.1.13 The following exercise may clarify for the inexperienced the role of characteristic functions.

1. Given $a, b \in \mathbb{R}$, prove that the only real solutions p, q for the equation

$$pe^{i\theta a} + re^{i\theta b} = 0, \ \forall \theta \in \mathbb{R}$$

can be classified as
$$\begin{cases} p = 0, \ r = 0; \ \text{if } a \neq b, \\ p = -r; \ \text{if } a = b. \end{cases}$$

Note that the above statement is linked to the linear independence of functions.

2. Let X, Y be two random variables which take two different values a and b. Prove that if $\varphi_X(\theta) = \varphi_Y(\theta)$ for all $\theta \in \mathbb{R}$ then $\mathbb{P}(X = a) = \mathbb{P}(Y = a)$ and therefore the laws of X and Y are equal.
3. Suppose that X takes two values a and b and Y take two values a and c. Prove that if $\varphi_X(\theta) = \varphi_Y(\theta)$ for all $\theta \in \mathbb{R}$ then $c = b$ and $\mathbb{P}(X = a) = \mathbb{P}(Y = a)$ and therefore the laws of X and Y are equal.

Notice that for all the above statements to be valid one does not need to require the validity for all $\theta \in \mathbb{R}$. Propose a statement with weaker requirements. Think about the generalization to discrete random variables X taking n different values respectively and Y taking m different values.

In particular, note the relation with Theorem 1.1.10. In fact, given the characteristic function $\varphi_X(\theta)$ of a random variable X taking the values a and b, then multiplying it by the function $e^{-i\theta a}$ will give that the constant term of $e^{-i\theta a}\varphi_X(\theta)$ corresponds to $\mathbb{P}(X = a)$. This is therefore linked with the residue theorem of complex analysis. On this last point, one has to be careful in the calculation as the point 0 which is a singular point lies on the path of the integral in (1.1).

1.2 Conditional Expectation

X is a random variable on the probability triple $(\Omega, \mathcal{F}, \mathbb{P})$ with $\mathbb{E}[|X|] < \infty$. Let $\mathcal{G}$ be a sub σ-algebra of $\mathcal{F}$. Then there exists a random variable Y such that: (i) $Y \in \mathcal{M}(\mathcal{G})$, (ii) $\mathbb{E}[|Y|] < \infty$, (iii) for every $G \in \mathcal{G}$ with $\int_G Y d\mathbb{P} = \int_G X d\mathbb{P}$.

Definition 1.2.1 (*Conditional expectation*) In the discussion above, the random variable Y satisfying properties (i)–(iii) is called a version of the conditional expectation $\mathbb{E}[X|\mathcal{G}]$.

From the definition, one obtains the a.s. uniqueness of the conditional expectation. That is, for two random variables X_1, X_2 such that $X_1 = X_2$, a.s. then their corresponding conditional expectations are also equal a.s. In this sense the word *version* is used in the above definition.

Theorem 1.2.2 (Some properties of conditional expectation) *Let $\mathscr{G}$ and $\mathscr{H}$ be sub σ-algebras of $\mathscr{F}$. Then all the following statements are true supposing that all random variables appearing in the corresponding statement satisfy the conditions for the existence of conditional expectations:*

1. *If Y is any version of $\mathbb{E}[X|\mathscr{G}]$, then $\mathbb{E}[Y] = \mathbb{E}[X]$.*
2. *If X is $\mathscr{G}$-measurable, then $\mathbb{E}[X|\mathscr{G}] = X$, a.s.*
3. *$\mathbb{E}[a_1 X_1 + a_2 X_2|\mathscr{G}] = a_1\mathbb{E}[X_1|\mathscr{G}] + a_2\mathbb{E}[X_2|\mathscr{G}]$.*
4. *If $X \geq 0$, a.s., then $\mathbb{E}[X|\mathscr{G}] \geq 0$ a.s.*
5. *If $0 \leq X_n \uparrow X$, then $\mathbb{E}[X_n|\mathscr{G}] \uparrow \mathbb{E}[X|\mathscr{G}]$, a.s.*
6. *If $X_n \geq 0$, then $\mathbb{E}[\liminf X_n|\mathscr{G}] \leq \liminf \mathbb{E}[X_n|\mathscr{G}]$.*
7. *If $\mathscr{H}$ is a sub σ-algebra of $\mathscr{G}$, then $\mathbb{E}[\mathbb{E}[X|\mathscr{G}]|\mathscr{H}] = \mathbb{E}[X|\mathscr{H}]$, a.s.*
8. *If Z is a bounded $\mathscr{G}$-measurable r.v., then $\mathbb{E}[ZX|\mathscr{G}] = Z\mathbb{E}[X|\mathscr{G}]$, a.s.*
9. *If $\mathscr{H}$ is independent of $\sigma(\sigma(X), \mathscr{G})$, then $\mathbb{E}[X|\sigma(\mathscr{G}, \mathscr{H})] = \mathbb{E}[X|\mathscr{G}]$, a.s.*

Lemma 1.2.3 *Let X, Y be two random variables on the probability space $(\Omega, \mathscr{F}, \mathbb{P})$. Furthermore, let $\mathscr{G}$ be a sub σ-algebra of $\mathscr{F}$ independent of $\sigma(X)$ and Y be a $\mathscr{G}$-measurable. If $f : \mathbb{R}^2 \to \mathbb{R}$ is a bounded measurable function, then*

$$\mathbb{E}[f(X, Y)|\mathscr{G}] = g_f(Y), \text{ a.s.},$$

where $g_f(y) := \mathbb{E}[f(X, y)]$.

Proof Define a class of functions $\mathscr{H}$ by

$$\mathscr{H} := \{f \in \mathscr{M}_b(\mathbb{R}^2);\ \mathbb{E}[f(X, Y)|\mathscr{G}] = g_f(Y), \text{ a.s.}\}.$$

Here $\mathscr{M}_b(A)$ stands for the set of bounded measurable functions defined in $\mathbb{R}^2$. Then we will show that $\mathscr{H} = \mathscr{M}_b(\mathbb{R}^2)$ by the monotone class theorem (see e.g. [60], Sect. 3.14).

(i) For $f, f_1, f_2 \in \mathscr{H}$, $\alpha \in \mathbb{R}$, since

$$\begin{aligned}
\mathbb{E}[(f_1 + f_2)(X, Y)|\mathscr{G}] &= \mathbb{E}[f_1(X, Y)|\mathscr{G}] + \mathbb{E}[f_2(X, Y)|\mathscr{G}] \\
&= g_{f_1}(Y) + g_{f_2}(Y) = g_{f_1+f_2}(Y), \\
\mathbb{E}[(\alpha f)(X, Y)|\mathscr{G}] &= \alpha g_f(Y) = g_{\alpha f}(Y),
\end{aligned}$$

we have $f_1 + f_2, \alpha f \in \mathscr{H}$, so $\mathscr{H}$ is a vector space on $\mathbb{R}$.

(ii) Since $\mathbb{E}[\mathbf{1}_{\mathbb{R}^2}(X, Y)|\mathscr{G}] = 1 = g_{\mathbf{1}_{\mathbb{R}^2}}(Y)$, $\mathbf{1}_{\mathbb{R}^2} \in \mathscr{H}$.

(iii) If (f_n) is a sequence of non-negative functions in $\mathscr{H}$ such that $f_n \uparrow f$, where f is a bounded function, then using Theorem 1.2.2 (5) we have

$$\begin{aligned}
\mathbb{E}[f(X, Y)|\mathscr{G}] &= \lim_{n\to\infty} \mathbb{E}[f_n(X, Y)|\mathscr{G}] = \lim_{n\to\infty} \mathbb{E}[f_n(X, y)|\mathscr{G}]|_{y=Y} \\
&= \mathbb{E}[f(X, y)|\mathscr{G}]|_{y=Y} = g_f(Y),
\end{aligned}$$

hence $f \in \mathscr{H}$.
(iv) For $(a, b) \times (c, d) \in \mathscr{I} := \{(a, b) \times (c, d); a, b, c, d \in \mathbb{R}\}$,

$$\begin{aligned}\mathbb{E}[\mathbf{1}_{(a,b)\times(c,d)}(X, Y)|\mathscr{G}] &= \mathbf{1}_{(c,d)}(Y)\mathbb{E}[\mathbf{1}_{(a,b)}(X)|\mathscr{G}] = \mathbf{1}_{(c,d)}(y)\mathbb{E}[\mathbf{1}_{(a,b)}(X)]|_{y=Y}\\ &= \mathbb{E}[\mathbf{1}_{(a,b)\times(c,d)}(X, y)]|_{y=Y} = g_{\mathbf{1}_{(a,b)\times(c,d)}}(Y),\end{aligned}$$

hence $\mathbf{1}_{(a,b)\times(c,d)} \in \mathscr{H}$. Therefore $\mathscr{H} = \mathscr{M}_b(\mathbb{R}^2)$.

Definition 1.2.4 The convolution μ of two distributions (or finite measures) on $\mathbb{R}^d$ is a distribution (finite measure) denoted by $\mu = \mu_1 * \mu_2$ and defined by

$$\mu(B) = \int_{\mathbb{R}^d}\int_{\mathbb{R}^d} \mathbf{1}_B(x + y)\mu_1(dx)\mu_2(dy), \quad B \in \mathscr{B}(\mathbb{R}^d).$$

Definition 1.2.5 The convolution of two functions f and g is denoted by $f * g$ and defined as

$$f * g(z) = \int f(z - x)g(x)dx.$$

Here we suppose that the above integral is finite for all z.

Exercise 1.2.6 Prove that if the measures μ_1 and μ_2 are absolutely continuous with respect to the Lebesgue measure with densities f and g respectively then the convolution measure is absolutely continuous and its density is given by the convolution of the two densities.[7]

[7]Note that we are already assuming that equality as functions means equality a.e. Hint: If you want more information you can check also Proposition 6.1.1.

Part I
Construction of Lévy Processes and Their Stochastic Calculus

Chapter 2
Simple Poisson Process and Its Corresponding SDEs

2.1 Introduction and Poisson Process

Poisson processes are generalizations of the Poisson distribution which are often used to describe the random behavior of some counting random quantities such as the number of arrivals to a queue, the number of hits to a webpage etc.

2.1.1 Preliminaries: The Poisson and the Exponential Distribution

As a preliminary, before introducing the Poisson process, we give basic properties of some related distributions.

Definition 2.1.1 (*Exponential distribution*) A positive random variable T is said to follow an *exponential distribution* with parameter $\lambda > 0$ if it has a probability density function of the form

$$\lambda e^{-\lambda t}\mathbf{1}_{[0,\infty)}(t).$$

We denote this fact with $T \sim \mathrm{Exp}(\lambda)$.

Theorem 2.1.2 *Let $T \sim Exp(\lambda)$. Then the distribution function of T is given by*

$$\forall y \in [0,\infty),\quad F_T(y) = \mathbb{P}(T \leq y) = 1 - \exp(-\lambda y).$$

F_T is invertible and its inverse is given by

$$\forall t \in [0,1),\ \ {F_T}^{-1}(t) = -\frac{1}{\lambda}\log(1-t).$$

© Springer Nature Singapore Pte Ltd. 2019

A. Kohatsu-Higa and A. Takeuchi, *Jump SDEs and the Study of Their Densities*, Universitext, https://doi.org/10.1007/978-981-32-9741-8_2

The mean and variance of T is given by

$$\mathbb{E}[T] = \frac{1}{\lambda}, \ Var[T] = \frac{1}{\lambda^2}.$$

Corollary 2.1.3 *If S is an exponential random variable with parameter* 1 *then* S/λ *is an exponential random variable with parameter* λ.

Proof We just check that the distribution function of S/λ forms an exponential distribution. In fact, $\mathbb{P}\left(\frac{S}{\lambda} \le t\right) = \mathbb{P}(S \le \lambda t) = 1 - e^{-\lambda t}$.

Lemma 2.1.4 (Absence of memory) *If T is an exponential random variable then*

$$\forall t, s > 0, \mathbb{P}(T > t + s | T > t) = \mathbb{P}(T > s).$$

Proof We calculate as below:

$$\mathbb{P}(T > t + s | T > t) = \frac{\mathbb{P}(T > t + s, T > t)}{\mathbb{P}(T > t)} = \exp(-\lambda s) = \mathbb{P}(T > s).$$

The following proposition tells us that the only continuous[1] distribution which possesses the *abscence of memory*[2] property is the exponential distribution.

Proposition 2.1.5 *Let T be a positive random variable such that* $P(T > t) > 0$ *for all* $t > 0$ *and*

$$\forall t, s > 0, \ \mathbb{P}(T > t + s | T > t) = \mathbb{P}(T > s).$$

Then T has an exponential distribution.

Proof Let $g(t) := \mathbb{P}(T > t) > 0$. Since $1 - g$ is a distribution function, g is a decreasing and right-continuous function. Since $g(t + s) = g(s)g(t)$ for all $s, t > 0$, we can get different representations for all $n, m \in \mathbb{N}$,

$$g\left(\frac{n}{m}\right) = g\left(\frac{1}{m} + \cdots \frac{1}{m}\right) = g\left(\frac{1}{m}\right)^n$$

$$g\left(\frac{n}{m}\right) = g\left(\frac{n}{m} - 1 + 1\right) = g\left(\frac{n}{m} - 1\right) g(1) = g\left(\frac{1}{m}\right)^{n-m} g(1) \ (n \ge m).$$

Hence we have

$$g\left(\frac{1}{m}\right) = g(1)^{\frac{1}{m}}, \ g\left(\frac{n}{m}\right) = g(1)^{\frac{n}{m}}, \ \forall n, m \in \mathbb{N}.$$

---

[1]The geometric distribution has the property of *abscence of memory* if we request it to be satisfied only for 0, 1, ...

[2]Other books call this the memoryless property.

Let t_n be a rational sequence such that $t_n \downarrow t \in \mathbb{R}^+$. Since g is decreasing and right-continuous,

$$g(t_n) \uparrow g(t),\ g(1)^{t_n} \uparrow g(1)^t.$$

Therefore $g(t) = g(1)^t$. Let $\lambda := \log\left(\frac{1}{g(1)}\right)$ which is positive by hypothesis. Then we get $P(T > t) = \exp(-\lambda t)$, so T has an exponential distribution.

Lemma 2.1.6 *Let $T \sim Exp(\lambda)$ and $S \sim Exp(\mu)$ be independent. Then* $\min(T, S) \sim$ *$Exp(\lambda + \mu)$.*

Proof

$$\mathbb{P}(\min(T, S) > t) = \mathbb{P}(T > t, S > t) = \mathbb{P}(T > t)\mathbb{P}(S > t) = e^{-(\lambda+\mu)t}.$$

Lemma 2.1.7 *Let $\{\tau_j;\ 1 \le j \le n\}$ be independent random variables such that $\tau_j \sim$ $Exp(\lambda_j)$. Then*

$$\mathbb{P}\big(\tau_i = \min(\tau_1, \ldots, \tau_n)\big) = \frac{\lambda_i}{\lambda_1 + \cdots + \lambda_n},\quad 1 \le i \le n.$$

Proof Let $T \sim \mathrm{Exp}(\lambda)$ and $S \sim \mathrm{Exp}(\mu)$ be independent. Denote the joint density function of (S, T) by $f_{S,T}$. Then

$$\begin{aligned}\mathbb{P}(S \le T) &= \mathbb{E}[\mathbf{1}_{\{S \le T\}}]\\ &= \int_0^\infty \int_0^\infty \mathbf{1}_{\{s \le t\}}(s, t) f_{S,T}(s, t)\,dsdt\\ &= \int_0^\infty \mu e^{-\mu s} \int_s^\infty \lambda e^{-\lambda t}\,dtds\\ &= \frac{\mu}{\lambda + \mu}.\end{aligned}$$

We know from Lemma 2.1.6 that $\min\{\tau_j; j \ne i\} \sim Exp(\sum_{j \ne i} \lambda_j)$. Hence

$$\mathbb{P}(\tau_i = \min(\tau_1, \ldots, \tau_n)) = \frac{\lambda_i}{\lambda_1 + \cdots + \lambda_n}.$$

Let $\{\tau_i;\ i \in \mathbb{N}\}$ be independent exponential random variables with parameter λ. Define $T_n := \sum_{i=1}^{n} \tau_i$.

Exercise 2.1.8 Prove that $\sigma(T_1, \ldots, T_n) = \sigma(\tau_1, \ldots, \tau_n)$.

Hint: Prove that there is a bijective application $f : \mathbb{R}^n \to \mathbb{R}^n$ such that $f(\tau_1, \ldots, \tau_n) = (T_1, \ldots, T_n)$.

Lemma 2.1.9 *$(T_1, \ldots, T_n)$ has a probability density on $\mathbb{R}^n$ given by*

$$f_{T_1,\ldots,T_n}(t_1, \ldots, t_n) = \lambda^n e^{-\lambda t_n} \mathbf{1}_{\{(t_1,\ldots,t_n)\,;\,0 \le t_1 < \cdots < t_n\}}(t_1, \ldots, t_n).$$

Proof We define a posteriori the function f_{n+1} as

$$\begin{aligned} f_{n+1}(t_1, \ldots, t_{n+1}) : &= f_{T_1,\ldots,T_n}(t_1, \ldots, t_n) f_{\tau_{n+1}}(\eta) \quad (t_{n+1} = t_n + \eta,\ \eta \ge 0) \\ &= \lambda^n e^{-\lambda t_n} \mathbf{1}_{\{(t_1,\ldots,t_n)\,;\,0 \le t_1 < \cdots < t_n\}}(t_1, \ldots, t_n) \lambda e^{-\lambda \eta} \mathbf{1}_{[0,\infty)}(\eta) \\ &= \lambda^{n+1} e^{-\lambda t_{n+1}} \mathbf{1}_{\{(t_1,\ldots,t_{n+1})\,;\,0 \le t_1 < \cdots < t_{n+1}\}}(t_1, \ldots, t_{n+1}) \text{ a.e.,} \end{aligned}$$

where $f_{T_1,\cdots,T_n}$ is a density of $(T_1, \cdots, T_n)$. Then for $a_1, \ldots, a_{n+1} \in \mathbb{R}$,

$$\begin{aligned} &\mathbb{P}(T_1 \le a_1, \ldots, T_{n+1} \le a_{n+1}) \\ &\quad = \mathbb{P}(T_1 \le a_1, \ldots, T_n + \tau_{n+1} \le a_{n+1}) \\ &\quad = \int \mathbf{1}_{\{t_1 \le a_1, \ldots, t_n \le a_n, t_n + \eta \le a_{n+1}\}} \mathbf{1}_{\{\eta \ge 0\}} f_{T_1,\ldots,T_n}(t_1, \ldots, t_n) f_{\tau_{n+1}}(\eta) dt_1 \cdots dt_n d\eta \\ &\quad = \int \mathbf{1}_{\{t_1 \le a_1, \ldots, t_n \le a_n, t_{n+1} \le a_{n+1}\}} f_{n+1}(t_1, \ldots, t_{n+1}) dt_1 \cdots dt_{n+1}. \end{aligned}$$

Therefore the proof follows by induction on n.

Definition 2.1.10 (*Gamma distribution*) A positive random variable X is said to follow a *gamma distribution* with *rate* parameter θ and *shape* parameter $r > 0$ if it has a probability density function of the form

$$f_X(x) = \frac{\theta^r x^{r-1} e^{-\theta x}}{\Gamma(r)}.$$

We denote this fact by $X \sim Gamma(\theta, r)$. Recall that $\Gamma(r) := \int_0^\infty x^{r-1} e^{-x} dx$. This function satisfies that $\Gamma(r+1) = r\Gamma(r)$ for $r > 0$ and $\Gamma(n) = (n-1)!$ for $n \in \mathbb{N}$.

Corollary 2.1.11 $T_n \sim Gamma(\lambda, n)$.

Proof By direct integration of Lemma 2.1.9 we have $\mathbb{P}(T_n \le t) = \displaystyle\int_0^t \frac{\lambda^n t_n^{n-1} e^{-\lambda t_n}}{(n-1)!} dt_n$. Therefore one obtains the probability density function of T_n by differentiation of $P(T_n \le t)$ with respect to t. This gives

$$f_{T_n}(t) = \frac{d}{dt} \mathbb{P}(T_n \le t) = \frac{\lambda^n t^{n-1} e^{-\lambda t}}{(n-1)!}.$$

Definition 2.1.12 (*Poisson distribution*) A positive integer-valued random variable N is said to follow a *Poisson distribution* with parameter $\lambda > 0$ if

$$\forall n \in \mathbb{Z}^+, \mathbb{P}(N = n) = e^{-\lambda}\frac{\lambda^n}{n!},$$

and we write $N \sim \text{Poisson}(\lambda)$.

The following properties are an immediate consequence of the above definition:

Corollary 2.1.13 *The mean and variance of $N \sim \text{Poisson}(\lambda)$ are λ. That is,*

$$\mathbb{E}[N] = \lambda, \ Var[N] = \lambda.$$

Furthermore, the characteristic function of the Poisson distribution is given by $\exp[\lambda(e^{i\theta} - 1)]$.

Remark 2.1.14 In order to make the relationship with the general theory that will follow in subsequent chapters, we remark here that the above characteristic function can be written by using the Dirac point mass measure at one, denoted by $\delta_1(dx)$ (i.e. the measure that characterizes the distribution of the random variable which takes the value 1, a.s.), [3] as

$$\mathbb{E}[e^{i\theta N}] = \exp\left[\int_{\mathbb{R}} (e^{i\theta x} - 1)\lambda\delta_1(dx)\right]. \tag{2.1}$$

Therefore the interpretation of the measure $\lambda\delta_1(dx)$ is that at every time that the process jumps, the size of the jump is one and that the average frequency of these jumps is λ. We will see about this interpretation shortly. The measure $\lambda\delta_1(dx)$ is usually called the Lévy measure associated with the random variable N.

Exercise 2.1.15 Suppose that there exists another finite measure μ such that for all $\theta \in \mathbb{R}$,

$$\int_{\mathbb{R}} (e^{i\theta x} - 1)\lambda\delta_1(dx) = \int_{\mathbb{R}} (e^{i\theta x} - 1)\mu(dx).$$

Prove that[4] $\mu = \lambda\delta_1$.

Proposition 2.1.16 *Let $\{\tau_i;\, i \in \mathbb{N}\}$ be a sequence of independent exponential random variables with parameter λ. Then, for any $t > 0$ the random variable*

$$N_t = \sum_{n\geq 1} \mathbf{1}_{\{T_n \leq t\}}$$

---

[3] Another way of saying the same thing is to say that δ_1 is a probability measure so that $\delta_1(A) = 1$ if $1 \in A$ and zero otherwise.

[4] This measure is essentially unique, although we have not yet discussed its uniqueness. This will follow because the exponential function is a generating family. Recall the discussion after (1.1) and before Exercise 1.1.11. That is, exponential functions generate indicators and therefore the corresponding measures have to be equal.

follows a Poisson distribution with parameter λt. *That is,*

$$\forall n \in \mathbb{N},\ \mathbb{P}(N_t = n) = e^{-\lambda t}\frac{(\lambda t)^n}{n!}.$$

Therefore we see that the previous interpretation is reinforced in this probabilistic representation. The random variable N_1 is the count of how many times the increasing sequence of random times T_i satisfy that $T_i \leq 1$. This count is done in units of one (every time the indicator function is equal to one) and the associated frequency is related to τ_i whose mean is λ^{-1}, which means that on average we will have λ jumps on the interval [0, 1].

Proof By Lemma 2.1.9, the density of $(T_1, \ldots, T_n)$ is given by

$$\lambda^n e^{-\lambda t_n}\mathbf{1}_{\{(t_1,\ldots,t_n)\,;\,0<t_1<\cdots<t_n\}}.$$

Hence, as in the proof of Corollary 2.1.11 and Lemma 2.1.9, we have

$$\begin{aligned}\mathbb{P}(N_t = n) &= \mathbb{P}(T_n \leq t < T_{n+1})\\ &= \int \lambda^{n+1} e^{-\lambda t_{n+1}}\mathbf{1}_{\{(t_1,\ldots,t_n)\,;\,0<t_1<\cdots<t_n<t<t_{n+1}\}}dt_1 \ldots dt_{n+1} = e^{-\lambda t}\frac{(\lambda t)^n}{n!}.\end{aligned}$$

One may wonder if there exists a sample space supporting the infinite sequence of random variables $\{\tau_i;\, i \in \mathbb{N}\}$. This is a classical mathematical result that requires the infinite product of sample spaces. This may be done using the Carathéodory extension theorem for measures.

2.1.2 Definition of the Poisson Process

Definition 2.1.17 (*point process*) Let $T = \{T_n;\, n \in \mathbb{N}\}$ be a discrete time stochastic process on $(\Omega, \mathscr{F}, \mathbb{P})$. Then T is called a *point process* on $\mathbb{R}^+$ if

$$0 < T_1 < \cdots < T_n < \cdots \text{ and } T_n \uparrow \infty.$$

Sometimes we use the notation[5] $T_0 = 0$.

Definition 2.1.18 (*counting process*) $\{N_t;\, t \geq 0\}$ is called a *counting process* of the point process $T = \{T_n;\, n \in \mathbb{N}\}$ if

$$N_t = \sum_{n\geq 1}\mathbf{1}_{\{T_n \leq t\}}.$$

[5]In some advanced texts this definition is considered in greater generality, without the condition that there are a finite number of counted events in any finite interval.

The following lemma is a immediate consequence of the above definitions.

Lemma 2.1.19 *The counting process* $\{N_t; t \geq 0\}$ *satisfies the following conditions:*

1. $\{N_t \geq n\} = \{T_n \leq t\}$,
2. $\{N_t = n\} = \{T_n \leq t < T_{n+1}\}$,
3. $\{N_s < n \leq N_t\} = \{s < T_n \leq t\}$,
4. *There exists the left limit* $N_{t-} := \lim_{s \uparrow t} N_s$, *and* N_t *is a right-continuous process (i.e.* $\lim_{s \downarrow t} N_s = N_t$*).*

Property **(4)** is usually called the *càdlàg property*.

Definition 2.1.20 (*Poisson process*) Let $\{\tau_i; i \geq 1\}$ be a sequence of independent exponential random variables with parameter λ and $T_n = \sum_{i=1}^{n} \tau_i$. The process $\{N_t, t \geq 0\}$ defined by

$$N_t = \sum_{n \geq 1} \mathbf{1}_{\{T_n \leq t\}}$$

is called a Poisson process with parameter λ.

In some texts λ is called the intensity of the Poisson parameter as it measures the rate of jumps. Many properties of Poisson processes have already been given in the previous section in the form of properties of Poisson random variables.

Proposition 2.1.21 *Let* $\{N_t\}$ *be a Poisson process with parameter* λ. *Given* $N_t = n$, $n \geq 1$, *then the jump times of the Poisson process are given by the following density:*

$$\mathbb{P}\big((T_1, T_2, \ldots, T_n) \in A | N_t = n\big) = \int_{A \cap \{0 \leq t_1 < t_2 < \cdots < t_n < t\}} dt_1 dt_2 \cdots dt_n \frac{n!}{t^n}, \; A \in \mathscr{B}(\mathbb{R}^n).$$

That is, given $N_t = n$ *the jump times* $(T_1, \cdots, T_n)$ *are distributed like the order statistics of n uniform random variables in the interval*[6] $[0, t]$.

Proof From Lemma 2.1.9, we have that

$$\begin{aligned}
&\mathbb{P}((T_1, T_2, \ldots, T_n) \in A | N_t = n) \\
=&\frac{\mathbb{P}((T_1, T_2, \ldots, T_n) \in A, N_t = n)}{\mathbb{P}(N_t = n)} \\
=&\mathbb{P}((T_1, T_2, \ldots, T_n) \in A, T_n \leq t < T_{n+1}) \frac{n!}{(\lambda t)^n} e^{\lambda t} \\
=&\int_{(A \cap \mathbb{R}^{n-1} \times [0,t]) \times (t, \infty)} \lambda^{n+1} e^{-\lambda t_{n+1}} \mathbf{1}_{\{0 \leq t_1 < \cdots < t_{n+1}\}}(t_1, \ldots, t_{n+1}) dt_1 \cdots dt_{n+1} \frac{n!}{(\lambda t)^n} e^{\lambda t}
\end{aligned}$$

[6] Given n independent random variables $U_1, \cdots, U_n$ each with the uniform distribution in $[0, t]$, the order statistic distribution is the n-dimensional distribution of the n random variables once they have been ordered.

$$= \int_{A\cap\mathbb{R}^{n-1}\times[0,t]} \int_t^\infty \lambda e^{-\lambda t_{n+1}} \mathbf{1}_{\{0\le t_1<\cdots<t_n\}}(t_1,\ldots,t_n)dt_1\cdots dt_{n+1}\frac{n!}{t^n}e^{\lambda t}$$
$$= \int_{A\cap\{0\le t_1<t_2<\cdots<t_n<t\}} \int_t^\infty \lambda e^{-\lambda t_{n+1}} dt_{n+1}dt_1\cdots dt_n\frac{n!}{t^n}e^{\lambda t}$$
$$= \int_{A\cap\{0\le t_1<t_2<\cdots<t_n<t\}} e^{-\lambda t}dt_1\cdots dt_n\frac{n!}{t^n}e^{\lambda t}$$
$$= \int_{A\cap\{0\le t_1<t_2<\cdots<t_n<t\}} dt_1dt_2\cdots dt_n\frac{n!}{t^n}.$$

Lemma 2.1.22 *If $\{N_t; t\ge 0\}$ is a Poisson process with parameter λ, then, for $t\ge 0$:*

(i) $T_{N_t+1}-t$ *is independent of* $\sigma(\tau_{N_t+k}: k\ge 2)$.
(ii) $T_{N_t+1}-t$ *is an exponentially distributed r.v. with parameter* λ.
(iii) $T_{N_t+1}-t$ *is independent of* $\sigma(N_t)$.

Proof First, we show (ii) and (iii). For $s_1\in\mathbb{R}$ and $k\in\mathbb{N}$, we have

$$\begin{aligned}\mathbb{P}(T_{N_t+1}-t\le s_1, N_t=k) &= \mathbb{P}(T_{k+1}-t\le s_1, N_t=k)\\ &= \mathbb{E}(\mathbf{1}_{\{T_{k+1}\le s_1+t, T_k\le t<T_{k+1}\}}).\end{aligned}$$

From Lemma 1.2.3, we obtain

$$\begin{aligned}\mathbb{E}[\mathbf{1}_{\{T_{k+1}\le s_1+t, T_k\le t<T_{k+1}\}}] =&\mathbb{E}[\mathbf{1}_{\{T_k\le t\}}\mathbb{E}[\mathbf{1}_{\{t<T_k+\tau_{k+1}\le s_1+t\}}|T_k]]\\ =&\mathbb{E}[\mathbf{1}_{\{T_k\le t\}}(e^{-\lambda(t-T_k)}-e^{-\lambda(s_1+t-T_k)})] = (1-e^{-\lambda s_1})\mathbb{P}(N_t=k).\end{aligned}$$

Next, we compute the following distribution function in order to prove (i). It is easy to check that, for $\{s_i\}_{i\ge 2}\subset\mathbb{R}$,

$$\mathbb{P}\left(\bigcap_{k\ge 2}\{\tau_{N_t+k}\le s_k\}\right) = \mathbb{P}\left(\bigcap_{k\ge 2}\{\tau_k\le s_k\}\right).$$

Therefore, we have

$$\begin{aligned}&\mathbb{P}\left(\{T_{N_t+1}-t\le s_1\}\cap\bigcap_{k\ge 2}\{\tau_{N_t+k}\le s_k\}\right)\\ &= \sum_{l=0}^\infty \mathbb{P}\left(\{T_{N_t+1}-t\le s_1\}\cap\bigcap_{k\ge 2}\{\tau_{N_t+k}\le s_k\}\cap\{N_t=l\}\right)\\ &= \sum_{l=0}^\infty \mathbb{P}\Big(\{T_{N_t+1}-t\le s_1\}\cap\{N_t=l\}\Big)\mathbb{P}\left(\bigcap_{k\ge 2}\{\tau_{l+k}\le s_k\}\right)\\ &= \mathbb{P}(T_{N_t+1}-t\le s_1)\mathbb{P}\left(\bigcap_{k\ge 2}\{\tau_{N_t+k}\le s_k\}\right).\end{aligned}$$

The proof is complete.

Proposition 2.1.23 *Let $\{N_t; t \geq 0\}$ be a Poisson process. Then for any $s < t$, $N_t - N_s$ and N_s, are independent.*

Proof Let $a, b \in \mathbb{N}$,

$$\begin{aligned}
\mathbb{P}(N_t - N_s \leq a, N_s \leq b) &= \sum_{k=0}^{b} \mathbb{P}(N_t \leq a+k, N_s = k) = \sum_{k=0}^{b}\sum_{l=k}^{a+k} \mathbb{P}(N_t = l, N_s = k) \\
&= \sum_{k=0}^{b}\sum_{l=k}^{a+k} \mathbb{P}(T_l \leq t < T_{l+1}, T_k \leq s < T_{k+1}) \\
&= \sum_{k=0}^{b}\sum_{j=0}^{a} \mathbb{P}(T_{k+j} \leq t < T_{k+j+1}, T_k \leq s < T_{k+1}). \qquad (2.2)
\end{aligned}$$

First for $j = 0$, we have

$$\mathbb{P}(T_k \leq s < t < T_{k+1}) = \int_0^s \mathbb{P}(t < u + \tau_{k+1}) e^{-\lambda u} \frac{\lambda^k u^{k-1}}{(k-1)!} du = \frac{(\lambda s)^k}{k!} e^{-\lambda t}.$$

For general $j \geq 1$, we have, applying conditioning twice (first $T_k = u$ and then $\tau_{k+1} = s_1$),

$$\begin{aligned}
&\mathbb{P}(T_{k+j} \leq t < T_{k+j+1}, T_k \leq s < T_{k+1}) \\
&= \int_0^s \int_{s-u}^{t-u} \mathbb{P}(u + s_1 + S_j \leq t < u + s_1 + S_{j+1}) \lambda e^{-\lambda s_1} e^{-\lambda u} \frac{\lambda^k u^{k-1}}{(k-1)!} ds_1 du.
\end{aligned}$$

Here in order to simplify the notation we have set $S_j = \sum_{l=2}^{j} \tau_{k+l}$. Now using the previous step, we have that

$$\begin{aligned}
&\mathbb{P}(T_{k+j} \leq t < T_{k+j+1}, T_k \leq s < T_{k+1}) \\
&= \int_0^s \int_{s-u}^{t-u} \frac{\lambda^{j-1}(t-u-s_1)^{j-1}}{(j-1)!} \lambda e^{-\lambda t} \frac{\lambda^k u^{k-1}}{(k-1)!} ds_1 du \\
&= \frac{(\lambda(t-s))^j}{j!} \frac{(\lambda s)^k}{k!} e^{-\lambda t}.
\end{aligned}$$

Thus, we have that

$$\begin{aligned}
(2.2) &= \sum_{k=0}^{b} \{ \frac{(\lambda s)^k}{k!} e^{-\lambda t} + \sum_{j=1}^{a} \frac{(\lambda s)^k}{k!} \frac{(\lambda(t-s))^j}{j!} e^{-\lambda t} \} \\
&= \sum_{k=0}^{b}\sum_{j=0}^{a} \frac{(\lambda s)^k}{k!} \frac{(\lambda(t-s))^j}{j!} e^{-\lambda t}
\end{aligned}$$

$$= \sum_{k=0}^{b} \{ \frac{(\lambda s)^k}{k!} e^{-\lambda s} \sum_{j=0}^{a} \frac{(\lambda(t-s))^j}{j!} e^{-\lambda(t-s)} \}$$
$$= \mathbb{P}(N_t - N_s \le a)\mathbb{P}(N_s \le b).$$

In fact, we have the following generalization.

Proposition 2.1.24 *The Poisson process $\{N_t; t \ge 0\}$ has independent increments. That is, for any partition $0 < t_1 < \cdots < t_n$, the random variables $N_{t_1}, N_{t_2} - N_{t_1}, \ldots, N_{t_n} - N_{t_{n-1}}$ are independent.*

Lemma 2.1.25 *Let $\{N_t; t \ge 0\}$ be a counting process of the point process $\{T_n; n \in \mathbb{N}\}$ with stationary independent increments. That is:*

(a) for all $n \ge 2$, $0 \le t_0 < t_1 < \cdots < t_n$, the increments $\{N_{t_j} - N_{t_{j-1}}; 1 \le j \le n\}$ are mutually independent,

(b) for all $0 \le s < t$, the law of $N_t - N_s$ depends upon the pair (s, t) only through the difference $t - s$.

We define that $\tau_1 := T_1, \tau_i := T_i - T_{i-1}$ $(i \ge 2)$. If, for any $t \ge 0$, N_t follows a Poisson distribution with parameter λt, then $\{\tau_i; i \in \mathbb{N}\}$ is a sequence of independent exponential random variables with parameter λ.

Proof First note that for $i < j$ and $s < u$

$$\mathbb{P}(T_i < s, T_j > u) = \mathbb{P}(N_s \ge i, N_u \le j-1) = e^{-\lambda u} \sum_{k=i}^{j-1} \sum_{l=0}^{j-1-k} \frac{s^k (u-s)^l \lambda^{k+l}}{k!l!}.$$

From here one obtains all the required properties. In fact, one proves first by taking $j = i + 1$, differentiation with respect to s and integration for $s \in [0, u]$ that $T_i \sim Gamma(\lambda, i)$ and similarly that $\tau_{i+1} = T_{i+1} - T_i$ is independent of T_i and that it is distributed exponentially with parameter λ.

Lemma 2.1.26 *Let X and Y be $\mathbb{Z}^+$-valued random variables with $\mathbb{E}[X^m], \mathbb{E}[Y^m] < \infty$ for any $m \in \mathbb{N}$. Then X and Y are identical in law if and only if for any u in a neighborhood of 0,*

$$\mathbb{E}[u^X] = \mathbb{E}[u^Y].$$

The function $G(u) := \mathbb{E}[u^X] = \sum_{i \ge 0} u^i \mathbb{P}(X = i)$ is usually called the probability generating function of X. In other texts, is called the moment generating function. In the particular case that $u = e^{-\lambda}$ and $\lambda \ge 0$ then it is called the Laplace transform, which is, loosely speaking, a transformation of the characteristic function.

Proof Consider $G(u)$, $u \in [0, 1]$. Since $\mathbb{E}[X^m], \mathbb{E}[Y^m] < \infty$, $G(u)$ is termwise differentiable. Then we have

$$G'(u) = \sum_{i \geq 0} iu^{i-1}\mathbb{P}(X = i) = \sum_{i \geq 1} iu^{i-1}\mathbb{P}(X = i).$$

Hence $G'(0) = \lim_{u \downarrow 0} G'(u) = 1 \cdot \mathbb{P}(X = 1)$. In the same fashion, we have

$$G^{(n)}(0) = n!\mathbb{P}(X = n).$$

Suppose that for some $\varepsilon > 0$ and for all u in $[0, \varepsilon \wedge 1]$, $\mathbb{E}[u^X] = \mathbb{E}[u^Y]$. Then we have

$$\mathbb{P}(X = n) = \frac{G^{(n)}(0)}{n!} = \mathbb{P}(Y = n).$$

Lemma 2.1.27 *Let X be a Poisson random variable with parameter λ. Then, for any $n \in \mathbb{N}$, $\mathbb{E}[X^n] < \infty$.*

Proof For all $n \in \mathbb{N}$, we have that

$$\mathbb{E}[X^n] = \sum_{k \geq 0} k^n \frac{\lambda^k}{k!} e^{-\lambda}.$$

Also,

$$\frac{k^n \frac{\lambda^k}{k!} e^{-\lambda}}{(k-1)^n \frac{\lambda^{k-1}}{(k-1)!} e^{-\lambda}} = \frac{\lambda}{k}\left(\frac{k}{k-1}\right)^n.$$

Since $\lim_{k \to \infty} \frac{\lambda}{k}\left(\frac{k}{k-1}\right)^n = 0$, from the ratio test, we get $\mathbb{E}[X^n] < \infty$.

In Lemma 2.1.25, we assumed that the marginal distribution of N follows a Poisson distribution. In the next result we generalize this result.

Proposition 2.1.28 *Let $\{N_t; t \geq 0\}$ be a counting process of the point process $T = \{T_n; n \in \mathbb{N}\}$ with stationary independent increments. That is:*

(a) for all $n \geq 2$, $0 \leq t_0 < t_1 < \cdots < t_n$, the increments $\left\{N_{t_j} - N_{t_{j-1}}; 1 \leq j \leq n\right\}$ are mutually independent,

(b) for all $0 \leq s < t$, the law of $N_t - N_s$ depends upon the pair (s, t) only through the difference $t - s$.

In such a case $\{N_t; t \geq 0\}$ is a Poisson process.

Proof For any $t \geq 0$, define the moment generating function $f_t : [0, 1] \to [0, 1]$ as follows:

$$f_t(u) := \mathbb{E}[u^{N_t}] = \sum_{k \geq 0} u^k \mathbb{P}(N_t = k) \ (t > 0),$$

$$f_0(u) := 1.$$

We remark that the above functions are well defined and that $f_t(0) = \mathbb{P}(N_t = 0)$. From Lemmas 2.1.26 and 2.1.27, we will prove in Step 1 that there exists $\lambda \geq 0$ such that $f_t(u) = e^{-\lambda t(1-u)}$ and in Step 2 that the process $T = \{T_n; n \in \mathbb{N}\}$ satisfies that each T_n follows a Gamma distribution with scale λ and shape n .

Step 1.

From assumptions (a) and (b), we have

$$\begin{aligned} f_{t+s}(u) &= \mathbb{E}[u^{N_{t+s}}] \\ &= \mathbb{E}[u^{N_s} u^{N_{t+s}-N_s}] \\ &= \mathbb{E}[u^{N_s}]\mathbb{E}[u^{N_{t+s}-N_s}] \\ &= f_s(u) f_t(u). \end{aligned}$$

Therefore for any $n, m \in \mathbb{N}$,

$$\begin{aligned} f_{\frac{n}{m}}(u) &= f_{\frac{1}{m}+\cdots+\frac{1}{m}}(u) = \left\{f_{\frac{1}{m}}(u)\right\}^n, \\ f_{\frac{n}{m}}(u) &= f_{\frac{n-m}{m}+1}(u) = f_{\frac{n-m}{m}}(u) f_1(u) = \left\{f_{\frac{1}{m}}(u)\right\}^{n-m} f_1(u), \ n \geq m. \end{aligned}$$

Hence we have

$$f_{\frac{1}{m}}(u) = \{f_1(u)\}^{\frac{1}{m}}, \ f_{\frac{n}{m}}(u) = \{f_1(u)\}^{\frac{n}{m}} \ \forall n, m \in \mathbb{N}.$$

Let $\{t_n; n \in \mathbb{N}\}$ be a sequence of rational numbers such that $t_n \downarrow t \in \mathbb{R}^+$. Then as $f_1(u) \geq 0$, we have

$$\{f_1(u)\}^{t_n} \uparrow \{f_1(u)\}^t .$$

Since N_t is a right-continuous function,

$$N_{t_n} \downarrow N_t, \text{ a.s.}, \ u^{N_{t_n}} \uparrow u^{N_t}, \text{ a.s.}$$

From the monotone convergence theorem, we have

$$f_{t_n}(u) = \mathbb{E}[u^{N_{t_n}}] \uparrow \mathbb{E}[u^{N_t}] = f_t(u).$$

Therefore $f_t(u) = \{f_1(u)\}^t$.

Let $\{s_n; n \in \mathbb{N}\}$ be a real sequence such that $s_n \downarrow 0$. Then

$$\begin{aligned} \{f_1(u)\}^{s_n} &= f_{s_n}(u) \\ &= \sum_{k \geq 0} u^k \mathbb{P}(N_{s_n} = k) \\ &\geq \mathbb{P}(N_{s_n} = 0) \end{aligned}$$

$$= \mathbb{P}(T_1 > s_n).$$

Since $\mathbb{P}(T_1 > s_n) \uparrow 1$, as $n \uparrow \infty$ then for some $N \in \mathbb{N}$, we have

$$\{f_1(u)\}^{s_N} \geq \mathbb{P}(T_1 > s_N) > \frac{1}{2}.$$

Therefore $f_1(u) \neq 0$. Let $\lambda(u) := \log \dfrac{1}{f_1(u)}$. Then we have

$$\mathbb{E}[u^{N_t}] = f_t(u) = \{f_1(u)\}^t = e^{-\lambda(u)t}.$$

In particular, $\lambda(0) = \log f_1(0)^{-1} = \log \mathbb{P}(N_1 = 0)^{-1} \geq 0$. Hence it just remains to show that for any $u \in [0, 1]$,

$$\lambda(u) = \lambda(0)(1 - u).$$

Clearly,

$$\begin{aligned}
\lambda(u) &= \lim_{t \downarrow 0} \frac{1 - e^{-\lambda(u)t}}{t} = \lim_{t \downarrow 0} \frac{1 - \mathbb{E}[u^{N_t}]}{t} \\
&= \lim_{t \downarrow 0} \frac{1}{t} \left\{ 1 - \sum_{k \geq 0} u^k \mathbb{P}(N_t = k) \right\} \\
&= \lim_{t \downarrow 0} \sum_{k \geq 1} \frac{1}{t} (1 - u^k) \mathbb{P}(N_t = k). \qquad (2.3)
\end{aligned}$$

Since $0 \leq u \leq 1$,

$$0 \leq \sum_{k \geq 2} \frac{1}{t} (1 - u^k) \mathbb{P}(N_t = k) \leq \frac{\mathbb{P}(N_t \geq 2)}{t}.$$

Now, we will prove that

$$\frac{\mathbb{P}(N_t \geq 2)}{t} \to 0, \ as\ t \downarrow 0.$$

We divide this into two cases: First, if $\lambda(0) = 0$ then $f_t(0) = 1$ for any t. Since $1 = \mathbb{E}[0^{N_t}] = \mathbb{P}(N_t = 0)$, $\mathbb{P}(N_t \geq 2) = 0$. Next, if $\lambda(0) > 0$ then $e^{-\lambda(0)t} < 1$, and we have

$$\begin{aligned}\bigcup_{n\geq 0}\{N_{nt}=0, N_{nt+t}\geq 2\} &= \bigcup_{n\geq 0}\{nt<T_1, T_2\leq nt+t\}\\ &= \bigcup_{n\geq 0}\{nt<T_1, T_2\leq nt+t<T_1+t\}\\ &\subseteq \{T_2<T_1+t\}\,.\end{aligned}$$

We deduce from the above and assumptions (a) and (b) that

$$\begin{aligned}\mathbb{P}(\bigcup_{n\geq 0}\{N_{nt}=0, N_{nt+t}\geq 2\}) &\leq \mathbb{P}(T_2<T_1+t)\\ &\to \mathbb{P}(T_2\leq T_1)=0\ as\ t\downarrow 0.\end{aligned}$$

Hence we have

$$0\leq \frac{\mathbb{P}(N_t\geq 2)}{\lambda(0)t}\leq \frac{\mathbb{P}(N_t\geq 2)}{1-e^{-\lambda(0)t}}\leq \mathbb{P}(T_2<T_1+t)\to 0.$$

So from (2.3), we conclude that for all $u\in[0,1]$,

$$\lambda(u)=\lim_{t\downarrow 0}\frac{\mathbb{P}(N_t=1)}{t}(1-u)=C(1-u).$$

Taking $u=0$, we obtain that $\lambda(u)=\lambda(0)(1-u)$.

Step 2.

Let $\tau_1:=T_1, \tau_i:=T_i-T_{i-1}\ (i\geq 2)$. We prove that $\{\tau_i; i\in\mathbb{N}\}$ is a sequence of independent exponential random variables with parameter $\lambda(0)$. From Lemma 2.1.25, it is clear.

Exercise 2.1.29 Let N be a Poisson process. Prove the following:

- $\lim_{s\uparrow t}N_s=N_t$ except for $t=T_i$, $i\in\mathbb{N}$.
- For any fixed $t>0$, we have $\mathbb{P}(N_t-\lim_{s\uparrow t}N_s=1)=0$.
- Let T be a random time which is independent of N, then $\mathbb{P}(N_T-\lim_{s\uparrow T}N_s=1)=0$.

Corollary 2.1.30 *The moment generating function of a Poisson process with parameter λ is*

$$\mathbb{E}[u^{N_t}]=e^{-\lambda(1-u)t}.$$

Theorem 2.1.31 (Sums of independent Poisson processes; super-position property) *If $\{N_t^{(1)}; t\geq 0\}$, $\{N_t^{(2)}; t\geq 0\}$ are two independent*[7] *Poisson process with intensities λ_1,λ_2, then $\{N_t=N_t^{(1)}+N_t^{(2)}; t\geq 0\}$ is a Poisson process with intensity $\lambda_1+\lambda_2$.*

---

[7]Recall that independence of random process means that the σ-fields generated by these process are independent.

Proof We prove that N_t is a counting process, N_t has stationary independent increments and N_t follows a Poisson distribution with parameter $\lambda_1 + \lambda_2$. Therefore the result will follow from Proposition 2.1.28.

Step 1. N_t is a counting process. Since $N_t^{(i)}$ is Poisson process, there exists a point process $\{T_n^{(i)}\}_{n\in\mathbb{N}}$ such that

$$N_t^{(i)} = \sum_{n\geq 1} \mathbf{1}_{\{T_n^{(i)}\leq t\}}.$$

Hence we have $N_t = \sum_{n\geq 1}(\mathbf{1}_{\{T_n^{(1)}\leq t\}} + \mathbf{1}_{\{T_n^{(2)}\leq t\}})$. We define a point process $\{T_n; n \in \mathbb{N}\}$ which is relabeled $\{T_n^{(1)}; n \in \mathbb{N}\}$, $\{T_n^{(2)}; n \in \mathbb{N}\}$ in an increasing order in that

$$T_1 := \inf\{t > 0; \Delta N_t \neq 0\},$$
$$T_i := \inf\{t > T_{i-1}; \Delta N_t \neq 0\},$$

where $\Delta N_t := N_t - N_{t-}$. Then we have $N_t = \sum_{n\geq 1} \mathbf{1}_{\{T_n\leq t\}}$. Hence N_t is counting process.

Step 2. N_t has the stationary increment property

$$\begin{aligned} N_{t+h} - N_t &= (N_{t+h}^{(1)} + N_{t+h}^{(2)}) - (N_t^{(1)} + N_t^{(2)}) \\ &= (N_{t+h}^{(1)} - N_t^{(1)}) + (N_{t+h}^{(2)} - N_t^{(2)}) \\ &\overset{\mathrm{d}}{=} N_h^{(1)} + N_h^{(2)} = N_h. \end{aligned}$$

Step 3. N_t has the independent increments property. We only check the following equation:

$$\mathbb{E}[\exp(i\sum_{j=1}^{n} \alpha_j(N_{t_j} - N_{t_{j-1}}))] = \prod_{j=1}^{n} \mathbb{E}[\exp(i\alpha_j(N_{t_j} - N_{t_{j-1}}))].$$

We decompose $N_{t_j} - N_{t_{j-1}}$ to $N_{t_j}^{(1)} - N_{t_{j-1}}^{(1)} + N_{t_j}^{(2)} - N_{t_{j-1}}^{(2)}$ as in Step 2. Since $N^{(1)}$ and $N^{(2)}$ are independent, we get

$$(LHS) = \mathbb{E}[\exp(i\sum_{j=1}^{n} \alpha_j(N_{t_j}^{(1)} - N_{t_{j-1}}^{(1)})]\mathbb{E}[i\sum_{j=1}^{n} \alpha_j(N_{t_j}^{(2)} - N_{t_{j-1}}^{(2)}))],$$

$$(RHS) = \prod_{j=1}^{n} \mathbb{E}[\exp(i\alpha_j(N_{t_j}^{(1)} - N_{t_{j-1}}^{(1)}))]\mathbb{E}[\exp(i\alpha_j(N_{t_j}^{(2)} - N_{t_{j-1}}^{(2)}))].$$

Therefore $(LHS) = (RHS)$.

Step 4. N_t follows the Poisson distribution with parameter $\lambda_1 + \lambda_2$. It is clear by checking that $\mathbb{E}[e^{i\theta N_t}] = \exp((\lambda_1 + \lambda_2)t(e^{i\theta} - 1))$.

Remark 2.1.32 Step 4 in the above proof should reinforce your understanding of the representation (2.1). We note that as in Remark 2.1.14 we can rewrite the above characteristic function as $\exp(\int (e^{iux} - 1)\nu(dx))$, where $\nu(dx) = \lambda_1\delta_1(dx) + \lambda_2\delta_1(dx)$. Therefore we understand that this may lead to a general form of some characteristic functions where the distribution is now characterized by a measure ν.

Exercise 2.1.33 Use characteristic functions (see e.g. Theorem 1.1.6) in order to prove the following two results for a sequence of independent Poisson random variables N^i, $i \in \mathbb{N}$ with parameter λ:

- $\sum_{i=1}^n N^i$ is a Poisson r.v. with parameter λn. In particular, prove that $\sum_{i=1}^n N^i \to \infty$, a.s. The following renormalization result gives the asymptotic behavior of this sum of random variables:
- Prove that $\lim_{n\to\infty} \frac{1}{n} \sum_{i=1}^n N^i = \lambda$, a.s.
- $\frac{1}{\sqrt{n}} \sum_{i=1}^n (N^i - \lambda) \Rightarrow Z$, where $Z \sim N(0, 1)$.

Notice that this last result is interesting. A discrete distribution like the Poisson random variable after a limit renormalization becomes a continuous distribution!

Exercise 2.1.34 In this exercise, we would like to briefly discuss some basic properties on the modulus of continuity of the Poisson process $\{N_t; t \geq 0\}$ with parameter λ. First, we recall that the paths of Poisson processes are càdlàg. Still, prove the following properties:

- $\lim_{t\downarrow 0} \frac{\mathbb{E}[N_t]}{t} = \lambda$.
- $\lim_{t\downarrow 0} \frac{N_t}{t} = 0$, a.s.
- $\lim_{t\downarrow 0} \mathbb{P}(\frac{N_t}{t} > 0) = 0$.

These results can also be stated for $t \uparrow \infty$. Write and prove those results and compare with Exercise 2.1.33. For the moment, it looks that not much of interest is going on as these paths always increase by one unit. Starting in the next chapter, we will see much more interesting behavior.

Problem 2.1.35 This exercise is designed for people that may have some experience with Markov chains. Prove that a Poisson process is a time homogeneous Markov chain with state space $\mathbb{N} \cup \{0\}$. In particular prove that for $t_n < \ldots < t_1$ and $a_i \in \mathbb{N}$, $i = 1, \ldots, n$, $a_n \leq a_{n-1} \leq \ldots \leq a_1$ one has

$$\mathbb{P}(N_{t_1} = a_1 / N_{t_j} = a_j, j = 2, \ldots, n) = \mathbb{P}(N_{t_1} = a_1 / N_{t_2} = a_2).$$

Compute its transition probabilities and prove its transience.

Exercise 2.1.36 One of the properties which separates the Poisson process from the Wiener process is its lack of scaling property. In fact, prove that although

$$\mathbb{E}[t^{-1}N_t] = \lambda,$$

in general, $\mathbb{E}[(t^{-1}N_t)^p]$ depends on t. This can also be seen from the moment generating function; that is, the fact that the generating function of $t^{-1}N_t$ is not independent of t.

Exercise 2.1.37 The following properties are related with the long-time behavior of the Poisson process:

1. Compute $\lim_{t\uparrow\infty} t^{-2}\int_0^t \mathbb{E}[N_s]ds$.
2. Prove that $\int_0^t N_s ds = \sum_{i=0}^{T_{N_t}} i(T_{i+1} - T_i) + N_t(t - T_{N_t})$.
3. Prove that $\lim_{t\uparrow\infty} N_t = \infty$, a.s. and therefore $\lim_{t\uparrow\infty} T_{N_t} = \infty$, a.s.
4. Prove that $\lim_{n\uparrow\infty} \sum_{i=0}^{n} i(T_{i+1} - T_i) = \infty$.
5. Try to guess the value of the limit $\lim_{t\uparrow\infty} t^{-2}\int_0^t N_s ds$.[8]

Exercise 2.1.38 This exercise will help you get acquainted with σ-algebras which will be needed later. Define $\mathscr{F}_t := \sigma(N_s; s \le t)$. Recall that we always consider that this means that $\mathscr{F}_t$ is the smallest σ-algebra for which N_s is measurable for all $s \le t$ and which is completed with all the subsets of sets of probability zero.

1. Prove that $\mathscr{F}_t = \sigma(\{T_i \le s\}; s \le t, i \in \mathbb{N}) = \sigma(T_i \wedge t; i \in \mathbb{N})$. In particular, T_i is a stopping time for any $i \in \mathbb{N}$.
2. Let $g : \Omega \times \mathbb{R} \times [0, T] \to \mathbb{R}$ be a function that is measurable with respect to the σ-field $\mathscr{F}_T \otimes \mathscr{B} \otimes \mathscr{B}[0, T]$. Prove that there exists a sequence of functions of the type
$$g_n(\omega, z, t) = \sum_{i=1}^{n} X_i(\omega)a_i(z)b_i(t)$$
such that $g_n \to g$, $a.s.$ Here $X_i : \Omega \to \mathbb{R}$ is a $\mathscr{F}_T$ random variable and $a_i : \mathbb{R} \to \mathbb{R}$ together with $b_i : [0, T] \to \mathbb{R}$ are measurable functions.
3. Now suppose that for each $t \in [0, T]$, $g(\omega, z, s)1(s \le t)$ is measurable with respect to $\mathscr{F}_t \otimes \mathscr{B} \otimes \mathscr{B}[0, t]$. Prove that $g(\cdot, \cdot, t)$ is measurable with respect to $\mathscr{F}_t \otimes \mathscr{B}$.

Exercise 2.1.39 This exercise is about the extension of the stationary increment property of N to a weaker property and is strongly related to Exercise 2.1.38.

1. Let $s, t > 0$; prove that $N_{t+s} - N_s$ is independent of $\mathscr{F}_s$. Hint: Prove this by using a simple set in $\mathscr{F}_s$ then extend this result by taking limits.

[8] Recall results related with the law of large numbers.

2. We say that τ is a stopping time if for any $t > 0$, $\{\tau \le t\} \in \mathscr{F}_t$. Prove that given any stopping time τ then $N_{\tau+t} - N_\tau$ is independent of $\mathscr{F}_\tau = \{A \in \mathscr{F}_\infty; A \cap \{\tau \le t\} \in \mathscr{F}_t, t > 0\}$. Furthermore prove that the laws of $N_{\tau+t} - N_\tau$ and N_t are the same. That is, $\mathbb{P}(N_{\tau+t} - N_\tau = k) = \mathbb{P}(N_t = k)$ for any $k \in \mathbb{N}$. This property is called the strong Markov property. When τ is non-random it is called the Markov property.
Hint: Prove this property by approximations. That is, assume that τ takes a finite number of values. Then proceed by taking limits.

Exercise 2.1.40 This exercise is about the simplest form of Girsanov's theorem (change of measure) for Poisson processes. It will appear later in a much more complex form.

1. Let N be a Poisson process with parameter $\lambda > 0$; prove the following equality for any f bounded and measurable function:

$$\mathbb{E}[f(N_t)] = \mathbb{E}[f(N_t') \exp((\lambda' - \lambda)t - N_t' \log(\frac{\lambda'}{\lambda}))].$$

Here N' is a Poisson process of parameter $\lambda' > 0$ and we assume without loss of generality that the probability space supports these two processes.
2. Prove that one may define a new probability measure by considering

$$Q_t(A) = \mathbb{E}[\mathbf{1}_A \exp((\lambda' - \lambda)t - N_t' \log(\frac{\lambda'}{\lambda}))].$$

3. Prove that if $A \in \mathscr{F}_s$ for $s < t$ then $Q_t(A) = Q_s(A)$.

Problem 2.1.41 This exercise is about the large deviation principle for the Poisson process.

1. Prove that $\mathbb{P}(N_t > x) \le f(x)^{-1}\mathbb{E}[f(N_t)]$ for any increasing function f.
2. Using $f(x) = e^{ax}$ prove that $\mathbb{P}(N_t > x) \le \exp(-ax)\exp((e^a - 1)\lambda t)$.
3. Prove that

$$\limsup_{x\to+\infty} \frac{\log \mathbb{P}(N_t > x)}{x} = -\infty.$$

4. State and prove an improvement of the above result[9].

Problem 2.1.42 (An alternative method of defining the Poisson law. The infinitesimal definition[10]) Suppose that a stochastic process[11] $N_t, t \ge 0$, satisfies the following relations as $h \downarrow 0$ for $k = 0, \dots$:

---

[9]This is an exercise to test your understanding.

[10]Recall that $o(1)$ stands for any function that converges to zero as the related parameter (which should be clear from the statement) approaches a certain limit. In this case, the parameter is h.

[11]A stochastic process is a family of random variables $\{N_t\}_{t\in[0,\infty)}$ such that $N : \Omega \times [0, \infty) \to \mathbb{R}$ is jointly measurable. I suppose that you interpreted this in a similar way in Definition 2.1.17.

1. $\mathbb{P}(N_{t+h} = k + 1 | N_t = k) = \lambda h + o(h)$.
2. $\mathbb{P}(N_{t+h} > k + 1 | N_t = k) = o(h)$. [12].
3. $\mathbb{P}(N_{t+h} = k | N_t = k) = 1 - \lambda h + o(h)$
4. $t \to N_t(\omega)$ (a.s.) is increasing and N has independent increments.

Prove the following statements:

1. $\mathbb{P}(N_t = 0) - \mathbb{P}(N_{t-h} = 0) = -\lambda h \mathbb{P}(N_{t-h} = 0) + o(h)$.
2. $\partial_t \mathbb{P}(N_t = 0) = -\lambda \mathbb{P}(N_t = 0)$ and therefore $\mathbb{P}(N_t = 0) = e^{-\lambda t}$.
3. For $k \geq 1$, $\mathbb{P}(N_t = k) - \mathbb{P}(N_{t-h} = k) = \lambda h(\mathbb{P}(N_{t-h} = k - 1) - \mathbb{P}(N_{t-h} = k)) + o(h)$.
4. Prove that the probabilities for N_t should satisfy the differential equation $\partial_t \mathbb{P}(N_t = k) = \lambda(\mathbb{P}(N_t = k - 1) - \mathbb{P}(N_t = k))$.
5. Provide boundary conditions for the above system of ordinary differential equations and prove that the unique solution is given by $\mathbb{P}(N_t = k) = e^{-\lambda t} \frac{(\lambda t)^k}{k!}$.

Note that this exercise does not tell you how to construct the process N, which is what we have done in this section. Still this is not impossible to do. In a general setting, one uses the independence increment property together with the laws of the increments to obtain the joint law of any vector $(N_{t_1}, \ldots, N_{t_n})$ for any sequence $0 < t_1 < \cdots < t_n$ and then applies the so-called Kolmogorov extension theorem.

Problem 2.1.43 The following exercise is about the behavior of the simple Poisson process when $\lambda \to \infty$. Prove that

$$\lim_{\lambda \to \infty} \frac{N_T}{\lambda} = 1, (a.s).$$

This is also known as the renewal theorem when one considers large time. That is,

$$\lim_{T \to \infty} \frac{N_T}{T} = \lambda, (a.s).$$

Hint: Use characteristic functions.

Exercise 2.1.44 This exercise will help us to understand the definitions of stochastic integrals to appear in the chapters that follow.

1. Prove that for $\omega \in \Omega$, the function $t \to N_t(\omega)$ is of bounded variation on compact intervals.
2. Given the previous result prove that for any function $g \in C[0, T]$, one can define the integral $\int_0^T g(s) dN_s$.
3. In fact, give an alternative expression for the above integral using the jump times $\{T_i\}_{i \in \mathbb{N}}$.

---

[12] Recall that $o(h)$ is any term such that $\lim_{h \to 0} \frac{o(h)}{h} = 0$.

Chapter 3
Compound Poisson Process and Its Associated Stochastic Calculus

In this chapter, we enlarge on the previous chapter by considering processes whose jumps may be random but independent between them. We also give some further definitions on general theory of stochastic processes and stochastic analysis which may be easier to understand given that they are used in this particular application.

3.1 Compound Poisson Process

Definition 3.1.1 (*compound Poisson process*) A compound Poisson process $\{Z_t; t \geq 0\}$ with intensity $\lambda > 0$ and jump size distribution F is a stochastic process defined as

$$Z_t := \begin{cases} \sum_{j=1}^{N_t} Y_j & \text{if } N_t > 0, \\ 0 & \text{if } N_t = 0, \end{cases} \tag{3.1}$$

where jump sizes $\{Y_j; j \geq 1\}$ are i.i.d.r.v.s with distribution F and $\{N_t; t \geq 0\}$ is a Poisson process with intensity λ, independent from $\{Y_j; j \geq 1\}$.

One can prove the following theorem by easy computation.

Theorem 3.1.2 *The compound Poisson process has a characteristic function given by*

$$\mathbb{E}[e^{i\theta Z_t}] = \exp\{\lambda t(\mathbb{E}[e^{i\theta Y_j} - 1])\}.$$

In the particular case that the random variables Y_i have a density f we have that

$$\mathbb{E}[e^{i\theta Z_t}] = \exp\{\lambda t \int (e^{i\theta x} - 1)f(x)dx\}.$$

© Springer Nature Singapore Pte Ltd. 2019
A. Kohatsu-Higa and A. Takeuchi, *Jump SDEs and the Study of Their Densities*, Universitext, https://doi.org/10.1007/978-981-32-9741-8_3

Proof

$$\begin{aligned}
\mathbb{E}[e^{i\theta Z_t}] &= \mathbb{E}[e^{i\theta \sum_{j=1}^{N_t} Y_j}] = \sum_{n\geq 0} \mathbb{E}[e^{i\theta \sum_{j=1}^{n} Y_j} \mathbf{1}_{\{N_t=n\}}] \\
&= \sum_{n\geq 0} \mathbb{E}[e^{i\theta Y_j}]^n \frac{(\lambda t)^n}{n!} e^{-\lambda t} = \sum_{n\geq 0} \frac{(\lambda t \mathbb{E}[e^{i\theta Y_j}])^n}{n!} e^{-\lambda t} \\
&= e^{-\lambda t} \exp(\lambda t \mathbb{E}[e^{i\theta Y_j}]) = \exp\{\lambda t(\mathbb{E}[e^{i\theta Y_j} - 1])\}.
\end{aligned}$$

Corollary 3.1.3 *Let $\{N_t^{(1)}; t \geq 0\}$ and $\{N_t^{(2)}; t \geq 0\}$ be two independent Poisson processes with parameters λ_1 and λ_2, then $N^{(1)} + N^{(2)}$ is a Poisson process with parameter $\lambda = \lambda_1 + \lambda_2$. In fact, any linear combination of $N^{(1)}$ and $N^{(2)}$ is a compound Poisson process.*

Proof By Lemma 2.1.25, it is enough to use Theorem 1.1.12 to check that $\mathbb{E}[e^{iu(N_t^{(1)}+N_t^{(2)})}]$ is equal to the characteristic function of the stated Poisson process. In the general case the proof is similar but one has to be careful about the law of the jump size. We leave this as an exercise for the reader.

Exercise 3.1.4 In this exercise, we study the properties of the process $N_t^{(1)} - N_t^{(2)}$ in Corollary 3.1.3. In fact, we will prove later on that $N^{(1)} - N^{(2)}$ is itself has the structure of a compound Poisson process using a method which is slightly different from using characteristic functions. Here we give some exercises to point in this direction.

1. Prove that the following representation is valid:

$$N_t^{(1)} - N_t^{(2)} = \sum_{i=0}^{N_t^{(1)}+N_t^{(2)}} Y_i$$

   In particular, give the explicit definition for Y_i.
2. Prove that $Z_t := N_t^{(1)} - N_t^{(2)}$ has the property of independent increments.
3. Define τ as the first jump time of Z. Prove that $N_{\tau+t}^{(i)} - N_\tau^{(i)}$ is a Poisson process.
4. Prove that Y_i is a sequence of i.i.d.r.v.s with $\mathbb{P}(Y_1 = 1) = \frac{\lambda_1}{\lambda}$ and $\mathbb{P}(Y_1 = -1) = 1 - \mathbb{P}(Y_1 = 1)$.

For a hint, see Chap. 14.

Remark 3.1.5 (i) Note that this last result matches with the structure in Remarks 2.1.14 and 2.1.32. In the case of Corollary 3.1.3, we have that $\nu(dx) = \frac{\lambda_1}{\lambda}\delta_1(dx) + \frac{\lambda_2}{\lambda}\delta_{-1}(dx)$. In this case of Theorem 3.1.2, we have that $\nu(dx) = \lambda f(x)dx$. Recall that ν is the so-called Lévy measure associated with the process Z and which is unique.[1]

[1] State and prove the equivalent statement of Exercise 2.1.15.

(ii) The property revealed in Corollary 3.1.3 will be important in what follows as it will be used many times in more general settings.
(iii) Note that the time variable appears separately in the exponent. One may always claim it to be part of the definition of ν (note that we are not requiring ν to be a probability measure) if one considers only one random variable with fixed time. That is, one may think that the compound Poisson process is always observed at time 1 and that instead its intensity is given by λt. As this will no longer be the case in the study of processes, we prefer to consider it as a separate component. Still, such a interpretation is sometimes useful.

Theorem 3.1.6 (Wald's Lemma) *Let* $\{Y_j; j \geq 1\}$ *be i.i.d. r.v.'s with* $\mathbb{E}[|Y_j|^2] < \infty$, N *be a* $\mathbb{Z}^+$*-valued r.v. independent from* $(Y_j)_{j\geq 1}$. *Let*

$$S := \begin{cases} 0 & \text{if } (N = 0), \\ \sum_{i=1}^{N} Y_i & \text{if } (N \geq 1). \end{cases}$$

Then:

(i) If $\mathbb{E}[N] < \infty$, *then* $\mathbb{E}[S] = \mathbb{E}[N]\mathbb{E}[Y_1]$.
(ii) If $\mathbb{E}[N^2] < \infty$, *then* $\mathrm{Var}[S] = \mathbb{E}[N]\mathrm{Var}[Y_1] + \mathrm{Var}[N]\mathbb{E}[Y_1]^2$.
(iii) If $N \sim$ *Poisson*(λ), *then* $\mathrm{Var}[S] = \lambda\mathbb{E}[Y_1^2]$.

Result (ii) above states that the variance of a random sum has two components, one coming from the variance of the random variables being summed and the other due to the number of terms in the sum.

Proof (i) $S = \sum_{i=0}^{N} Y_i = \sum_{n\geq 0}\sum_{j=1}^{n} Y_j \mathbf{1}_{\{N=n\}}$, so that

$$\mathbb{E}[S] = \mathbb{E}\left[\sum_{n\geq 0}\sum_{j=1}^{n} Y_j \mathbf{1}_{\{N=n\}}\right] = \sum_{n\geq 0}\sum_{j=1}^{n} \mathbb{E}\left[Y_j \mathbf{1}_{\{N=n\}}\right]$$
$$= \sum_{n\geq 0}\sum_{j=1}^{n} \mathbb{E}[Y_j]\mathbb{E}\left[\mathbf{1}_{\{N=n\}}\right] = \sum_{n\geq 0} n \cdot \mathbb{E}[Y_1]\mathbb{P}(N = n) = \mathbb{E}[N]\mathbb{E}[Y_1].$$

Note that the above interchange of sums and expectations is possible due to the Lebesgue convergence theorem. The needed hypotheses follow from the assumptions.

(ii)

$$\mathbb{E}[S^2] = \mathbb{E}\left[\sum_{n\geq 0}\left(\sum_{j=1}^{n} Y_j\right)^2 \mathbf{1}_{\{N=n\}}\right]$$
$$= \sum_{n\geq 0}\mathbb{E}\left[\left(\sum_{j=1}^{n} Y_j\right)^2\right]\mathbb{P}(N = n)$$

$$= \sum_{n \geq 0} \left\{ \sum_{j=1}^{n} \mathbb{E}[Y_j^2] + \sum_{l \neq m} \mathbb{E}[Y_l]\mathbb{E}[Y_m] \right\} \mathbb{P}(N = n)$$

$$= \sum_{n \geq 0} \{n \cdot \mathbb{E}[Y_1^2] + (n^2 - n)\mathbb{E}[Y_1]^2\}\mathbb{P}(N = n)$$

$$= \mathbb{E}[Y_1^2]\mathbb{E}[N] + \mathbb{E}[Y_1]^2\{\mathbb{E}[N^2] - \mathbb{E}[N]\}.$$

So we get $\mathrm{Var}[S] = \mathbb{E}[N]\mathrm{Var}[Y_1] + \mathrm{Var}[N]\mathbb{E}[Y_1]^2$.

(iii) Since $N \sim Poisson(\lambda)$, $\mathbb{E}[N] = \lambda, \mathrm{Var}[N] = \lambda$. Therefore

$$\mathrm{Var}[S] = \lambda(\mathrm{Var}[Y_1] + \mathbb{E}[Y_1]^2) = \lambda\mathbb{E}[Y_1^2].$$

We remark that in the above proof the following useful identity (which was also discussed in Lemma 2.1.19) is hidden in the calculations. This identity may be useful in order to prove various properties:

$$\sum_{i=1}^{N_t} Y_i = \sum_{i=1}^{\infty} Y_i \mathbf{1}_{\{T_i \leq t\}}.$$

Property (ii) above states that the variance of a sum of i.i.d. random variables whose number of terms is random has two components, one due to the variability of X and the other due to the variability of N.

Theorem 3.1.7 *Let Z be a compound Poisson process. Then the law of $Z_t - Z_s$ is equal to the law of Z_{t-s}. In other words,*

$$\mathbb{P} \circ Z_{t-s}^{-1} = \mathbb{P} \circ (Z_t - Z_s)^{-1}, \quad (t > s).$$

Proof We prove that for any θ, $\mathbb{E}[e^{i\theta(Z_t - Z_s)}] = \mathbb{E}[e^{i\theta Z_{t-s}}]$. From Definition 3.1.1 and Proposition 2.1.23, we have

$$\mathbb{E}[e^{i\theta(Z_t - Z_s)}] = \mathbb{E}[e^{i\theta \sum_{j=N_s+1}^{N_t} Y_j} \mathbf{1}_{\{N_t > N_s\}}] + \mathbb{E}[e^{i\theta \cdot 0} \mathbf{1}_{\{N_t = N_s\}}]$$

$$= \sum_{k \geq 0} \sum_{l \geq k+1} \mathbb{E}[e^{i\theta \sum_{j=k+1}^{l} Y_j} \mathbf{1}_{\{N_t = l, N_s = k\}}] + \mathbb{P}(N_t = N_s)$$

$$= \sum_{k \geq 0} \sum_{l \geq k+1} \mathbb{E}[e^{i\theta Y_1}]^{l-k} \mathbb{P}(N_t = l, N_s = k) + \mathbb{P}(N_t = N_s)$$

$$= \sum_{k \geq 0} \sum_{j \geq 1} \mathbb{E}[e^{i\theta Y_1}]^j \mathbb{P}(N_t = k + j, N_s = k) + \sum_{k \geq 0} \mathbb{P}(N_t = N_s = k)$$

$$= \sum_{k \geq 0} \sum_{j \geq 0} \mathbb{E}[e^{i\theta Y_1}]^j \mathbb{P}(N_t = k + j, N_s = k)$$

$$= \sum_{k\geq 0}\sum_{j\geq 0} \mathbb{E}[e^{i\theta Y_1}]^j \mathbb{P}(N_t - N_s = j, N_s = k)$$

$$= \sum_{k\geq 0}\sum_{j\geq 0} \mathbb{E}[e^{i\theta Y_1}]^j \mathbb{P}(N_t - N_s = j)\mathbb{P}(N_s = k)$$

$$= \sum_{k\geq 0} \mathbb{P}(N_s = k) \sum_{j\geq 0} \mathbb{E}[e^{i\theta Y_1}]^j \mathbb{P}(N_t - N_s = j)$$

$$= \mathbb{E}[e^{i\theta Z_{t-s}}].$$

Proposition 3.1.8 (Sum of independent compound Poisson processes) *If* $\{Z_t^{(1)}; t \geq 0\}$, $\{Z_t^{(2)}; t \geq 0\}$ *are independent compound Poisson processes such that jumps sizes* $\{Y_j^{(1)}; j \geq 1\}$ *and* $\{Y_j^{(2)}; j \geq 1\}$ *are i.i.d.r.v.s with distribution* F*, then for any* $t \geq 0$,

$$\mathbb{E}[e^{i\theta(Z_t^{(1)}+Z_t^{(2)})}] = \exp\{\lambda t \mathbb{E}[e^{i\theta Y} - 1]\}.$$

Here $\lambda = \lambda_1 + \lambda_2$ and Y has the distribution F.

Exercise 3.1.9 Extend and prove the above result to the case that the distribution of $Y_j^{(1)}$ is F and the distribution of $Y_j^{(2)}$ is [2] G.

For a hint, see Chap. 14.

Exercise 3.1.10 Extend the results in Exercises 2.1.33 and 2.1.34 to the setting of compound Poisson processes. Notice that from a qualitative viewpoint there is no change in the properties. This will change in the next section!

Exercise 3.1.11 This exercise states a "converse" of Proposition 3.1.8. Let $Z_t = \sum_{i=1}^{N_t} Y_i$ be a compound Poisson process. For two measurable subsets A_1, A_2 such that $A_1 + A_2 = \mathbb{R}$, define the following processes for $k = 1, 2$:

$$N_t^k = \sum_{i=1}^{\infty} \mathbf{1}_{Y_i \in A_k} \mathbf{1}_{\{T_i \leq t\}},$$

$$Z_t^k = \sum_{i=1}^{N_t} Y_i \, \mathbf{1}_{Y_i \in A_k}.$$

Prove the following assertions:

1. $N_t = N_t^1 + N_t^2$ and similarly $Z_t = Z_t^1 + Z_t^2$.
2. $Z_t^k = \sum_{i=1}^{N_t} Y_i^k$ with $Y_i^1 = Y_i \, \mathbf{1}_{Y_i \in A_1}$ and $Y_i^2 = Y_i \, \mathbf{1}_{Y_i \in A_1^c}$.
3. Prove that $N^k, k = 1, 2$ are independent Poisson processes. Find their corresponding parameters.
4. Prove that Z^k is a compound Poisson process and its corresponding jump counting process is N^k.

---

[2] The new distribution of Y should be the mixture of the distributions of $Y^{(1)}$ and $Y^{(2)}$ with weights $\frac{\lambda_1}{\lambda}$ and $1 - \frac{\lambda_1}{\lambda}$.

5. Deduce from the above that Z^1 is independent of the jump times of Z^2. Also the jump times of Z^1 are independent of the jump times of Z^2.

   For a hint, see Chap. 14.

3.2 Lévy Process

Lévy processes are natural generalizations of compound Poisson processes. This generalization is done through its basic properties studied in Proposition 2.1.28. This leads to the following definition.

Definition 3.2.1 (*Lévy process*) A stochastic process $\{Z_t; t \geq 0\}$ is a Lévy process if the following conditions are satisfied:

(i) For any $n \in \mathbb{N}$ and $0 = t_0 < t_1 < \cdots < t_n$, the random variables $Z_{t_1} - Z_{t_0}, \cdots, Z_{t_n} - Z_{t_{n-1}}$ are independent.
(ii) $Z_0 = 0$, a.s.
(iii) The distribution of $Z_{t+h} - Z_t$ does not depend on t.
(iv) For any $a > 0$ and $t \geq 0$, $\lim_{h \to 0} P(|Z_{t+h} - Z_t| > a) = 0$.
(v) There exists $\Omega_0 \in \mathcal{F}$ with $\mathbb{P}(\Omega_0) = 1$ such that, for every $\omega \in \Omega_0$, $Z_t(\omega)$ is right-continuous in $t \geq 0$ and has left limits in $t > 0$, i.e. $Z_t(\omega)$ is a càdlàg function on $[0, \infty)$.

Problem 3.2.2 (*Brownian motion*) One of the most famous examples of Lévy process is Brownian motion. The above definition is trivially satisfied by the definition of Brownian motion whose increments are distributed according to a Gaussian law with mean zero and variance given by the corresponding time difference. This important example will only be marginally treated in this text.

We note that both the Poisson process and the compound Poisson process defined previously are examples of Lévy processes.

Corollary 3.2.3 *A Poisson process is a Lévy process.*

Corollary 3.2.4 *A compound Poisson process is a Lévy process.*

Exercise 3.2.5 Prove the above two corollaries recalling Proposition 2.1.28.

Remark 3.2.6 One theoretically interesting question which has to be dealt with is if the properties stated in Definition 3.2.1 uniquely characterize all probability laws of functionals of the Lévy process. We do not discuss these issues, just stating that these issues require first the proof of the specific construction of the corresponding probability space which shows that the process we are discussing exists. Secondly, we establish that the process has the path properties stated in (v) so that all measurable functionals are uniquely defined.[3]

---

[3]Actually this is the contents of the Kolmogorov extension theorem. The measurability has to be defined in the space of càdlàg functions with an appropriate topology. To make it simple, we may use for the moment the topology given by the supremum norm on compact intervals of time.

As we have already, sums of independent Poisson processes are Poisson processes (cf. Theorem 2.1.31). Similar properties repeat in Corollary 3.1.3 and Proposition 3.1.8. We can check easily that sums of independent Lévy processes are Lévy processes.

Theorem 3.2.7 *Sums of independent Lévy processes are Lévy processes.*

Exercise 3.2.8 Prove Theorem 3.2.7.

For the next theorem, which deals with the independent increment property for stopping times, we need to define $\mathscr{F}_t = \sigma(Z_s; s \le t)$ and $\mathscr{F}_\tau = \{A \in \mathscr{F}_\infty; A \cap \{\tau \le t\} \in \mathscr{F}_t, t > 0\}$.

Theorem 3.2.9 *Any Lévy process satisfies the following strong Markov property for any bounded function f and any set $A_\tau \in \mathscr{F}_\tau$:*

$$\mathbb{E}[f(Z_{\tau+s} - Z_\tau); A_\tau] = \mathbb{E}[f(Z_{\tau+s} - Z_\tau)]\mathbb{P}(A_\tau) = \mathbb{E}[f(Z_s)]\mathbb{P}(A_\tau).$$

We leave the proof of this statement as an exercise, noting that the proof is similar to Exercise 2.1.39 which is obtained by considering an appropriate simpler approximation of the quantity under consideration and then taking limits.

Exercise 3.2.10 Let Z be a compound Poisson process defined as

$$Z_t := \sum_{i=1}^{N_t} Y_i,$$

such that N is a Poisson process of parameter λ and $\{Y_i\}_i$ is a sequence of *i.i.d.* random variables such that $\mathbb{P}(Y_i = 0) = 1/2$, $\mathbb{P}(Y_i = 1) = 1/8$ and $\mathbb{P}(Y_i = -2) = 3/8$.

Prove that Z is "equivalent" to the stochastic process

$$Z'_t =:= \sum_{i=1}^{N'_t} Y'_i,$$

where N' is a Poisson process of parameter $\lambda/2$ and $\{Y'_i\}_i$ is a sequence of *i.i.d.* random variables such that $\mathbb{P}(Y'_i = 1) = 1/4$ and $\mathbb{P}(Y'_i = -2) = 3/4$.

Exercise 3.2.11 This is a continuation of Exercise 2.1.44. Let Z be a compound Poisson process.

1. Prove that for $\omega \in \Omega$, the function $t \to Z_t(\omega)$ is of bounded variation on compact intervals.
2. Given the previous result prove that for any function $g \in C[0, T]$, one can define the integral $\int_0^T g(s)dZ_s$.
3. In fact, give an alternative expression for the above integral using the jump times and jump sizes $\{(T_i, Y_i)\}_{i\in\mathbb{N}}$.

3.3 Poisson Random Measure

We will now try to discuss in detail the structure of Lévy processes from the point of view of extensions of compound Poisson processes. This will also tie the notation introduced in Remarks 2.1.14, 2.1.32 and 3.1.5 which gives the structure of the process from the point of view of its characteristic function. This could not be seen directly in Theorem 3.2.7.

Throughout the following sections we will assume without loss of generality the following hypothesis for any jump process: That the probability that a jump of size 0 is 0. That is, e.g. in the case of a compound Poisson process, we assume that for any Y_i, $\mathbb{P}(Y_i = 0) = 0$ (recall Exercise 3.2.10).

This hypothesis is a general hypothesis that is always assumed in the study of Lévy processes. We will instead make a much stronger hypothesis in order to simplify our calculations. We will assume that Y_i has a density function given by $f(y)$, $y \in \mathbb{R}$.

Definition 3.3.1 (*Dirac point mass measure*) For any a $\in \mathbb{R}$, the Dirac point mass measure δ_a at a is defined as

$$\delta_a(A) = \begin{cases} 1 \text{ if } a \in A, \\ 0 \text{ if } a \notin A. \end{cases}$$

Definition 3.3.2 (*Poisson random measure*) Let $\{Z_t = \sum_{i=1}^{\infty} Y_i \mathbf{1}_{\{T_i \le t\}}; t \geq 0\}$ be a compound Poisson process. Then $\mathcal{N}$ is called a Poisson random measure associated with $\{Z_t; t \geq 0\}$, if

$$\mathcal{N}(F) := \sum_{i=1}^{\infty} \delta_{Y_i} \otimes \delta_{T_i}(F), \quad F \in \mathscr{B}(\mathbb{R}) \otimes \mathscr{B}[0, \infty).$$

Exercise 3.3.3 Check that the Poisson random measure is a measure.

It should be clear from the above definition that the goal of the Poisson random measure is to count how many times events associated with Z have happened in the product space $\mathbb{R} \times [0, \infty)$.

Lemma 3.3.4 *Let* (S, Σ) *be a measurable space and let* $(\mu_n)_n$ *and* μ *be measures on* (S, Σ) *with* $\mu = \sum_{n \geq 1} \mu_n$. *Then for any* $f \in L^1(S, \Sigma, \mu)$,

$$\int_S f(s)\mu(ds) = \sum_{n=1}^{\infty} \int_S f(s)\mu_n(ds).$$

Proposition 3.3.5 *Let $\mathcal{N}$ be a Poisson random measure associated with $\{Z_t;\, t \geq 0\}$. Then*

(i) $\displaystyle\int_{\mathbb{R}\times[0,\infty)} \mathbf{1}_{\mathbb{R}\times[0,t]}(z,s)\mathcal{N}(dz,ds)$ *is a Poisson process.*

(ii) *For any $A \in \mathscr{B}(\mathbb{R})$, $B \in \mathscr{B}[0,\infty)$ and for any $g \in L^1(\mathbb{R}\times[0,\infty), \mathscr{B}(\mathbb{R}) \otimes \mathscr{B}[0,\infty), \mathcal{N})$,*

$$\int_{\mathbb{R}\times[0,\infty)} g(z,s)\mathbf{1}_{A\times B}(z,s)\mathcal{N}(dz,ds) = \sum_{i=1}^{\infty} g(Y_i,T_i)\mathbf{1}_{\{Y_i\in A\}}\mathbf{1}_{\{T_i\in B\}}. \tag{3.2}$$

In particular, if $g : \mathbb{R} \to \mathbb{R}$, then

$$\int_{\mathbb{R}\times[0,\infty)} g(z)\mathbf{1}_{\mathbb{R}\times[0,t]}(z,s)\mathcal{N}(dz,ds)$$

is a compound Poisson process with each jump size following the same distribution as that of $g(Y)$.

Proof **(i)** Trivial.

(ii) From Lemma 3.3.4, we have

$$\int_{\mathbb{R}\times[0,\infty)} g(z,s)\mathbf{1}_{A\times B}(z,s)\mathcal{N}(dz,ds) = \sum_{i=1}^{\infty} g(Y_i,T_i)\mathbf{1}_{\{Y_i\in A\}}\mathbf{1}_{\{T_i\in B\}}.$$

Remark 3.3.6 We remark that in general this last integral, (3.2), is no longer a compound Poisson process as the jump size may now depend on the time at which the jump happens. Therefore one of the goals of this new notation is to allow for greater generality. On the other hand, also note that this integral is a càdlàg function in t for $B = [0,t]$. For more on this, see Sect. 5.4.

Proposition 3.3.7 *The characteristic function of the r.v.*

$$\int_{\mathbb{R}\times[0,\infty)} g(z,s)\mathbf{1}_{A\times B}(z,s)\mathcal{N}(dz,ds)$$

is given by

$$\exp\left(\lambda \int_{\mathbb{R}\times[0,\infty)} (e^{i\theta g(z,s)} - 1)\mathbf{1}_{A\times B}(z,s)f(z)dzds\right).$$

Here we suppose that the above integral is finite.

Proof It is enough to consider the case of $\int_{\mathbb{R}\times[0,t]} g(z)\mathbf{1}_A(z)\mathcal{N}(dz,ds)$. We leave the details of generalizing these arguments to simple functions in the time variable and their corresponding limits. The characteristic function in this case is given by

$$\mathbb{E}[\prod_{j=1}^{N_t} \exp(i\theta g(Y_j)\, \mathbf{1}_{\{Y_j \in A\}})] = \exp\left(\lambda t(\mathbb{E}[\exp(i\theta g(Y)\, \mathbf{1}_{\{Y \in A\}})] - 1)\right)$$
$$= \exp\left(\lambda t(\mathbb{E}[\, \mathbf{1}_{\{Y \in A\}}(\exp(i\theta g(Y)) - 1)])\right).$$

From here the result follows.

Definition 3.3.8 In concordance with previous definitions, notice that if $g(z, s) = z$ then

$$Z_t = \int_{\mathbb{R}\times[0,t]} z\mathcal{N}(dz, ds).$$

Its characteristic function is given by the above expression which is an extension of Theorem 3.1.2. From now on, we define $\nu(dx) = \lambda f(x)dx$ as the associated Lévy measure. See also Remark 3.1.5.

Corollary 3.3.9 *For any $A, B \in \mathscr{B}(\mathbb{R})$ and $0 \leq a < b \leq c < d$, $\mathcal{N}(A \times (a, b])$ and $\mathcal{N}(B \times (c, d])$ are independent random variables.*

Proof Either compute $\mathcal{N}(A \times (a, b])$ and $\mathcal{N}(B \times (c, d])$ directly to check that the statement holds or use Proposition 3.3.7.

From this result, one can already see the formation of the parallel results to Theorem 3.2.7 and Corollary 3.1.3 using the results in Corollary 3.3.9 and Proposition 3.3.7. That is, whenever one wants to construct a Lévy process, one may as well build it first in particular subsets of the jump space and then adding independent copies of each to obtain that the final result will give the desired process.

This issue will be important for the constructions to follow in the next two chapters. For the moment, we will now move to the study of the filtration generated by a compound Poisson process, its martingale properties etc.

Exercise 3.3.10 This exercise is about the approximation of the Gaussian distribution using compound Poisson process. Let f be the density of the jump size distribution. Assume that f follows the Laplace density function with mean zero and scale parameter $b > 0$. Prove that if $\lambda = cb^{-2} \to \infty$ for fixed $c > 0$ then the characteristic function of the corresponding compound Poisson process at time t converges to the characteristic function of a Gaussian random variable with variance $2ct$. This means that the corresponding random variables converge in law according to Theorem 1.1.6. Furthermore notice that one may interpret this as a Central Limit type Theorem with random number of summands. Hint: Recall that the characteristic function of the Laplace density function is $\int e^{itx} f(x)dx = \frac{1}{1+t^2b^2}$.

Exercise 3.3.11 This is an extension of the previous exercise.[4] Assume that the density of the jump size of a compound Poisson process with parameter λ is a gen-

[4]This is an extension in the sense that we can measure the error of the characteristic function associated with approximating the compound Poisson process using Gaussian laws.

eral density function, denoted by f, which is an even function and which satisfies that $\sum_{k=0}^{\infty} \frac{C^k}{k!} \int |x|^k f(x)dx < \infty$ for any $C > 0$.

Denote by $\varphi_{Z_t}(\theta)$ the characteristic function of the compound Poisson process at time t. Similarly, denote by $\varphi_{X_t}(\theta)$ the characteristic function of a Gaussian distribution with mean zero and variance $\sigma^2 t := \lambda t \int x^2 f(x)dx$. Prove the following approximation estimate for the error between characteristic functions:

$$\varphi_{Z_t}(\theta) = \exp\left(\sum_{j=3}^{\infty} \lambda t \frac{(i\theta)^k}{k!} \mu_k\right) \varphi_{X_t}(\theta).$$

Here $\mu_k = \int x^k f(x)dx$.[5]

3.4 Stochastic Calculus for Compound Poisson Processes

Definition 3.4.1 Let $\{Z_t; t \geq 0\}$ be a compound Poisson process. Then we define the increment of Z at s (also called the jump of Z at s) as follows:

$$\Delta Z_s := Z_s - Z_{s-}.$$

Note that ΔZ satisfies the following properties:

1. For any time interval $[0, t]$, $\Delta Z_s \neq 0$ for a finite number of values of s.
2. From the above, we may abuse the sum notation to have that

$$\sum_{s \leq t} \Delta Z_s := \sum_{\Delta Z_s \neq 0,\, s \leq t} \Delta Z_s \equiv \sum_{s \leq t} \Delta Z_s.$$

3. From the above one has that

$$\begin{aligned} Z_t &= \sum_{s \leq t} \Delta Z_s = \sum_{i=1}^{N_t} Y_i \\ &= \sum_{i=1}^{\infty} (Z_{T_i} - Z_{T_i-}) \mathbf{1}_{\{T_i \leq t\}}. \end{aligned}$$

The above abuse of the summation notation where a non-denumerable number of terms are summed which only a countable number are different from zero will be used in the future.

[5] If you have some experience with Fourier transforms you may even realize that this implies an expansion of the distribution of Z_t using the Gaussian distribution. But one has to be careful as Z_t in general does not have a density in the case that $N_t = 0$.

We now define the filtration[6] $\mathscr{F} = \{\mathscr{F}_t; t \geq 0\}$ associated with the compound Poisson process Z as

$$\begin{aligned}
\mathscr{F}_t &:= \sigma(Z_s; s \leq t) = \sigma(\Delta Z_s; s \leq t) \\
&= \sigma\Big(\sum_{i=1}^{\infty} Y_i \mathbf{1}_{\{T_i \leq s\}}; s \leq t\Big) \\
&\overset{*}{=} \sigma(Y_i \mathbf{1}_{\{T_i \leq s\}}; s \leq t, i \in \mathbb{N}). \qquad (3.3)
\end{aligned}$$

Note that the inclusion $\sigma(\sum_{i=1}^{\infty} Y_i \mathbf{1}_{\{T_i \leq s\}}; s \leq t) \subseteq \sigma(Y_i \mathbf{1}_{\{T_i \leq s\}}; s \leq t, i \in \mathbb{N})$ is obvious. Before giving the ideas for proving the reverse inclusion let us give a preparatory lemma.

The filtration above is the smallest filtration generated by the processes $Y_i\ \mathbf{1}_{\{T_i \leq s\}}$ and $\mathbf{1}_{\{T_i \leq s\}}$ and completed with the null sets. That is,

Lemma 3.4.2 $\sigma(Y_i \mathbf{1}_{\{T_i \leq s\}}; s \leq t, i \in \mathbb{N}) = \sigma(Y_i \mathbf{1}_{\{T_i \leq s\}}, \mathbf{1}_{\{T_i \leq s\}}; s \leq t, i \in \mathbb{N})$.

Proof Let $\mathscr{G}_t := \sigma(Y_i \mathbf{1}_{\{T_i \leq s\}}, \mathbf{1}_{\{T_i \leq s\}}; s \leq t, i \in \mathbb{N})$ and $\mathscr{H}_t := \sigma(Y_i \mathbf{1}_{\{T_i \leq s\}}; s \leq t, i \in \mathbb{N})$.

One clearly has that $\mathscr{H}_t \subset \mathscr{G}_t$. Since for any $s \leq t$, $\mathscr{H}_s \subset \mathscr{H}_t$, it is enough to prove that

$$\{T_i > t\} \in \mathscr{H}_t, \quad \forall t \geq 0.$$

It is obvious that

$$\{T_i > t\} = \Big(\bigcap_{s \leq t} \{\omega \in \Omega; Y_i(\omega)\mathbf{1}_{\{T_i \leq s\}}(\omega) = 0\} \cap \{Y_i \neq 0\}\Big) \cup \Big(\{T_i > t\} \cap \{Y_i = 0\}\Big).$$

Since $\mathbb{P}(Y_i = 0) = 0$ and all σ-fields are completed with respect to the probability space $(\Omega, \mathscr{F}, \mathbb{P})$ we have $\{T_i > t\} \cap \{Y_i = 0\} \in \mathscr{H}_t$. For notational convenience, let us denote two sets,

$$\begin{aligned}
L &:= \bigcap_{s \leq t} \{\omega \in \Omega; Y_i(\omega)\mathbf{1}_{\{T_i \leq s\}}(\omega) = 0\}, \\
R &:= \bigcap_{\substack{s \leq t, \\ s \in \mathbb{Q} \cup \{t\}}} \{\omega \in \Omega; Y_i(\omega)\mathbf{1}_{\{T_i \leq s\}}(\omega) = 0\}.
\end{aligned}$$

We will now prove that $L = R$.

$L \subset R$ is obvious. By contradiction, suppose that $\omega \in R$, $\omega \notin L$. Then there exists $s_0 \notin \mathbb{Q} \cup \{t\}$ such that

[6] For a much more formal and general definition see Definition 3.4.13. But note that a filtration and a filtered space are not the same thing. A filtration, in general, is just an increasing sequence of σ algebras.

$$Y_i(\omega)\mathbf{1}_{\{T_i \le s_0\}}(\omega) \neq 0.$$

Define $g(s) := Y_i(\omega)\mathbf{1}_{\{T_i \le s\}}(\omega)$. Since g is a right-continuous function, for any rational sequence $\{y_n; n \in \mathbb{N}\}$ such that $y_n \downarrow s_0$,

$$0 \neq g(s_0) = \lim_{n\to\infty} g(y_n) = \lim_{n\to\infty} Y_i(\omega)\mathbf{1}_{\{T_i \le y_n\}}(\omega) = 0.$$

This is a contradiction. Hence $L = R$. Since for any $s \le t$, $Y_i\mathbf{1}_{\{T_i \le s\}}$ is an $\mathscr{H}_t$-measurable function, $\{T_i > t\} \in \mathscr{H}_t$.

Note that in the above proof we have used the fact that $P(Y_i = 0) = 0$. Lemma 3.4.2 is a prelude to issues of right-continuity of filtrations, completeness etc. which will not be treated in this text in full generality as they are quite technical. An advanced student can look at [34] for a glance at what we mean here.

But to start with, recall that we have always assumed that all filtrations are completed. Then one can verify that these filtrations are right-continuous.[7]

If you wish to be reassured, one can make any filtration right-continuous, by defining $\mathscr{F}_{t+} = \cap_{s>t}\mathscr{F}_s$. The main reason for doing this is that any martingale [8] in continuous time defined over such a filtration will have a version.[9] which is càdlàg[10]

Exercise 3.4.3 Prove the equality $\stackrel{*}{=}$ in (3.3). We suggest that the reader think of making the following statements rigorous:

- For $\mathbf{1}_{\{T_1 > s\}}$ with $s \le t$, note that this indicator function is one if and only if for any positive rational number $s_1 \le s$, $\sum_{i=1}^{\infty} Y_i\mathbf{1}_{\{T_i \le s_1\}} = 0$ with the exception of sets of measure zero.
- Let $s \le t$ and $A \in \mathscr{B}(\mathbb{R})$ then $Y_1\mathbf{1}_{\{T_1 \le s\}} \in A$ if and only if there exists $s_1 \le s$ with $s_1 \in \mathbb{Q}$ such that for all $s_2 \le s_1$, $\sum_{i=1}^{\infty} Y_i\mathbf{1}_{\{T_i \le s_2\}} = 0$ and $\sum_{i=1}^{\infty} Y_i\mathbf{1}_{\{T_i \le s_1\}} \in A$ with the exception of a set of probability zero.
- Finally, one has to generalise the above statements by induction for general $\mathbf{1}_{\{T_i > s\}}$ and $Y_i\mathbf{1}_{\{T_i \le s\}}$.

Corollary 3.4.4 *Let $g \in L^1(\mathbb{R}, \mathscr{B}(\mathbb{R}), \mathscr{N}(\cdot, [0, t]))$ for all $t > 0$ and let $A \in \mathscr{B}(\mathbb{R})$. Define*

$$\tilde{Z}_t := \int_{\mathbb{R}\times[0,\infty)} g(z)\mathbf{1}_{A\times[0,t]}(z, s)\mathscr{N}(dz, ds).$$

Then $\tilde{Z}_t$ is a $\mathscr{F}_t$-measurable function and for any $c \ge 0$, $\tilde{Z}_t - \tilde{Z}_c$ is independent of $\mathscr{F}_c$.

---

[7] This property follows for Lévy processes.

[8] See Definition 3.4.17.

[9] We say that a stochastic process X is a version of the stochastic process Y if $\mathbb{P}(X_t = Y_t) = 1$ for all $t \ge 0$.

[10] As we will see this will become useful when considering stochastic integrals.

Proof Let $\hat{g} = g\mathbf{1}_A$. From Proposition 3.3.5-(ii),

$$\tilde{Z}_t = \sum_{i=1}^{\infty} \hat{g}(Y_i)\mathbf{1}_{\{T_i \le t\}} = \sum_{i=1}^{\infty} \hat{g}(Y_i\, \mathbf{1}_{\{T_i \le t\}})\mathbf{1}_{\{T_i \le t\}} \in \mathscr{M}(\mathscr{F}_t).$$

The other conclusions follow due to Corollary 3.2.4.[11]

Proposition 3.4.5 *Let* $g \in \mathscr{M}_b(\mathbb{R})$. *Then* $\mathbb{E}\left[\left|\int_{(a,b]\times(c,d]} g(z)\mathcal{N}(dz, ds)\right|\right] < \infty$ *and*

$$\mathbb{E}\Big[\int_{(a,b]\times(c,d]} g(z)\mathcal{N}(dz, ds)|\mathscr{F}_c\Big] = \int_c^d \int_a^b \lambda g(z)f(z)dzds$$

where f *is a probability density function of* Y.

Proof Since Y_i, T_i are independent and $\{Y_i;\, i \in \mathbb{N}\}$ are i.i.d.r.v.s,

$$\begin{aligned}
(LHS) &= \sum_{i=1}^{\infty} \mathbb{E}[g(Y_i)\mathbf{1}_{\{Y_i\in(a,b]\}}\mathbf{1}_{\{T_i\in(c,d]\}}] \\
&= \mathbb{E}[g(Y)\mathbf{1}_{\{Y\in(a,b]\}}] \sum_{i=1}^{\infty}\{\mathbb{P}(N_d \ge i) - \mathbb{P}(N_c \ge i)\} \\
&= \mathbb{E}[g(Y)\mathbf{1}_{\{Y\in(a,b]\}}]\{\mathbb{E}[N_d] - \mathbb{E}[N_c]\} \\
&= \int_a^b g(z)f(z)dz\, \lambda(d-c) = (RHS).
\end{aligned}$$

Remark 3.4.6 (Important remark about integrability) Note that in the above proof integrability of the (RHS), the exchange of limits, sums and expectations has not been discussed in detail. This is due to the fact that g is bounded and we hope that you are already at ease with this argument. This issue will not be discussed in detail in the proofs that follow. In fact, the other method to prove integrability is to first assume without loss of generality that g only takes positive values and then deal separately with the positive and negative part. This will also be used in what follows.

Exercise 3.4.7 Suppose that A is a bounded subset of $\mathbb{R}$. Prove that the random variable $\int_{A\times[0,t]} \mathcal{N}(dz, ds)$ is a Poisson random variable. Compute its parameter.

3.4.1 Compensated Poisson Random Measure

As we will start taking limits on the paths of various càdlàg processes, we will need some path properties that are valid for any càdlàg function.

[11] Note that $\tilde{Z}$ can be rewritten using Z.

Lemma 3.4.8 *Let f be a càdlàg function. Then, for any $\varepsilon > 0$, there exists a partition $t_0 = 0 < t_1 < \ldots < t_n = T$ and $\delta > 0$ such that $t_{i+1} - t_i < \delta$ and*

$$\max_i \sup_{s,t \in (t_i, t_{i+1}]} |f(t) - f(s)| < \varepsilon.$$

Exercise 3.4.9 Prove the above lemma.[12] Also prove that a càdlàg function can only have at most a countable number of discontinuities. The solution is an extension of a classical result for continuous functions which states that continuous functions on compacts are uniformly continuous.

For a hint, see Chap. 14.

Definition 3.4.10 (*Compensator, compensated Poisson random measure*) Let $\mathcal{N}$ be a Poisson random measure. Then we define the compensator of $\mathcal{N}$ as the σ-finite measure

$$\widehat{\mathcal{N}}(F) := \lambda \int_F f(z) dz ds, \quad F \in \mathscr{B}(\mathbb{R}) \otimes \mathscr{B}[0, \infty)$$

and the compensated Poisson random measureas

$$\widetilde{\mathcal{N}} := \mathcal{N} - \widehat{\mathcal{N}}.$$

Theorem 3.4.11 *Let g be a bounded $\mathscr{B}(\mathbb{R}) \otimes \mathscr{B}[0, \infty)$-measurable function where $g(z, \cdot)$ is a càdlàg function for any $z \in \mathbb{R}$. Then for any partition : $0 \le t_0 < t_1 < \cdots < t_n = t$,*

$$\mathbb{E}\Big[\sum_{i=0}^{n-1} \int_{\mathbb{R}\times(t_i, t_{i+1}]} g(z, t_i) \mathcal{N}(dz, ds) | \mathscr{F}_{t_0}\Big] = \sum_{i=0}^{n-1} \int_{\mathbb{R}\times(t_i, t_{i+1}]} g(z, t_i) \widehat{\mathcal{N}}(dz, ds).$$

Furthermore if[13] $\max\{|t_{i+1} - t_i|;\ i = 0, \ldots, n-1\} \to 0$ *as $n \to \infty$ then by taking limits we also have*

$$(LHS) \to \mathbb{E}\Big[\int_{\mathbb{R}\times(t_0, t]} g(z, s-) \mathcal{N}(dz, ds) | \mathscr{F}_{t_0}\Big] \ (n \to \infty),$$

$$(RHS) \to \int_{\mathbb{R}\times(t_0, t]} g(z, s-) \widehat{\mathcal{N}}(dz, ds) \ (n \to \infty).$$

Moreover if $\int_{\mathbb{R}\times(t_0, t]} g(z, s-) \mathcal{N}(dz, ds) \in L^1(\Omega, \mathscr{F}, \mathbb{P})$, then we get

$$\mathbb{E}\left[\int_{\mathbb{R}\times(t_0, t]} g(z, s-) \mathcal{N}(dz, ds) | \mathscr{F}_{t_0}\right] = \int_{\mathbb{R}\times(t_0, t]} g(z, s-) \widehat{\mathcal{N}}(dz, ds).$$

---

[12]Hint: Recall the argument to prove that a continuous function on a bounded interval is uniformly continuous.

[13]This is usually called the norm of the partition.

Proof **Step 1.** $(LHS) = (RHS)$. It is obvious from Proposition 3.4.5.

Step 2. Convergence of (RHS). From properties of Lebesgue integrals, we have

$$\sum_{i=0}^{n-1} \int_{t_i}^{t_{i+1}} g(z, t_i)ds = \sum_{i=0}^{n-1} g(z, t_i)(t_{i+1} - t_i)$$
$$\overset{n\uparrow\infty}{\to} \int_{t_0}^{t} g(z, s-)ds \left(= \int_{t_0}^{t} g(z, s)ds \right).$$

Using this result, we obtain that

$$\begin{aligned}(RHS) &= \sum_{i=1}^{n-1} \int_{t_i}^{t_{i+1}} \int_{\mathbb{R}} \lambda g(z, t_i)f(z)dzds \\ &= \int_{\mathbb{R}} \lambda \sum_{i=1}^{n-1} \int_{t_i}^{t_{i+1}} g(z, t_i)ds f(z)dz \\ &\to \lambda \int_{\mathbb{R}} \int_{t_0}^{t} g(z, s-)dsf(z)dz.\end{aligned}$$

Hence the (RHS) converges to $\int_{\mathbb{R}\times(t_0,t]} g(z, s-)\widehat{\mathcal{N}}(dz, ds)$ due to the bounded convergence theorem, as $g(z, t_i)$ is uniformly bounded.

Step 3. Convergence of (LHS). We will carry the proof of convergence in two steps: First, we will prove the convergence of the corresponding random variables a.s. for all $\omega \in \Omega$ and then we will consider the limit in expectation.[14]

We define

$$F(\omega) := \{k \in \mathbb{N};\ T_k \in (t_0, t]\}.$$

Since $T_k \uparrow \infty$, a.s., $|F| < \infty$, a.s. Also, from Proposition 3.3.5, we get that

$$\sum_{i=0}^{n-1} \int_{\mathbb{R}\times(t_i,t_{i+1}]} g(z, t_i)\mathcal{N}(dz, ds) = \sum_{k=1}^{\infty} \sum_{i=0}^{n-1} g(Y_k, t_i)\, \mathbf{1}_{\{T_k \in (t_i, t_{i+1}]\}},$$
$$\int_{\mathbb{R}\times(t_0,t]} g(z, s-)\mathcal{N}(dz, ds) = \sum_{k=1}^{\infty} g(Y_k, T_k-)\, \mathbf{1}_{\{T_k \in (t_0, t]\}}.$$

[14]This is done with the purpose of exercising the use of the Poisson random point measure.

Hence,

$$
\begin{aligned}
&\left|\sum_{i=0}^{n-1}\int_{\mathbb{R}\times(t_i,t_{i+1}]} g(z,t_i)\mathcal{N}(dz,ds) - \int_{\mathbb{R}\times(t_0,t]} g(z,s-)\mathcal{N}(dz,ds)\right| \\
&= \left|\sum_{k=1}^{\infty}\left(\sum_{i=0}^{n-1} g(Y_k,t_i)\,\mathbf{1}_{\{T_k\in(t_i,t_{i+1}]\}} - g(Y_k,T_k-)\,\mathbf{1}_{\{T_k\in(t_0,t]\}}\right)\right| \\
&\le \sum_{k\in F}\left|\sum_{i=0}^{n-1} g(Y_k,t_i)\,\mathbf{1}_{\{T_k\in(t_i,t_{i+1}]\}} - g(Y_k,T_k-)\right|.
\end{aligned}
$$

Using Lemma 3.4.8, we have

$$
\left|\sum_{i=0}^{n-1}\int_{\mathbb{R}\times(t_i,t_{i+1}]} g(z,t_i)\mathcal{N}(dz,ds) - \int_{\mathbb{R}\times(t_0,t]} g(z,s-)\mathcal{N}(dz,ds)\right| \to 0 \text{ as } n\to\infty, \text{ a.s.}
$$

Since g is a bounded function, we obtain the convergence of (LHS) in $L^1(\Omega)$ by the dominated convergence theorem. Therefore the convergence of conditional expectations follows.

Exercise 3.4.12 Carry out the above proof without using the finiteness of the number of jumps. Instead use the dominated convergence theorem. The reason why we preferred the above proof is so that the reader practices the pathwise calculation used in the proof.

Definition 3.4.13 (*Filtered probability space*) Let $(\Omega, \mathcal{F}, \mathbb{P})$ be a probability space. If the family of sub σ-algebras $\mathcal{F}_t$ of $\mathcal{F}$ satisfy the following conditions, $(\Omega, \mathcal{F}, \{\mathcal{F}_t\}_{t\ge 0}, \mathbb{P})$ is called a filtered probability space:

1. Monotonicity: If $0 \le s < t$, then $\mathcal{F}_s \subset \mathcal{F}_t$.
2. Right-continuity: For all $t \ge 0$, $\mathcal{F}_t = \mathcal{F}_{t+}\left(:=\bigcap_{s>t}\mathcal{F}_s\right)$.
3. Completeness: Let $\mathcal{N}$ be a set of all $\mathbb{P}$-negligible sets,[15] then $\mathcal{N} \subset \mathcal{F}_0$.

Definition 3.4.14 (*Adapted*) If the stochastic process $X := \{X_t; t \ge 0\}$ satisfies

$$
X_t \in \mathcal{M}(\mathcal{F}_t), \forall t \ge 0,
$$

then we say that X is a $\{\mathcal{F}_t\}_{t\ge 0}$-adapted process.

Definition 3.4.15 (*Progressively measurable*)[16] Given the stochastic process $X := \{X_t; t \ge 0\}$, we say that it is progressively measurable if for any $t > 0$, $X : [0,t] \times \Omega \to \mathbb{R}$ is measurable with respect to the σ-field $\mathcal{B}[0,t] \times \mathcal{F}_t$.

---

[15] Recall that a negligible set is any subset of a measurable set with probability zero.

[16] Note the link between this definition and Exercise 2.1.38.

Exercise 3.4.16 1. Prove that a progressively measurable process is an adapted process.
2. Prove that if X^n is a sequence of progressively measurable processes such that $\lim_{n\to\infty} X_t^n(\omega) = X_t$ for all $t > 0$ and $\omega \in \Omega$ then X is progressively measurable.
3. Given a stochastic process X such that it has right-limits at each point $t > 0$ for all $\omega \in \Omega$, prove that X is progressively measurable. Hint: Use an approximation argument. That is, find a sequence of progressively measurable processes X^n such that $X^n \to X$.

Definition 3.4.17 (*Martingale*) Let $X := \{X_t; t \geq 0\}$ be a stochastic process and $(\Omega, \mathscr{F}, \{\mathscr{F}_t\}_{t\geq 0}, \mathbb{P})$ be a filtered probability space. If X satisfies:

1. X is $\{\mathscr{F}_t\}_{t\geq 0}$-adapted,
2. $E[|X_t|] < \infty, \forall t \geq 0$,
3. $E[X_t|\mathscr{F}_s] = X_s$, a.s., $\forall 0 \leq s < t$,

then we say that X is a $\{\mathscr{F}_t\}_{t\geq 0}$-martingale.

Note that under the current situation $\mathscr{N}$ and $\widetilde{\mathscr{N}}$ are finite measures on finite intervals, therefore integrals with respect to these random measures can be defined naturally. Therefore for a measurable bounded function g, $\int_{\mathbb{R}\times[0,t]} g(z, s-)\widetilde{\mathscr{N}}(dz, ds)$ is always well defined.

Remark 3.4.18 Note that in the present situation $\mathscr{N}$ is a point measure and therefore one may have that the measure of $\mathscr{N}(\mathbb{R} \times \{0\})$ is not zero for a general point measure $\mathscr{N}$. Still, in the present case of a compound Poisson process and in all cases to follow, $\mathscr{N}(\mathbb{R} \times \{0\}) = 0$, a.s. Still, we remind the reader of this possibility and that in general, we have to be careful when we integrate open or closed intervals when $\omega \in \Omega$ is fixed. In other books you may see that the author may prefer the notation $\int_{\mathbb{R}\times(0,t]} g(z, s-)\widetilde{\mathscr{N}}(dz, ds)$, which is used to remind the reader of this possibility.

Theorem 3.4.19 *Let g be a $\mathscr{B}(\mathbb{R}) \otimes \mathscr{B}[0, \infty)$-measurable càdlàg function and integrable*[17] *with respect to $\mathscr{N}$ and $\widehat{\mathscr{N}}$. Then the process*

$$\left\{\int_{\mathbb{R}\times[0,t]} g(z, s-)\widetilde{\mathscr{N}}(dz, ds), t \geq 0\right\}$$

is a martingale relative to the filtration $\{\mathscr{F}_t\}_{t\geq 0}$. Furthermore the above integral is a càdlàg function in t.

Proof First, without loss of generality we assume that $g(z, s-) \geq 0$ for all $(z, s) \in \mathbb{R} \times [0, \infty)$. Taking limits[18] in Theorem 3.4.11 we obtain that for any $0 < u < t$,

[17]In the sense that $\mathbb{E}[\int_{\mathbb{R}\times[0,t]} |g(z, s-)|\mathscr{N}(dz, ds)] < \infty$ and $\mathbb{E}[\int_{\mathbb{R}\times[0,t]} |g(z, s-)|\widehat{\mathscr{N}}(dz, ds)] < \infty$.

[18]This is an exercise. This also proves the integrability needed in order to obtain the martingale property.

$$\mathbb{E}\Big[\int_{\mathbb{R}\times(0,u]} g(z,s-)\mathscr{N}(dz,ds)|\mathscr{F}_u\Big] + \mathbb{E}\Big[\int_{\mathbb{R}\times(u,t]} g(z,s-)\mathscr{N}(dz,ds)|\mathscr{F}_u\Big]$$
$$= \int_{\mathbb{R}\times(0,u]} g(z,s-)\mathscr{N}(dz,ds) + \int_{\mathbb{R}\times(u,t]} g(z,s-)\widehat{\mathscr{N}}(dz,ds).$$

From the definition of a compensated Poisson random measure,

$$\mathbb{E}\Big[\int_{\mathbb{R}\times[0,t]} g(z,s-)\widetilde{\mathscr{N}}(dz,ds)|\mathscr{F}_u\Big] = \int_{\mathbb{R}\times(0,u]} g(z,s-)\widetilde{\mathscr{N}}(dz,ds).$$

Hence $\int_{\mathbb{R}\times[0,t]} g(z,s-)\widetilde{\mathscr{N}}(dz,ds)$ is a martingale relative to the filtration $\{\mathscr{F}_t\}_{t\geq 0}$. The general case follows, considering $g = g_+ - g_-$. The càdlàg property follows trivially because of the explicit expression of the integral as a sum.

As remarked before, the reduction of the arguments to the case that $g \geq 0$ will be done in other proofs without necessarily mentioning it.

Corollary 3.4.20 *Assume the same conditions as in Theorem 3.4.19. Then, for any $t \geq 0$, the following equality holds:*

$$\mathbb{E}\left[\int_{\mathbb{R}\times[0,t]} g(z,s-)\mathscr{N}(dz,ds)\right] = \int_{\mathbb{R}\times[0,t]} g(z,s-)\widehat{\mathscr{N}}(dz,ds).$$

Corollary 3.4.21 *Let $\{Z_t^{(1)}\ ;\ t \geq 0\}$, $\{Z_t^{(2)}\ ;\ t \geq 0\}$ be two independent compound Poisson processes; for i=1,2, $\mathscr{N}_i$ is the Poisson random measure associated with $\{Z_t^{(i)}\ ;\ t \geq 0\}$ and $\widehat{\mathscr{N}_i}$ is the compensator of $\mathscr{N}_i$. We define a random measure $\mathscr{N}$ and a measure $\widehat{\mathscr{N}}$ as follows:*

$$\mathscr{N} := \mathscr{N}_1 + \mathscr{N}_2,\ \widehat{\mathscr{N}} := \widehat{\mathscr{N}_1} + \widehat{\mathscr{N}_2}.$$

Furthermore, we assume that g is a $\mathscr{B}(\mathbb{R}) \otimes \mathscr{B}[0,\infty)$-measurable càdlàg function and integrable with respect to $\mathscr{N}$ and $\widehat{\mathscr{N}}$. Then,

$$\mathbb{E}\left[\int_{\mathbb{R}\times[0,t]} g(z,s-)\mathscr{N}(dz,ds)\right] = \int_{\mathbb{R}\times[0,t]} g(z,s-)\widehat{\mathscr{N}}(dz,ds).$$

Corollary 3.4.22 *Let g be a bounded measurable càdlàg function and $\widetilde{\mathscr{N}}$ be a compensated Poisson random measure. Then the process*

$$\left\{\int_{\mathbb{R}\times(t_0,t]} g(z,s-)\widetilde{\mathscr{N}}(dz,ds);\ t > t_0\right\}$$

is a martingale relative to the filtration $\{\mathscr{F}_t\}_{t\in[t_0,\infty)}$ with càdlàg paths.

Exercise 3.4.23 Let $g : \mathbb{R} \times [0, T] \to \mathbb{R}$ be measurable and bounded. Give an upper bound for

$$\mathbb{E}\left[\left|\int_{\mathbb{R}\times[0,t]} g(z, s-)\mathscr{N}(dz, ds)\right|^2\right],$$

so that the upper bound depends explicitly on a deterministic norm of g. Such a bound is useful when taking limits.

3.4.2 *The Itô Formula*

The Itô formula[19] is a famous result in stochastic analysis which explains the value of a smooth function evaluated at a process through is time-dynamical change of value. The proof in the compound Poisson process case is easy.

Theorem 3.4.24 (Itô formula for jump process) *Let $Z = \{Z_t; t \geq 0\}$ be a compound Poisson process. Then for any $\mathscr{B}(\mathbb{R})$-measurable function h, and $t \geq 0$,*

$$\begin{aligned} h(Z_t) &= h(Z_0) + \sum_{s\leq t}\{h(Z_s) - h(Z_{s-})\} \\ &= h(Z_0) + \int_{\mathbb{R}\times[0,t]} \{h(Z_{s-} + z) - h(Z_{s-})\}\mathscr{N}(dz, ds). \end{aligned}$$

This formula which at first glance may appear as a triviality[20] describes the dynamical evolution of Z. This formula will be generalized to many situations and it is the cornerstone for many other calculations.

Proof Since $h(Z_{T_i}) = h(Z_{T_{i+1}-})$,

$$\sum_{s\leq t}\{h(Z_s) - h(Z_{s-})\} = \sum_{i=1}^{N_t}\{h(Z_{T_i}) - h(Z_{T_i-})\} = h(Z_t) - h(Z_0).$$

Hence $h(Z_t) = h(Z_0) + \sum_{s\leq t}\{h(Z_s) - h(Z_{s-})\}$. From Proposition 3.3.5 and $Z_{T_j-} + Y_j = Z_{T_j}$,

[19]For any continuous semi-martingale X in $\mathbb{R}^d$ and function $f \in C^2(\mathbb{R})$, we have

$$h(X) = h(X_0) + \sum_{i=1}^{d} h_i'(X)\cdot X^i + \frac{1}{2}\sum_{i,j=1}^{d} h_{ij}''(X)\cdot\langle X^i, X^j\rangle, \text{ a.s.},$$

where $X \cdot Y$ means the stochastic integral of X with respect to Y, and $\langle X, Y\rangle$ means the cross variation between X and Y.

[20]Usually it is called the telescopic sum formula.

$$\int_{\mathbb{R}\times[0,t]} \{h(Z_{s-}+z) - h(Z_{s-})\}\mathcal{N}(dz, ds) = \sum_{s\le t}\{h(Z_s) - h(Z_{s-})\}.$$

Hence $h(Z_t) = h(Z_0) + \int_{\mathbb{R}\times[0,t]} \{h(Z_{s-}+z) - h(Z_{s-})\}\mathcal{N}(dz, ds)$.

Corollary 3.4.25 *Let $\{Z_t^{(1)}; t \ge 0\}$, $\{Z_t^{(2)}; t \ge 0\}$ be two compound Poisson processes and $\mathcal{N}_i$ denotes the Poisson random measure associated with $\{Z_t^{(i)}; t \ge 0\}$. We define a stochastic process $\{Z_t; t \ge 0\}$ and a random measure $\mathcal{N}$ as follows:*

$$Z_t = Z_t^{(1)} + Z_t^{(2)}, \quad \mathcal{N} = \mathcal{N}_1 + \mathcal{N}_2.$$

Then for any $\mathcal{B}(\mathbb{R})$-measurable function h,

$$\begin{aligned} h(Z_t) &= h(Z_0) + \sum_{s\le t}\{h(Z_s) - h(Z_{s-})\} \\ &= h(Z_0) + \int_{\mathbb{R}\times[0,t]} \{h(Z_{s-}+z) - h(Z_{s-})\}\mathcal{N}(dz, ds). \end{aligned}$$

Theorem 3.4.26 *Let $a : \mathbb{R} \to \mathbb{R}$ be a bounded measurable function, and $b : \mathbb{R} \times [0, \infty) \to \mathbb{R}$ be a measurable càdlàg function and integrable with respect to $\mathcal{N}$ and $x \in \mathbb{R}$. We define*

$$X_t := x + \int_0^t a(s)ds + \int_{\mathbb{R}\times[0,t]} g(z, s-)\mathcal{N}(dz, ds).$$

Then, for any $h \in C_b^1(\mathbb{R})$,

$$h(X_t) = h(x) + \int_0^t h'(X_s)a(s)ds + \int_{\mathbb{R}\times[0,t]} \{h(X_{s-} + g(z, s-)) - h(X_{s-})\}\mathcal{N}(dz, ds).$$

Proof Note that the integral $\int_0^t h'(X_s)a(s)ds$ exists because h' and a are bounded and that X_t is a càdlàg function. Let $0 \le t < T_1$. Since

$$\int_{\mathbb{R}\times[0,t]} g(z, s-)\mathcal{N}(dz, ds) = 0,$$

we have that

$$X_t = x + \int_0^t a(s)ds.$$

Also a is bounded, so that X_t is differentiable and

$$\frac{d}{dt}X_t = a(t).$$

Therefore we get that

$$\begin{aligned}h(X_t) &= h(X_0) + \int_0^t \frac{d}{ds} h(X_s) ds \\ &= h(x) + \int_0^t h'(X_s)a(s)ds + \int_{\mathbb{R}\times[0,t]} \{h(X_{s-} + g(z, s-)) - h(X_{s-})\}\mathcal{N}(dz, ds),\end{aligned}$$

because $\int_{\mathbb{R}\times[0,t]} \left\{h(X_{s-} + g(z, s-)) - h(X_{s-})\right\}\mathcal{N}(dz, ds) = 0$. We assume that for $t < T_i$,

$$h(X_t) = h(x) + \int_0^t h'(X_s)a(s)ds + \int_{\mathbb{R}\times[0,t]} \{h(X_{s-} + g(z, s-)) - h(X_{s-})\}\mathcal{N}(dz, ds).$$

Let $T_i \leq t < T_{i+1}$. Then we have that

$$\begin{aligned}&h(x) + \int_0^t h'(X_s)a(s)ds + \int_{\mathbb{R}\times[0,t]} \{h(X_{s-} + g(z, s-)) - h(X_{s-})\}\mathcal{N}(dz, ds) \\ &= h(X_{T_i-}) + \int_{T_i}^t h'(X_s)a(s)ds + \int_{\mathbb{R}\times[T_i,t]} \{h(X_{s-} + g(z, s-)) - h(X_{s-})\}\mathcal{N}(dz, ds) \\ &= h(X_{T_i-}) + \int_{T_i}^t \frac{d}{ds} h(X_s)ds + h(X_{T_i-} + g(Y_i, T_i-)) - h(X_{T_i-}) \\ &= h(X_t) - h(X_{T_i}) + h(X_{T_i}) \\ &= h(X_t).\end{aligned}$$

Thus, for any $t \geq 0$,

$$h(X_t) = h(x) + \int_0^t h'(X_s)a(s)ds + \int_{\mathbb{R}\times[0,t]} \{h(X_{s-} + g(z, s-)) - h(X_{s-})\}\mathcal{N}(dz, ds).$$

Exercise 3.4.27 1. For all $h \in C^1(\mathbb{R})$, prove Theorem 3.4.26.
2. Prove that the result in Theorem 3.4.26 is also valid for coefficients a and g which are random adapted processes.

Exercise 3.4.28 Another way of specifying Markov processes is through their associated Itô formula. This is somewhat out of the scope of this book. But we can state the following exercise, which shows the elements that are used for such a description.

1. Define for $h \in C_b^1(\mathbb{R})$ the operator $P_t h(x) = \mathbb{E}[h(X_t)]$ where X is given in Theorem 3.4.26 with $a(s) = a$ and $g(z, s-) = g(z)$ (that is, the coefficients do not depend on time). Then prove that $P_0 h(x) = h(x)$ and that the semigroup property is satisfied. That is, for any $0 < s < t$ we have

$$P_{s+t} h(x) = P_s(P_t h)(x).$$

2. Find the generator of the semigroup using Itô's formula. That is,

$$Ah(x) = \lim_{t\downarrow 0} \frac{P_t h(x) - P_0 h(x)}{t}.$$

3. Prove that $P_t h(x)$ satisfies the differential equation

$$\frac{dP_t h(x)}{dt} = AP_t h(x).$$

Try to examine the relationship between the above exercises with Exercise 2.1.42. For a hint, see Chap. 14.

For a proof of the following well-known result, see [26], Chap. 4.

Theorem 3.4.29 (Doob's maximal inequality) *Let $\{X_t; t \geq 0\}$ be a positive submartingale such that every path of X is a càdlàg function. Let $[s_1, s_2]$ be a subinterval of $[0, \infty)$, and $p > 1$. If $\mathbb{E}[X_t^p] < \infty$ for any $t \geq 0$, then*

$$\mathbb{E}\left[\left(\sup_{s_1 \leq t \leq s_2} X_t\right)^p\right] \leq \left(\frac{p}{p-1}\right)^p \mathbb{E}[X_{s_2}^p].$$

Lemma 3.4.30 *Let g be a bounded $\mathscr{B}(\mathbb{R}) \otimes \mathscr{B}([0, \infty))$-measurable function such that $g(x, \cdot)$ is càdlàg function for any $x \in \mathbb{R}$. Also, let $\{X_t; t \geq 0\}$ be a $L^2(\Omega)$ martingale, integrable with respect to $\mathcal{N}$ and such that its sample paths are càdlàg. Then for any partition: $0 \leq t_0 < t_1 < \cdots < t_n = t$,*

$$\mathbb{E}\left[\sum_{i=0}^{n-1} \int_{\mathbb{R}\times(t_i, t_{i+1}]} g(z, t_i) X_{t_i} \mathcal{N}(dz, ds) \middle| \mathscr{F}_{t_0}\right] = \sum_{i=0}^{n-1} \int_{\mathbb{R}\times(t_i, t_{i+1}]} g(z, t_i) X_{t_0} \widehat{\mathcal{N}}(dz, ds).$$

Therefore if $\max\{|t_{i+1} - t_i|; i = 0, \ldots, n-1\} \to 0$, as $n \to \infty$ then by taking limits we also have

$$(LHS) \overset{n\uparrow\infty}{\to} \mathbb{E}\left[\int_{\mathbb{R}\times(t_0, t]} g(z, s-) X_{s-} \mathcal{N}(dz, ds) \middle| \mathscr{F}_{t_0}\right], \text{ a.s.,}$$

$$(RHS) \overset{n\uparrow\infty}{\to} \int_{\mathbb{R}\times(t_0, t]} g(z, s-) X_{t_0} \widehat{\mathcal{N}}(dz, ds), \text{ a.s.}$$

Proof For any $i \geq 1$, we have

$$\begin{aligned}
&\mathbb{E}\left[\int_{\mathbb{R}\times(t_i, t_{i+1}]} g(z, t_i) X_{t_i} \mathcal{N}(dz, ds) \middle| \mathscr{F}_{t_0}\right] \\
&= \mathbb{E}\left[\mathbb{E}\left[X_{t_i} \int_{\mathbb{R}\times(t_i, t_{i+1}]} g(z, t_i) \mathcal{N}(dz, ds) \middle| \mathscr{F}_{t_i}\right] \middle| \mathscr{F}_{t_0}\right] \\
&= \mathbb{E}\left[X_{t_i} \mathbb{E}\left[\int_{\mathbb{R}\times(t_i, t_{i+1}]} g(z, t_i) \mathcal{N}(dz, ds) \middle| \mathscr{F}_{t_i}\right] \middle| \mathscr{F}_{t_0}\right].
\end{aligned}$$

From Proposition 3.4.5, we get that

$$\mathbb{E}\left[\int_{\mathbb{R}\times(t_i,t_{i+1}]} g(z,t_i)\mathcal{N}(dz,ds)\,\middle|\,\mathcal{F}_{t_i}\right] = \int_{\mathbb{R}\times(t_i,t_{i+1}]} g(z,t_i)\widehat{\mathcal{N}}(dz,ds).$$

Also, since X is a martingale, we get that

$$\begin{aligned}\mathbb{E}\left[\int_{\mathbb{R}\times(t_i,t_{i+1}]} g(z,t_i)X_{t_i}\mathcal{N}(dz,ds)\,\middle|\,\mathcal{F}_{t_0}\right] &= \mathbb{E}\left[X_{t_i}\mathbb{E}\left[\int_{\mathbb{R}\times(t_i,t_{i+1}]} g(z,t_i)\mathcal{N}(dz,ds)\right]\,\middle|\,\mathcal{F}_{t_0}\right]\\ &= \mathbb{E}[X_{t_i}|\mathcal{F}_{t_0}]\int_{\mathbb{R}\times(t_i,t_{i+1}]} g(z,t_i)\widehat{\mathcal{N}}(dz,ds)\\ &= \int_{\mathbb{R}\times(t_i,t_{i+1}]} g(z,t_i)X_{t_0}\widehat{\mathcal{N}}(dz,ds).\end{aligned}$$

Also, X_{t_0} is a $\mathcal{F}_{t_0}$-measurable random variable. Hence,

$$\begin{aligned}\mathbb{E}\left[\int_{\mathbb{R}\times(t_0,t_1]} g(z,t_0)X_{t_0}\mathcal{N}(dz,ds)\,\middle|\,\mathcal{F}_{t_0}\right] &= X_{t_0}\;\mathbb{E}\left[\int_{\mathbb{R}\times(t_0,t_1]} g(z,t_0)\mathcal{N}(dz,ds)\,\middle|\,\mathcal{F}_{t_0}\right]\\ &= \int_{\mathbb{R}\times(t_0,t_1]} g(z,t_0)X_{t_0}\widehat{\mathcal{N}}(dz,ds).\end{aligned}$$

Therefore we have that

$$\mathbb{E}\left[\sum_{i=0}^{n-1}\int_{\mathbb{R}\times(t_i,t_{i+1}]} g(z,t_i)X_{t_i}\mathcal{N}(dz,ds)\,\middle|\,\mathcal{F}_{t_0}\right] = \sum_{i=0}^{n-1}\int_{\mathbb{R}\times(t_i,t_{i+1}]} g(z,t_i)X_{t_0}\widehat{\mathcal{N}}(dz,ds).$$

From Theorem 3.4.11, we get

$$\begin{aligned}(RHS) &= X_{t_0}\sum_{i=0}^{n-1}\int_{\mathbb{R}\times(t_i,t_{i+1}]} g(z,t_i)\widehat{\mathcal{N}}(dz,ds)\\ &\to X_{t_0}\int_{\mathbb{R}\times(t_0,t]} g(z,s-)\widehat{\mathcal{N}}(dz,ds)\\ &= \int_{\mathbb{R}\times(t_0,t]} g(z,s-)X_{t_0}\widehat{\mathcal{N}}(dz,ds).\end{aligned}$$

Hence the (RHS) converges to $\int_{\mathbb{R}\times(t_0,t]} g(z,s-)X_{t_0}\widehat{\mathcal{N}}(dz,ds)$.

Also, using an argument similar to the proof of Proposition 3.4.11, we can prove that

$$\sum_{i=0}^{n-1}\int_{\mathbb{R}\times(t_i,t_{i+1}]} g(z,t_i)X_{t_i}\mathcal{N}(dz,ds) \to \int_{\mathbb{R}\times(t_0,t]} g(z,s-)X_{s-}\mathcal{N}(dz,ds)\; n\to\infty.$$

Note that a similar argument also gives that $\int_{\mathbb{R}\times(t_i,t_{i+1}]} g(z,t_i)X_{t_i}\mathcal{N}(dz,ds) \in L^1(\Omega)$. In fact, since g is a bounded function, we get

$$\begin{aligned}\left|\sum_{i=0}^{n-1}\int_{\mathbb{R}\times(t_i,t_{i+1}]} g(z,t_i)X_{t_i}\mathcal{N}(dz,ds)\right| &\le \sum_{i=0}^{n-1}\int_{\mathbb{R}\times(t_i,t_{i+1}]} |g(z,t_i)||X_{t_i}|\mathcal{N}(dz,ds)\\ &\le K\sup_{0\le s\le T}|X_s|\int_{\mathbb{R}\times(t_0,t]}\mathcal{N}(dz,ds)\\ &= K\sup_{0\le s\le T}|X_s|(N_t-N_{t_0}).\end{aligned}$$

Using Schwartz's inequality and Doob's inequality, we have

$$\begin{aligned}\mathbb{E}[\sup_{0\le s\le T}|X_s|(N_t-N_{t_0})] &\le (\mathbb{E}[(\sup_{0\le s\le T}|X_s|)^2])^{\frac{1}{2}}(\mathbb{E}[(N_t-N_{t_0})^2])^{\frac{1}{2}}\\ &\le 2(\mathbb{E}[|X_T|^2])^{\frac{1}{2}}(\mathbb{E}[(N_{t-t_0})^2])^{\frac{1}{2}}\\ &< \infty.\end{aligned}$$

Hence, from the dominated convergence theorem,

$$(LHS) \to \mathbb{E}\left[\int_{\mathbb{R}\times(t_0,t]} g(z,s-)X_{s-}\mathcal{N}(dz,ds)\,\middle|\,\mathcal{F}_{t_0}\right].$$

Therefore,

$$\mathbb{E}\left[\int_{\mathbb{R}\times(t_0,t]} g(z,s-)X_{s-}\mathcal{N}(dz,ds)\,\middle|\,\mathcal{F}_{t_0}\right] = \int_{\mathbb{R}\times(t_0,t]} g(z,s-)X_{t_0}\widehat{\mathcal{N}}(dz,ds).$$

Next, we prove the convergence of the (LHS) by the monotone class theorem. Define $\mathcal{H}$ as the class of functions in $\mathcal{M}_b(\mathbb{R}\times[0,\infty))$ such that they are càdlàg functions of time and that they satisfy that

$$\sum_{i=0}^{n-1}\int_{\mathbb{R}\times(t_i,t_{i+1}]} h(z,t_i)X_{t_i}\mathcal{N}(dz,ds) \overset{n\to\infty}{\to} \int_{\mathbb{R}\times(t_0,t]} h(z,s-)X_{s-}\mathcal{N}(dz,ds),\ \text{a.s.}$$

Firstly, for any $\alpha,\beta\in\mathbb{R}$ and $h_1,h_2\in\mathcal{H}$,

$$\begin{aligned}&\sum_{i=0}^{n-1}\int_{\mathbb{R}\times(t_i,t_{i+1}]}(\alpha h_1(z,t_i)+\beta h_2(z,t_i))X_{t_i}\mathcal{N}(dz,ds)\\ &\overset{n\to\infty}{\to}\int_{\mathbb{R}\times(t_0,t]}(\alpha h_1(z,s-)+\beta h_2(z,s-))X_{s-}\mathcal{N}(dz,ds),\ \text{a.s.}\end{aligned}$$

Therefore $\mathcal{H}$ is a vector space on $\mathbb{R}$.

Secondly, let $T > 0$. We remark that

$$\sum_{i=0}^{n-1} \int_{\mathbb{R}\times(t_i,t_{i+1}]} X_{t_i} \mathscr{N}(dz, ds) = \sum_{i=0}^{n-1} \sum_{k\in\mathbb{N}} X_{t_i} \mathbf{1}_{\{T_k \in (t_i, t_{i+1}]\}}.$$

In addition,

$$\begin{aligned}\int_{\mathbb{R}\times(t_0,t]} X_{s-} \mathscr{N}(dz, ds) &= \sum_{k\in\mathbb{N}} X_{T_k-} \mathbf{1}_{\{T_k \in (t_0, t]\}} \\ &= \sum_{k\in\mathbb{N}} \sum_{i=0}^{n-1} X_{T_k-} \mathbf{1}_{\{T_k \in (t_i, t_{i+1}]\}}.\end{aligned}$$

Since the sample paths of X are càdlàg, we have from Lemma 3.4.8 that for $k \in \mathbb{N}$

$$\left| \sum_{i=0}^{n-1} (X_{t_i} - X_{T_{k-}}) \mathbf{1}_{\{T_k \in (t_i, t_{i+1}]\}} \right| \overset{n\to\infty}{\to} 0, \text{ a.s.}$$

Hence,

$$\sum_{i=0}^{n-1} \int_{\mathbb{R}\times(t_i,t_{i+1}]} X_{t_i} \mathscr{N}(dz, ds) \overset{n\to\infty}{\to} \int_{\mathbb{R}\times(t_0,t]} X_{s-} \mathscr{N}(dz, ds), \text{ a.s.}$$

That is, $\mathbf{1}_{\mathbb{R}\times[0,\infty)} \in \mathscr{H}$.

Thirdly, let $\{h_m; m \in \mathbb{N}\}$ be a sequence of non-negative functions in $\mathscr{H}$ such that $h_m \uparrow h$, where h is a bounded $\mathscr{B}(\mathbb{R}) \otimes \mathscr{B}[0, \infty)$-measurable function which is càdlàg with respect to the time variable. By splitting X_s as $X_s = X_s^+ - X_s^-$, we see that it is enough to deal with the case when $X_s \geq 0$. Then, if we take limits in the a.s. sense we have

$$\begin{aligned}\lim_{n\to\infty} \sum_{i=0}^{n-1} \int_{\mathbb{R}\times(t_i,t_{i+1}]} h(z, t_i) X_{t_i} \mathscr{N}(dz, ds) &= \lim_{n\to\infty} \lim_{m\to\infty} \sum_{i=0}^{n-1} \int_{\mathbb{R}\times(t_i,t_{i+1}]} h_m(z, t_i) X_{t_i} \mathscr{N}(dz, ds) \\ &= \lim_{m\to\infty} \lim_{n\to\infty} \sum_{i=0}^{n-1} \int_{\mathbb{R}\times(t_i,t_{i+1}]} h_m(z, t_i) X_{t_i} \mathscr{N}(dz, ds) \\ &= \lim_{m\to\infty} \int_{\mathbb{R}\times(t_0,t]} h_m(z, s-) X_{s-} \mathscr{N}(dz, ds) \\ &= \int_{\mathbb{R}\times(t_0,t]} h(z, s-) X_{s-} \mathscr{N}(dz, ds).\end{aligned}$$

Hence $h \in \mathscr{H}$.

Lastly, let

$$\mathscr{A} := \{(a,b) \times (c,d);\, a,b \in \mathbb{R},\, c,d \in [0,\infty)\} \cup \{(a,b) \times \{0\};\, a,b \in \mathbb{R}\}.$$

Then $\mathscr{A}$ is a π-system on $\mathscr{B}(\mathbb{R}) \otimes \mathscr{B}[0,\infty)$ and $\sigma(\mathscr{A}) = \mathscr{B}(\mathbb{R}) \otimes \mathscr{B}[0,\infty)$. For any $A = (a,b) \times (c,d) \in \mathscr{A}$,

$$\begin{aligned}
&\lim_{n\to\infty} \sum_{i=0}^{n-1} \int_{\mathbb{R}\times(t_i,t_{i+1}]} \mathbf{1}_A(z,t_i) X_{t_i} \mathscr{N}(dz,ds) \\
&= \sum_{j\geq 1} \mathbf{1}_{(a,b)}(Y_j) \lim_{n\to\infty} \sum_{i=0}^{n-1} \mathbf{1}_{(c,d)}(t_i) X_{t_i}\, \mathbf{1}_{\{T_j \in (t_i,t_{i+1}]\}}, \\
&\int_{\mathbb{R}\times(t_0,t]} \mathbf{1}_A(z,s-) X_{s-} \mathscr{N}(dz,ds) \\
&= \sum_{j\geq 1} \mathbf{1}_{(a,b)}(Y_j)\, \mathbf{1}_{(c,d)}(T_j-) X_{T_j-}\, \mathbf{1}_{\{T_j \in (t_0,t]\}}.
\end{aligned}$$

Hence, in order to prove that $\mathbf{1}_A \in \mathscr{H}$ for any $A \in \mathscr{A}$ we only need to prove

$$\sum_{i=0}^{n-1} \mathbf{1}_{(c,d)}(t_i) X_{t_i}\, \mathbf{1}_{\{T_j \in (t_i,t_{i+1}]\}} \overset{n\to\infty}{\to} \mathbf{1}_{(c,d)}(T_j-) X_{T_j-}\, \mathbf{1}_{\{T_j \in (t_0,t]\}}, \text{ a.s.}$$

We consider a sequence of strictly increasing real numbers $\{y_m; m \in \mathbb{N}\}$ such that $y_m \uparrow T_j$.

(i) If $T_j \leq c$, for any natural number m, $y_m \notin (c,d)$. So

$$\mathbf{1}_{(c,d)}(T_j-) = \lim_{m\to\infty} \mathbf{1}_{(c,d)}(y_m) = 0.$$

For a partition with sufficiently small norm, there exists t_k such that $t_k < T_j \leq t_{k+1} \leq c$. So

$$\sum_{i=0}^{n-1} \mathbf{1}_{(c,d)}(t_i) X_{t_i}\, \mathbf{1}_{\{T_j \in (t_i,t_{i+1}]\}} = 0.$$

(ii) If $T_j > d$, there exists a natural number N such that for any $m \geq N$, $d < y_m < T_j$. So

$$\mathbf{1}_{(c,d)}(T_j-) = \lim_{m\to\infty} \mathbf{1}_{(c,d)}(y_m) = 0.$$

For a small partition, there exists t_k such that $t_k < d \leq t_{k+1} < T_j$. So

$$\sum_{i=0}^{n-1} \mathbf{1}_{(c,d)}(t_i) X_{t_i} \mathbf{1}_{\{T_j \in (t_i, t_{i+1}]\}} = 0.$$

(iii) If $T_j = d$, there exists a natural number N such that for any $m \geq N$, $c < y_m < d = T_j$. So

$$\mathbf{1}_{(c,d)}(T_j-) = \lim_{m\to\infty} \mathbf{1}_{(c,d)}(y_m) = 1.$$

For a partition with sufficiently small norm, there exists t_k such that $c < t_k < d = T_j \leq t_{k+1}$. So

$$\begin{aligned}\sum_{i=0}^{n-1} \mathbf{1}_{(c,d)}(t_i) X_{t_i} \mathbf{1}_{\{T_j \in (t_i, t_{i+1}]\}} &= X_{t_k} \\ &\to X_{T_j-}.\end{aligned}$$

(iv) If $c < T_j < d$, there exists a natural number N such that for any $m \geq N$, $c < y_m < T_j < d$. So

$$\mathbf{1}_{(c,d)}(T_j-) = \lim_{m\to\infty} \mathbf{1}_{(c,d)}(y_m) = 1.$$

For a partition with sufficiently small norm, there exists t_k such that $c < t_k < T_j \leq t_{k+1} < d$. So

$$\sum_{i=0}^{n-1} \mathbf{1}_{(c,d)}(t_i) X_{t_i} \mathbf{1}_{\{T_j \in (t_i, t_{i+1}]\}} = X_{t_k} \to X_{T_j-}.$$

Hence we have

$$\lim_{n\to\infty} \sum_{i=0}^{n-1} \mathbf{1}_{(c,d)}(t_i) X_{t_i} \mathbf{1}_{\{T_j \in (t_i, t_{i+1}]\}} = \mathbf{1}_{(c,d)}(T_j-) X_{T_j-} \mathbf{1}_{\{T_j \in (t_0, t]\}}.$$

That is, $\mathbf{1}_A \in \mathscr{H}$. From the monotone class theorem, we get that

$$\lim_{n\to\infty} \sum_{i=0}^{n-1} \int_{\mathbb{R}\times(t_i, t_{i+1}]} g(z, t_i) X_{t_i} \mathscr{N}(dz, ds) = \int_{\mathbb{R}\times(t_0, t]} g(z, s-) X_{s-} \mathscr{N}(dz, ds).$$

Since g is a bounded function, by the dominated convergence theorem, we have that

$$(LHS) \to \mathbb{E}\left[\int_{\mathbb{R}\times(t_0, t]} g(z, s-) X_{s-} \mathscr{N}(dz, ds) \middle| \mathscr{F}_{t_0}\right].$$

Therefore,

$$\mathbb{E}\left[\int_{\mathbb{R}\times(t_0,t]} g(z,s-)X_{s-}\mathscr{N}(dz,ds)\,\middle|\,\mathscr{F}_{t_0}\right] = \int_{\mathbb{R}\times(t_0,t]} g(z,s-)X_{t_0}\widehat{\mathscr{N}}(dz,ds).$$

Theorem 3.4.31 *Let g be a bounded measurable function. Then*

$$\mathbb{E}\left[\left(\int_{\mathbb{R}\times[0,t]} g(z,s-)\widetilde{\mathscr{N}}(dz,ds)\right)^2\right] = \int_{\mathbb{R}\times[0,t]} g^2(z,s-)\widehat{\mathscr{N}}(dz,ds).$$

Proof Let

$$\begin{aligned}
X_t &:= \int_{\mathbb{R}\times[0,t]} g(z,s-)\widetilde{\mathscr{N}}(dz,ds)\\
&= -\int_{\mathbb{R}\times[0,t]} g(z,s-)\widehat{\mathscr{N}}(dz,ds) + \int_{\mathbb{R}\times[0,t]} g(z,s-)\mathscr{N}(dz,ds)\\
&= -\int_0^t \lambda \int_{\mathbb{R}} g(z,s-)f(z)dzds + \int_{\mathbb{R}\times[0,t]} g(z,s-)\mathscr{N}(dz,ds)\\
&= -\int_0^t \lambda\mathbb{E}[g(Y,s-)]ds + \int_{\mathbb{R}\times[0,t]} g(z,s-)\mathscr{N}(dz,ds).
\end{aligned}$$

Clearly the above is integrable. From Theorem 3.4.26, we get that

$$\begin{aligned}
X_t^2 &= X_0^2 + \int_0^t 2X_s(-\lambda\mathbb{E}[g(Y,s-)])ds + \int_{\mathbb{R}\times[0,t]} \{(X_{s-}+g(z,s-))^2 - X_{s-}^2\}\mathscr{N}(dz,ds)\\
&= -2\lambda\int_0^t X_s\mathbb{E}[g(Y,s-)]ds + \int_{\mathbb{R}\times[0,t]} (g^2(z,s-)+2g(z,s-)X_{s-})\mathscr{N}(dz,ds).
\end{aligned}$$

From Lemma 3.4.30 one obtains that the above is integrable and therefore X_t is a $L^2(\Omega)$-martingale, so that $\mathbb{E}[X_t]=0$. Therefore we have from Corollary 3.4.22 that

$$\begin{aligned}
\mathbb{E}[X_t^2] =& -2\lambda\mathbb{E}\left[\int_0^t X_s\mathbb{E}[g(Y,s-)]ds\right]\\
&+\mathbb{E}\left[\int_{\mathbb{R}\times[0,t]} g^2(z,s-)\mathscr{N}(dz,ds)\right] + 2\mathbb{E}\left[\int_{\mathbb{R}\times[0,t]} g(z,s-)X_{s-}\mathscr{N}(dz,ds)\right]\\
=& -2\lambda\int_0^t \mathbb{E}[X_s]\mathbb{E}[g(Y,s-)]ds\\
&+\int_{\mathbb{R}\times[0,t]} g^2(z,s-)\widehat{\mathscr{N}}(dz,ds) + 2\mathbb{E}\left[\int_{\mathbb{R}\times[0,t]} g(z,s-)X_{s-}\mathscr{N}(dz,ds)\right]\\
=& \int_{\mathbb{R}\times[0,t]} g^2(z,s-)\widehat{\mathscr{N}}(dz,ds) + 2\mathbb{E}\left[\int_{\mathbb{R}\times[0,t]} g(z,s-)X_{s-}\mathscr{N}(dz,ds)\right].
\end{aligned}$$

From Lemma 3.4.30, we get that

$$\mathbb{E}\left[\int_{\mathbb{R}\times[0,t]} g(z,s-)X_{s-}\mathcal{N}(dz,ds)\right] = \int_{\mathbb{R}\times[0,t]} g(z,s-)\mathbb{E}[X_0]\widehat{\mathcal{N}}(dz,ds)$$
$$= 0.$$

Hence,

$$\mathbb{E}\left[\left(\int_{\mathbb{R}\times[0,t]} g(z,s-)\widetilde{\mathcal{N}}(dz,ds)\right)^2\right] = \int_{\mathbb{R}\times[0,t]} g^2(z,s-)\widehat{\mathcal{N}}(dz,ds).$$

Exercise 3.4.32 Generalize the above result using approximations to the case that g may not be bounded but instead

$$\int_{\mathbb{R}\times[0,t]} g^2(z,s-)\widehat{\mathcal{N}}(dz,ds) < \infty.$$

Use Theorem 3.4.29 in order to prove that the approximations converge uniformly in time and finally obtain that the stochastic integral has càdlàg paths.

Exercise 3.4.33 For $g : \mathbb{R}\times[0,T] \to \mathbb{R}$, define $X_t := \int_{\mathbb{R}\times[0,t]} g(z,s-)\mathcal{N}(dz, ds)$. Apply Itô's formula with $h(x) = |x|^p$ with $p \in \mathbb{N}$. Finally, obtain an explicit bound for $\mathbb{E}[|X_t|^p]$ which depends explicitly on a deterministic norm of g.

3.5 Stochastic Integrals

All the previous sections were a preparation to ask the following question. Is it possible to define an integral with respect to integrands which are random?

As our goal is to go as fast as possible to stochastic differential equations, we will only briefly explain how the general construction is done. We refer the reader to [2] or [48] for the details.

To start, let us just suppose that $g : \Omega\times\mathbb{R}\times[0,T] \to \mathbb{R}$ is a jointly measurable function. Then as $\mathcal{N}$ is a finite measure in the present case, one can define

$$\int_{\mathbb{R}\times[0,t]} g(z,s)\mathcal{N}(dz,ds) = \sum_{i=1}^{\infty} g(Y_i,T_i)\,\mathbf{1}_{\{T_i\le t\}}.$$

This definition is valid for any random function g, but the properties in Corollary 3.4.4 and Theorem 3.4.11 will not be satisfied, the main reason being that now g may depend on all the variables $\{Y_i, T_i;\ i\in\mathbb{N}\}$.

Giving up on the property in Corollary 3.4.4, we would like to retain a property similar to the result in Theorem 3.4.11 because this leads to the martingale property. An instructive example, related to this definition, is to consider the following two particular situations: $g_1(z,s) = Z_s z$ and $g_2(z,s) = Z_{s-}z$. In fact, one has

$$\int_{\mathbb{R}\times[0,t]} g_1(z,s)\mathcal{N}(dz,ds) = \sum_{i=1}^{\infty} Z_{T_i} Y_i \mathbf{1}_{\{T_i \le t\}} = \sum_{i=1}^{\infty}\sum_{j=1}^{i} Y_j Y_i \mathbf{1}_{\{T_i \le t\}},$$

$$\int_{\mathbb{R}\times[0,t]} g_2(z,s)\mathcal{N}(dz,ds) = \sum_{i=1}^{\infty} Z_{T_i -} Y_i \mathbf{1}_{\{T_i \le t\}} = \sum_{i=2}^{\infty}\sum_{j=1}^{i-1} Y_j Y_i \mathbf{1}_{\{T_i \le t\}}.$$

Now if we compute the expectations in each case, we see that the calculation in the case of g_1 above involves the variance of Y, while in the case of g_2 it only involves the expectation of Y. In fact,

$$\begin{aligned}\mathbb{E}\left[\int_{\mathbb{R}\times[0,t]} g_2(z,s)\mathcal{N}(dz,ds)\right] &= \sum_{i=2}^{\infty}\sum_{j=1}^{i-1} \mathbb{E}[Y]^2 \mathbb{P}(N_t \ge i) \\ &= \lambda \int_{\mathbb{R}\times[0,t]} \mathbb{E}[g_2(z,s)] f(z) dz ds. \qquad (3.4)\end{aligned}$$

On the other hand, note that $\mathbb{E}[Z_{s-}] = \mathbb{E}[Z_s]$. Therefore this means that

$$\mathbb{E}\left[\int_{\mathbb{R}\times[0,t]} g_1(z,s)\mathcal{N}(dz,ds)\right] \neq \lambda \int_{\mathbb{R}\times[0,t]} \mathbb{E}[g_1(z,s)] f(z) dz ds.$$

Therefore there is a significant property that allows us to obtain (3.4), which is not satisfied by g_1. This property is called predictability.[21].

Definition 3.5.1 Let $\mathcal{P}$ be the σ-algebra generated on $[0,T]\times\Omega$ by processes of the type

$$g_0(\omega)\,\mathbf{1}_{\{t=0\}}(t) + \sum_{i=0}^{n-1} g_i(\omega)\,\mathbf{1}_{(t_i,t_{i+1}]}(t).$$

Here $\{0 = t_0 < \ldots < t_n = T\}$ is a partition of the interval $[0,T]$ and g_i is a bounded $\mathcal{F}_{t_i}$-measurable function.

$\mathcal{P}$ is called the $\mathcal{F}$ predictable σ-algebra and any process measurable with respect to it is called predictable.

Exercise 3.5.2 Write the above definition using mathematical symbols and formulae.

[21] The reason for this name should be clear. The random process g_2 is left-continuous, that means that its value at any time can be "predicted" by the values at times before it. In fact, the example g_1 is called anticipating because it uses information that cannot be predicted as in g_2. The terms associated with the terms $\mathbb{E}[Y^2]$ are called trace terms for obvious reasons.

That is, $\mathscr{P}$ is defined the smallest filtration generated by all adapted processes which are left continuous.[22] Then, we define the stochastic integral for predictable processes, which in the present case is trivial and one has the following result[23]:

Theorem 3.5.3 *Let* $g : \Omega \times \mathbb{R} \times [0, T] \to \mathbb{R}$ *be a predictable process such that*

$$\mathbb{E}[\int_{\mathbb{R}\times[0,T]} |g(z,s)|\widehat{\mathcal{N}}(dz, ds)] < \infty.$$

Then for any $t \in [0, T]$,

$$\mathbb{E}[\int_{\mathbb{R}\times[0,t]} g(z,s)\mathcal{N}(dz, ds)] = \mathbb{E}[\int_{\mathbb{R}\times[0,t]} g(z,s)\widehat{\mathcal{N}}(dz, ds)].$$

Furthermore $\{\int_{\mathbb{R}\times[0,t]} g(z,s)\widetilde{\mathcal{N}}(dz, ds) : t \in [0, T]\}$ *is a martingale which has càdlàg paths.*

The proof of this fact follows through techniques used in the proofs of Theorem 3.4.11 or Lemma 3.4.30 which essentially ask you to first prove the statement for a basic generating class of processes and then extend the result through the monotone class theorem.

Exercise 3.5.4 Prove the above theorem for the following basic class of processes, the set of processes of the type

$$g_0(\omega, z)\, \mathbf{1}_{\{t=0\}} + \sum_{i=0}^{n-1} g_i(\omega, z)\, \mathbf{1}_{(t_i,t_{i+1}]}(t).$$

Here $\{0 = t_0 < ... < t_n = T\}$ is a partition of the interval $[0, T]$ and g_i is a bounded $\mathscr{F}_{t_i} \otimes \mathscr{B}$-measurable function. In order to prove this result recall the result in Theorem 3.4.11.

Exercise 3.5.5 Prove the following extension to Theorem 3.4.31. Let g be a bounded predictable process such that $\int_{\mathbb{R}\times[0,t]} \mathbb{E}[g^2(z, s-)]\widehat{\mathcal{N}}(dz, ds) < \infty$. Then

$$\mathbb{E}\left[\left(\int_{\mathbb{R}\times[0,t]} g(z, s-)\widetilde{\mathcal{N}}(dz, ds)\right)^2\right] = \int_{\mathbb{R}\times[0,t]} \mathbb{E}[g^2(z, s-)]\widehat{\mathcal{N}}(dz, ds).$$

---

[22] This also explains why in the previous section one always takes $g(z, s-)$ as the integrand function, that is, just to try to make you aware that when integrating a process like Z_s one really needs to use Z_{s-} rather than Z_s. Another way of looking at this notation is that it reminds you of the need of predictability in the integrand.

[23] Note that this problem does not appear at all with Brownian motion. For this just consider the uniform partition $t_i = \frac{Ti}{n}$ and the Riemann sum $\sum_{i=0}^{n-1} B_{t_i}(B_{t_{i+1}} - B_{t_i})$ and $\sum_{i=0}^{n-1} B_{t_i-\varepsilon_n}(B_{t_{i+1}} - B_{t_i})$ with $\varepsilon_n \to 0$ as $n \to \infty$. Prove that the limit of the difference converges to zero.

One can also prove the corresponding Itô formulas parallel to the ones in Theorems 3.4.24 and 3.4.26.

Theorem 3.5.6 *Assume the same conditions as in Theorem 3.5.3 and that* $h \in \mathscr{M}_b(\mathbb{R})$*, then for* $Y_t = \int_{\mathbb{R}\times[0,t]} g(z,s)\mathscr{N}(dz,ds)$*,*

$$h(Y_t) = h(0) + \int_{\mathbb{R}\times[0,t]} (h(Y_{s-} + g(z,s)) - h(Y_{s-}))\mathscr{N}(dz,ds). \tag{3.5}$$

Furthermore

$$\mathbb{E}[h(Y_t)] = h(0) + \int_{\mathbb{R}\times[0,t]} \mathbb{E}[h(Y_{s-} + g(z,s)) - h(Y_{s-})]\widehat{\mathscr{N}}(dz,ds).$$

Exercise 3.5.7 Prove the above theorem. Generalize the Itô formula in the case that the condition $\mathbb{E}[\int_{\mathbb{R}\times[0,T]} |g(z,s)|\widehat{\mathscr{N}}(dz,ds)] < \infty$ is not satisfied.

In fact, the predictability is not even needed if we only want the formula (3.5) to be satisfied.

For a hint, see Chap. 14.

Exercise 3.5.8 Find conditions that assure that the moments of stochastic integrals are finite. That is, let $g : \Omega \times \mathbb{R} \times [0,T] \to \mathbb{R}$ be predictable and fix $p \in \mathbb{N}$. Find conditions under which[24]

$$\mathbb{E}\left[\left|\int_{\mathbb{R}\times[0,t]} g(z,s)\mathscr{N}(dz,ds)\right|^p\right] < \infty.$$

Notice that for $p = 1$, the required condition is

$$\mathbb{E}\left[\int_{\mathbb{R}\times[0,T]} |g(z,s)|\widehat{\mathscr{N}}(dz,ds)\right] < \infty.$$

In particular, prove that the following equality is true whenever the right-hand side is finite:

$$\mathbb{E}\left[\left|\int_{\mathbb{R}\times[0,t]} g(z,s)\widetilde{\mathscr{N}}(dz,ds)\right|^2\right] = \mathbb{E}\left[\int_{\mathbb{R}\times[0,t]} |g(z,s)|^2\widehat{\mathscr{N}}(dz,ds)\right].$$

With these exercises we wanted to make the reader understand that many statements that are valid for functions g can then be properly extended to predictable processes if they are bounded in the appropriate norm.

In the following exercises we try to reinforce the correct understanding of Proposition 3.1.8, Exercises 3.1.11 and 3.1.4, Theorem 3.2.7 and Corollaries 3.1.3, 3.3.9, 3.4.21.

[24] As before, this has to be some norm on g.

Exercise 3.5.9 Let $Z_t^i = \sum_{j=1}^{N_t^i} Y_j^i$, $i = 1, 2$ be two independent compound Poisson processes with respective Poisson random measures $\mathcal{N}^i$.

1. Let $f : \mathbb{R}^2 \to \mathbb{R}$ be a bounded measurable function. Is the following process a compound Poisson process?

$$\sum_{i=1}^{N_t^1} \sum_{j=1}^{N_t^2} f(Y_i^1, Y_j^2).$$

2. Consider N to be a Poisson process with intensity given by $t^{-2}\mathbb{E}[N_t^1]\mathbb{E}[N_t^2]$. Let (X_i^1, X_i^2) be a sequence of i.i.d.r.v.s such that its joint law is the same as the law of (Y^1, Y^2). Is the following process a compound Poisson process?

$$\sum_{i=1}^{N_t} f(X_i^1, X_j^2).$$

   We will denote its corresponding random measure as $\mathcal{N}$.
3. Note that the above question can also be reformulated using random point measures. In particular, determine if the laws of the processes $\int_{\mathbb{R}\times[0,t]} \int_{\mathbb{R}\times[0,t]} f(z_1, z_2)\mathcal{N}^1(dz_1, ds_1)\mathcal{N}^2(dz_2, ds_2)$ and $\int_{\mathbb{R}^2\times[0,t]} f(z)\mathcal{N}(dz, ds)$ are the same.
4. If you have noticed, the intention of this exercise is to find an interpretation of $\int_{\mathbb{R}^2\times[0,t]} f(z)\mathcal{N}(dz, ds)$ as a doubly iterated integral. Can you provide such a construction?

For a hint, see Chap. 14.

3.6 Stochastic Differential Equations

In this section, we consider the trivial but yet essential case of stochastic differential equations driven by compound Poisson processes. This understanding will be important in future chapters.

Definition 3.6.1 Y is a modification of X if, for every $t \geq 0$, we have $\mathbb{P}(X_t = Y_t) = 1$.

Definition 3.6.2 X and Y are called indistinguishable if almost all their sample paths agree:

$$\mathbb{P}(X_t = Y_t, \ t \geq 0) = 1.$$

Proposition 3.6.3 *Let X and Y be two right-continuous processes (or left-continuous processes) and Y be a modification of X. Then X and Y are indistinguishable.*

Proof Since Y is a modification of X,

$$\mathbb{P}(X_t = Y_t,\ t \in \mathbb{Q}_+) = 1.$$

For any $t \geq 0$, let q_n be a rational sequence such that $q_n \downarrow t$. For every $\omega \in \{X_t = Y_t,\ t \in \mathbb{Q}_+\}$, $X_{q_n}(\omega) = Y_{q_n}(\omega)$. Since X and Y are right-continuous processes, we get

$$\lim_{n\to\infty} X_{q_n}(\omega) = X_t(\omega), \quad \lim_{n\to\infty} Y_{q_n}(\omega) = Y_t(\omega).$$

Therefore we have $X_t(\omega) = Y_t(\omega)$. Then,

$$\{X_t = Y_t,\ t \in \mathbb{Q}_+\} \subset \{X_t = Y_t,\ t \geq 0\}.$$

This proposition completes the proof.

Definition 3.6.4 Let $b \in \mathscr{M}_b(\mathbb{R}^2)$. A solution of the stochastic equation

$$X_t = X_0 + \int_{\mathbb{R}\times[0,t]} b(z, X_{s-}) \mathscr{N}(dz, ds)$$

on the given probability space $(\Omega, \mathscr{F}, \mathbb{P})$ with initial condition ξ is a process $X = \{X_t\ ;\ t \geq 0\}$ with càdlàg sample paths with the following properties:
(i) $\mathbb{P}(X_0 = \xi) = 1$.
(ii) $\mathbb{P}(\int_{\mathbb{R}\times[0,t]} |b(z, X_{s-})| \mathscr{N}(dz, ds) < \infty) = 1,\ t \geq 0$.
(iii) $X_t = X_0 + \int_{\mathbb{R}\times[0,t]} b(z, X_{s-}) \mathscr{N}(dz, ds)\ t \geq 0$, a.s.
Moreover, a solution of the stochastic equation is unique if, for any two solutions X and $\widetilde{X}$ satisfying the above conditions, $\mathbb{P}(X_t = \widetilde{X}_t,\ t \geq 0) = 1$.

Equations of the above type are generally called stochastic differential equations or in abbreviated form sdes. In many cases that will follow, we will stop writing the above definitions assuming that the reader can build the definition for solution and uniqueness (usually called pathwise uniqueness).

Theorem 3.6.5 *Let Z be a compound Poisson process and $\mathscr{N}$ be the Poisson random measure associated with Z. Furthermore, let $b \in \mathscr{M}_b(\mathbb{R}^2)$ and $x \in \mathbb{R}$. Then there exists a unique solution to the stochastic equation*

$$X_t = x + \int_{\mathbb{R}\times[0,t]} b(X_{s-}, z) \mathscr{N}(dz, ds). \tag{3.6}$$

Proof We define

$$X_t := \sum_{i\in\mathbb{N}} \xi_i\, \mathbf{1}_{\{T_{i-1}\leq t<T_i\}}, \quad T_0 := 0,$$

where

$$\xi_1 := x$$
$$\xi_i := \xi_{i-1} + b(\xi_{i-1}, Y_{i-1}),\ i = 2, 3, \cdots .$$

From the definition of X_t, for almost every ω, $X_t(\omega)$ is a càdlàg function and $\xi_i = X_{T_i-}$. Using Proposition 3.3.5 we get

$$\begin{aligned}\int_{\mathbb{R}\times[0,t]} |b(X_{s-}, z)| \mathcal{N}(dz, ds) &= \sum_{i\in\mathbb{N}} |b(X_{T_i-}, Y_i)|\ \mathbf{1}_{\{T_i\le t\}} \\ &< \infty,\ \text{a.s.}\end{aligned}$$

Thus,

$$\begin{aligned}x + \int_{\mathbb{R}\times[0,t]} b(X_{s-}, z)\mathcal{N}(dz, ds) &= x + \sum_{i\in\mathbb{N}} b(X_{T_i-}, Y_i)\ \mathbf{1}_{\{T_i\le t\}} \\ &= x + \sum_{k=1}^{i_0-1} b(X_{T_k-}, Y_k)\ (\exists i_0 \text{ s.t. } T_{i_0-1} \le t < T_{i_0}) \\ &= X_{T_1-} + b(X_{T_1-}, Y_1) + \sum_{k=2}^{i_0-1} b(X_{T_k-}, Y_k) \\ &= X_{T_2-} + \sum_{k=2}^{i_0-1} b(X_{T_k-}, Y_k) \\ &= \cdots \\ &= X_{T_{i_0}-} \\ &= \sum_{i\in\mathbb{N}} X_{T_i-}\ \mathbf{1}_{\{T_{i-1}\le t<T_i\}} \\ &= X_t.\end{aligned}$$

We assume that a process $\widetilde{X}_t$ exists such that

$$\widetilde{X}_t = x + \int_{\mathbb{R}\times[0,t]} b(\widetilde{X}_{s-}, z)\mathcal{N}(dz, ds).$$

Then, we have

$$\widetilde{X}_t = x + \sum_{i\in\mathbb{N}} b(\widetilde{X}_{T_i-}, Y_i)\ \mathbf{1}_{\{T_i\le t\}}.$$

Hence, from Proposition 3.6.3, we only have to show $X_{T_i-} = \widetilde{X}_{T_i-}$ for any $i \in \mathbb{N}$. Clearly, $X_{T_1-} = \widetilde{X}_{T_1-}$. If $X_{T_i-} = \widetilde{X}_{T_i-}$, then

$$\begin{aligned}\widetilde{X}_{T_{i+1}-} &= x + \sum_{k\in\mathbb{N}} b(\widetilde{X}_{T_k-}, Y_k)\, \mathbf{1}_{\{T_k \le T_{i+1}-\}} \\ &= x + \sum_{k=1}^{i} b(X_{T_k-}, Y_k) \\ &= x + \sum_{k\in\mathbb{N}} b(X_{T_k-}, Y_k)\, \mathbf{1}_{\{T_k \le T_{i+1}-\}} \\ &= X_{T_{i+1}-}.\end{aligned}$$

Therefore, X is a unique solution to the stochastic equation (3.6).

Corollary 3.6.6 *Let $b \in \mathscr{M}_b(\mathbb{R})$ and $x \in \mathbb{R}$. Then there exists a unique solution to the stochastic equation*

$$X_t = x + \int_0^t b(X_{s-})dZ_s \equiv x + \int_{\mathbb{R}\times[0,t]} b(X_{s-})z\mathscr{N}(dz, ds).$$

Theorem 3.6.7 *Let $\{Z_t^{(1)}; t \ge 0\}$, $\{Z_t^{(2)}; t \ge 0\}$ be independent compound Poisson processes and $\mathscr{N}_i$ be the Poisson random measure associated with $\{Z_t^{(i)}; t \ge 0\}$. We define a random measure $\mathscr{N}$ as follows:*

$$\mathscr{N} = \mathscr{N}_1 + \mathscr{N}_2.$$

Let $b \in \mathscr{M}_b(\mathbb{R}^2)$ and $x \in \mathbb{R}$. Then there exists a unique solution to the stochastic equation

$$X_t = x + \int_{\mathbb{R}\times[0,t]} b(X_{s-}, z)\mathscr{N}(dz, ds).$$

Proposition 3.6.8 *The solution of the linear equation*

$$X_t = x + \int_{\mathbb{R}\times[0,t]} X_{s-}b(z)\mathscr{N}(dz, ds)$$

is given by

$$X_t = x\prod_{s\le t}(1 + \mathbf{1}_{\{Z_s - Z_{s-}\neq 0\}}b(Z_s - Z_{s-})).$$

Proof From Theorem 3.6.5, the solution of the stochastic equation

$$X_t = x + \int_{\mathbb{R}\times[0,t]} X_{s-}b(z)\mathscr{N}(dz, ds)$$

is given by

$$X_t = x + \sum_{i\in\mathbb{N}} X_{T_i-}b(Y_i)\, \mathbf{1}_{\{T_i\le t\}}.$$

Let $t < T_1$, then $X_t = x$. For any $s \le t$, $\mathbf{1}_{\{Z_s - Z_{s-} \neq 0\}} = 0$, so that

$$X_t = x \prod_{s \le t} (1 + \mathbf{1}_{\{Z_s - Z_{s-} \neq 0\}} b(Z_s - Z_{s-})).$$

We assume that

$$X_t = x \prod_{s \le t} (1 + \mathbf{1}_{\{Z_s - Z_{s-} \neq 0\}} b(Z_s - Z_{s-})), \; t < T_i.$$

Then, for all $T_i \le t < T_{i+1}$,

$$\begin{aligned}
X_t &= x + \sum_{k=1}^{i} X_{T_k -} b(Y_k) \\
&= x + \sum_{k=1}^{i-1} X_{T_k -} b(Y_k) + X_{T_i -} b(Y_i) \\
&= X_{T_i -} + X_{T_i -} b(Y_i) \\
&= X_{T_i -} (1 + b(Z_{T_i} - Z_{T_i -})) \\
&= (1 + b(Z_{T_i} - Z_{T_i -})) x \prod_{s \le T_i -} (1 + \mathbf{1}_{\{Z_s - Z_{s-} \neq 0\}} b(Z_s - Z_{s-})) \\
&= x \prod_{s \le t} (1 + \mathbf{1}_{\{Z_s - Z_{s-} \neq 0\}} b(Z_s - Z_{s-})).
\end{aligned}$$

Therefore, for any $t \ge 0$,

$$X_t = x \prod_{s \le t} (1 + \mathbf{1}_{\{Z_s - Z_{s-} \neq 0\}} b(Z_s - Z_{s-})).$$

Exercise 3.6.9 Suppose that the function $b(x, z)$ satisfies the inequality

$$|b(x, z)| \le C(1 + |x|)|z|.$$

Furthermore suppose that $\mathbb{E}[|Y|] < \infty$. Find conditions that assure that the solution X to the Eq. (3.6) has a finite first absolute moment. Write the upper bound explicitly.

Exercise 3.6.10 Use Itô's formula to prove finiteness of $\mathbb{E}[|X_t|^p]$ for $p \in \mathbb{N}$ under appropriate conditions on b.

Exercise 3.6.11 This important exercise serves to study the solution X of the stochastic equation (3.6) as a function of x. That is, in this exercise, we denote this solution by $X_t(x)$, noting that this random variable is defined for each $t > 0$ and $x \in \mathbb{R}$.

1. Prove that $X_{t+s}(x) = X_{s,t+s}(X_s(x))$, where $X_{s,s+\cdot}(x)$ is the unique solution to the equation

$$X_{s,s+t}(x) = x + \int_{\mathbb{R}\times[s,s+t]} b(X_{s,u-}(x), z)\mathcal{N}(dz, du).$$

2. Prove that the law of $X_{s,s+t}(x)$ is the same as the law of $X_t(x)$ for fixed values of $s, t > 0$ and $x \in \mathbb{R}$.
3. Prove that if $b \in C^1$ then $X_t(x)$ is differentiable with respect to $x \in \mathbb{R}$ with $t > 0$ fixed. Furthermore prove that the derivative denoted by $X_t'(x)$ satisfies the equation

$$X_t'(x) = 1 + \int_{\mathbb{R}\times[0,t]} b'(X_{u-}(x), z)X_{u-}'(x)\mathcal{N}(dz, du).$$

   Extend the above result for second derivatives under appropriate conditions on b.
4. Noting that the equation satisfied by $X_t'(x)$ is a linear equation write its solution in an explicit form as in Proposition 3.6.8.

Problem 3.6.12 (*An example from mathematical insurance*) Consider the risk process defined as

$$S_t = u + ct - \sum_{i=1}^{N_t} Y_i.$$

Here N is a Poisson process with parameter λ and Y is a sequence of i.i.d. random variables with density function $f : [0, \infty) \to [0, \infty)$. Consider $\Psi(u) := P(\inf_{s>0} S_s < 0)$ which is called the probability of ruin and assume that Ψ is differentiable. Condition on the first jump of the process S in order to obtain an integro-differential equation associated with Ψ. Use this idea in order to prove that Ψ satisfies the following linear equation:

$$c\Psi'(u) + \lambda\left(\int_0^u (\Psi(u-y) - \Psi(u))f(y)dy + (1 - \Psi(u))\bar{F}(u)\right) = 0.$$

Compute $\frac{\mathbb{E}[\Psi(S_t)]-\Psi(u)}{t}$ as $t \to 0$. Recall from Exercise 3.4.28 that this quantity is the generator of S applied to Ψ. Compare with the previous result.

Note that proving that the derivative of Ψ exists is related to the regularity of the law of the process $\inf_{s>0} S_s$.

Problem 3.6.13 Suppose that f is differentiable such that $f' \in L^1(\mathbb{R})$. Prove that in the case that $c = 0$ then $\Psi(u) := \mathbb{P}(\inf_{s>0} S_s < 0)$ is a differentiable function.

Chapter 4
Construction of Lévy Processes and Their Corresponding SDEs: The Finite Variation Case

In this chapter, we will generalize the previous construction of compound Poisson processes and allow the possibility of a infinite number of jumps on a fixed interval. The stochastic process constructed in this section will satisfy that the number of jumps whose absolute size is larger than any fixed positive value is finite in any fixed interval. Therefore the fact that there are infinite number of jumps is due to the fact that most of these jumps are small in size. The conditions imposed will also imply that the generated stochastic process has paths of bounded variation and therefore Stiltjes integration can be used to give a meaning to stochastic integrals. We also introduce the associated stochastic calculus.

4.1 Construction of Lévy Processes: The Finite Variation Case

In this section, we will construct a Lévy process Z such that

$$\mathbb{E}[e^{i\theta Z_t}] = \exp[ct \int_{-\infty}^{\infty} (e^{i\theta x} - 1) f(x) dx]. \tag{4.1}$$

where $c > 0$ is some constant.

Here the function $f : \mathbb{R} \to \mathbb{R}_+$ satisfies the following conditions:

$$\begin{aligned} f(0) &= 0, \\ \int_{-\infty}^{\infty} (|x| \wedge 1) f(x) dx &< \infty. \end{aligned} \tag{4.2}$$

As one can see, we do not require f to be a density function. Furthermore in comparison with Remarks 2.1.14, 2.1.32 and 3.1.5 we do not see the appearance of λ. In fact, from now on, what appeared as λf in Remark 3.1.5 is now just cf in

© Springer Nature Singapore Pte Ltd. 2019
A. Kohatsu-Higa and A. Takeuchi, *Jump SDEs and the Study of Their Densities*, Universitext, https://doi.org/10.1007/978-981-32-9741-8_4

the above expression. The reason for this is that f is not necessarily an integrable function anymore (and therefore one cannot build a density out of it) and for this reason λ is eventually used for the purpose of proving the existence of such a process. This should be somewhat clear in the proofs that follow. We may use without loss of generality that $c = 1$.

A beginner may wonder what is the use of such a procedure as one may think that this will not correspond to any physical process such as the previous compound Poisson process. This is far from the truth! In fact, there are physical processes, usually known as Lévy flights, which are characterized by the above characteristic function. We will later discuss further in detail the properties of such a process.

One classical example of this is $f(x) = Cx^{-(1+\alpha)}$ for $x > 0$ and $\alpha \in (0, 1)$. This function cannot give a density function as it is not integrable. Still, we will see that it can be used for (4.1).

Lemma 4.1.1 *If $f : \mathbb{R} \to \mathbb{R}_+$ satisfies $\int_{-\infty}^{\infty}(|x| \wedge 1)f(x)dx < \infty$, then f also satisfies:*

1. $\int_{\varepsilon}^{\infty} f(x)dx < \infty$ *for any* $\varepsilon > 0$.
2. $|\int_{-\infty}^{\infty}(e^{i\theta x} - 1)f(x)dx| < \infty, \theta \in \mathbb{R}$.

The above lemma shows that although f is not a density when integrated on the whole space, it is integrable when restricted to a set that does not include a neighborhood around zero. Therefore using the approach described after Corollary 3.3.9, we will construct this process in pieces and then put all of them together with a limit procedure.

For this, we define

$$f_\varepsilon(x) := \frac{f(x)\mathbf{1}_{\{x>\varepsilon\}}}{\lambda_\varepsilon},$$

where $\lambda_\varepsilon := \int_{\varepsilon}^{\infty} f(x)dx < \infty$. Notice that in our example $f(x) = Cx^{-(1+\alpha)}$ we will have that $\lambda_\varepsilon = C\alpha^{-1}\varepsilon^{-\alpha}$.

Step 1. Using this we construct the process with jumps larger than one. That is, we define the following compound Poisson process:

$$Z_t^{(1,+)} := \sum_{i=1}^{\infty} Y_i^{(1,+)}\mathbf{1}_{\{T_i^{(1,+)}\leq t\}} = \int_{\mathbb{R}\times[0,t]} x\mathcal{N}^{(1,+)}(dx, ds),$$

where $Y_i^{(1,+)}$ is a $[1, \infty)$-valued random variable with $\mathbb{P}(Y_i^{(1,+)} \in A) = \int_A f_1(x)dx$. Next, $T_i^{(1,+)} = s_1^{(1,+)} + \cdots + s_i^{(1,+)}$, and $\{s_i^{(1,+)}\}$ are i.i.d.r.v.s each with the $Exp(c\lambda_1)$ distribution.

Step 2. For $0 < \varepsilon' < \varepsilon$, we define

$$f_{\varepsilon',\varepsilon}(x) := \frac{f(x)\mathbf{1}_{\{\varepsilon'<x\leq\varepsilon\}}}{\lambda_{\varepsilon',\varepsilon}},$$

where $\lambda_{\varepsilon',\varepsilon} = \int_{\varepsilon'}^{\varepsilon} f(x)dx$. Then we construct

$$Z_t^{(\varepsilon',\,\varepsilon,+)} := \sum_{i=1}^{\infty} Y_i^{(\varepsilon',\varepsilon)} \mathbf{1}_{\{T_i^{(\varepsilon',\varepsilon)} \le t\}} = \int_0^t \int_{\mathbb{R}} x \mathcal{N}^{(\varepsilon',\varepsilon,+)}(dx, ds),$$

where $Y_i^{(\varepsilon',\varepsilon)}$ is a $(\varepsilon', \varepsilon]$-valued random variable with $\mathbb{P}(Y_i^{(\varepsilon',\varepsilon)} \in A) = \int_A f_{(\varepsilon',\varepsilon)}(x)dx$, $T_i^{(\varepsilon',\varepsilon)} = s_1^{(\varepsilon',\varepsilon)} + \cdots + s_i^{(\varepsilon',\varepsilon)}$, and $\{s_i^{(\varepsilon',\varepsilon)}\}$ are i.i.d.r.v.s each with the $Exp(c\lambda_{\varepsilon',\varepsilon})$ distribution. All the r.v.s here are independent.

To construct Z, we take $\varepsilon_n = \dfrac{1}{2^{n-1}}$. We define

$$0 \le Z_t^{(\varepsilon_n,+)} := Z_t^{(1,+)} + \sum_{i=1}^{n-1} Z_t^{(\varepsilon_{i+1},\varepsilon_i,+)} \le Z_t^{(\varepsilon_{n+1},+)},$$

$$Z_t^+ := \lim_{n\to\infty} Z_t^{(\varepsilon_n,+)} = Z_t^{(1,+)} + \sum_{i=1}^{\infty} Z_t^{(\varepsilon_{i+1},\varepsilon_i,+)}.$$

Note that $Z_t^{(\varepsilon_1,+)} \equiv Z_t^{(1,+)}$ and that Z_t^+ is well defined as a limit of an increasing positive sequence, although it may be infinite. Sometimes when the meaning is clear, we will simplify the notation, using instead $Z_t^{(\varepsilon_{i+1},\varepsilon_i)} \equiv Z_t^{(\varepsilon_{i+1},\varepsilon_i,+)}$ or $Z_t^{(\varepsilon_n)} \equiv Z_t^{(\varepsilon_n,+)}$. Similarly, for the Poisson random measure we may use $\mathcal{N}^{(\varepsilon',\varepsilon)} \equiv \mathcal{N}^{(\varepsilon',\varepsilon,+)}$. For fixed n, we have $Z_t^{(1,+)}, Z_t^{(\varepsilon_2,\varepsilon_1,+)}, \cdots, Z_t^{(\varepsilon_n,\varepsilon_{n-1},+)}$ are independent compound Poisson processes. From Corollary 3.2.4 and Theorem 3.2.7 we have $Z_t^{(\varepsilon_n,+)}$ is a Lévy process. Moreover, we have that its characteristic function is given by

$$\begin{aligned}
&\mathbb{E}[e^{i\theta Z_t^{(\varepsilon_n,+)}}]\\
&= \mathbb{E}[e^{i\theta(Z_t^{(1,+)}+\sum_{i=1}^{n-1} Z_t^{(\varepsilon_{i+1},\varepsilon_i,+)})}] = \mathbb{E}[e^{i\theta Z_t^{(1,+)}} \prod_{j=1}^{n-1} e^{i\theta Z_t^{(\varepsilon_{i+1},\varepsilon_i,+)}}]\\
&= \mathbb{E}[e^{i\theta Z_t^{(1,+)}}] \prod_{j=1}^{n-1} \mathbb{E}[e^{i\theta Z_t^{(\varepsilon_{i+1},\varepsilon_i,+)}}]\\
&= \exp[c\lambda_{\varepsilon_1} t \int_{\mathbb{R}} (e^{i\theta x} - 1) f_{\varepsilon_1}(x)dx] \prod_{j=1}^{n-1} \exp[c\lambda_{\varepsilon_{j+1},\varepsilon_j} t \int_{\mathbb{R}} (e^{i\theta x} - 1) f_{\varepsilon_{j+1},\varepsilon_j}(x)dx]\\
&= \exp[ct \int_{\mathbb{R}} (e^{i\theta x} - 1)\Big\{\lambda_{\varepsilon_1} f_{\varepsilon_1}(x) + \sum_{j=1}^{n-1} c\lambda_{\varepsilon_{j+1},\varepsilon_j} f_{\varepsilon_{j+1},\varepsilon_j}(x)\Big\}dx]\\
&= \exp[ct \int_{\mathbb{R}} (e^{i\theta x} - 1) f(x) \mathbf{1}_{\{x>\varepsilon_n\}} dx]. \qquad (4.3)
\end{aligned}$$

Lastly, we define a random measure $\mathcal{N}^{(\varepsilon_n,+)}$ as follows:

$$\mathcal{N}^{(\varepsilon_n,+)} := \sum_{i=1}^{n} \mathcal{N}^{(\varepsilon_i,\varepsilon_{i-1},+)}, \tag{4.4}$$

where $\varepsilon_0 = \infty$ and $\mathcal{N}^{(\varepsilon_1,\varepsilon_0,+)} \equiv \mathcal{N}^{(\varepsilon_1,\infty,+)} \equiv \mathcal{N}^{(\varepsilon_1,+)}$.

Proposition 4.1.2 *$Z_t^{(\varepsilon_n,+)}$ and Z_t^+ satisfies the following properties:*

(i) $Z_t^+ < \infty$, *a.s.,*
(ii) (finite variation property) $\Delta Z_t^{(\varepsilon_n)} = \Delta Z_t^+ \mathbf{1}_{\{\Delta Z_t^+ > \varepsilon_n\}}$, a.s.,
(iii) $Z_t^+ = \sum_{s \le t} \Delta Z_s^+$,
(iv) $\lim_{n\to\infty} \sup_{s \le t} |Z_s^{(\varepsilon_n,+)} - Z_s^+| = 0.$

Note that property (i) above implies that Z is of bounded variation because its paths are non-decreasing. Therefore, again stochastic integrals for g measurable and bounded with respect to $\mathcal{N}$ and $\widetilde{\mathcal{N}}$ can be naturally defined.

Proof (i) Note that $Z^{(1,+)}$ is a compound Poisson process and therefore it is well defined and finite. Note that $Z_t^+ - Z_t^{(1,+)}$ is a positive process and furthermore

$$\mathbb{E}[Z_t^+ - Z_t^{(1,+)}] = \lim_{n\to\infty} \mathbb{E}[Z_t^{(\varepsilon_n,+)} - Z_t^{(1,+)}] = t \int_0^1 x f(x) dx < \infty.$$

Therefore $Z_t^+ < \infty$, a.s..
(ii) Since Z_t^+ is monotone non-decreasing in t then Z_t^+ has left limits a.s. and

$$\lim_{s \uparrow t} Z_s^+ \le Z_t^+ < \infty, \text{ a.s.}$$

Define the following measurable subset of Ω,

$$\Omega_0 := \{T_j^{(\varepsilon_{i+1},\varepsilon_i,+)} \neq T_{j'}^{(\varepsilon_{i'+1},\varepsilon_{i'},+)},\ i = 0, \cdots, n-1, i' \ge n, j, j' \in \mathbb{N}\},$$

where $T_j^{(\varepsilon_1,\varepsilon_0,+)} := T_j^{(\varepsilon_1,+)}$.

The complement of this set characterizes the possibility that different jumps may happen at the same time. We now prove that this is not possible. Since for any $i = 0, \cdots, n-1$, $i' \ge n$ and $j, j' \in \mathbb{N}$, $T_j^{(\varepsilon_{i+1},\varepsilon_i,+)}$ and $T_{j'}^{(\varepsilon_{i'+1},\varepsilon_{i'},+)}$ are independent, we have $\mathbb{P}(T_j^{(\varepsilon_{i+1},\varepsilon_i,+)} \neq T_{j'}^{(\varepsilon_{i'+1},\varepsilon_{i'},+)}) = 1$. Hence, $\mathbb{P}(\Omega_0) = 1$.

Now, we get

$$\Delta Z_t^+ = \Delta Z_t^{(\varepsilon_1,+)} + \sum_{i=1}^{\infty} \Delta Z_t^{(\varepsilon_{i+1},\varepsilon_i,+)}$$

$$= \Delta Z_t^{(\varepsilon_1,+)} + \sum_{i=1}^{n-1} \Delta Z_t^{(\varepsilon_{i+1},\varepsilon_i,+)} + \sum_{i=n}^{\infty} \Delta Z_t^{(\varepsilon_{i+1},\varepsilon_i,+)}$$
$$= \Delta Z_t^{(\varepsilon_n,+)} + \sum_{i=n}^{\infty} \Delta Z_t^{(\varepsilon_{i+1},\varepsilon_i,+)}.$$

For every $i \geq n$, $\Delta Z_t^{(\varepsilon_{i+1},\varepsilon_i,+)} \leq \varepsilon_n$. Therefore, we have $\Delta Z_t^+ = \Delta Z_t^{(\varepsilon_n,+)}$ on $\{\Delta Z_t^+ > \varepsilon_n\} \cap \Omega_0$. For all $\omega \in \{\Delta Z_t^+ \leq \varepsilon_n\} \cap \Omega_0$, we have $\Delta Z_t^{(\varepsilon_n,+)}(\omega) = 0$. Hence we have $\Delta Z_t^{(\varepsilon_n,+)} = \Delta Z_t^+ \mathbf{1}_{\{\Delta Z_t^+ > \varepsilon_n\}}$, a.s.

(iii) From the monotone convergence theorem, we have

$$Z_t^+ = \lim_{n\to\infty} Z_t^{(\varepsilon_n,+)} = \lim_{n\to\infty} \sum_{s\leq t} \Delta Z_s^{(\varepsilon_n,+)} = \sum_{s\leq t} \lim_{n\to\infty} (Z_s^{(\varepsilon_n,+)} - Z_{s-}^{(\varepsilon_n,+)})$$
$$= \sum_{s\leq t} (Z_s^+ - Z_{s-}^+) = \sum_{s\leq t} \Delta Z_s^+.$$

(iv) From (ii),

$$\sup_{s\leq t} |Z_s^{(\varepsilon_n,+)} - Z_s^+|$$
$$= \sup_{s\leq t} |\sum_{u\leq s} \Delta Z_u^{(\varepsilon_n,+)} - \sum_{u\leq s} \Delta Z_u^+| = \sup_{s\leq t} |\sum_{u\leq s} (\Delta Z_u^+ \mathbf{1}_{\{\Delta Z_u^+ > \varepsilon_n\}} - \Delta Z_u^+)|$$
$$\leq \sup_{s\leq t} \sum_{u\leq s} |\Delta Z_u^+ \mathbf{1}_{\{\Delta Z_u^+ > \varepsilon_n\}} - \Delta Z_u^+| = \sup_{s\leq t} \sum_{u\leq s} \Delta Z_u^+ \mathbf{1}_{\{\Delta Z_u^+ \leq \varepsilon_n\}}$$
$$= \sum_{u\leq t} \Delta Z_u^+ \mathbf{1}_{\{\Delta Z_u^+ \leq \varepsilon_n\}} = \sum_{u\leq t} (\Delta Z_u^+ - \Delta Z_u^+ \mathbf{1}_{\{\Delta Z_u^+ > \varepsilon_n\}})$$
$$= \sum_{u\leq t} (\Delta Z_u^+ - \Delta Z_u^{(\varepsilon_n,+)}) = Z_t - Z_t^{(\varepsilon_n,+)} \to 0 \text{ as } (n \to \infty).$$

Exercise 4.1.3 This exercise is related to Proposition 4.1.2 (i).

1. Give an explicit function f satisfying that $\int |x| \wedge 1 f(x)dx < \infty$ but so that $\mathbb{E}[Z_t^{(1,+)}] = \infty$.
2. In general, note that under the condition $\int |x| \wedge 1 f(x)dx < \infty$, $\mathbb{E}[Z_t^{(1,+)}]$ may not be finite, but we always have that $\mathbb{E}[|Z_t^+ - Z_t^{(1,+)}|^2]$ is finite.[1]

Lemma 4.1.4 *Let* $T > 0$ *and* f_n *be a càdlàg function on* $[0, T]$. *If* $\lim_{n\to\infty} \sup_{x\in[0,T]} |f_n(x) - f(x)| = 0$, *then* f *is a càdlàg function on* $[0, T]$.

[1]Hint: Note that the process $Z_t^+ - Z_t^{(1,+)}$ does not include any jump of size bigger than one. Also, see Exercise 4.1.26 .

Proof We prove for any $c \in [0, T]$, $f(c)$ is a càdlàg function. Since $\lim_{n\to\infty} \sup_{x\in[0,T]} |f_n(x) - f(x)| = 0$, we have for any $\varepsilon > 0$, there exists $N \in \mathbb{N}$ such that for any $x \in [0, T]$,

$$|f_N(x) - f(x)| < \varepsilon/3.$$

(i) Right-continuity. Since f_N is càdlàg at c, there exists $\delta > 0$ such that

$$0 < x - c < \delta,\ x \in [0, T] \Rightarrow |f_N(x) - f_N(c)| < \varepsilon/3.$$

Hence we have

$$|f(x) - f(c)| \le |f(x) - f_N(x)| + |f_N(x) - f_N(c)| + |f_N(c) - f(c)| < \varepsilon,$$

that is, f is a right-continuous function.
(ii) Existence of left limit. We take a sequence $\{c_n\}_{n\in\mathbb{N}}$ with $c_n \uparrow c$. Since f_N has a left limit at c, there exists $\lim_{n\to\infty} f_N(c_n)$. Because a convergent sequence is a Cauchy sequence, there exists $M \in \mathbb{N}$ such that

$$n, k \ge M \Rightarrow |f_N(c_n) - f_N(c_k)| < \varepsilon/3.$$

Hence we have

$$|f(c_n) - f(c_k)| \le |f(c_n) - f_N(c_n)| + |f_N(c_n) - f_N(c_k)| + |f_N(c_k) - f(c_k)| < \varepsilon,$$

so $\{f(c_n)\}$ is also a Cauchy sequence. Therefore there exists a left limit $\lim_{n\to\infty} f(c_n) = \lim_{s\uparrow c} f(s) = f(c-)$.

Theorem 4.1.5 *$\{Z_t^+; t \ge 0\}$ is a Lévy process.*

Proof (i) We prove for any $\theta, \eta \in \mathbb{R}$,

$$\mathbb{E}[e^{i\theta(Z_t^+ - Z_s^+) + i\eta Z_s^+}] = \mathbb{E}[e^{i\theta(Z_t^+ - Z_s^+)}]\mathbb{E}[e^{i\eta Z_s^+}].$$

From the dominated convergence theorem, we have

$$\begin{aligned}\mathbb{E}[e^{i\theta(Z_t^+ - Z_s^+) + i\eta Z_s^+}] &= \lim_{n\to\infty} \mathbb{E}[e^{i\theta(Z_t^{(\varepsilon_n,+)} - Z_s^{(\varepsilon_n,+)}) + i\eta Z_s^{(\varepsilon_n,+)}}] \\ &= \lim_{n\to\infty} \mathbb{E}[e^{i\theta(Z_t^{(\varepsilon_n,+)} - Z_s^{(\varepsilon_n,+)})}]\mathbb{E}[e^{i\eta Z_s^{(\varepsilon_n,+)}}] \\ &= \mathbb{E}[e^{i\theta(Z_t^+ - Z_s^+)}]\mathbb{E}[e^{i\eta Z_s^+}].\end{aligned}$$

(ii) $Z_0^+ = \lim_{n\to\infty} Z_0^{(\varepsilon_n,+)} = 0$.
(iii) For any θ,

$$\mathbb{E}[e^{i\theta(Z_t^+ - Z_s^+)}] = \lim_{n\to\infty} \mathbb{E}[e^{i\theta(Z_t^{(\varepsilon_n,+)} - Z_s^{(\varepsilon_n,+)})}] = \lim_{n\to\infty} \mathbb{E}[e^{i\theta Z_{t-s}^{(\varepsilon_n,+)}}] = \mathbb{E}[e^{i\theta Z_{t-s}^+}].$$

(iv) We take any $\varepsilon > 0$, then

$$\begin{aligned}\mathbb{P}\big[|Z_t^+ - Z_s^+| > \varepsilon\big] &= \mathbb{P}\big[Z_{t-s}^+ > \varepsilon\big] \\ &= \mathbb{P}\left[Z_{t-s}^+ > \varepsilon,\ Z_{t-s}^{(1,+)} > \frac{\varepsilon}{2}\right] + \mathbb{P}\left[Z_{t-s}^+ > \varepsilon,\ Z_{t-s}^{(1,+)} \le \frac{\varepsilon}{2}\right].\end{aligned}$$

Now, we prove that each term on the right-hand side of the above equality converges to zero as $t - s \downarrow 0$:

$$\begin{aligned}(\text{1st term of RHS}) &\le \mathbb{P}\left[Z_{t-s}^{(1,+)} > \frac{\varepsilon}{2}\right] \to 0, \\ (\text{2nd term of RHS}) &\le \mathbb{P}\left[Z_{t-s}^+ - Z_{t-s}^{(1,+)} > \frac{\varepsilon}{2}\right] \le \frac{2}{\varepsilon}\mathbb{E}\big[Z_{t-s}^+ - Z_{t-s}^{(1,+)}\big] \\ &= \frac{2(t-s)}{\varepsilon}\int_0^1 x\, f(x)\, dx \to 0.\end{aligned}$$

Hence $\lim_{s\to t} \mathbb{P}(|Z_t^+ - Z_s^+| > \varepsilon) = 0$.
(v) From Proposition 4.1.2 (iv) and Lemma 4.1.4, $t \mapsto Z_t^+$ is a càdlàg function.

Now we define the random measure $\mathcal{N}^+$ which corresponds to Z^+. From (4.4),

$$\mathcal{N}^{(\varepsilon_n,+)} = \sum_{i=1}^{n} \mathcal{N}^{(\varepsilon_i,\varepsilon_{i-1},+)} \le \mathcal{N}^{(\varepsilon_{n+1},+)},$$

hence we define the random measure $\mathcal{N}^+$ as the following σ-finite measure:

$$\mathcal{N}^+ = \lim_{n\to\infty} \mathcal{N}^{(\varepsilon_n,+)} = \sum_{i=1}^{\infty} \mathcal{N}^{(\varepsilon_i,\varepsilon_{i-1},+)}.$$

We will now give some properties on the number of jumps of Z in closed time intervals. First, we consider the number of jumps away from zero.

Corollary 4.1.6 *$\mathcal{N}^+$ is a finite measure on $((a, b] \times [0, t],\ \mathscr{B}(a, b] \otimes \mathscr{B}[0, t])$ for any $b > a > 0$.*

That is, the number of jumps of size $(a, b]$ in the time interval $[0, t]$ is finite a.s.

Proof Since for any $a > 0$, there exists $n \in \mathbb{N}$ such that $\varepsilon_n < a \le \varepsilon_{n-1}$, we have

$$\begin{aligned}\mathcal{N}^+((a, b] \times [0, t]) =& \mathcal{N}^{(\varepsilon_n,+)}((a, b] \times [0, t]) \\ \le& \mathcal{N}^{(\varepsilon_n,+)}([0, \infty) \times [0, t]) = N_t^{(\varepsilon_n,+)} < \infty, \text{ a.s.}\end{aligned}$$

Theorem 4.1.7 $Z_t^+ = \int_{(0,\infty)\times[0,t]} x\mathcal{N}^+(dx, ds).$

Proof From Lemma 3.3.4, we have

$$\begin{aligned} Z_t^+ &= \lim_{n\to\infty} Z_t^{(\varepsilon_n,+)} = \lim_{n\to\infty} \int_{(0,\infty)\times[0,t]} x\mathcal{N}^{(\varepsilon_n,+)}(dx, ds) \\ &= \lim_{n\to\infty} \sum_{i=1}^{n} \int_{(0,\infty)\times[0,t]} x\mathcal{N}^{(\varepsilon_i,\varepsilon_{i-1},+)}(dx, ds) \\ &= \sum_{i=1}^{\infty} \int_{(0,\infty)\times[0,t]} x\mathcal{N}^{(\varepsilon_i,\varepsilon_{i-1},+)}(dx, ds) = \int_{(0,\infty)\times[0,t]} x\mathcal{N}^+(dx, ds). \end{aligned}$$

To finish the construction, one needs to repeat all the previous steps for

$$\begin{aligned} f_{-\varepsilon}(x) &= \lambda_{-\varepsilon}^{-1} f(x)\mathbf{1}_{\{x<-\varepsilon\}} \\ \lambda_{-\varepsilon} &= \int_{-\infty}^{-\varepsilon} f(x)dx. \end{aligned}$$

This construction will also lead to a Poisson random measure $\mathcal{N}^-$.

Theorem 4.1.8 $Z_t^- = \int_{(-\infty,0)\times[0,t]} x\mathcal{N}^-(dx, ds).$

The following representation is sometimes called the Lévy-Itô representation of the corresponding Lévy process Z.

Theorem 4.1.9 *Let $\mathcal{N} := \mathcal{N}^+ + \mathcal{N}^-$. Then we have*

$$Z_t = \int_{\mathbb{R}\times[0,t]} x\mathcal{N}(dx, ds).$$

Exercise 4.1.10 Prove the above theorems.

Then we can define Z^- which is statistically independent Z^+. Adding the processes for positive and negative jumps we get the following result:

Theorem 4.1.11 *Let $Z_t := Z_t^+ + Z_t^-$. Then $\{Z_t\ ;\ t \geq 0\}$ is a Lévy process which satisfies that $|Z_t| \leq \sum_{s\leq t} |\Delta Z_s| < +\infty$ a.s. That is, Z is a process whose paths are a.s. of finite variation on any closed interval.*

From this theorem stems the title of this chapter.

Exercise 4.1.12 Prove that $\sum_{s\leq t} |\Delta Z_s|^p < +\infty$ a.s. for any $p \geq 1$.
For a hint, see Chap. 14.

Corollary 4.1.13 *The characteristic function of Z_t is given by*

$$\mathbb{E}[e^{i\theta Z_t}] = \exp[ct \int_{\mathbb{R}} (e^{i\theta x} - 1) f(x) dx].$$

Due to the above result, one says that the Lévy measure associated with Z is $cf(x)dx$.

Proof Let $Z_t^{(\varepsilon)} := Z_t^{(\varepsilon,+)} + Z_t^{(\varepsilon,-)}$. From (4.3),

$$\begin{aligned}\mathbb{E}[e^{i\theta Z_t}] = \lim_{n\to\infty} \mathbb{E}[e^{i\theta Z_t^{(\varepsilon_n)}}] &= \lim_{n\to\infty} \exp[ct \int_{\mathbb{R}} (e^{i\theta x} - 1) f(x) \left(\mathbf{1}_{\{x>\varepsilon_n\}} + \mathbf{1}_{\{x<-\varepsilon_n\}}\right) dx] \\ &= \exp[ct \int_{\mathbb{R}} (e^{i\theta x} - 1) f(x) dx].\end{aligned}$$

The fact that Z is a process of bounded variation also assures that the measure $\mathcal{N}(dx, ds)$ can be defined away from zero. Therefore all integrals can be defined as the usual integrals with respect to σ-finite measures.

Definition 4.1.14 (*Definition of stochastic integral*) Let g be a $\mathcal{F} \otimes \mathcal{B}(\mathbb{R}) \otimes \mathcal{B}[0, \infty)$-measurable càdlàg function such that

$$\int_{\mathbb{R}\times[0,t]} |g(x, s-)| \mathcal{N}(dx, ds) := \lim_{\varepsilon\downarrow 0} \int_{\mathbb{R}\times[0,t]} |g(x, s-)| \mathcal{N}^{(\varepsilon)}(dx, ds) < \infty.$$

Here $\mathcal{N}^{(\varepsilon)}(dx, ds) := \mathbf{1}_{|x|>\varepsilon} \mathcal{N}(dx, ds)$. In such a case the integral

$$\int_{\mathbb{R}\times[0,t]} g(x, s-) \mathcal{N}(dx, ds)$$

is called the stochastic integral with respect to $\mathcal{N}$.

Similarly to the notation $Z_t^{(a,b)}$, we will also use the following notation:

$$\mathcal{N}^{(a,b)}(dx, ds) := \mathbf{1}_{|x|\in(a,b]} \mathcal{N}(dx, ds).$$

Exercise 4.1.15 Prove that the above integral has càdlàg paths. That is, for almost all $\omega \in \Omega$, $t \to \int_{\mathbb{R}\times[0,t]} g(x, s-) \mathcal{N}(dx, ds)$ is a càdlàg function.

Exercise 4.1.16 Due to Theorem 4.1.11, the Lévy process Z has paths of finite variation. Suppose that in Definition 4.1.14, we consider the particular case that $g(x, s) = g_1(s)x$ for a step function $g_1 : \mathbb{R}_+ \to \mathbb{R}$. Prove that

$$\int_0^t g_1(s) dZ_s = \int_{\mathbb{R}\times[0,t]} g(x, s-) \mathcal{N}(dx, ds).$$

Here, the integral on the left side of the above equation is understood as an integral with respect to functions of bounded variation. Finally, taking limits (i.e. using the monotone class theorem) prove that the above result is valid for continuous functions g_1.

Note that Corollary 3.4.21 gives the following theorem:

Theorem 4.1.17 *Define the σ-finite measure $\widehat{\mathcal{N}}(dx, ds) := cf(x)dxds$. Assume that g is a $\mathcal{B}(\mathbb{R}) \otimes \mathcal{B}[0, \infty)$-measurable càdlàg function and that the random variable $\int_{\mathbb{R}\times[0,t]} |g(x, s-)|\mathcal{N}(dx, ds) \in L^1(\Omega, \mathcal{F}, \mathbb{P})$ and $\int_{\mathbb{R}\times[0,t]} |g(x, s-)|\widehat{\mathcal{N}}(dx, ds) < \infty$. Then*

$$\mathbb{E}\left[\int_{\mathbb{R}\times[0,t]} g(x, s-)\mathcal{N}(dx, ds)\right] = \int_{\mathbb{R}\times[0,t]} g(x, s-)\widehat{\mathcal{N}}(dx, ds).$$

Proof From Corollary 3.4.21,[2]

$$\begin{aligned}\mathbb{E}\left[\int_{\mathbb{R}\times[0,t]} g(x, s-)\mathcal{N}(dx, ds)\right] &= \lim_{n\to\infty} \mathbb{E}\left[\int_{\mathbb{R}\times[0,t]} g(x, s-)\mathcal{N}^{(\varepsilon_n)}(dx, ds)\right] \\ &= \lim_{n\to\infty} c \int_{\mathbb{R}\times[0,t]} g(x, s-)f(x)\left(\mathbf{1}(x > \varepsilon_n) + \mathbf{1}(x < -\varepsilon_n)\right) dxds \\ &= c \int_{\mathbb{R}\times[0,t]} g(x, s-)f(x)dxds \\ &= \int_{\mathbb{R}\times[0,t]} g(x, s-)\widehat{\mathcal{N}}(dx, ds).\end{aligned}$$

Definition 4.1.18 The measure $\widehat{\mathcal{N}}(dx, ds) := \nu(dx)ds = cf(x)dxds$ is called the compensator of $\mathcal{N}$. Note that under the current conditions f is not necessarily a density function.

Exercise 4.1.19 For a measurable set $A \subset \mathbb{R} - \{0\}$, such that $\int_A f(x)dx < \infty$, find the law of the random variable $C_t := \int_{A\times[0,t]} \mathcal{N}(dx, ds)$. Prove that C is an increasing càdlàg process. Note that this process counts how many jumps fall in the set A. Hint: Recall Exercise 3.4.9.

Exercise 4.1.20 Let $f(x) := \dfrac{e^{-\lambda|x|}}{|x|^{1+\alpha}} \mathbf{1}_{\{x\neq 0\}}$, where $\lambda > 0$ and $\alpha \in [0, 1)$. Prove that f satisfies conditions (4.2).

Exercise 4.1.21 Consider the case that $f(x) := \dfrac{e^{-\lambda|x|}}{|x|^{1+\alpha}} \mathbf{1}_{\{x>0\}}$. In this particular case the generated process Z is called the gamma process. Prove:

[2] Recall that $\mathcal{N}^{(\varepsilon_n)} = \mathcal{N}^{(\varepsilon_n,+)} + \mathcal{N}^{(\varepsilon_n,-)}$.

1. Z has increasing paths.
2. The distribution of Z_t follows a gamma law. Find its mean and variance.

   For a hint, see Chap. 14.

Exercise 4.1.22 This exercise is related with the jump transformation procedure introduced in Proposition 3.3.7.

For this, consider a Lévy process with $f(x) := \frac{1}{x^{1+\alpha}} \mathbf{1}_{\{x>0\}}$, $\alpha \in [0, 1)]$. As explained in Theorem 4.1.7, the associated Lévy process Z satisfies that $Z_t = \int_{(0,\infty)\times[0,t]} x\mathcal{N}^+(dx, ds)$. Prove that $W_t := \int_{(0,\infty)\times[0,t]} x^r \mathcal{N}^+(dx, ds)$ for $r > \alpha$ is also a Lévy process and find its Lévy measure.

For a hint, see Chap. 14.

Exercise 4.1.23 In this exercise, we perform some basic calculations for stable laws that will be used later.

1. If we denote by Z^λ the process corresponding to Exercise 4.1.20, prove that Z_t^λ converges weakly (in law) as $\lambda \downarrow 0$ for $t > 0$ fixed. Obtain the characteristic function of the limit random variable, which we denote Z_t.[3]
2. Let $f(x) := \dfrac{C}{|x|^{1+\alpha}}$, $\alpha \in (0, 1)$. Prove that

$$ct \int_{\mathbb{R}} (e^{i\theta x} - 1) f(x)dx = -2^{1-\alpha} Cct|\theta|^\alpha \int_{\mathbb{R}_+} \sin^2(u)u^{-(1+\alpha)}du.$$

   In particular, prove the finiteness of the above integral. This law is known as the stable law and its characteristic function is given by $\exp(-c|\theta|^\alpha)$ for some positive constant $c > 0$. In particular, this law is symmetric.[4]
3. Prove that the random variable Z_t satisfies that the law of $t^{-1/\alpha} Z_t$ is independent of t.[5]
4. Prove that the density of Z_t exists and is smooth.[6]

   For a hint, see Chap. 14.

Remark 4.1.24 We remark that the above construction also reveals the nature of the process just defined. This becomes more clear if we consider the example $f(x) = C|x|^{-(1+\alpha)} \mathbf{1}_{\{x\neq 0\}}$ for $\alpha \in (0, 1)$. Then it is clear that $\lambda_{\varepsilon,\varepsilon'}$ is blowing to infinity. Therefore the average frequency of small jumps is getting larger and blowing to infinity. Still, the approximating process converges to Z. This is the essential nature of the above construction. One may in fact, even prove that for almost every $\omega \in \Omega$ there are a infinite number of jumps.

---

[3] Recall Theorem 1.1.6.

[4] One has to be careful with other texts as there are generalized versions of stable laws which include a parameter which measures symmetry.

[5] Recall Exercise 2.1.36.

[6] Recall Exercise 1.1.11.

Exercise 4.1.25 In this exercise, we show that, in general, the number of jumps of any size in any time interval may be infinite. For this, let us consider the explicit example of $f(x) = \frac{C}{x^{1+\alpha}} \mathbf{1}_{\{x \in (0,1)\}}$ for $\alpha \in [0, 1)$, $C > 0$. Prove the following statements:

1. Compute $\mathbb{E}[\mathcal{N}^{(\varepsilon_i,\varepsilon_{i-1})}((0, 1] \times [0, t])]$, $i \in \mathbb{N}$. Note that in this case, $\mathcal{N}^{(\varepsilon_i,\varepsilon_{i-1})} = \mathcal{N}^{(\varepsilon_i,\varepsilon_{i-1},+)}$.
2. Prove that the following law of large numbers is satisfied:

$$\frac{1}{n}\sum_{i=1}^{n} 2^{-i\alpha} \mathcal{N}^{(\varepsilon_i,\varepsilon_{i-1})}((0, 1] \times [0, t]) \to \frac{Ct}{\alpha} 2^{-\alpha}(1 - 2^{-\alpha}), \text{ a.s.}$$

3. Prove that $\mathcal{N}^+((0, 1] \times [0, t]) = \infty$, a.s.
4. Prove that for any $s < t$, one also has $\mathcal{N}^+((0, 1] \times (s, t]) = \infty$, a.s.

For a hint, see Chap. 14.

Exercise 4.1.26 This exercise is related to martingale properties of the process Z. Suppose that $f(x) = f(x)\, \mathbf{1}_{\{|x| \leq 1\}}$ with $\int_{[-1,1]} |x| f(x)dx < \infty$.

1. Compute $\mathbb{E}[Z_t]$ and $\mathbb{E}[Z_t^2]$
2. Find deterministic functions μ_t and σ_t such that $Z_t - \mu_t$ and $(Z_t - \mu_t)^2 - \sigma_t^2$ are martingales. These functions μ and σ^2 are usually called the compensators of Z_t and $(Z_t - \mu_t)^2$, respectively.

For a hint, see Chap. 14.

Exercise 4.1.27 Consider $f(x) = \frac{\alpha}{x^{1+\alpha}} \mathbf{1}_{\{x>1\}}, \alpha \in (0, 1)$. Let Z be the Lévy process with Lévy measure $f(x)dx$.

1. Prove that in this case Z is an increasing compound Poisson process and state explicitly the law of the jumps and the associated frequency of jumps.[7]
2. Prove that the expectation of Z_t^k does not exist (i.e. is not finite) for any $k \in \mathbb{N}$.
3. Prove that the expectation of Z_t^α does not exist.
4. Prove that $\mathbb{E}[Z_t^\beta] < \infty$ for all $0 < \beta < \alpha$.[8]
5. Prove that $\mathbb{E}[Z_t^\beta] = \infty$ in the case that $\beta < 0$. In fact, prove that $\mathbb{P}(Z_t = 0) > 0$ for any $t > 0$.

For a hint, see Chap. 14.

Many other interesting properties of Lévy processes are well known and determined by its Lévy measure. For more on this, see [51].

Exercise 4.1.28 Propose and prove the equivalent of Exercise 2.1.36 in the present setting. For example, compute the moments of $t^{-1}Z_t$ or prove that the law of $t^{-1}Z_t$ depends on t.

---

[7]This is trivial.

[8]Hint: Decompose the expectation according to the number of jumps of Z_t. Then use repeatedly the inequality $(x + y)^\beta \leq x^\beta y^\beta$ for $x, y > 0$.

Exercise 4.1.29 Define $\mathscr{F}_t^{(\varepsilon_n)} := \sigma(Z_s^{(\varepsilon_n)}; s \le t)$. Prove that

$$\mathscr{F}_t^{(\varepsilon_n)} := \sigma(Z_s^{(\varepsilon_k,\varepsilon_{k-1})}; s \le t, k = 1, \cdots n).$$

Exercise 4.1.30 Let $b_n, b : \mathbb{R} \times [0, T] \to \mathbb{R}_+$ be measurable uniformly bounded functions such that they satisfy that $|b_n(z, s)| \le C|b(z, s)|$ for some deterministic positive constant C. Suppose that $b_n \to b$ a.e. in $\mathbb{R} \times [0, T]$ with $\mathbb{E}[\int_{\mathbb{R}\times[0,t]} |b(x, s)| \widehat{\mathcal{N}}(dx, ds)] < \infty$. Prove that

$$\lim_{n\to\infty} \mathbb{E}\left[\int_{\mathbb{R}\times[0,t]} b_n(z, s)\mathcal{N}^{(\varepsilon_n)}(dz, ds)\right] = \mathbb{E}\left[\int_{\mathbb{R}\times[0,t]} b(z, s)\mathcal{N}(dz, ds)\right].$$

Exercise 4.1.31 (*The strong rate of convergence of the* $Z^{(\varepsilon)}$) Prove that for $\varepsilon < 1$:

1. $\mathbb{E}\left[\left|Z_T - Z_T^{(\varepsilon)}\right|\right] \le cT \int_{|z|\le\varepsilon} |z| f(z)dz$.
2. $\lim_{\varepsilon\downarrow 0} \int_{|z|\le\varepsilon} |z| f(z)dz = 0$.
3. Prove that the rate of convergence $\int_{|z|\le\varepsilon} |z| f(z)dz$ is optimal in the sense that there exists a Lévy process Z such that

$$\lim_{\varepsilon\downarrow 0}\left(\int_{|z|\le\varepsilon} |z| f(z)dz\right)^{-1}\mathbb{E}\left[\left|Z_T - Z_T^{(\varepsilon)}\right|\right] > 0.$$

4. Find an upper bound for the rate of convergence of $\mathbb{E}\left[\left|Z_T - Z_T^{(\varepsilon)}\right|\right]$ to zero in the case that $f(x) = \frac{1}{x^{1+\alpha}}, x > 0, \alpha \in (0, 1)$.

Exercise 4.1.32 (*The weak rate of convergence of* $Z^{(\varepsilon)}$) Prove that for $\varepsilon < 1$:

1. $\mathbb{E}[Z_T - Z_T^{(\varepsilon)}] = T \int_{|z|\le\varepsilon} z f(z)dz$.
2. Give an example of a function f satisfying the conditions (4.2) such that the strong rate of convergence $\int_{|z|\le\varepsilon} |z| f(z)dz$ is bigger than the weak rate of convergence $\int_{|z|\le\varepsilon} z f(z)dz$.
3. Suppose that $g : \mathbb{R} \to \mathbb{R}$ is a bounded differentiable function with bounded derivative. Assume furthermore that $f(x) = \frac{1}{|x|^{1+\alpha}} \mathbf{1}_{\{|x|<1\}}, \alpha \in [0, 1]$. Use a Taylor expansion of first order on g to obtain that

$$\lim_{\varepsilon\downarrow 0}\left(\int_{|z|\le\varepsilon} |z|^2 f(z)dz\right)^{-1}|\mathbb{E}[g(Z_T) - g(Z_T^{(\varepsilon)})]| = 0.$$

   Therefore the weak rate of convergence is faster than the strong rate of convergence.

   For a hint, see Chap. 14.

Exercise 4.1.33 This exercise is about real functions. Recall that in Exercise 3.4.9 we have proved that if $f : [0, 1] \to \mathbb{R}$ is a function such that the following right limit

$\lim_{x \downarrow a} f(x)$ exists at every point $a \in [0, 1)$ then f may have at most a countable number of discontinuity points.

Prove that if f is an increasing function then the number of discontinuity points is at most countable. Conclude that a function of bounded variation also has at most a countable number of discontinuity points.

4.1.1 Itô Formula

Theorem 4.1.34 (Itô formula for finite variation jump process) *Let $g \in C_b^2(\mathbb{R})$. Then we have the following Itô formula:*

$$\begin{aligned} g(Z_t) &= g(Z_0) + \sum_{s \le t} \{g(Z_s) - g(Z_{s-})\} \\ &= g(Z_0) + \int_{\mathbb{R} \times [0,t]} \{g(Z_{s-} + z) - g(Z_{s-})\} \mathscr{N}(dz, ds). \end{aligned}$$

Proof From Corollary 3.4.25 we have

$$g(Z_t^{(\varepsilon_n)}) = g(Z_0^{(\varepsilon_n)}) + \int_{\mathbb{R} \times [0,t]} \left\{ g(Z_{s-}^{(\varepsilon_n)} + z) - g(Z_{s-}^{(\varepsilon_n)}) \right\} \mathscr{N}^{(\varepsilon_n)}(dz, ds).$$

Since g is a continuous function, we have $g(Z_t^{(\varepsilon_n)}) \to g(Z_t)$ and $g(Z_0^{(\varepsilon_n)}) \to g(Z_0)$ as $n \to \infty$. So we show that

$$\begin{aligned} &\int_{\mathbb{R} \times [0,t]} \left\{ g(Z_{s-}^{(\varepsilon_n)} + z) - g(Z_{s-}^{(\varepsilon_n)}) \right\} \mathscr{N}^{(\varepsilon_n)}(dz, ds) \\ &\longrightarrow \int_{\mathbb{R} \times [0,t]} \{g(Z_{s-} + z) - g(Z_{s-})\} \mathscr{N}(dz, ds). \end{aligned}$$

The integral $\int_{\mathbb{R} \times [0,t]} \{g(Z_{s-} + z) - g(Z_{s-})\} \mathscr{N}(dz, ds)$ exists because

$$\begin{aligned} &\left| \int_{\mathbb{R} \times [0,t]} \{g(Z_{s-} + z) - g(Z_{s-})\} \mathscr{N}(dz, ds) \right| \\ &\le \int_{\mathbb{R} \times [0,t]} |g(Z_{s-} + z) - g(Z_{s-})| \mathscr{N}(dz, ds) \\ &\le C \int_{\mathbb{R} \times [0,t]} |z| \mathscr{N}(dz, ds) = C(Z_t^+ - Z_t^-) < \infty. \end{aligned}$$

Let $Z_t^{(\varepsilon_n)} := Z_t^{(\varepsilon_n,+)} + Z_t^{(\varepsilon_n,-)}$. From $\mathscr{N} - \mathscr{N}^{(\varepsilon_n)} \ge 0$, we get

$$\begin{aligned}
&\left|\int_{\mathbb{R}\times[0,t]} \{g(Z_{s-}+z)-g(Z_{s-})\}\,(\mathscr{N}-\mathscr{N}^{(\varepsilon_n)})(dz,ds)\right| \\
&\le \int_{\mathbb{R}\times[0,t]} |g(Z_{s-}+z)-g(Z_{s-})|\,(\mathscr{N}-\mathscr{N}^{(\varepsilon_n)})(dz,ds) \\
&\le C\int_{\mathbb{R}\times[0,t]} |z|\,(\mathscr{N}-\mathscr{N}^{(\varepsilon_n)})(dz,ds) \\
&= C\left\{(Z_t^+-Z_t^-)-(Z_t^{(\varepsilon_n,+)}-Z_t^{(\varepsilon_n,-)})\right\} \to 0, \text{ a.s.}
\end{aligned}$$

Hence we have

$$\begin{aligned}
&\int_{\mathbb{R}\times[0,t]} \{g(Z_{s-}+z)-g(Z_{s-})\}\,\mathscr{N}^{(\varepsilon_n)}(dz,ds) \\
&\to \int_{\mathbb{R}\times[0,t]} \{g(Z_{s-}+z)-g(Z_{s-})\}\,\mathscr{N}(dz,ds), \text{ a.s.}
\end{aligned}$$

Since if $g\in C^1(\mathbb{R})$ then $g(z+x)-g(z)=x\int_0^1 g'(z+\alpha x)d\alpha$, we obtain that

$$\begin{aligned}
&\left|\int_{\mathbb{R}\times[0,t]} \left\{g(Z_{s-}^{(\varepsilon_n)}+z)-g(Z_{s-}^{(\varepsilon_n)})\right\}-\{g(Z_{s-}+z)-g(Z_{s-})\}\,\mathscr{N}^{(\varepsilon_n)}(dz,ds)\right| \\
&=\left|\int_{\mathbb{R}\times[0,t]} z\int_0^1 \left\{g'(Z_{s-}^{(\varepsilon_n)}+\alpha z)-g'(Z_{s-}+\alpha z)\right\}d\alpha\,\mathscr{N}^{(\varepsilon_n)}(dz,ds)\right| \\
&\le \int_{\mathbb{R}\times[0,t]} |z|\int_0^1 \left|g'(Z_{s-}^{(\varepsilon_n)}+\alpha z)-g'(Z_{s-}+\alpha z)\right|d\alpha\,\mathscr{N}^{(\varepsilon_n)}(dz,ds) \\
&\le K\int_{\mathbb{R}\times[0,t]} |z|\left|Z_{s-}^{(\varepsilon_n)}-Z_{s-}\right|\mathscr{N}^{(\varepsilon_n)}(dz,ds) \\
&\le K\sup_{0\le s\le t}\left|Z_{s-}^{(\varepsilon_n)}-Z_{s-}\right|\int_{\mathbb{R}\times[0,t]} |z|\mathscr{N}^{(\varepsilon_n)}(dz,ds) \\
&\le K\sup_{0\le s\le t}\left|Z_{s-}^{(\varepsilon_n)}-Z_{s-}\right|\left(Z_t^{(\varepsilon_n,+)}-Z_t^{(\varepsilon_n,-)}\right) \to 0\cdot(Z_t^+-Z_t^-)=0, \text{ a.s.}
\end{aligned}$$

Therefore we have

$$\begin{aligned}
&\int_{\mathbb{R}\times[0,t]} \left\{g(Z_{s-}^{(\varepsilon_n)}+z)-g(Z_{s-}^{(\varepsilon_n)})\right\}\mathscr{N}^{(\varepsilon_n)}(dz,ds) \\
&\to \int_{\mathbb{R}\times[0,t]} \{g(Z_{s-}+z)-g(Z_{s-})\}\,\mathscr{N}(dz,ds).
\end{aligned}$$

Remark 4.1.35 We have to note that the random function $|g(Z_{s-}+z)-g(Z_{s-})|\le \|g'\|_\infty|z|$ is integrable with respect to $\mathscr{N}$.

As in Sect. 3.5, we denote by $(\mathscr{F}_t)_{t\in[0,T]}$ the right-continuous filtration completed with all sets of probability zero. We also define the class of predictable stochastic processes as the ones that are generated by the class of left-continuous processes. Then we have as before:

Theorem 4.1.36 *Define* $X_t = x + \int_0^t a(s)ds + Z_t$, *where* $a \in \mathscr{M}_b(\mathbb{R})$ *and* $x \in \mathbb{R}$. *Then, for any* $g \in C_b^2(\mathbb{R})$,

$$g(X_t) = g(x + Z_0) + \int_0^t g'(X_s)a(s)ds + \int_{\mathbb{R}\times[0,t]} \{g(X_{s-} + z) - g(X_{s-})\}\mathscr{N}(dz, ds).$$

Proof We define

$$X_t^{(n)} := x + \int_0^t a(s)ds + Z_t^{(\varepsilon_n)}.$$

It is clear that $X_t^{(n)} \to X_t$ as $n \to \infty$. From Theorem 3.4.26, we get

$$g(X_t^{(n)}) = g(x + Z_0^{(\varepsilon_n)}) + \int_0^t g'(X_s^{(n)})a(s)ds \\ + \int_{\mathbb{R}\times[0,t]} \{g(X_{s-}^{(n)} + z) - g(X_{s-}^{(n)})\}\mathscr{N}^{(\varepsilon_n)}(dz, ds).$$

Since g is a continuous function, we get $g(X_t^{(n)}) \to g(X_t)$ and $g(x + Z_0^{(\varepsilon_n)}) \to g(x + Z_0)$ as $n \to \infty$. So we show that

$$\int_0^t g'(X_s^{(n)})a(s)ds + \int_{\mathbb{R}\times[0,t]} \{g(X_{s-}^{(n)} + z) - g(X_{s-}^{(n)})\}\mathscr{N}^{(\varepsilon_n)}(dz, ds) \\ \to \int_0^t g'(X_s)a(s)ds + \int_{\mathbb{R}\times[0,t]} \{g(X_{s-} + z) - g(X_{s-})\}\mathscr{N}(dz, ds).$$

Since g' is a continuous function, $g'(X_s^{(n)})a(s) \to g'(X_s)a(s)$ as $n \to \infty$. Also, g' and a are bounded functions, so that, from theb dominated convergence theorem, we have

$$\int_0^t g'(X_s^{(n)})a(s)ds \to \int_0^t g'(X_s)a(s)ds \; n \to \infty.$$

Using a similar argument to the proof of Theorem 4.1.34, we can get

$$\int_{\mathbb{R}\times[0,t]} \{g(X_{s-}^{(n)} + z) - g(X_{s-}^{(n)})\}\mathscr{N}^{(\varepsilon_n)}(dz, ds) \to \int_{\mathbb{R}\times[0,t]} \{g(X_{s-} + z) - g(X_{s-})\}\mathscr{N}(dz, ds).$$

Therefore we have

$$\int_0^t g'(X_s^{(n)})a(s)ds + \int_{\mathbb{R}\times[0,t]} \{g(X_{s-}^{(n)} + z) - g(X_{s-}^{(n)})\}\mathcal{N}^{(\varepsilon_n)}(dz, ds)$$

$$\to \int_0^t g'(X_s)a(s)ds + \int_{\mathbb{R}\times[0,t]} \{g(X_{s-} + z) - g(X_{s-})\}\mathcal{N}(dz, ds).$$

Exercise 4.1.37 1. Prove an extension of the above result in the case that $g \in C^2(\mathbb{R})$.
2. Prove an extension of the above result in the case that a is an adapted process.

Theorem 4.1.38 *Let $b : \Omega \times \mathbb{R} \times [0, T] \to \mathbb{R}$ be a predictable process such that $\mathbb{E}[\int_{\mathbb{R}\times[0,T]} |b(z, s)|\widehat{\mathcal{N}}(dz, ds)] < \infty$. Then the stochastic integral $\int_{\mathbb{R}\times[0,T]} |b(z, s)|\mathcal{N}(dz, ds)$ is well defined and*

$$\mathbb{E}[\int_{\mathbb{R}\times[0,T]} b(z, s)\mathcal{N}(dz, ds)] = \mathbb{E}[\int_{\mathbb{R}\times[0,T]} b(z, s)\widehat{\mathcal{N}}(dz, ds)].$$

Exercise 4.1.39 Consider first any monotone approximation for b^+ and prove the above statement. Finalize it by considering $b = b^+ - b^-$.

Theorem 4.1.40 *Let $g : \Omega \times \mathbb{R} \times [0, T] \to \mathbb{R}$ be a stochastic process such that g is predictable*[9] *and such that*

$$\int_{\mathbb{R}\times[0,t]} |g(z, s)|\widehat{\mathcal{N}}(dz, ds) < \infty, \text{ a.s.}$$

Then the stochastic integral $\int_{\mathbb{R}\times[0,t]} g(z, s)\mathcal{N}(dz, ds) < \infty$ is well defined and $\int_{\mathbb{R}\times[0,t]} g(z, s)\widetilde{\mathcal{N}}(dz, ds)$ is a local martingale.[10]

As in Sect. 3.5, you may try to write and prove the statement of Itô's formula for the martingale process $Y_t = \int_{\mathbb{R}\times[0,t]} g(z, s)\widetilde{\mathcal{N}}(dz, ds)$.

4.2 Differential Equations

Following our results in Sect. 3.6, we now consider the corresponding stochastic differential equations for finite variation jump processes with an infinite number of jumps. Before this, we need to consider the following inequality.

---

[9]Recall the notation issues discussed in footnote 22.

[10]We have taken the advantage of assuming that the reader knows about local martingales. Otherwise you may try to prove the same statement assuming that $\mathbb{E}[\int_{\mathbb{R}\times[0,t]} |g(z, s)|\widehat{\mathcal{N}}(dz, ds)] < \infty$.

Lemma 4.2.1 (Gronwall's inequality) *Let u be a non-negative function that satisfies the integral inequality*

$$u(t) \le c + \int_{t_0}^{t} a(s)u(s)ds, \; c \ge 0,$$

where $a(t)u(t)$ is a continuous non-negative function for $t \ge t_0$. Then,

$$u(t) \le c \exp\left\{ \int_{t_0}^{t} a(s)ds \right\}.$$

Proof Let

$$v(s) := \exp\left\{ -\int_{t_0}^{s} a(r)dr \right\} \int_{t_0}^{s} a(r)u(r)dr, \; s \ge t_0.$$

Then,

$$\begin{aligned} v'(s) &= -a(s)\exp\left\{ -\int_{t_0}^{s} a(r)dr \right\} \int_{t_0}^{s} a(r)u(r)dr + \exp\left\{ -\int_{t_0}^{s} a(r)dr \right\} a(s)u(s) \\ &= \left(u(s) - \int_{t_0}^{s} a(r)u(r)dr \right) a(s)\exp\left\{ -\int_{t_0}^{s} a(r)dr \right\}. \end{aligned}$$

By assumption,

$$u(s) - \int_{t_0}^{s} a(r)u(r)dr \le c.$$

Thus, we get

$$v'(s) \le ca(s)\exp\left\{ -\int_{t_0}^{s} a(r)dr \right\}.$$

Since $v(t_0) = 0$, integrating this inequality from t_0 to t gives

$$v(t) \le c \int_{t_0}^{t} a(s)\exp\left\{ -\int_{t_0}^{s} a(r)dr \right\} ds.$$

From the definition of v, we have

$$\begin{aligned} \int_{t_0}^{t} a(s)u(s)ds &= \exp\left\{ \int_{t_0}^{t} a(s)ds \right\} v(t) \\ &\le c\exp\left\{ \int_{t_0}^{t} a(s)ds \right\} \int_{t_0}^{t} a(s')\exp\left\{ -\int_{t_0}^{s'} a(r)dr \right\} ds' \\ &= c\exp\left\{ \int_{t_0}^{t} a(s)ds \right\} \int_{t_0}^{t} \frac{d}{ds'}\left(-\exp\left\{ -\int_{t_0}^{s'} a(r)dr \right\} \right) ds' \end{aligned}$$

$$= c \exp\left\{\int_{t_0}^{t} a(s)ds\right\}\left(-\exp\left\{-\int_{t_0}^{t} a(s)ds\right\}+1\right)$$
$$= c \exp\left\{\int_{t_0}^{t} a(s)ds\right\} - c.$$

Therefore,

$$u(t) \leq c \exp\left\{\int_{t_0}^{t} a(s)ds\right\}.$$

Theorem 4.2.2 *Let $a : \mathbb{R} \to \mathbb{R}$ be a Lipschitz function (i.e. $|a(x) - a(y)| \leq C|x - y|$ for all $x, y \in \mathbb{R}$ and for some $C > 0$) and b is a Borel function such that $I(b) := \int_{\mathbb{R}} |b(z)| f(z)dz < \infty$. Then, the following stochastic equation has a unique solution:*

$$X_t = x + \int_0^t a(X_s)ds + \int_{\mathbb{R}\times[0,t]} X_{s-}b(z)\mathcal{N}(dz, ds),\ t \geq 0. \tag{4.5}$$

Exercise 4.2.3 Give the definition of the solution and uniqueness for the above equation after reading the proof of the theorem. Hint: Remember Definition 3.6.4.

Exercise 4.2.4 Prove that the solution of the Eq. (4.5) is also the solution of the stochastic equation

$$X_t = x + \int_0^t a(X_{s-})ds + \int_{\mathbb{R}\times[0,t]} X_{s-}b(z)\mathcal{N}(dz, ds).$$

Write its solution[11] as explicitly as possible in the case that $a(x) = a_0 x$ for some constant $a_0 \in \mathbb{R}$. Recall and discuss the relation between this problem and footnote 22.

Proof of Theorem 4.2.2. Without any further mention, we note first that all stochastic integrals will be well defined due to Theorem 4.1.38. For any $T > 0$, $t \in [0, T]$, we define

$$X_t^{(0)} := x$$
$$X_t^{(k)} := x + \int_0^t a(X_s^{(k-1)})ds + \int_{\mathbb{R}\times[0,t]} X_{s-}^{(k-1)}b(z)\mathcal{N}(dz, ds),\ \ k \in \mathbb{N}.$$

By induction, we get[12]

$$\int_{\mathbb{R}\times[0,t]} |X_{s-}^{(k-1)}b(z)|\mathcal{N}(dz, ds) < \infty \text{ a.s.},\ k \in \mathbb{N}. \tag{4.6}$$

[11]This is just a consequence of the following property: Any càdlàg function has at most a countable number of discontinuities. Recall Exercise 3.4.9.

[12]We leave this as an exercise for the reader. One can do this using an argument similar to the one in the rest of the proof.

Also, for all $k \in \mathbb{Z}_{\geq 0}$, it is clear that $X_t^{(k)}$ is a càdlàg function on $[0, T]$ a.s. For every $k \geq 2$,

$$
\begin{aligned}
&\sup_{0\leq t\leq T} |X_t^{(k)} - X_t^{(k-1)}| \\
&\leq \sup_{0\leq t\leq T} \int_0^t |a(X_s^{(k-1)}) - a(X_s^{(k-2)})|ds + \sup_{0\leq t\leq T} \int_{\mathbb{R}\times[0,t]} |X_{s-}^{(k-1)} - X_{s-}^{(k-2)}||b(z)|\mathcal{N}(dz, ds) \\
&= \int_0^T |a(X_s^{(k-1)}) - a(X_t^{(k-2)})|ds + \int_{\mathbb{R}\times[0,T]} |X_{s-}^{(k-1)} - X_{s-}^{(k-2)}||b(z)|\mathcal{N}(dz, ds) \\
&\leq C \int_0^T |X_s^{(k-1)} - X_s^{(k-2)}|ds + \int_{\mathbb{R}\times[0,T]} |X_{s-}^{(k-1)} - X_{s-}^{(k-2)}||b(z)|\widetilde{\mathcal{N}}(dz, ds) \\
&\quad + \int_{\mathbb{R}\times[0,T]} |X_{s-}^{(k-1)} - X_{s-}^{(k-2)}||b(z)|\widehat{\mathcal{N}}(dz, ds) \\
&= C \int_0^T |X_s^{(k-1)} - X_s^{(k-2)}|ds + \int_{\mathbb{R}\times[0,T]} |X_{s-}^{(k-1)} - X_{s-}^{(k-2)}||b(z)|\widetilde{\mathcal{N}}(dz, ds) \\
&\quad + \int_0^T \int_{\mathbb{R}} |X_{s-}^{(k-1)} - X_{s-}^{(k-2)}||b(z)|f(z)dzds \\
&= C \int_0^T |X_s^{(k-1)} - X_s^{(k-2)}|ds \\
&\quad + \int_{\mathbb{R}\times[0,T]} |X_{s-}^{(k-1)} - X_{s-}^{(k-2)}||b(z)|\widetilde{\mathcal{N}}(dz, ds) + I(b) \int_0^T |X_{s-}^{(k-1)} - X_{s-}^{(k-2)}|ds \\
&= (C + I(b)) \int_0^T |X_s^{(k-1)} - X_s^{(k-2)}|ds + \int_{\mathbb{R}\times[0,T]} |X_{s-}^{(k-1)} - X_{s-}^{(k-2)}||b(z)|\widetilde{\mathcal{N}}(dz, ds).
\end{aligned}
$$

Let $L := C + I(b)$, $I(b) := \int_{\mathbb{R}} |b(z)|f(z)dz < \infty$. From Theorem 4.1.38, we have

$$
\begin{aligned}
&\mathbb{E}[\sup_{0\leq t\leq T} |X_t^{(k)} - X_t^{(k-1)}|] \\
&\leq L\mathbb{E}\left[\int_0^T |X_s^{(k-1)} - X_s^{(k-2)}|ds\right] + \mathbb{E}\left[\int_{\mathbb{R}\times[0,T]} |X_{s-}^{(k-1)} - X_{s-}^{(k-2)}||b(z)|\widetilde{\mathcal{N}}(dz, ds)\right] \\
&= L \int_0^T \mathbb{E}[|X_s^{(k-1)} - X_s^{(k-2)}|]ds.
\end{aligned}
$$

Applying this inequality successively, we find that

$$
\begin{aligned}
&\mathbb{E}[\sup_{0\leq t\leq T} |X_t^{(k)} - X_t^{(k-1)}|] \\
&\leq L^{k-1} \int_0^T \int_0^{s_1} \cdots \int_0^{s_{k-2}} \mathbb{E}[\sup_{0\leq s_k\leq s_{k-1}} |X_{s_k}^{(1)} - X_{s_k}^{(0)}|]ds_{k-1} \cdots ds_1.
\end{aligned}
$$

Since

$$|X_{s_k}^{(1)} - X_{s_k}^{(0)}| \leq \int_0^{s_k} |a(x)|ds + \int_{\mathbb{R}\times[0,s_k]} |x||b(z)|\mathcal{N}(dz, ds),$$

we get

$$\begin{aligned}\mathbb{E}[\sup_{0\leq s_k\leq s_{k-1}} |X_{s_k}^{(1)} - X_{s_k}^{(0)}|] &\leq (|a(x)| + |x|I(b))\, s_{k-1} \\ &=: K s_{k-1}.\end{aligned}$$

Therefore,

$$\begin{aligned}\mathbb{E}[\sup_{0\leq t\leq T} |X_t^{(k)} - X_t^{(k-1)}|] &\leq L^{k-1} \int_0^T \int_0^{s_1} \cdots \int_0^{s_{k-2}} K s_{k-1} ds_{k-1} \cdots ds_1 \\ &= K L^{k-1} \cdot \frac{T^{k-1}}{(k-1)!} \\ &= \frac{K(LT)^{k-1}}{(k-1)!}.\end{aligned}$$

By Markov's inequality,

$$\mathbb{P}\left(\sup_{0\leq t\leq T} |X_t^{(k)} - X_t^{(k-1)}| > \frac{1}{2^{k-1}}\right) \leq 2^{k-1}\mathbb{E}[\sup_{0\leq t\leq T} |X_t^{(k)} - X_t^{(k-1)}|] \leq \frac{K(2LT)^{k-1}}{(k-1)!}.$$

Thus,

$$\sum_{k\in\mathbb{N}} \mathbb{P}\left(\sup_{0\leq t\leq T} |X_t^{(k)} - X_t^{(k-1)}| > \frac{1}{2^{k-1}}\right) \leq \sum_{k\in\mathbb{N}} \frac{K(2LT)^{k-1}}{(k-1)!} < \infty.$$

From the Borel–Cantelli lemma, we get

$$\mathbb{P}\left(\limsup_k \left\{\sup_{0\leq t\leq T} |X_t^{(k)} - X_t^{(k-1)}| > \frac{1}{2^{k-1}}\right\}\right) = 0.$$

That is,

$$\mathbb{P}\left(\liminf_k \left\{\sup_{0\leq t\leq T} |X_t^{(k)} - X_t^{(k-1)}| \leq \frac{1}{2^{k-1}}\right\}\right) = 1.$$

Therefore,

$$\mathbb{P}\left(\lim_{k,l\to\infty} \sup_{0\leq t\leq T} |X_t^{(l)} - X_t^{(k)}| = 0\right) = 1.$$

Since $X_t^{(k)}$ is a Cauchy sequence almost surely, there exists X_t such that $X_t^{(k)}$ uniformly converges to X_t on $[0, T]$ as $k \to \infty$. From Lemma 4.1.4, $X_t(\omega)$ is a càdlàg function. Also, since a is Lipschitz continuous, $a(X_t^{(k)})$ uniformly converges to $a(X_t)$ on $[0, T]$ as $k \to \infty$. Thus,

$$\begin{aligned} X_t &= \lim_{k\to\infty} X_t^{(k)} \\ &= x + \lim_{k\to\infty} \int_0^t a(X_s^{(k-1)})ds + \lim_{k\to\infty} \int_{\mathbb{R}\times[0,t]} X_{s-}^{(k-1)} b(z)\mathcal{N}(dz, ds) \\ &= x + \int_0^t a(X_s)ds + \int_{\mathbb{R}\times[0,t]} X_{s-}b(z)\mathcal{N}(dz, ds). \end{aligned}$$

T is arbitrary, so that the following stochastic equation has a solution

$$X_t = x + \int_0^t a(X_s)ds + \int_{\mathbb{R}\times[0,t]} X_{s-}b(z)\mathcal{N}(dz, ds).$$

Next, we shall prove the uniqueness of solutions. We assume that $\widetilde{X}_t$ satisfies the stochastic equation:

$$\widetilde{X}_t = x + \int_0^t a(\widetilde{X}_s)ds + \int_{\mathbb{R}\times[0,t]} \widetilde{X}_{s-}b(z)\mathcal{N}(dz, ds)$$

Then,

$$\begin{aligned} \mathbb{E}[|X_t - \widetilde{X}_t|] &\le \mathbb{E}\left[\int_0^t |a(X_s) - a(\widetilde{X}_s)|ds\right] + \mathbb{E}\left[\int_{\mathbb{R}\times[0,t]} |X_{s-} - \widetilde{X}_{s-}||b(z)|\mathcal{N}(dz, ds)\right] \\ &\le \mathbb{E}\left[C\int_0^t |X_s - \widetilde{X}_s|ds\right] + \mathbb{E}\left[\int_{\mathbb{R}\times[0,t]} |X_{s-} - \widetilde{X}_{s-}||b(z)|\widetilde{\mathcal{N}}(dz, ds)\right] \\ &\quad + \mathbb{E}\left[\int_{\mathbb{R}\times[0,t]} |X_{s-} - \widetilde{X}_{s-}||b(z)|\widehat{\mathcal{N}}(dz, ds)\right] \\ &= C\int_0^t \mathbb{E}[|X_s - \widetilde{X}_s|]ds + \mathbb{E}\left[I(b)\int_0^t |X_{s-} - \widetilde{X}_{s-}|ds\right] \\ &= C\int_0^t \mathbb{E}[|X_s - \widetilde{X}_s|]ds + I(b)\mathbb{E}\left[\int_0^t |X_{s-} - \widetilde{X}_{s-}|ds\right] \\ &= C\int_0^t \mathbb{E}[|X_s - \widetilde{X}_s|]ds + I(b)\int_0^t \mathbb{E}[|X_s - \widetilde{X}_s|]ds \\ &= (C + I(b))\int_0^t \mathbb{E}[|X_s - \widetilde{X}_s|]ds. \end{aligned} \tag{4.7}$$

Applying Gronwall's inequality in Lemma 4.2.1, we have

$$\mathbb{E}[|X_t - \widetilde{X}_t|] = 0.$$

Thus,

$$\mathbb{P}(X_t = \widetilde{X}_t) = 1, \quad t \geq 0.$$

Since X_t and $\widetilde{X}_t$ have càdlàg paths,[13] from Proposition 3.6.3, we get

$$\mathbb{P}(X_t = \widetilde{X}_t \; ; \; t \geq 0) = 1.$$

Remark 4.2.5 (i) In the proof of results like the one in Theorem 4.2.2, inequalities like (4.7) are one of the goals when trying to find solutions for equations through the fixed point theorem.[14]

(ii) Another way of proceeding in our proof would be to consider

$$\mathbb{E}[\sup_{0\leq t\leq T} |X_t^{(k)} - X_t^{(k-1)}|] \leq \frac{K(LT)^{k-1}}{(k-1)!},$$

so that

$$\begin{aligned}
\mathbb{E}[|X_t^{(m)} - X_t^{(n)}|] &\leq \sum_{k=n+1}^{m} \mathbb{E}[|X_t^{(k)} - X_t^{(k-1)}|] \\
&\leq \sum_{k=n+1}^{m} \mathbb{E}[\sup_{0\leq t\leq T} |X_t^{(k)} - X_t^{(k-1)}|] \\
&\leq \sum_{k=n+1}^{m} \frac{K(LT)^{k-1}}{(k-1)!} \\
&\to 0 \;\; (n, m \to \infty).
\end{aligned}$$

Therefore we get

$$\sup_{0\leq t\leq T} \mathbb{E}[|X_t^{(m)} - X_t^{(n)}|] \to 0 \; (n, m \to \infty).$$

Thus, $X_t^{(k)}$ is a uniformly Cauchy sequence in $L^1(\Omega)$, so that there exists $\widehat{X}_t$ such that $X_t^{(k)}$ uniformly converges to $\widehat{X}_t$ in $L^1(\Omega)$ as $k \to \infty$ (i.e. $\sup_{0\leq t\leq T} \mathbb{E}[|X_t^{(k)} - \widehat{X}_t|] \to 0$ as $k \to \infty$). But, we cannot assure that $\widehat{X}_t(\omega)$ is a càdlàg function. Thus $\{\widehat{X}_t; t \in [0, T]\}$ is not a solution of the stochastic equation if we want the solution to have càdlàg paths.[15] This notion would belong to a weaker notion of a solution which may be used when necessary. In general, it should hold that

[13]That is, $X_t(\omega)$ and $\widetilde{X}_t(\omega)$ are càdlàg functions.

[14]For example, let $f : \mathbb{R} \to \mathbb{R}$ such that f is Lipschitz with a Lipschitz constant smaller than 1. Then the equation $f(x) = x$ has a unique solution. In our setting one has to choose a time small enough so that this idea can be applied.

[15]Of course, in the case that you may not need the càdlàg property then you can get by with this kind of solution.

weaker notions of solutions should require weaker conditions on the coefficients in order to obtain existence and uniqueness, otherwise their interest will be limited.

Exercise 4.2.6 Use the proof of Theorem 4.2.2 to prove that the following rate of convergence is also satisfied for any $p > 0$ and $t > 0$:

$$\lim_{k\to\infty} k^p(X_t^{(k)} - X_t) = 0, \text{ a.s.}$$

Think about the possibility of using functions that explode faster than k^p.

For a hint, see Chap. 14.

Exercise 4.2.7 Prove the statement of (4.6). This statement may not be the optimal one to prove. So you also need to make an effort to clarify in your own way what needs to be proved in order to make the proof efficient.

For a hint, see Chap. 14.

Exercise 4.2.8 Consider the same problem as in Exercise 3.6.11. In particular, prove first that under enough conditions on the coefficients a and b of (4.5), we have:

1. $\lim_{k\uparrow\infty} \sup_x |X_t^{(k)}(x) - X_t(x)| = 0$.
2. Use induction to prove that $X_t^{(k)}(x)$ is differentiable with respect to $x \in \mathbb{R}$ for all $k \in \mathbb{N}$.
3. Prove then that $X_t(x)$ is differentiable and find the equation satisfied by $X_t'(x)$.

Now we move towards the case when the equations are driven by the generalized process defined in Sect. 4.1.

Theorem 4.2.9 *Suppose that $\int |z| f(z)dz < \infty$ and let $a : \mathbb{R} \to \mathbb{R}$ be a Lipschitz bounded function. Then there exists a unique solution to the stochastic equation*

$$X_t = x + \int_{\mathbb{R}\times[0,t]} a(X_{s-})z\mathcal{N}(dz, ds), \tag{4.8}$$

in the space of càdlàg predictable stochastic processes.

Recall that in the above statement the notion of a solution is also being defined. Before giving the proof we need to give the following exercise.

Exercise 4.2.10 Prove that the space $D[0, T] = \{x : [0, T] \to \mathbb{R}; x \text{ is càdlàg}\}$ is a Banach space with the norm of uniform convergence.

Proof From Corollary 3.6.6, for any $n \in \mathbb{N}$, there exists a unique solution to the stochastic equation

$$X_t^{(n)} = x + \int_{\mathbb{R}\times[0,t]} a(X_{s-}^{(n)})z\mathcal{N}^{(\varepsilon_n)}(dz, ds).$$

Consider for $m \geq n$

$$X_t^{(m)} - X_t^{(n)} = \int_{\mathbb{R}\times[0,t]} (a(X_{s-}^{(m)}) - a(X_{s-}^{(n)}))z\mathcal{N}^{(\varepsilon_n)}(dz, ds)$$
$$+ \int_{\mathbb{R}\times[0,t]} a(X_{s-}^{(m)})z\mathcal{N}^{(\varepsilon_m,\varepsilon_n)}(dz, ds).$$

Therefore if K denotes the Lipschitz constant of a, we have

$$\sup_{t\in[0,T]} |X_t^{(m)} - X_t^{(n)}| \leq K \sup_{t\in[0,T]} |X_t^{(m)} - X_t^{(n)}| \int_{\mathbb{R}\times[0,T]} |z|\mathcal{N}^{(\varepsilon_n)}(dz, ds)$$
$$+ \left| \int_{\mathbb{R}\times[0,t]} a(X_{s-}^{(m)})z\mathcal{N}^{(\varepsilon_m,\varepsilon_n)}(dz, ds) \right|.$$

Note that

$$\left| \int_{\mathbb{R}\times[0,t]} a(X_{s-}^{(m)})z\mathcal{N}^{(\varepsilon_m,\varepsilon_n)}(dz, ds) \right| \leq \|a\|_\infty \int_{\mathbb{R}\times[0,t]} |z|\mathcal{N}^{(\varepsilon_m,\varepsilon_n)}(dz, ds) \to 0, \text{ a.s.},$$

as $n, m \to \infty$. Consider $T \equiv T(\omega)$ small enough so that $\int_{\mathbb{R}\times[0,T]} |z|\mathcal{N}(dz, ds) < 1/K$. Then one has that

$$\sup_{t\in[0,T]} |X_t^{(m)} - X_t^{(n)}| \leq \left(1 - K \int_{\mathbb{R}\times[0,T]} |z|\mathcal{N}(dz, ds)\right)^{-1} \left| \int_{\mathbb{R}\times[0,t]} a(X_{s-}^{(m)})z\mathcal{N}^{(\varepsilon_m,\varepsilon_n)}(dz, ds) \right|.$$

Therefore one obtains that $\lim_{m,n\to\infty} \sup_{t\in[0,T]} |X_t^{(m)} - X_t^{(n)}| = 0$. Therefore by the completeness of the space of càdlàg functions with the uniform norm (cf. Exercise 4.2.10), one obtains that the limit X exists and is càdlàg. Also,

$$\begin{aligned}
&\left| \int_{\mathbb{R}\times[0,t]} a(X_{s-}^{(n)})z\mathcal{N}^{(\varepsilon_n)}(dz, ds) - \int_{\mathbb{R}\times[0,t]} a(X_{s-})z\mathcal{N}(dz, ds) \right| \\
&\leq \left| \int_{\mathbb{R}\times[0,t]} a(X_{s-}^{(n)})z\mathcal{N}^{(\varepsilon_n)}(dz, ds) - \int_{\mathbb{R}\times[0,t]} a(X_{s-})z\mathcal{N}^{(\varepsilon_n)}(dz, ds) \right| \\
&\quad + \left| \int_{\mathbb{R}\times[0,t]} a(X_{s-})z\mathcal{N}^{(\varepsilon_n)}(dz, ds) - \int_{\mathbb{R}\times[0,t]} a(X_{s-})z\mathcal{N}(dz, ds) \right| \\
&\leq \int_{\mathbb{R}\times[0,t]} |a(X_{s-}^{(n)}) - a(X_{s-})||z|\mathcal{N}^{(\varepsilon_n)}(dz, ds) \\
&\quad + \int_{\mathbb{R}\times[0,t]} |a(X_{s-})||z|(\mathcal{N} - \mathcal{N}^{(\varepsilon_n)})(dz, ds) \\
&\leq K \int_{\mathbb{R}\times[0,t]} |X_{s-}^{(n)} - X_{s-}||z|\mathcal{N}^{(\varepsilon_n)}(dz, ds) + K \int_{\mathbb{R}\times[0,t]} |z|(\mathcal{N} - \mathcal{N}^{(\varepsilon_n)})(dz, ds) \\
&\leq K \sup_{0\leq s\leq t} |X_s^{(n)} - X_s| \int_{\mathbb{R}\times[0,t]} |z|\mathcal{N}^{(\varepsilon_n)}(dz, ds) + K((Z_t^+ - Z_t^-) - (Z_t^{(\varepsilon_n,+)} - Z_t^{(\varepsilon_n,-)})) \\
&\to 0, \text{ a.s.}
\end{aligned}$$

Thus, we get

$$\int_{\mathbb{R}\times[0,t]} a(X_{s-}^{(n)})z\mathcal{N}^{(\varepsilon_n)}(dz,ds) \to \int_{\mathbb{R}\times[0,t]} a(X_{s-})z\mathcal{N}(dz,ds), \text{ a.s.}$$

Then, X_t has been constructed on the interval $t \in [0,T]$ with T satisfying $\int_{\mathbb{R}\times[0,T]} |z|\mathcal{N}(dz,ds) < 1/K$. Now, we may continue constructing the solution using that X_T is known and starting again the above argument in an interval starting at T. As $\int_{\mathbb{R}\times[T,2T]} |z|\mathcal{N}(dz,ds) = \int_{\mathbb{R}\times[0,T]} |z|\mathcal{N}(dz,ds) < 1/K$, we obtain that the solution can be constructed for any time t. Therefore, the stochastic equation (4.8) has a solution. Also,

$$\begin{aligned}
\mathbb{E}\left[\int_{\mathbb{R}\times[0,t]} |a(X_{s-})||z|\widehat{N}(dz,ds)\right] &= \mathbb{E}\left[\int_{\mathbb{R}}\int_0^t |a(X_{s-})||z|f(z)dzds\right] \\
&= \int_0^t \mathbb{E}[|a(X_s)|]ds \int_{\mathbb{R}} |z|f(z)dz \\
&\le Kt \int_{\mathbb{R}} |z|f(z)dz \\
&< \infty.
\end{aligned}$$

Therefore X_t satisfies (4.8) and is integrable. To prove uniqueness we use a fixed point argument. Let X and Y be two solutions of (4.8). Then using the Lipschitz property and Theorem 4.1.17 we obtain that

$$\begin{aligned}
\mathbb{E}[|X_t - Y_t|] &= \mathbb{E}\left[\left|\int_{\mathbb{R}\times[0,t]} (a(X_{s-}) - a(Y_{s-}))z\mathcal{N}(dz,ds)\right|\right] \\
&\le \mathbb{E}\left[\int_{\mathbb{R}\times[0,t]} |a(X_{s-}) - a(Y_{s-})||z|\mathcal{N}(dz,ds)\right] \\
&\le K\mathbb{E}\left[\int_{\mathbb{R}\times[0,t]} |X_{s-} - Y_{s-}||z|\mathcal{N}(dz,ds)\right] \\
&= K\mathbb{E}\left[\int_{\mathbb{R}\times[0,t]} |X_{s-} - Y_{s-}||z|\widehat{\mathcal{N}}(dz,ds)\right] \\
&= K\mathbb{E}\left[\int_0^t |X_{s-} - Y_{s-}| \int_{\mathbb{R}} |z|f(z)dzds\right] \\
&= K\int_{\mathbb{R}} |z|f(z)dz \int_0^t \mathbb{E}[|X_s - Y_s|]ds.
\end{aligned}$$

Applying Gronwall's inequality in Lemma 4.2.1, we have $\mathbb{E}[|X_t - Y_t|] = 0$. That is $\mathbb{P}(X_t = Y_t) = 1$, for any $t \ge 0$. Therefore we get $\mathbb{P}(X_t = Y_t,\ t \ge 0)$ from Proposition 3.6.3.

Exercise 4.2.11 Suppose that $\text{supp}(f) \subseteq \mathbb{R}_+$. Note that (4.8) will have a unique solution if we consider the driving process $Z - Z^{(1,+)}$. Use the fact that $Z^{(1,+)}$ is a compound Poisson process to prove that there exists a unique solution to (4.8) for the driving process Z even if the condition $\int_0^\infty xf(x)dx < \infty$ is not satisfied.

Theorem 4.2.12 *Suppose that* $\int |z|f(z)dz < \infty$*. Let* X *be a solution to* $X_t = x + \int_{\mathbb{R}\times[0,t]} a(X_{s-})z\mathcal{N}(dz, ds)$*, where* a *is a bounded Lipschitz function and* $x \in \mathbb{R}$*. Then, for any* $g \in C_b^2(\mathbb{R})$*,*

$$g(X_t) = g(x) + \int_{\mathbb{R}\times[0,t]} \{g(X_{s-} + a(X_{s-})z) - g(X_{s-})\}\mathcal{N}(dz, ds).$$

Proof From Theorem 3.5.6, we get

$$g(X_t^{(n)}) = g(x) + \int_{\mathbb{R}\times[0,t]} \{g(X_{s-}^{(n)} + a(X_{s-}^{(n)})z) - g(X_{s-}^{(n)})\}\mathcal{N}^{(\varepsilon_n)}(dz, ds).$$

Since g is a continuous function and $X_t^{(n)} \to X_t$ as $n \to \infty$, we have $g(X_t^{(n)}) \to g(X_t)$ as $n \to \infty$. So we need to show that

$$\begin{aligned} &\int_{\mathbb{R}\times[0,t]} \{g(X_{s-}^{(n)} + a(X_{s-}^{(n)})z) - g(X_{s-}^{(n)})\}\mathcal{N}^{(\varepsilon_n)}(dz, ds) \\ \to &\int_{\mathbb{R}\times[0,t]} \{g(X_{s-} + a(X_{s-})z) - g(X_{s-})\}\mathcal{N}(dz, ds). \end{aligned}$$

We prove this in two parts:

$$\left|\int_{\mathbb{R}\times[0,t]} \{g(X_{s-} + a(X_{s-})z) - g(X_{s-})\}(\mathcal{N} - \mathcal{N}^{(\varepsilon_n)})(dz, ds)\right| \to 0,$$

$$\left|\int_{\mathbb{R}\times[0,t]} \{(g(X_{s-} + a(X_{s-})z) - g(X_{s-})) - (g(X_{s-}^{(n)} + a(X_{s-}^{(n)})z) - g(X_{s-}^{(n)}))\}\mathcal{N}^{(\varepsilon_n)}(dz, ds)\right| \to 0.$$

Using the Lipschitz property of g and the boundedness of a, we get that

$$\begin{aligned} &\left|\int_{\mathbb{R}\times[0,t]} \{g(X_{s-} + a(X_{s-})z) - g(X_{s-})\}(\mathcal{N} - \mathcal{N}^{(\varepsilon_n)})(dz, ds)\right| \\ &\leq \int_{\mathbb{R}\times[0,t]} |g(X_{s-} + a(X_{s-})z) - g(X_{s-})|(\mathcal{N} - \mathcal{N}^{(\varepsilon_n)})(dz, ds) \\ &\leq K \int_{\mathbb{R}\times[0,t]} |a(X_{s-})||z|(\mathcal{N} - \mathcal{N}^{(\varepsilon_n)})(dz, ds) \\ &\leq K^2 \int_{\mathbb{R}\times[0,t]} |z|(\mathcal{N} - \mathcal{N}^{(\varepsilon_n)})(dz, ds) \\ &= K^2(Z_t^+ - Z_t^- - (Z_t^{(\varepsilon_n,+)} - Z_t^{(\varepsilon_n,-)})) \\ &\to 0\ (n \to \infty). \end{aligned}$$

Hence, we have that

$$\left|\int_{\mathbb{R}\times[0,t]} \{g(X_{s-}+a(X_{s-})z)-g(X_{s-})\}(\mathcal{N}-\mathcal{N}^{(\varepsilon_n)})(dz,ds)\right| \to 0.$$

Since if $g \in C^1$ then $g(x+z)-g(x) = z\int_0^1 g'(x+\alpha z)d\alpha$, we obtain that

$$\begin{aligned}
&\{(g(X_{s-}+a(X_{s-})z)-g(X_{s-}))-(g(X_{s-}^{(n)}+a(X_{s-}^{(n)})z)-g(X_{s-}^{(n)}))\}\\
&= z\int_0^1 \{a(X_{s-})g'(X_{s-}+\alpha a(X_{s-})z)-a(X_{s-}^{(n)})g'(X_{s-}^{(n)}+\alpha a(X_{s-}^{(n)})z)\}d\alpha\\
&= a(X_{s-})z\int_0^1 \{g'(X_{s-}+\alpha a(X_{s-})z)-g'(X_{s-}^{(n)}+\alpha a(X_{s-}^{(n)})z)\}d\alpha\\
&\qquad + (a(X_{s-})-a(X_{s-}^{n}))z\int_0^1 g'(X_{s-}^{(n)}+\alpha a(X_{s-}^{(n)})z)d\alpha.
\end{aligned}$$

Also, g' and a are bounded Lipschitz functions, so that we get that

$$\begin{aligned}
&\left|\int_{\mathbb{R}\times[0,t]} \{(g(X_{s-}+a(X_{s-})z)-g(X_{s-}))-(g(X_{s-}^{(n)}+a(X_{s-}^{(n)})z)-g(X_{s-}^{(n)}))\}\mathcal{N}^{(\varepsilon_n)}(dz,ds)\right|\\
&\le \int_{\mathbb{R}\times[0,t]} |a(X_{s-})||z|\int_0^1 |g'(X_{s-}+\alpha a(X_{s-})z)-g'(X_{s-}^{(n)}+\alpha a(X_{s-}^{(n)})z)|d\alpha\mathcal{N}^{(\varepsilon_n)}(dz,ds)\\
&\quad + \int_{\mathbb{R}\times[0,t]} |a(X_{s-})-a(X_{s-}^{(n)})||z|\int_0^1 |g'(X_{s-}^{(n)}+\alpha a(X_{s-}^{(n)})z)|d\alpha\mathcal{N}^{(\varepsilon_n)}(dz,ds)\\
&\le K\int_{\mathbb{R}\times[0,t]} |z|\int_0^1 |X_{s-}+\alpha a(X_{s-})z-X_{s-}^{(n)}-\alpha a(X_{s-}^{(n)})z|d\alpha\mathcal{N}^{(\varepsilon_n)}(dz,ds)\\
&\quad + K\int_{\mathbb{R}\times[0,t]} |X_{s-}-X_{s-}^{(n)}||z|\int_0^1 d\alpha\mathcal{N}^{(\varepsilon_n)}(dz,ds)\\
&\le K\int_{\mathbb{R}\times[0,t]} |z|\int_0^1 (|X_{s-}-X_{s-}^{(n)}|+\alpha|z||a(X_{s-})-a(X_{s-}^{(n)})|)d\alpha\mathcal{N}^{(\varepsilon_n)}(dz,ds)\\
&\quad + K\sup_{0\le s\le t}|X_s-X_s^{(n)}|\int_{\mathbb{R}\times[0,t]} |z|\mathcal{N}^{(\varepsilon_n)}(dz,ds)\\
&= 2K\sup_{0\le s\le t}|X_s-X_s^{(n)}|\int_{\mathbb{R}\times[0,t]} |z|\mathcal{N}^{(\varepsilon_n)}(dz,ds)\\
&\quad + \frac{1}{2}K^2\sup_{0\le s\le t}|X_s-X_s^{(n)}|\int_{\mathbb{R}\times[0,t]} |z|^2\mathcal{N}^{(\varepsilon_n)}(dz,ds).
\end{aligned}$$

Since $X_t^{(n)}$ uniformly converges to X_t, we have using Theorem 4.1.11 and Exercise 4.1.12 that $\sup_{0\le s\le t}|X_s-X_s^{(n)}| \to 0$ as $n\to\infty$. Therefore we have

$$\left|\int_{\mathbb{R}\times[0,t]} \{(g(X_{s-}+a(X_{s-})z)-g(X_{s-}))-(g(X_{s-}^{(n)}+a(X_{s-}^{(n)})z)-g(X_{s-}^{(n)}))\}\mathcal{N}^{(\varepsilon_n)}(dz,ds)\right| \to 0.$$

Hence we get

$$\int_{\mathbb{R}\times[0,t]} \{g(X_{s-}^{(n)} + a(X_{s-}^{(n)})z) - g(X_{s-}^{(n)})\}\mathcal{N}^{(\varepsilon_n)}(dz, ds)$$
$$\to \int_{\mathbb{R}\times[0,t]} \{g(X_{s-} + a(X_{s-})z) - g(X_{s-})\}\mathcal{N}(dz, ds).$$

Exercise 4.2.13 Take further limits on the function g in Theorem 4.2.12 to prove that the Itô formula is also valid for $g \in C_b^1(\mathbb{R})$. Try to further extend it to $g \in C^1(\mathbb{R})$.

Exercise 4.2.14 1. Prove that under the conditions of Theorem 4.2.9 the unique solution of (4.8) satisfies that $\mathbb{E}[|X_t|] < \infty$.
2. Find conditions that assure that $\mathbb{E}[|X_t|^2] < \infty$.
3. Find conditions that assure that $\mathbb{E}[\sup_{s\in[0,t]} |X_s|] < \infty$.

Exercise 4.2.15 Use Itô's formula and Theorem 4.1.38 to prove that for any $b : \Omega \times \mathbb{R} \times [0, T] \to \mathbb{R}$ predictable process such that $\mathbb{E}[\int_{\mathbb{R}\times[0,T]} |b(z, s)|^i \widehat{\mathcal{N}}(dz, ds)] < \infty$, $i = 1, 2$. Then the stochastic integral $\int_{\mathbb{R}\times[0,T]} |b(z, s)|\mathcal{N}(dz, ds)$ is well defined and

$$\mathbb{E}\left[\left(\int_{\mathbb{R}\times[0,T]} b(z, s)\mathcal{N}(dz, ds)\right)^2\right]$$
$$\le C\mathbb{E}\left[\int_{\mathbb{R}\times[0,T]} |b(z, s)|^2\widehat{\mathcal{N}}(dz, ds) + \left(\int_{\mathbb{R}\times[0,T]} b(z, s)\widehat{\mathcal{N}}(dz, ds)\right)^2\right].$$

Note that this result is the natural extension of $\mathbb{E}[N_T^2] = \lambda T + (\lambda T)^2$ in the Poisson case.

For a hint, see Chap. 14.

Problem 4.2.16 (*Taylor expansions for solutions of stochastic equations*) Consider X to be the solution of Eq. (4.8). Suppose that a is a smooth function with bounded derivatives. Find the first terms of a Taylor expansion for X in terms of multiple integrals of $\mathcal{N}(dz, ds)$ where the coefficients are derivatives of a.

Hint: Use Itô's formula to find some of the coefficients associated with such an expansion.

Note that another version of this problem is to consider coefficients of the type $a(x) = a_0(rx)$ for a parameter $r \in \mathbb{R}$ and then the problem is to find an expansion in terms of the powers of r.

Problem 4.2.17 (*Taylor expansions for expectations of solutions of stochastic equations*) Assume that $\int e^z f(z)dz < \infty$. Find the first three terms in the Taylor expansion for $\mathbb{E}[X_t]$ in terms of powers of t^j for $j = 0, 1, 2$.

Notice that the Taylor expansion is a very general name referring to the expansion of a certain quantity around a fixed value. This fixed value may vary depending on the objective. It is in this sense that the above expression "Taylor expansion" is being used. The reader should invest some effort in understanding what is playing the role of fixed value in each case.

Chapter 5
Construction of Lévy Processes and Their Corresponding SDEs: The Infinite Variation Case

In this chapter, we consider a class of Lévy processes which are not of bounded variation as in the preceding chapter but instead they are processes with paths of infinite variation. From the pedagogical point of view, this chapter provides the construction of the Lévy process, leaving for the reader most of the developments related to the construction of the stochastic integral, the Itô formula and the associated stochastic differential equations. This is done in the exercises in order to let you test your understanding of the subject. This is done on two levels. You will find the ideas written in words in the proofs. If you do not understand them you may try a further description that may be given in Chap. 14. It is a good exercise to try to link the words and the equations so that you understand the underlying meaning. This is also a chapter that may be used for promoting discussion between students and the guiding lecturer.

5.1 Construction of Lévy Processes: The Infinite Variation Case

So far, the Lévy processes we have defined have been originated in natural extensions using Poisson random measures and their corresponding integrals in the measure theoretic sense. That is, we have given the definition of the simple Poisson process which is a non-decreasing process with jumps of size one. Next, we extended this definition to the compound Poisson process which allows jump sizes to be random. Still the number of jumps in any interval is finite. In a first non-trivial extension we defined using limits an increasing process for which the average number of jumps in any interval is infinite.[1] In order to do this we first assumed that all jumps are

[1] In fact, one can prove that in any interval the number of jumps is infinite. This will left as an exercise for the reader.

© Springer Nature Singapore Pte Ltd. 2019

A. Kohatsu-Higa and A. Takeuchi, *Jump SDEs and the Study of Their Densities*, Universitext, https://doi.org/10.1007/978-981-32-9741-8_5

positive so that we could use the non-decreasing property in order to take limits on a sequence of compound Poisson processes. Similarly, one can define the limit of non-increasing compound Poisson processes and the final combination leads to the generalization that we have introduced in the previous section. This construction was possible because the constructed process paths had the finite variation property in any finite interval. That is, $\sum_{s\leq t}|\Delta Z_s| < \infty$.

In this section, we will introduce the last generalization step of this procedure. That is, we will consider cases where the process Z to be defined does not have the finite variation property. The idea to treat this case is to define the approximations as in the previous chapter, but now we will take away the terms that will tend to explode. This procedure is usually called "compensation". In fact, suppose that we want to consider the case $f(x) = C|x|^{-(1+\alpha)}\,\mathbf{1}_{|x|\neq 0}$ for $\alpha \in [1, 2)$, $C > 0$. In this case, if we follow the procedure in Proposition 4.1.2 the process $Z^{(\varepsilon_n,+)}$ will blow up,[2] which implies that the searched process if it exists will have infinite variation in any finite interval.

First, we will apply the following simplification to the problem. According to the procedure used before, we may first construct the Lévy process associated with $f_1(x) = C|x|^{-(1+\alpha)}\,\mathbf{1}_{|x|>1}$. This will lead to a compound Poisson process because the function f_1 is integrable.[3]

Now, we continue to the heart of the problem, which is how to construct a Lévy process associated with the Lévy measure $(f - f_1)(x) = C|x|^{-(1+\alpha)}\,\mathbf{1}_{0<|x|\leq 1}$. Clearly, the monotone limit idea used in the previous chapter cannot be used. Instead, we will use the following trick: consider an appropriate sequence $\mu^{(\varepsilon_n)}$ of real numbers which will blow up as $\varepsilon_n \downarrow 0$. Then find the correct sequence so that $Z_t^{(\varepsilon_n,+)} - \mu^{(\varepsilon_n)}$ will converge.[4] This may deceive some, as it means that clearly we are not taking limits of the process $Z^{(\varepsilon_n,+)}$ but of a "compensated" version, meaning that we hope that the way this quantity blows up can be estimated in a deterministic fashion.

This is exactly what happens and this mean compensation will appear in all the formulas to follow. In fact, the choice of $\mu^{(\varepsilon_n)}$ is not unique and this is always a point of distraction for students as the definition of the compensation is different according to the book you may be looking at. In this case we will consider the compensation $\mu^{(\varepsilon_n)} = \lambda_{\varepsilon_n}\int_{\varepsilon_n}^{1} xf(x)dx$.

To resume all the above ideas, our first goal in this chapter is then to extend the results of the previous chapter in order to construct a Lévy process Z such that

$$\mathbb{E}[e^{i\theta Z_t}] = \exp[ct\int_{-\infty}^{\infty}(e^{i\theta x} - 1 - i\theta x\,\mathbf{1}_{|x|\leq 1})f(x)dx]. \tag{5.1}$$

where $c > 0$ is some constant.[5]

---

[2] In fact, as an exercise prove that in this case, the integral $\int_0^\infty (e^{\theta x} - 1)f(x)dx$ is not finite.

[3] Although some moments will not be finite. Recall Exercise 4.1.27.

[4] Note that this construction is done using $f - f_1$ in Sect. 4.1.

[5] In many situations we will assume that $c = 1$ without further mentioning it.

In order for (5.1) to make sense we need to have that $f : \mathbb{R} \to \mathbb{R}_+$ is given such that

$$\left| \int_{-\infty}^{\infty} (e^{i\theta x} - 1 - i\theta x \, \mathbf{1}_{|x|\leq 1}) f(x) dx \right| < \infty, \text{ for all } \theta. \tag{5.2}$$

In fact, we will shortly see (Exercise 5.1.1 below) that instead of (5.2) we just need to assume

$$\int_{\mathbb{R}} (|x|^2 \wedge 1) f(x) dx < \infty \tag{5.3}$$

in order to assure that (5.2) is satisfied for all $\theta \in \mathbb{R}$.

We note here that the condition above has been weakened in comparison with the previous section (from $|x| \wedge 1$ to $|x|^2 \wedge 1$). This in fact implies that now the Example 4.1.20 can be extended from $\alpha \in [0, 1)$ to $\alpha \in [0, 2)$. The main difference between this section and the previous one is that in the previous section our main tool to construct such a process was to construct first an approximation using compound Poisson processes and then taking monotone limits. In the general case which we will consider here, this technique cannot be applied. In fact, one can prove that the limits taken in the previous section will diverge in the present case. But fortunately enough, this divergence in the sum can be correctly estimated. That is, by subtracting a diverging expression (related with a local mean), one finds that using the martingale convergence theorem is possible and therefore proving that a modification of the sum in Proposition 4.1.2 (ii) will converge. This modification is related with the term $i\theta x 1(|x| \leq 1)$ in the characteristic function (5.2).[6]

Exercise 5.1.1 Prove that condition (5.3) implies that

$$\int_{-\infty}^{\infty} |(e^{i\theta x} - 1 - i\theta x \, \mathbf{1}_{|x|\leq 1})| f(x) dx < \infty, \text{ for all } \theta.$$

This is done by dividing the integral into two regions: $[-1, 1]$ and its complement. Then apply Taylor's theorem with residue in each region to obtain (5.2). That is, this calculation divided in two regions will be carried out in different forms through this chapter which explains that the characteristic of the infinite variation processes is usually different for jumps in a neighbourhood of zero and for jumps away from zero.

Finally note that we are proving a stronger claim than (5.2).

Exercise 5.1.2 Let $f(x) := \dfrac{e^{-\lambda|x|}}{|x|^{1+\alpha}} \mathbf{1}_{\{x \neq 0\}}$ where $\lambda \geq 0$ and $\alpha \in [0, 2)$. Then f satisfies conditions (5.2).

[6] In fact, recall that if $\varphi(\theta)$ is the characteristic function of some random variable X then $\varphi(\theta)e^{-icIt\theta}$ is the characteristic function of $X - ctI$ with $I = \int_{|x|\leq 1} x f(x) dx$. For more on this, see Exercise 5.1.3.

Exercise 5.1.3 In order to realize that the process that we define in this chapter does not differ too much from the one in the previous chapter, consider the particular case that $\int_{\mathbb{R}}(|x| \wedge 1)f(x)dx < \infty$. Then to avoid confusion, we denote by Z' the Lévy process defined in Chap. 4.

1. Prove using characteristic functions that the random variable Z_t is related with Z'_t. That is, there is a constant A such that
$$Z'_t + At \overset{\mathscr{L}}{=} Z_t.$$
Find the value of A.
2. Let us focus now on the value of A. Therefore in this part, we stop assuming that $\int_{\mathbb{R}}(|x| \wedge 1)f(x)dx < \infty$. Consider then the case that $f_n(x) := \frac{e^{-\lambda|x|}}{|x|^{1+\alpha}}\mathbf{1}_{x>n^{-1}}$ with $\alpha \in [1, 2)$. Prove that $\lim_{n\to\infty} A_n = -\infty$.

Therefore we will be "taking away" an infinite amount from the process Z' in order to make it converge so that Z can be defined.

Now, let us continue with the construction of Z. As in the previous section, we define
$$f_\varepsilon(x) := \frac{f(x)\mathbf{1}_{\{x>\varepsilon\}}}{\lambda_\varepsilon}, \tag{5.4}$$
where $\lambda_\varepsilon := \int_\varepsilon^\infty f(x)dx < \infty$ and again for the example $f(x) = Cx^{-(1+\alpha)}$ we have that $\lambda_\varepsilon = C\alpha^{-1}\varepsilon^{-\alpha}$. We also let
$$f_{\varepsilon',\varepsilon}(x) := \frac{f(x)\mathbf{1}_{\{\varepsilon'<x\leq\varepsilon\}}}{\lambda_{\varepsilon',\varepsilon}}$$
where $\lambda_{\varepsilon',\varepsilon} = \int_{\varepsilon'}^{\varepsilon} f(x)dx$.

Step 1. First, we start constructing the compound Poisson process associated with the jumps bigger than one. That is,[7] for $\varepsilon_1 = 1$
$$Z_t^{(\varepsilon_1,+)} := \sum_{i=1}^\infty Y_i^{(\varepsilon_1)}\mathbf{1}_{\{T_i^{(\varepsilon_1)}\leq t\}} = \int_{\mathbb{R}\times[0,t]} x\mathscr{N}^{(\varepsilon_1,+)}(dx, ds).$$

Then the second step used the processes where the jump sizes are confined in an interval $(\varepsilon_{i+1}, \varepsilon_i]$. In general, the process $Z^{(\varepsilon_n,+)}$ will blow up. The measures $\mathscr{N}^{(\varepsilon_n,\pm)}$, $\mathscr{N}^{(\varepsilon_n)}$ can still be defined as an increasing sequence and therefore we will not pursue the line of discussion about the construction of the measure. For the small jumps, we define the following modification:

[7]Recall the arguments in Sect. 4.1.

$$\bar{Z}_t^{(\varepsilon,+)} := \sum_{i=1}^{\infty} Y_i^{(\varepsilon,1)} \mathbf{1}_{\{T_i^{(\varepsilon)} \le t\}} - c\mu^{(\varepsilon)} t,$$

where $\mu^{(\varepsilon)} = \lambda_{\varepsilon,1} \int_\varepsilon^1 x f_{\varepsilon,1}(x)dx$. The process $\bar{Z}^{(\varepsilon,+)}$ is a martingale in time.[8]

Exercise 5.1.4 Prove that $\{\bar{Z}_t^{(\varepsilon,+)}; t \in [0,T]\}$ is a martingale for fixed $\varepsilon > 0$. Specify the filtration.

Step 2. Now, we construct

$$\bar{Z}_t^{(\varepsilon',\varepsilon,+)} := \sum_{i=1}^{\infty} Y_i^{(\varepsilon',\varepsilon)} \mathbf{1}_{\{|Y_i^{(\varepsilon',\varepsilon)}| \le 1\}} \mathbf{1}_{\{T_i^{(\varepsilon',\varepsilon)} \le t\}} - \mu^{(\varepsilon',\varepsilon)} t,$$

where $\mu^{(\varepsilon',\varepsilon)} = \lambda_{\varepsilon',\varepsilon} \int_{\varepsilon'}^{\varepsilon} x f_{\varepsilon',\varepsilon}(x)dx$.

To construct Z^+, we take $\varepsilon_n = \dfrac{1}{2^{n-1}}$. We define the martingale sequence

$$\bar{Z}_t^{(\varepsilon_n,+)} := \sum_{i=1}^{n-1} \bar{Z}_t^{(\varepsilon_{i+1},\varepsilon_i,+)}.$$

The goal now is to prove the following result:

Theorem 5.1.5 *The following limit exists:*

$$\bar{Z}_t^+ := \lim_{n\to\infty} \bar{Z}_t^{(\varepsilon_n,+)} = \int_{[0,1]\times[0,t]} x \widetilde{\mathcal{N}}^{(+)}(dx, ds).$$

Proof By Doob's inequality we have that for any sequence of increasing times t_k, $k = 0, ..., l$ with $t_l = T$ that

$$\begin{aligned} P(\max_{k=1,...,l} |\bar{Z}_{t_k}^{(\varepsilon_n,+)} - \bar{Z}_{t_k}^{(\varepsilon_m,+)}| > r) &\le r^{-2}\mathbb{E}\left[|\bar{Z}_T^{(\varepsilon_n,+)} - \bar{Z}_T^{(\varepsilon_m,+)}|^2\right] \\ &= r^{-2} \int_{[0,T]\times[\varepsilon_m,\varepsilon_n]} |x|^2 f(x)dxds. \end{aligned}$$

Here, the last equality follows from an extension of Theorem 3.4.31.[9] Note that the bound we have obtained does not depend on l. Therefore we have that

$$\lim_{n,m\to\infty} P(\max_{t\in[0,T]} |\bar{Z}_t^{(\varepsilon_n,+)} - \bar{Z}_t^{(\varepsilon_m,+)}| > r) = 0.$$

[8]It is also a martingale sequence based on ε. Prove this as an exercise.

[9]One may also use Exercises 4.1.26 and 4.1.27.

Therefore the sequence of processes, Z^{ε_n} is a Cauchy sequence in probability[10] and therefore it converges.

As before, the associated Poisson random measure $\mathcal{N}$ is defined as the limit of the finite random measures $\mathcal{N}^{(\varepsilon_n,+)}(dx, ds)$ thus becoming a σ-finite measure. Therefore, integrals that are defined momentarily will be integrals with respect to σ-finite measures. We can define the compensated[11] Poisson random measure $\tilde{\mathcal{N}}^{+}(dx, ds) = \mathcal{N}^{(+)}(dx, ds) - \mathbf{1}_{\{0<x\leq 1\}} f(x)dxds$. We remark here that the fact that the limit exists is due to the compensation procedure which extracts the divergent component of $Z^{(\varepsilon_n,+)}$. For the same reason the process $\bar{Z}_t^+$ is no longer positive or increasing.

Proposition 5.1.6 *The characteristic function of $\bar{Z}_t^+$ is given by*

$$\mathbb{E}[e^{i\theta \bar{Z}_t^+}] = \exp[ct \int_{[0,1]} (e^{i\theta x} - 1 - i\theta x) f(x)dx]$$

Exercise 5.1.7 Prove the above proposition.[12]

To complete the construction of the process with positive jumps we only need to take an independent version of $Z_t^{(\varepsilon_1,+)}$ and let $Z_t^+ := Z_t^{(\varepsilon_1,+)} + \bar{Z}_t^+$, which[13] will have as characteristic function

$$\mathbb{E}[e^{i\theta Z_t^+}] = \exp[ct \int_{\mathbb{R}_+} (e^{i\theta x} - 1 - i\theta x\, \mathbf{1}_{\{|x|\leq 1\}}) f(x)dx].$$

The construction of Z^- can be done similarly and is left as an exercise for the reader.

Exercise 5.1.8 This exercise tries to discuss an alternative approximation process to obtain the same goal. Define

$$\tilde{Z}_t^{(\varepsilon,+)} := \sum_{i=1}^{N_t^{(\varepsilon,+)}} (Y_i^{(\varepsilon,1)} - c\tilde{\mu}^{(\varepsilon)}).$$

Here $\tilde{\mu}^{(\varepsilon)} := \int_\varepsilon^1 x f_{\varepsilon,1}(x)dx$ and $N_t^{(\varepsilon,+)} := \int_{[\varepsilon,1]\times[0,t]} \mathcal{N}^{(\varepsilon,+)}(dx, ds)$. Prove that the limit of this process exists and related this with the previous construction of $\bar{Z}^+$.[14]

---

[10]Exercise: Describe a stronger norm where this sequence is a Cauchy sequence.

[11]Compensated means that the mean is being taken from the process so that the resulting process has mean zero. This will imply that the process becomes a martingale.

[12]Recall that the characteristic function of $\bar{Z}_t^+$ can be computed as the limit of the characteristic functions of the sequence $\bar{Z}_t^{(\varepsilon_n,+)}$.

[13]Some may object to the usage of Z^+ as in the finite variation case this process is not compensated, while in the infinite variation case it becomes the compensated process.

[14]In fact, it may be easier to study the difference between the approximating processes.

Exercise 5.1.9 Continuing with Exercise 5.1.8 suppose that one uses a general density function f_ε instead of the explicit one given in (5.4). Give a proper distance function between f_ε and f so that the process similar to $\bar{Z}^+$ converges.

For a hint, see Chap. 14.

Exercise 5.1.10 Let $f(x) = \frac{C}{|x|^{1+\alpha}}\mathbf{1}_{\{|x|\leq 1\}}$, for $\alpha \in (1,2)$ and $C > 0$. Prove that for any $t > 0$,[15]

$$\lim_{n\uparrow\infty} \sum_{0\leq s\leq t} |\Delta \bar{Z}_s^{(\varepsilon_n,+)}| = \infty, \text{ a.s.}$$

This exercise, is linked with the title of the chapter which is not totally exact as the present theory also includes Lévy processes of finite variation.

Exercise 5.1.11 Let $f(x) = \frac{C}{|x|^{1+\alpha}}\mathbf{1}_{\{x\neq 0\}}$, for $\alpha \in (0,2)$ and $C > 0$. Recall the construction of $Z_t^{(\varepsilon_1,+)}$ in step 2 of 4.1 and prove that $\mathbb{E}[|Z_t^{(\varepsilon_1,+)}|]$ is infinite if $\alpha \in (0,1]$ and finite for $\alpha \in (1,2)$.[16]

Remark 5.1.12 Important consequences should be derived from the above exercise, if we recall that the construction given in Theorem 5.1.5 uses Doob's inequality and therefore $L^2(\Omega)$ estimates. These estimates, in general, cannot be valid for the process Z due to the above exercise. Therefore any construction may use $L^2(\Omega)$-estimates for jumps close to zero and a.s. constructions for jumps away from zero. This will be the case in most of the proofs that follow.

Exercise 5.1.13 Prove that for $f(x) = C|x|^{-(1+\alpha)}\,\mathbf{1}_{\{x\neq 0\}}$ for $\alpha \in [1,2)$. Characterize for which values of $p > 0$, $\mathbb{E}[(Z_t^+)^p] < \infty$. Recall the result in Exercise 4.1.27 and compare. In particular note that the behavior at zero of $f(x)$ is completely irrelevant in order to decide if the moments are finite or not.

As usual, one extends the construction to the negative jumps. The notation is also extended in a natural fashion. That is, one defines Z^- and $Z = Z^+ + Z^-$.

Theorem 5.1.14 $\{Z_t^+; t \geq 0\}$, $\{Z_t^-; t \geq 0\}$ *and* $\{Z_t; t \geq 0\}$ *are Lévy processes. The characteristic function of* Z_t *is given by*

$$\mathbb{E}[e^{i\theta Z_t}] = \exp[ct\int_{\mathbb{R}} (e^{i\theta x} - 1 - i\theta x\,\mathbf{1}_{\{|x|\leq 1\}})f(x)dx].$$

As usual we say that the Lévy measure associated with Z is $\nu(dx) = cf(x)dx$ but realize that the form of the characteristic function has an extra term in comparison with Corollary 4.1.13 which is due to compensation with the local mean (that is, considering only the jumps around zero) in order to gain integrability.

---

[15]Hint: Recall the solution of Exercise 4.1.25. But also there is a very short way of proving this by computing $\lim_{n\uparrow\infty} \mu^{(\varepsilon_n,1)}$.

[16]Therefore in general, variances of the process Z_t cannot be studied unless one adds extra conditions on f.

Proof The proof is similar to the one of Theorem 4.1.5 except for the càdlàg property, which is straightforward. We leave the details to the reader just noting that when proving the continuity in probability one has to realize that now the process Z^+ is neither positive nor increasing.

Theorem 5.1.15 *Z satisfies that for all $t \geq 0$*

$$\sum_{0 \leq s \leq t} (\Delta Z_s)^2 < \infty, \text{ a.s.}$$

Proof In fact, it is enough to prove this property for the jumps smaller than one as the process with jumps bigger than one is a compound Poisson process. For the jumps smaller than one, the property follows by taking limits and expectations. In fact, for $\bar{Z}$

$$\mathbb{E}[\sum_{0 \leq s \leq t} (\Delta \bar{Z}_s^+)^2] = ct \int_{[0,1]} z^2 f(z) dz < \infty.$$

The first exercise is about the distribution of measures of sets under the Poisson random measure.

Exercise 5.1.16 Let $A \subset \mathbb{R} - (-\varepsilon, \varepsilon)$ for some $\varepsilon > 0$ such that $\int_A f(x)dx > 0$. Prove that the random variable $\int_{A \times [0,t]} \mathcal{N}(dx, ds)$ is a well-defined finite a.s. Poisson distributed random variable with parameter $\int_{A \times [0,t]} f(x)dxds$. Hint: Recall Exercises 3.4.9 and 4.1.19.

We can now obtain also some results about stable laws.

Exercise 5.1.17 Recall Exercise 4.1.23 and check that the same results are satisfied for $\alpha \in [1, 2)$. That is, define the process Z^λ as the Lévy process associated with the Lévy density $f(x) := \frac{e^{-\lambda|x|}}{|x|^{1+\alpha}} \mathbf{1}_{\{x \neq 0\}}$, where $\lambda > 0$ and $\alpha \in [1, 2)$. Find the weak limit of Z_t^λ as $\lambda \to 0$. Prove that the density of the limit random variable is smooth. For a hint, see Chap. 14.

Remark 5.1.18 (i) We should remark that in the above exercise we are proving that the density of the limit process Z_t is infinitely smooth, which may be counterintuitive at first due to the introduction of the stable process as limit of a sequence of compound Poisson process.[17] But, in fact, this limit procedure makes the possibility of zero jumps disappear.

(ii) Also an important point in all the calculations is the symmetry of the Lévy measure which makes calculations possible.

Exercise 5.1.19 The following exercise will test your understanding of the given construction, as it is not strictly related with the previous results.

---

[17] Recall that in general, a compound Poisson process does not have a density.

1. Consider a discrete Lévy measure of the type

$$\nu(dx) = \sum_{i=0}^{\infty}(a_i - a_{i+1})\frac{\delta_{a_i}(dx) + \delta_{-a_i}(dx)}{a_i^{1+\alpha}}.$$

Here a_i is a positive sequence decreasing monotonically to zero. Give conditions on the sequence $\{a_i\}_i$ in order to construct the associated Lévy process by using compound Poisson processes with jumps having a discrete distribution. As before, you may have to consider first the case $\alpha \in (0, 1)$ and then $\alpha \in [1, 2)$.[18]
2. Find the characteristic function for the process defined above. Prove that the associated process has infinite differentiable densities. Here, you will need some knowledge of how to derive regularity properties of the densities from the characteristic function of the associated Lévy process.[19]

For a hint, see Chap. 14.

Exercise 5.1.20 In this exercise, we will try to understand the behavior of small jumps through the central limit theorem. Consider the process $a_n \bar{Z}_t^{(\varepsilon_n,+)}$ for some random sequence a_n and $f(x) = C|x|^{-(1+\alpha)}\,\mathbf{1}_{\{x\neq 0\}}$ for $\alpha \in [1, 2)$. Determine this random sequence so that the following result is valid:

$$\lim_{n\to\infty} \mathbb{P}(a_n \bar{Z}_t^{(\varepsilon_n,+)} \in A) = \mathbb{P}(Y \in A).$$

Here Y is a normal random variable. Determine the mean and variance of this random variable.[20]

For a hint, see Chap. 14.

Exercise 5.1.21 This exercise is about the jump structure of Lévy processes as stated by its Lévy measure and the sign of the associated Lévy process.

1. Prove that if the Lévy measure is given by $f(x) = f(x)\,\mathbf{1}_{\{x>0\}}$ with $\int(x \wedge 1)f(x)dx < \infty$ then the associated Lévy process Z is positive and increasing a.s.
2. Give an example in the general case with $f(x) = f(x)\,\mathbf{1}_{\{x>0\}}$ and $\int(x^2 \wedge 1)f(x)dx < \infty$, where $\mathbb{P}(Z_t < 0) > 0$ for all $t > 0$.

For a hint, see Chap. 14.

Exercise 5.1.22 Prove that if $\{T_i; i \in \mathbb{N}\}$ are the jump times associated with the process $Z^{(\varepsilon_1,+)}$ then the probability that the process $\bar{Z}^+$ jumps at any of these times is zero.

For a hint, see Chap. 14.

---

[18] Clearly, one may modify the Lévy measure so that asymptotically it is equivalent to this example. We prefer this way of writing as it makes it clear the relation with the concept of a Lévy measure.

[19] Hint: See the results in Chap. 9, in particular, Corollary 9.1.3.

[20] In fact, the limit is only determined by the behavior of f around zero. Therefore any change on the function f away from zero will not change the limit behavior.

Exercise 5.1.23 This exercise is the parallel of Exercise 4.1.21 together with 4.1.23.

1. Prove that the exponent of the characteristic function obtained in 2. of Exercise 4.1.23 does not converge as $\alpha \to 1$.
2. Consider the process constructed in Theorem 5.1.14 for $f(x) := \frac{1}{|x|^2} \mathbf{1}_{\{x \neq 0\}}$.
   Prove that the marginal laws are given by Cauchy laws.
3. Prove that the random variable Z_t satisfies that the law of $t^{-1}Z_t$ is independent of t.
4. Prove that the density of Z_t exists and is smooth.

   For a hint, see Chap. 14.

5.2 Itô Formula for the Lévy Process Z

We give a first Itô formula for the Lèvy process Z, which will be useful later in order to understand how to define stochastic integrals. This is also a short section that will serve as a test of the knowledge that you have accumulated so far. In particular, we do not give all details and ask you to fill them out using previous results or an extension of those results on which you may have to work on your own.

Recall that $\widehat{\mathcal{N}}(dx, ds) = cf(x)dxds$ is the σ-finite measure corresponding to the compensator of the Poisson random measure $\mathcal{N}(dx, ds)$. That is, the compensated Poisson random measure is given by $\widetilde{\mathcal{N}} := \mathcal{N} - \widehat{\mathcal{N}}$. Therefore, one has, for example $Z_t^+ = \int_{[0,1]\times[0,t]} z\widetilde{\mathcal{N}}(dz, ds) + \int_{(1,\infty)\times[0,t]} z\mathcal{N}(dz, ds)$.

Exercise 5.2.1 Prove the above formula.

Theorem 5.2.2 (Itô formula for jump process) *For $h \in C_b^2$, we have for $t \geq 0$*

$$h(Z_t) = h(Z_0) + \int_{\mathbb{R}\times[0,t]} \left\{h(Z_{s-} + z) - h(Z_{s-}) - h'(Z_{s-})z\mathbf{1}_{\{|z|\leq 1\}}\right\} \mathcal{N}(dz, ds)$$
$$+ \int_{[-1,1]\times[0,t]} h'(Z_s)z\widetilde{\mathcal{N}}(dz, ds).$$

Here, we interpret the first integral as

$$\int_{[-1,1]\times[0,t]} \left\{h(Z_{s-} + z) - h(Z_{s-}) - h'(Z_{s-})z\right\} \mathcal{N}(dz, ds)$$
$$:= \lim_{n\to\infty} \int_{[-1+\varepsilon_n,1-\varepsilon_n]\times[0,t]} \left\{h(Z_{s-} + z) - h(Z_{s-}) - h'(Z_{s-})z\right\} \mathcal{N}(dz, ds).$$

The second integral is interpreted as

$$\int_{[-1,1]\times[0,t]} h'(Z_s)z\widetilde{\mathcal{N}}(dz, ds) := \lim_{n\to\infty} \int_{[-1+\varepsilon_n,1-\varepsilon_n]\times[0,t]} h'(Z_s)z\widetilde{\mathcal{N}}(dz, ds).$$

Proof We will give the proof for Z^+ and $c = 1$. First, taking $n \geq 2$,

$$\bar{Z}_t^{(\varepsilon_n)} = \int_{[0,1]\times[0,t]} z\mathcal{N}^{(\varepsilon_n)}(dz, ds) - \int_{[\varepsilon_n,1]\times[0,t]} zf(z)dzds.$$

Then, derive an extension of Theorem 3.4.24 which gives that for $Y_t^n := \bar{Z}_t^{(\varepsilon_n)} + Z_t^{(\varepsilon_1)}$

$$\begin{aligned} h(Y_t^n) &= h(Y_0^n) + \int_{[0,1]\times[0,t]} \{h(Y_{s-}^n + z) - h(Y_{s-}^n)\} \mathcal{N}^{(\varepsilon_n)}(dz, ds) \\ &\quad + \int_{[1,\infty)\times[0,t]} \{h(Y_{s-}^n + z) - h(Y_{s-}^n)\} \mathcal{N}^{(\varepsilon_1)}(dz, ds) - \int_{[\varepsilon_n,1]\times[0,t]} h'(Y_s^n)zf(z)dzds \\ &= h(Y_0^n) + \int_{[0,\infty)\times[0,t]} \{h(Y_{s-}^n + z) - h(Y_{s-}^n) - h'(Y_{s-}^n)z\,\mathbf{1}_{|z|\leq 1}\} \mathcal{N}^{(\varepsilon_n)}(dz, ds) \\ &\quad + \int_{[0,1]\times[0,t]} h'(Y_s^n)z\widetilde{\mathcal{N}}^{(\varepsilon_n)}(dz, ds). \end{aligned}$$

The two terms above converge because $\int_{[1,\infty)\times[0,t]} |z|\mathcal{N}(dz, ds) < \infty$, a.s. and[21]

$$\int_{[0,1]\times[0,t]} |z|^2\widehat{\mathcal{N}}(dz, ds) < \infty.$$

In fact, for the first term note that

$$\begin{aligned} &\left|\int_{[0,1]\times[0,t]} \{h(Y_{s-}^n + z) - h(Y_{s-}^n) - h'(Y_{s-}^n)z\} \mathcal{N}^{(\varepsilon_n)}(dz, ds)\right| \\ &\leq \|h''\|_\infty \int_{[0,1]\times[0,t]} |z|^2\mathcal{N}^{(\varepsilon_n)}(dz, ds). \end{aligned}$$

For the second term, we have

$$\mathbb{E}\left[\left|\int_{[0,1]\times[0,t]} h'(Y_s^n)z\widetilde{\mathcal{N}}^{(\varepsilon_n)}(dz, ds)\right|\right] \leq C\|h'\|_\infty \int_{[0,1]\times[0,t]} |z|^2\widehat{\mathcal{N}}^{(\varepsilon_n)}(dz, ds).$$

Therefore the result follows after completing the details.

Remark 5.2.3 (i) Actually the last part in the previous proof is quite important in what follows because through it, one understands that the estimation of stochastic integrals in the case of infinite variation Lévy processes has to be done through estimates in $L^2(\Omega)$ for jumps close to zero while for jumps away from zero, the Itô formula is proven a.s. as $L^2(\Omega)$ estimates cannot be obtained in general. Also recall that the above estimates are obtained using the supremum norm for t in compact intervals.

[21] Prove this. Note that the first term converges in absolute variation, while the second only converges in the $L^2(\Omega)$ sense with the supremum norm in time. For this reason, for the first term one uses two derivatives, while for the second one only uses the first derivative.

(ii) Depending on the proof that you propose you may need to prove that any càdlàg function on a compact interval has a bounded image.
(iii) If you understood the proof correctly you will see that the jumps of size smaller than one are considered twice in the Itô formula above. This is done in order to achieve the finite integrability of each term. In the first term, the convergence is in the sense of finite variation a.s. In the second term the integrability is in the $L^2(\Omega)$ sense.

Exercise 5.2.4 For a possible line of proof of the above result, check in Chap. 14. For a hint, see Chap. 14.

5.3 Stochastic Integrals with Respect to Infinite Variation Lévy Processes and Their Corresponding SDEs

Note that up to now, the stochastic integrals with respect to compound Poisson processes were defined simply as finite sums. The case of finite variation was considered as integrals with respect to bounded variated functions. The integral with respect to a compound Poisson process can be considered as a particular case of an integral with respect to a bounded variated function. In the discussion just before Proposition 5.1.6, we have introduced the σ-finite measure $\mathcal{N}$ for which integrals can be defined which generalized all previously defined integrals.

We will now generalize these integrals into cases where the integral understood as an integral with respect to σ-finite measures cannot be defined. This will be related with our previous definitions of stochastic integrals. This will be done using the key idea in the last part of the proof of Theorem 5.2.2 and which is further highlighted in Remark 5.2.3. Before this, as before we will separate the jumps that are greater than one which can then be considered as a measure integral and therefore the integral can be defined without compensation. As before, we define $(\mathscr{F}_t)_{t\in[0,T]}$ as the smallest filtration such that the process Z is adapted to it, the filtration is complete and right-continuous. We also define the class of predictable processes which are generated by the left-continuous processes adapted to $(\mathscr{F}_t)_{t\in[0,T]}$. In particular, note that $\mathscr{F}_0$ is the trivial sigma field completed with all null sets.

Definition 5.3.1 Define the predictable σ-field as

$$\mathscr{P} = \sigma(F_0 \times \{0\}, F \times (s,t]; 0 \leq s < t \leq T, F \in \mathscr{F}_s, F_0 \in \mathscr{F}_0).$$

We say that a process $g : \Omega \times [0,T] \to \mathbb{R}$ is predictable if it is measurable with respect to $\mathscr{P}$.

Exercise 5.3.2 Prove that Definition 3.5.1 and the one above coincide.

Remark 5.3.3 At this point, note that we had to complete the filtration, making it right-continuous so that the filtration satisfies what is called "usual conditions".

This is done in order to avoid some situations where the proofs become awkwardly difficult. In general, when checking results in other sources, one always has to be careful to see if the stated results are valid with filtrations of this kind or otherwise. But happily for us in the case of Lévy processes, completed filtrations are right-continuous. See, for example, [48], Theorems 25 and 31 in the Preliminaries.[22]

Theorem 5.3.4 *Let $g : [0,1] \times [0,T] \to \mathbb{R}$ be a jointly measurable function so that $\int_{[0,1]\times[0,T]} |g(z,s)|^2 \widehat{\mathcal{N}}(dz,ds) < \infty$. Then the stochastic integral $\int_{[0,1]\times[0,T]} g(z,s)$ $\widetilde{\mathcal{N}}(dz,ds)$ is defined as the limit of the stochastic integrals $\int_{[0,1]\times[0,T]} g_n(z,s)\widetilde{\mathcal{N}}$ (dz,ds) for any sequence of step functions g_n such that $g_n(z,s) = g_n(z,s)\,\mathbf{1}_{|z|>\varepsilon_n}$ and*

$$\lim_{n\to\infty} \int_{[0,1]\times[0,T]} |(g_n - g)(z,s)|^2 \widehat{\mathcal{N}}(dz,ds) = 0.$$

Furthermore $\{\int_{[0,1]\times[0,t]} g(z,s)\widetilde{\mathcal{N}}(dz,ds); t \in [0,T]\}$ is a $(\mathscr{F}_t)_{t\in[0,T]}$-martingale and

$$\mathbb{E}\left[\left|\int_{[0,1]\times[0,t]} g(z,s)\widetilde{\mathcal{N}}(dz,ds)\right|^2\right] = \int_{[0,1]\times[0,T]} |g(z,s)|^2 \widehat{\mathcal{N}}(dz,ds).$$

Proof There are different ways of doing this. One of them is as follows: First, we note that we need to define the concept of the stochastic integral. Assume that g is positive and bounded. Then, there exists a sequence of simple functions g_n so that they are monotone in n and converge to g. Then one needs to slightly modify the sequence g_n so that jumps around zero do not appear in the definition of the stochastic integral. Then consider the $L^2(\Omega)$ norms and prove that the sequence of well-defined stochastic integrals of g_n converges by proving that the stochastic integrals are a Cauchy sequence in $L^2(\Omega)$. Next prove that for any other sequence of simple functions satisfying that it vanishes around zero and converges to g in the norm stated, their stochastic integrals converge, therefore obtaining that the stochastic integral is well defined.

For the martingale property, this is a repetition (and it was done on purpose so that now you should be well trained for this), of the proof of Theorem 3.4.11 if g were bounded. Note that all instances of a.s. convergence used in Theorem 3.4.11 have to be replaced now by convergence in $L^2(\Omega)$ which can be achieved due to the results in Theorem 3.4.31.

After this step, one starts by completing all other cases. That is, complete the arguments for the case that $g = g_+ - g_-$ for two positive bounded functions g_+ and g_-.

Once all integrals have been defined, one can take any sequence g_n in an even more general sense than the one in the statement to prove the convergence in $L^2(\Omega)$. This proves that the stochastic integral defined through the particular sequence is independent of the approximation sequence of functions g_n, $n \in \mathbb{N}$ taken.

[22] In fact, even the independent increment property will suffice.

You can also recall the proof you have provided for Theorem 5.2.2. Therefore the details of the proof are left as an exercise.

Exercise 5.3.5 If you have finished the proof, state exactly in what sense is the limit taken in the statement of Theorem 5.3.4.

For a hint, see Chap. 14.

Exercise 5.3.6 Prove that $\int_{[0,1]\times[0,T]} z\mathcal{N}(dz, ds) = \infty$, a.s. but $\int_{[0,1]\times[0,T]} z\widetilde{\mathcal{N}}(dz, ds) < \infty$, a.s. This explains that the extension of the integral is meaningful and that we need to understand the integral as a stochastic integral. Clearly, when both integrals (the one defined as an integral with respect to σ-measure and the stochastic integral) exist they coincide.

Exercise 5.3.7 Use the techniques introduced in the proof of Theorem 5.1.5 in order to prove that the stochastic integral just defined has càdlàg paths (a.s.).

Exercise 5.3.8 Use the independent increment property of the Lévy process Z to prove that the random variable $\int_{[0,1]\times(u,T]} g(z, s)\widetilde{\mathcal{N}}(dz, ds)$ is independent of $\mathcal{F}_u$. In fact, prove that $\int_{[0,1]\times(u,T]} g(z, s)\widetilde{\mathcal{N}}(dz, ds) \in \sigma(Z_s - Z_u; s \in (u, T])$.

Theorem 5.3.9 *Let $g : \Omega \times [0, 1] \times [0, T] \to \mathbb{R}$ be a predictable process with*

$$\mathbb{E}[\int_{[0,1]\times[0,T]} |g(z, s)|^2 \widehat{\mathcal{N}}(dz, ds)] < \infty.$$

Assume that g_n is a sequence of step processes of the type

$$g_n(\omega, z, s) = \mathbf{1}_{G_0\times\{0\}}(\omega, s)h^0(z) + \sum_{i=0}^{n-1} \mathbf{1}_{F_i\times(t_i,t_{i+1}]}(\omega, s)g^i(z).$$

Here, $\{0 = t_0 < ... < t_n = T\}$ is a partition of the interval $[0, T]$ such that $|\pi_n| := \max_{i=0,\dots,n-1}(t_{i+1} - t_i) \to 0$ as $n \to \infty$, $G_0 \in \mathcal{F}_0$, $F_i \in \mathcal{F}_{t_i}$ and $h^0, g^i : [0, 1] \to \mathbb{R}$ are measurable functions such that

$$\lim_{n\to\infty} \mathbb{E}[\int_{[0,1]\times[0,T]} |(g_n - g)(z, s)|^2 \widehat{\mathcal{N}}(dz, ds)] = 0.$$

Then the sequence of stochastic integrals $\int_{[0,1]\times[0,T]} g_n(z, s)\widetilde{\mathcal{N}}(dz, ds)$ converges in $L^2(\Omega)$. Its limit is called the stochastic integral of g with respect to Z and is denoted by $\int_{[0,1]\times[0,T]} g(z, s)\widetilde{\mathcal{N}}(dz, ds)$ and when considered as a process the stochastic integral is a $(\mathcal{F}_t)_{t\in[0,T]}$-martingale with càdlàg paths. In the particular case that g does not depend on Z, we may also use the notation $\int_0^T g(s)dZ_s$.

Proof The proof is just again a repetition of the proof in the previous theorem with appropriate changes. Therefore now the exercise is to think how to extend the previous proof to the current case using conditional expectations properly.

Exercise 5.3.10 Prove that for any $F \in \mathscr{F}_u$ and the stochastic integrals defined above satisfy the following formula:

$$\mathbf{1}_F \int_{[0,1]\times(u,T]} g(z,s)\widetilde{\mathcal{N}}(dz,ds) = \int_{[0,1]\times(u,T]} \mathbf{1}_F g(z,s)\widetilde{\mathcal{N}}(dz,ds).$$

Exercise 5.3.11 Prove that the sequence $g_n(\omega, z, s)$ satisfying the hypotheses of Theorem 5.3.9 can always be constructed under the hypothesis that

$$\mathbb{E}[\int_{[0,1]\times[0,T]} |g(z,s)|^2 \widehat{\mathcal{N}}(dz,ds)] < \infty.$$

Exercise 5.3.12 Give the corresponding construction for the stochastic integral considered in $[1,\infty) \times [0,T]$.

Exercise 5.3.13 In each of the two cases: stochastic integrals over $[0,1] \times [0,T]$ and stochastic integrals over $[1,\infty) \times [0,T]$, give bounds for the $L^2(\Omega)$ norms of the corresponding stochastic integrals. That is, prove that

$$\mathbb{E}\left[\left|\int_{[0,1]\times[0,T]} g(z,s)\widetilde{\mathcal{N}}(dz,ds)\right|^2\right] \le C\mathbb{E}\left[\int_{[0,1]\times[0,T]} |g(z,s)|^2 \widehat{\mathcal{N}}(dz,ds)\right],$$

$$\mathbb{E}\left[\left|\int_{[1,\infty)\times[0,T]} g(z,s)\mathcal{N}(dz,ds)\right|^2\right]$$
$$\le C\mathbb{E}\left[\int_{[1,\infty)\times[0,T]} |g(z,s)|^2 \widehat{\mathcal{N}}(dz,ds) + \left|\int_{[1,\infty)\times[0,T]} g(z,s)\widehat{\mathcal{N}}(dz,ds)\right|^2\right].$$

Furthermore prove the following bound in $L^1(\Omega)$:

$$\mathbb{E}\left[\left|\int_{[1,\infty)\times[0,T]} g(z,s)\mathcal{N}(dz,ds)\right|\right] \le C\mathbb{E}\left[\int_{[1,\infty)\times[0,T]} |g(z,s)| \widehat{\mathcal{N}}(dz,ds)\right].$$

Now we give the corresponding Itô formula for this case. The proof is left as an exercise.

Theorem 5.3.14 *Consider the process* $Y_t := \int_{[0,1]\times[0,t]} g(z,s)\widetilde{\mathcal{N}}(dz,ds)$, *where* $g : \Omega \times [0,1] \times [0,T] \to \mathbb{R}$ *satisfies the same conditions as in Theorem 5.3.9 and* $h \in C_b^2(\mathbb{R})$, *then for any* $t \le T$

$$h(Y_t) = h(Y_0) + \int_{[0,1]\times[0,t]} h(Y_{s-} + g(z,s)) - h(Y_{s-}) - h'(Y_{s-})g(z,s)\mathcal{N}(dz,ds)$$
$$+ \int_{[0,1]\times[0,t]} h'(Y_{s-})g(z,s)\widetilde{\mathcal{N}}(dz,ds).$$

Exercise 5.3.15 Supposing that $h \in C_b^2(\mathbb{R})$, prove the following Itô formula for $Z_t = \int_{[0,1]\times[0,t]} z\widetilde{\mathcal{N}}(dz, ds) + \int_{(1,\infty)\times[0,t]} z\mathcal{N}(dz, ds)$:

$$h(Z_t) = h(Z_0) + \int_{\mathbb{R}\times[0,t]} \{h(Z_{s-} + z) - h(Z_{s-}) - h'(Z_{s-})z\,\mathbf{1}_{\{|z|\le 1\}}\}\mathcal{N}(dz, ds)$$
$$+ \int_{[0,1]\times[0,t]} h'(Z_{s-})z\widetilde{\mathcal{N}}(dz, ds).$$

This exercise may confuse you if you compare with the result in Theorem 5.2.2. In fact, the problem here is to prove that the procedure used in the definition of Z^+ in Sect. 5.1 and the definition as stochastic integral given above coincide.

Remark 5.3.16 Generalizations similar to Theorem 4.1.36 can be obtained. We will somewhat assume in what follows that the proof of these generalizations is well understood.[23] We will in fact, use the following result, which you can try to prove as an exercise.

In order to understand the next theorem it is good to recall the discussion about predictability in Sect. 3.5. We also define the following σ field

$$\mathscr{F}_t := \sigma\left(\mathcal{N}(\mathbf{1}_{A\times B});\, B \subset [0,t],\, A \subset \mathbb{R} - (-\varepsilon, \varepsilon), \text{ for some } \varepsilon > 0\right).$$

We say that a process $a : \Omega \times \mathbb{R} \times [0, T] \to \mathbb{R}$ is progressively measurable if its restriction $a\,\mathbf{1}_{[0,t]}$ is $\mathscr{F}_t \otimes \mathscr{B}([0,t] \times \mathbb{R})$-measurable for any $t \ge 0$.

Theorem 5.3.17 *Assume that X satisfies the stochastic equation*

$$X_t = X_0 + \int_0^t b_s ds + \int_{(1,\infty)\times[0,t]} a_1(z,s)\mathcal{N}(dz, ds) + \int_{[0,1]\times[0,t]} a_2(z,s)\widetilde{\mathcal{N}}(dz, ds).$$

Here $a_1, a_2 : \Omega \times \mathbb{R} \times [0, T] \to \mathbb{R}$ *and* $b : \Omega \times [0, T] \to \mathbb{R}$ *are predictable processes such that they satisfy* $|a_2(z,s)|^2 \le C|z|^2$, $|b_s| + |a_1(z,s)| \le C$ *for some positive constant C and all* $z \in \mathbb{R},\ s \in [0, T]$. *Then the above integrals are finite almost surely and the following Itô formula is satisfied:*

$$h(X_t) = h(X_0) + \int_0^t h'(X_s)b_s ds + \int_{(1,\infty)\times[0,t]} \{h(X_{s-} + a_1(z,s)) - h(X_{s-})\}\mathcal{N}(dz, ds)$$
$$+ \int_{[0,1]\times[0,t]} \{h(X_{s-} + a_2(z,s)) - h(X_{s-}) - h'(X_{s-})a_2(z,s)\}\mathcal{N}(dz, ds)$$
$$+ \int_{[0,1]\times[0,t]} h'(X_{s-})a_2(z,s)\widetilde{\mathcal{N}}(dz, ds).$$

---

[23]Otherwise as an exercise you can try to think each time how to prove the needed version of the Itô formula.

Exercise 5.3.18 Write the Itô formula for the following case:

$$X_t = X_0 + \int_0^t b_s ds + \int_{(0,\infty)\times[0,t]} a_1(z,s)\mathcal{N}(dz,ds) + \int_{[0,1]\times[0,t]} a_2(z,s)\widetilde{\mathcal{N}}(dz,ds).$$

Here we assume the conditions $a_1, a_2 : \Omega \times \mathbb{R} \times [0,T] \to \mathbb{R}$ and $b : \Omega \times [0,T] \to \mathbb{R}$ are predictable processes such that they satisfy $|a_1(z,s)| \le C|z|^2$ $|a_2(z,s)|^2 \le C|z|^2$, $|b_s| \le C$ for some positive constant C and all $z \in \mathbb{R}$, $s \in [0,T]$.

We will now turn to the study of stochastic equations driven by Lévy processes of the type that we have introduced. Recall that we have already discussed certain stochastic differential equations with jumps in Theorems/Propositions 3.6.8, 4.2.9, 4.2.2 and 3.6.5. The techniques in each case were slightly different. Either through explicit calculation or through the definition of the correct norm in order to apply some type of approximation procedure or fixed-point-type theorem.

Theorem 5.3.19 *Suppose that $a : \mathbb{R} \to \mathbb{R}$ be a Lipschitz bounded function and that the Lévy measure satisfies that $\int |z|^2 f(z)dz < \infty$. Then there exists a unique solution to the stochastic equation*

$$X_t = x + \int_{[-1,1]\times[0,t]} a(X_{s-})z\widetilde{\mathcal{N}}(dz,ds) + \int_{[-1,1]^c\times[0,t]} a(X_{s-})z\mathcal{N}(dz,ds),$$

which satisfies that $\mathbb{E}[|X_t|^2] \le C$ for all $t \in [0,T]$.

Proof We will only give the ideas for uniqueness and leave the rest of the details to the reader. In case you need more help you may look at Theorem 6.2.3 in [2]. Remember that you need to define what is meant by solution of the above stochastic equation.

In order to be prepared to carry out this proof you have to remember all the previous methods used for proving existence and uniqueness for stochastic equations described in previous chapters.

Consider two solutions X and Y satisfying that $\mathbb{E}[|X_t|^2 + |Y_t|^2] \le C$ for all $t \in [0,T]$. Then using a "good" approximation for the function $g(x) = |x|^2$, one applies the Itô formula for $|X_t - Y_t|^2$; this gives

$$\begin{aligned}|X_t - Y_t|^2 = &\int_{[-1,1]^c\times[0,t]} 2(X_{s-} - Y_{s-})(a(X_{s-}) - a(Y_{s-}))z\mathcal{N}(dz,ds)\\ &+ \int_{\mathbb{R}\times[0,t]} (a(X_{s-}) - a(Y_{s-}))^2 z^2 \mathcal{N}(dz,ds)\\ &+ \int_{[-1,1]\times[0,t]} 2(X_{s-} - Y_{s-})(a(X_{s-}) - a(Y_{s-}))z\widetilde{\mathcal{N}}(dz,ds).\end{aligned}$$

Taking expectations and Gronwall's lemma will give the result if $\int |z|^2 f(z)dz < \infty$.

In general, if one only has that $\int 1 \wedge |z|^2 f(z)dz < \infty$. then one needs an extra argument.

Suppose that we restrict our study to the time interval up to the first jump of size bigger than one. Then the norm that we will use in order to prove existence and uniqueness will be $\mathbb{E}[|X_t - Y_t|^2 1(t < T_1)]$.[24] Therefore we obtain the uniqueness of solutions up to the first jump bigger than one. By iteration one obtains the general result.

Similarly, a localization argument will solve the case when a is not bounded but just Lipschitz. For that the following exercise is useful.

Exercise 5.3.20 Given a Lipschitz function a, prove that there exists a sequence of Lipschitz bounded functions a_n such that:

- $a_n(x) = a(x)$ for all $x \in [-n, n]$.
- The Lipschitz constant of a_n is uniformly bounded independent of n. That is, there exists a constant C such that $|a_n(x) - a_n(y)| \leq C|x - y|$ for all $x, y \in \mathbb{R}$.
- For each n there exists a constant C_n such that $|a_n(x)| \leq C_n$.

For a hint, see Chap. 14.

Exercise 5.3.21 Compute $\mathbb{E}[X_t^2]$ for the solution of the linear equation

$$X_t = x + \int_{[-1,1]\times[0,t]} X_{s-} z \widetilde{\mathcal{N}}(dz, ds).$$

For a hint, see Chap. 14.

Exercise 5.3.22 Consider the solution process X for the equation studied in Theorem 5.3.19 with $f(x) = \frac{c}{|x|^{1+\alpha}} \mathbf{1}_{|x|>1}, \alpha \in (1, 2)$. Prove that there exists a unique solution. Prove that $\mathbb{E}[|X_t|^\beta] < \infty$ for $0 < \beta < \alpha$.

For a hint, see Chap. 14.

5.4 Some Extensions of Jump Processes

In this section, we briefly discuss some extensions of jump processes which are used in applications.

In the following exercises, we will briefly discuss some extensions of jump processes.

Exercise 5.4.1 Consider over the set $[0, \infty) \times \mathbb{R}$ a σ-finite measure λ. That is, there exists a family of disjoint sets $A_n, n \in \mathbb{N}$ such that $\cup_{n\in\mathbb{N}} A_n = [0, \infty) \times \mathbb{R}$ with $\lambda(A_n) < \infty$. For $n \in \mathbb{N}$, define N^n to be a sequence of i.i.d. Poisson random variables with parameter $\lambda(A_n)$. Next, we define for each n, independent of all $\{N^n; n \in \mathbb{N}\}$ a

[24] Notice the difference with the norms used in the proofs of Theorems 4.2.2 and 4.2.9.

sequence of i.i.d. random variables Z_n^i, $i \in \mathbb{N}$ with the law $\mathbb{P}(Z_n^i \in dz) = \frac{\lambda(dz)}{\lambda(A_n)}$, $z \in A_n$. Therefore the family of random variables $\{N^n, Z_n^i; i, n \in \mathbb{N}\}$ is also independent. Define the following family of random variables for any A measurable subset of $[0, \infty) \times \mathbb{R}$,

$$\mathcal{N}^n(A) := \sum_{i=1}^{N^n} \mathbf{1}(Z_n^i \in A).$$

1. Find the law of $\mathcal{N}^n(A)$.
2. Prove that $\mathcal{N}^n(A)$ and $\mathcal{N}^n(B)$ are independent if A and B are disjoint.
3. Find the law of $\mathcal{A} = \sum_{n=1}^{\infty} \mathcal{N}^n(A)$.
4. Prove that $\mathcal{N}(A)$ and $\mathcal{N}(B)$ are independent if A and B are disjoint.

This defines a generalized form of Poisson random measure. In fact, we do not need that the measure λ has to be absolutely continuous with respect to the Lebesgue measure as in previous chapters.

For a hint, see Chap. 14.

Exercise 5.4.2 Suppose that $\lambda(A) = \int_A \lambda(t) f(x) dt dx$, where $\lambda(t) f(x) \geq 0$ and $\lambda([0, \infty) \times \mathbb{R}) < \infty$. Find the law of the first jump defined as $T_1 := \inf\{s > 0, \mathcal{N}([0, s] \times \mathbb{R}) > 0\}$. This generalization allows the definition of the so-called non-homogeneous Poisson process.[25] Hint: Recall Lemma 2.1.6.

Extrapolate the result for the case that $\lambda([0, \infty) \times \mathbb{R}) = \infty$.

For a hint, see Chap. 14.

Exercise 5.4.3 Let $\mathcal{N}^i, \mathcal{M}^i$, $i \in \mathbb{N}$ be a sequence of independent Poisson random measures associated with simple Poisson processes with parameters λ_i, μ_i, $i \in \mathbb{N}$. Define X as the process solution of the following stochastic equation:

$$X_t = 1 + \sum_{i=1}^{\infty} \int_{\{1\} \times [0,t]} \mathbf{1}_{\{X_{s-}=i\}} \mathcal{N}^i(dz, ds) - \sum_{i=1}^{\infty} \int_{\{1\} \times [0,t]} \mathbf{1}_{\{X_{s-}=i\}} \mathcal{M}^i(dz, ds).$$

Describe the movement of this process. This corresponds to the so-called birth and death process with birth rates given by $(\lambda_i)_i$ and death rates given by $(\mu_i)_i$.

1. Prove the existence and uniqueness of the process X solution of the above equation.
2. Prove that once $X_s = 0$ for some $s > 0$ then the process will always remain at zero.
3. Compute the conditional probabilities $\mathbb{P}(X_t = i + 1, X_u = i, u \in [s, t)/X_s = i)$ and $\mathbb{P}(X_t = i - 1, X_u = i, u \in [s, t)/X_s = i)$ for $s < t$.
4. Prove that X is a Markov process. That is, $\mathbb{E}[f(X_t)/\mathcal{F}_s^X] = \mathbb{E}[f(X_t)/X_s]$. Here $\mathcal{F}_s^X$ denotes the σ-algebra generated by the process $\{X_u; u \in [0, s]\}$.

[25] We hope that the abuse of notation using λ for the measure and the function that defines λ does not cause confusion.

Finally, compare the results above with similar results for the solution of the equation

$$X_t = 1 + \sum_{i=1}^{\infty} \int_{\{1\}\times[0,t]} \mathbf{1}_{\{X_{s-}\geq i\}} \mathcal{N}^i(dz, ds) - \sum_{i=1}^{\infty} \int_{\{1\}\times[0,t]} \mathbf{1}_{\{X_{s-}\geq i\}} \mathcal{M}^i(dz, ds).$$

In particular discuss the interpretation of $\{\lambda_i, \mu_i\}_i$ in each case.

5.5 Non-homogeneous Poisson Process

A counting process is any increasing process. The following three definitions are equivalent.

Definition 5.5.1 (*The infinitesimal definition*) A counting process $N = (N_t)_{t\geq 0}$ is a non-homogeneous Poisson process with rate function $\lambda : [0, \infty) \to \mathbb{R}_+$ if:

(i) N has independent increments.
(ii) $\mathbb{P}(N_{t+h} - N_t = 1) = \lambda(t)h + o(h)$ and $\mathbb{P}(N_{t+h} - N_t > 1) = o(h)$ as $h \to 0$.

Definition 5.5.2 (*The axiomatic definition*) A counting process $N = (N_t)_{t\geq 0}$ is a non-homogeneous Poisson process with integrable positive rate function $\lambda = (\lambda(t))_{t\geq 0}$ if:

(i) N has independent increments.
(ii) for any $t \geq 0$ and $h > 0$, $\mathbb{P}(N_{t+h} - N_t = k) = e^{m(t+h)-m(t)} \frac{(m(t+h)-m(t))^k}{k!}$, where $m(t) := \int_0^t \lambda(s)ds$.

Definition 5.5.3 (*The constructive definition*) A counting process $N = (N_t)_{t\geq 0}$ is a non-homogeneous Poisson process with rate function $\lambda = (\lambda(t))_{t\geq 0}$ if there exists a Poisson process $Y = (Y(t))_{t\geq 0}$ with rate $\lambda = 1$ such that[26]

$$N_t = Y\left(\int_0^t \lambda(s)ds\right) = Y(m(t)).$$

Remark 5.5.4 (i) By the independent increment property of Y, the process N satisfies the Markov property (recall Theorem 3.2.9). On the other hand, $\mathbb{P}(N_{t+h} - N_t = k) = \mathbb{P}(N_h = k)$ is not satisfied in general, so N is a non-homogeneous Markov process.
(ii) Note that the non-homogeneous Poisson process is used in models such as renewal population dynamics, reliability theory, biology and finance. This process satisfies:

$$\mathbb{P}(N_h = 0) = \exp\left(-\int_0^h \lambda(s)ds\right) = \exp(-m(h)).$$

[26]Note that we have slightly changed the notation so that it is easier to read. That is, $Y\left(\int_0^t \lambda(s)ds\right) \equiv Y(u)|_{u=\int_0^t \lambda(s)ds}$.

(iii) Let $M_t := N_t - \int_0^t \lambda(s)ds$, then M_t also satisfies the martingale property.
(iv) Let $(\tau_n^Y)_{n\in\mathbb{N}}$ and $(\tau_n^N)_{n\in\mathbb{N}}$ be jump times of Y and N, then it holds that $\tau_n^Y = \int_0^{\tau_n^N} \lambda(s)ds$.

Exercise 5.5.5 Prove the above properties using the constructive definition. Find the conditional distribution of jump times in the interval $[0, t]$ under the condition that $N_t = k$ for fixed $k > 0$.

Proposition 5.5.6 *Let $N = (N_t)_{t\geq 0}$ be a non-homogeneous Poisson process with positive integrable rate function $\lambda = (\lambda(t))_{t\geq 0}$. Then*

$$\mathbb{E}[N_t] = \mathbb{E}\left[Y\left(\int_0^t \lambda(s)ds\right)\right] = \int_0^t \lambda(s)ds,$$

$$\mathbb{E}[N_t^2] - \mathbb{E}[N_t]^2 = \mathbb{E}\left[Y^2\left(\int_0^t \lambda(s)ds\right)\right] - \mathbb{E}\left[Y\left(\int_0^t \lambda(s)ds\right)\right]^2 = \left(\int_0^t \lambda(s)ds\right)^2,$$

$$\mathbb{E}[e^{-i\theta N_t}] = \mathbb{E}\left[\exp\left(-i\theta Y\left(\int_0^t \lambda(s)ds\right)\right)\right] = \exp\left((e^{i\theta} - 1)\left(\int_0^t \lambda(s)ds\right)\right).$$

5.5.1 Stochastic Equation Driven by a Poisson Process

In this subsection, we will study the following stochastic equation:

$$N_t = N_0 + Y\left(\int_0^t \lambda(N_s)ds\right), \quad N_0 \in [0, \infty), \tag{5.5}$$

where λ is a positive measurable function which satisfies $\int_0^t \lambda(N_s)ds < \infty$. The solution process to the above equation N is also a counting process. This model may be interpreted as a self-exciting model. The fact that N increases makes the rate of the Poisson process increase (assuming that $\lambda \geq 0$) and therefore more jumps are likely. The reverse effect may be achieved if we accept that λ may take negative values but then we will have to modify the argument in Y to $\max\{\int_0^t \lambda(N_s)ds, 0\}$.

Proposition 5.5.7 *Let $Y = (Y(t))_{t\geq 0}$ be a Poisson process with rate 1 and $(\tau_n)_{n\in\mathbb{N}}$ be the jump times of Y. Then there exists a unique solution to the Eq. (5.5) which is given by*

$$N_t = N_0 + \sum_{n\geq 1} \mathbf{1}(T_n \leq t),\, t \in [0, T_\infty)$$

where $\tau_0 = T_0 = 0$ and for any $n \in \mathbb{N}$,

$$T_n := T_{n-1} + \frac{\tau_n - \tau_{n-1}}{\lambda(N_0 + n - 1)} = \sum_{k=1}^n \frac{\tau_k - \tau_{k-1}}{\lambda(N_0 + k - 1)},$$

and $T_\infty := \lim_{n\to\infty} T_n$.

Remark 5.5.8 Note that if λ is bounded, then $T_n \geq \tau_n/||\lambda||_\infty$, so $T_\infty = \infty$. If $\lambda(t) := t^p$ for $p > 1$, then

$$\mathbb{E}[T_n] = \sum_{k=1}^{n} \frac{1}{(N_0 + k - 1)^p}.$$

Therefore, by the monotone convergence theorem, we have

$$\mathbb{E}[T_\infty] = \sum_{k=1}^{\infty} \frac{1}{(N_0 + k - 1)^p} < \infty,$$

thus $\mathbb{P}(T_\infty < \infty) = 1$.

Proof (*Proof of Proposition* 5.5.7) We define $N_t = N_0 + \sum_{n\geq 1} \mathbf{1}(T_n \leq t)$. Then we prove N_t is a solution of (5.5).

Assume that $t \in [0, T_1)$. Then $N_t = N_0 = N_0 + Y(0) = N_0 + Y(\int_0^t \lambda(N_s)ds)$, so N_t is a solution to the SDE (5.5) on $[0, T_1)$.

Assume that $t \in [T_n, T_{n+1})$ for $n \geq 1$. Then $N_t = N_0 + n$. We first prove that $\int_0^t \lambda(N_s)ds \in [\tau_n, \tau_{n+1})$. From the definition of T_n, we have

$$\begin{aligned}
\int_0^t \lambda(N_s)ds &= \sum_{k=0}^{n-1} \left(\int_{T_k}^{T_{k+1}} + \int_{T_n}^{t} \right) \lambda(N_s)ds = \sum_{k=0}^{n-1} \int_{T_k}^{T_{k+1}} \lambda(N_0 + k)ds + \int_{T_n}^{t} \lambda(N_0 + n)ds \\
&= \sum_{k=0}^{n-1} \lambda(N_0 + k)(T_{k+1} - T_k) + \lambda(N_0 + n)(t - T_n) \\
&= \sum_{k=0}^{n-1} \lambda(N_0 + k)\frac{\tau_{k+1} - \tau_k}{\lambda(N_0 + k)} + \lambda(N_0 + n)(t - T_n) \\
&= \tau_n + \lambda(N_0 + n)(t - T_n).
\end{aligned}$$

Since $t \in [T_n, T_{n+1})$, we have $\int_0^t \lambda(N_s)ds \geq \tau_n$ and

$$\begin{aligned}
\int_0^t \lambda(N_s)ds &= \tau_n + \lambda(N_0 + n)(t - T_n) < \tau_n + \lambda(N_0 + n)(T_{n+1} - T_n) = \tau_n + (\tau_{n+1} - \tau_n) \\
&= \tau_{n+1},
\end{aligned}$$

thus we conclude $Y(\int_0^t \lambda(N_s)ds) = n$. Therefore, it holds that

$$N_t = N_0 + n = N_0 + Y\left(\int_0^t \lambda(N_s)ds \right),$$

which implies N_t is a solution to the SDE (5.5). The proof of uniqueness is left for the reader.

5.5.2 Generator: Time Homogeneous Case

We will now deal with Itô's formula. In order to learn about generators and their interpretation as an alternative form of Itô formula we give the following proposition.

Proposition 5.5.9 *Let N be a homogeneous Poisson process with rate $\lambda > 0$. Then for $h : \mathbb{N} \to \mathbb{R}$*

$$h(N_t) - h(N_0) - \int_0^t Ah(N_s)ds$$

is a $\mathscr{F}_t := \sigma(N_s; s \le t)$ martingale if $\mathbb{E}[\sup_{s\le t} |h(N_s)|] \vee \mathbb{E}[|Ah(N_t)|] < \infty$ for all $t > 0$, where

$$Ah(n) := \lambda(h(n+1) - h(n)).$$

A is usually called the generator of the process N. Note that previously (see Theorem 3.5.3) by assuming that f is bounded we proved the above proposition. Therefore one of the goals of the above statement is to say that the martingale property can be established under quite general conditions.

Proof Let $M_t := N_t - \lambda t$. It follows from Itô's formula that

$$\begin{aligned} h(N_t) &= h(N_0) + \int_0^t (h(N_{s-}+1) - h(N_{s-}))\, dN_s \\ &= h(N_0) + \int_0^t (h(N_{s-}+1) - h(N_{s-}))\, (dM_s + \lambda ds) \\ &= h(N_0) + \int_0^t Ah(N_s)ds + \int_0^t (h(N_{s-}+1) - h(N_{s-}))\, dM_s. \end{aligned}$$

The condition $\mathbb{E}[|Ah(N_t)|] < \infty$ implies that $\int_0^t \mathbb{E}[|Ah(N_s)|]ds < \infty$ (this we leave as an exercise). Therefore the above stochastic integral is a uniformly integrable martingale. Using the martingale property of M_t, the stochastic integral term above is a martingale.

Exercise 5.5.10 • Prove that $Ah(n) = \lim_{\varepsilon\downarrow 0} \frac{\mathbb{E}[h(N_{t+\varepsilon})|N_t=n]-h(n)}{\varepsilon}$.
- Prove that $\mathbb{E}[|Ah(N_t)|] < \infty$ implies that $\int_0^t \mathbb{E}[|Ah(N_s)|]ds < \infty$.[27]
- Prove that in the statement of Proposition 5.5.9, the following weaker condition: $\mathbb{E}[|h(N_t)|] \vee \mathbb{E}[|Ah(N_t)|] < \infty$, for all $t > 0$, is enough to obtain the same conclusion.[28]

Although the above proposition may seem too technical for a beginner (in fact, one may start the discussion just supposing that h is uniformly bounded), this proposition

---

[27]Hint: Prove that $\int_0^t \mathbb{E}[|Ah(N_s)|]ds \le te^{\lambda t}\mathbb{E}[|Ah(N_t)|]$.

[28]Hint: Use the fact that $\int_u^t h(N_{s-}+1) - f(N_{s-})dN_s = \sum_{k=1}^{\infty}(h(k) - h(k-1)\, \mathbf{1}_{\{u<T_k\le t\}}$.

and its exercises help the reader understand that the issue of extending the domain of the generator as much as possible is important because this tells us to what extent the Itô formula is valid between many other important conclusions.

5.5.3 Generator: Non-homogeneous Case

Proposition 5.5.11 *Let N be a non-homogeneous Poisson process with a bounded rate function $\lambda : (0, \infty) \to (0, \infty)$. Then for a bounded function h we have that*

$$h(N_t) - h(N_0) - \int_0^t A_s h(N_s) ds$$

is a $\{\mathscr{F}_t := \sigma(N_s; s \leq t); t \geq 0\}$-martingale where

$$A_s h(n) := \lambda(s)(h(n+1) - h(n)).$$

Furthermore if λ is a continuous function then

$$\lim_{\varepsilon \to 0} \frac{\mathbb{E}[h(N_{t+\varepsilon}) - f(N_t)]}{\varepsilon} = \mathbb{E}[\lambda(t)(h(N_t + 1) - h(N_t))].$$

Proof Let $M_t := N_t - \int_0^t \lambda(s) ds$. It follows from Itô's formula (Proposition 3.4.24) that

$$\begin{aligned} h(N_t) &= h(N_0) + \int_0^t (h(N_{s-} + 1) - h(N_{s-}))\, dN_s \\ &= h(N_0) + \int_0^t (h(N_{s-} + 1) - h(N_{s-}))\, (dM_s + \lambda(s) ds) \\ &= h(N_0) + \int_0^t A_s h(N_s) ds + \int_0^t (h(N_{s-} + 1) - h(N_{s-}))\, dM_s. \end{aligned}$$

Using a martingale property of M_t, the term of the stochastic integral is a martingale.

5.5.4 Generator for the Solution to the SDE (5.5)

Let N be the solution to the SDE (5.5).

Exercise 5.5.12 Compute $\mathbb{P}(N_t = 0)$ and $\mathbb{P}(N_t = 1)$.

Exercise 5.5.13 Fix $s > 0$. Prove that $N_t - N_s, t > s$ solves the equation

$$Z_{s,t}(y) = Y(\int_s^t \lambda(Z_{s,u}(y) + y)du).$$

Prove that the above equation has a unique solution.

Define $\mathbb{E}_{s,n}[h(t, N_t)] := \mathbb{E}[f(t, N_t)|N_s = n]$. Then we define the generator of N by

$$Ah(t, n) := \lim_{\varepsilon \downarrow 0} \frac{\mathbb{E}_{t,n}[h(t + \varepsilon, N_{t+h}) - h(t, N_t)]}{\varepsilon}.$$

Then one can prove the following proposition:

Proposition 5.5.14 *The generator of the solution of Eq. (5.5) is given by*

$$Ah(n) = \lambda(n)(h(n + 1) - h(n))$$

for any compactly supported function h.

5.5.5 SDE Driven by Poisson Process

5.5.5.1 ODE and Homogeneous Poisson Process

We first consider an SDE driven by homogeneous Poisson process. The result and its proof is a generalization of Theorem 3.6.5 in the particular case that $b(x, z) \equiv b(x)$.

Proposition 5.5.15 *Assume that the ODE*

$$x_t = x_0 + \int_0^t \mu(x_s)ds$$

has a unique solution for any $x_0 \in \mathbb{R}$. Let N be a homogeneous Poisson process with the rate $\lambda > 0$. Then the SDE

$$X_t = x_0 + \int_0^t \mu(X_s)ds + \int_0^t \beta(X_{s-})dN_s \tag{5.6}$$

has a unique solution.

Note that the generator of the solution (5.6) is given by

$$Ah(x) = \mu(x)h'(x) + \lambda(h(x + \beta(x)) - h(x)).$$

5.5.5.2 ODE and Non-homogeneous Poisson Process

Now we consider an SDE driven by a non-homogeneous Poisson process.

Proposition 5.5.16 *Assume that the ODE*

$$x_t = x_0 + \int_0^t \mu(x_s)ds$$

has a unique solution for any $x_0 \in \mathbb{R}$. *Let* $Y = (Y_t)_{t\geq 0}$ *be a homogeneous Poisson process with rate* 1. *Then the SDE*

$$X_t = x_0 + \int_0^t \mu(X_s)ds + Y\left(\int_0^t \lambda(X_s)ds\right) \tag{5.7}$$

has a unique solution on $[0, \tau_\infty)$, *where* $\tau_\infty = \inf\{t > 0 | X_t = \infty\}$.

Proof Let $(T_n)_n$ be the jump times of Y. Now we fix $\omega \in \Omega$.

(i) We solve the ODE:

$$X_t = x_0 + \int_0^t \mu(X_s)ds$$

on $t \in [0, \tau_1)$, where $\tau_1 := \inf\{t > 0 | \int_0^t \lambda(X_s)ds \geq T_1\}$, and let $X_{\tau_1 -} := \lim_{t\uparrow T_1} X_t$.

(ii) $X_{\tau_1} := X_{\tau_1 -} + 1$.

(iii) We solve the ODE:

$$X_t = X_{\tau_1} + \int_{\tau_1}^t \mu(X_s)ds$$

on $t \in [\tau_1, \tau_2)$, where $\tau_2 := \inf\{t > 0 | \int_{\tau_1}^t \lambda(X_s)ds \geq T_2 - T_1\}$. Then

$$\begin{aligned}
X_t = X_{\tau_1} + \int_{\tau_1}^t \mu(X_s)ds &= X_{\tau_1 -} + 1 + \int_{\tau_1}^t \mu(X_s)ds \\
&= x_0 + \int_0^t \mu(X_s)ds + 1 \\
&= x_0 + \int_0^t \mu(X_s)ds + Y\left(\int_0^t \lambda(X_s)ds\right).
\end{aligned}$$

Repeating the above procedure, we can solve the SDE (5.7).

There are many interesting questions related to these types of process. In fact, the interesting situation is when the function λ takes large values and therefore a large number of jumps happen. For more on this, the reader may search for keywords such as self-excited models or Hawkes process (see also Exercise 5.5.18 below as well as Refs. [23, 28, 29] or [32] between others).

Exercise 5.5.17 (*A non-linear equation*)

Consider $\mathcal{N}^n$ to be a sequence of independent Poisson random measures with compensator given by $\mathbf{1}(n \le z < n+1)dzds$. We define a σ-finite random measure as

$$\mathcal{N}(A) = \sum_{n=1}^{\infty} \mathcal{N}^n(A).$$

Here $A \subset [0,\infty) \times \mathbb{R}$ is any set of finite Lebesgue measure. With these definitions we study existence and uniqueness for the following equation for $x > 0$:

$$X_t = x + t - \int_{\mathbb{R}\times[0,t]} X_{s-}\, \mathbf{1}(z \le a(X_{s-}, \mathbb{E}[X_{s-}]))\mathcal{N}(dz, ds).$$

Here $a : \mathbb{R}^2 \to \mathbb{R}_+$ is a bounded Lipschitz function. Prove that the above equation has a unique solution.

Here we give just a brief argument which should be completed by the reader. In order to prove existence or uniqueness one has to be able to compare two solutions. Suppose that X and Y are two solutions of the equation. Then consider

$$\begin{aligned} X_t - Y_t &= \int_{\mathbb{R}\times[0,t]} (X_{s-} - Y_{s-})\, \mathbf{1}(z \le a(X_{s-}, \mathbb{E}[X_{s-}]))\mathcal{N}(dz, ds) \\ &\quad + \int_{\mathbb{R}\times[0,t]} Y_{s-}\left(\mathbf{1}(z \le a(X_{s-}, \mathbb{E}[X_{s-}])) - \mathbf{1}(z \le a(Y_{s-}, \mathbb{E}[Y_{s-}]))\right)\mathcal{N}(dz, ds) \\ &:= A_1 + A_2. \end{aligned}$$

If we consider the expectation of the absolute value of the second term above, we have

$$\mathbb{E}|A_2| \le \int_{\mathbb{R}\times[0,t]} \mathbb{E}\left[Y_{s-}\, |a(X_{s-}, \mathbb{E}[X_{s-}]) - a(Y_{s-}, \mathbb{E}[Y_{s-}])|\right] ds.$$

The argument can be completed by arguments that you can see in other stochastic equations treated previously and the fact that $Y_t \in (0, x+t)$.

Exercise 5.5.18 Let $T = \{T_n\,;\, n \in \mathbb{N}\}$ be a point process on $\mathbb{R}_+$, and denote by $N = \{N_t\}_{t\ge 0}$ a counting process of the point process T given by

$$N_t = \sum_{n\ge 1} \mathbb{I}_{\{T_n \le t\}}. \tag{5.8}$$

I. Suppose that there exists a non-negative bounded function λ_t, $t > 0$ such that

$$\lambda_t = \lim_{\delta\to 0} \frac{1}{\delta}\, \mathbb{E}\left[N_{t+\delta} - N_t \middle| \mathcal{F}_t\right]$$

for each $t \geq 0$, the process $\lambda = \{\lambda_t\}_{t\geq 0}$ is called the conditional intensity of the process N, where $\{\mathscr{F}_t\}_{t\geq 0}$ is the filtration associated with the process N. Define the process $\Lambda = \{\Lambda_t\}_{t\geq 0}$ by

$$\Lambda_t = \int_0^t \lambda_s \, \mathrm{d}s,$$

which is called the compensator of the process N. Prove that the process $\tilde{N} = \{\tilde{N}_t\}_{t\geq 0}$ given by

$$\tilde{N}_t = N_t - \int_0^t \lambda_s \, \mathrm{d}s$$

defines a $\{\mathscr{F}_t\}_{t\geq 0}$-martingale.

II. Let $\lambda_0 > 0$ be a constant, and $h : [0, +\infty) \to [0, +\infty)$ a bounded and Borel measurable function. Then, the process N is called a (linear) Hawkes process with the conditional intensity λ, if the process λ is determined by

$$\lambda_t = \lambda_0 + \sum_{n\geq 1} h(t - T_n)\, \mathbb{I}_{\{T_n \leq t\}}. \tag{5.9}$$

The constant λ_0 is called the background intensity, while the function h is called the exciting function.

1. Use an argument similar to the ones used in Sect. 5.5.1 in order to prove the existence and uniqueness of solutions for (5.8) and (5.9) so that $T_{n+1} - T_n$ follows an exponential distribution.
2. Consider the case $h(u) = e^{-u}$. Then, prove that the process λ satisfies the equation

$$\lambda_t = \lambda_0 - \int_0^t (\lambda_s - \lambda_0)\, ds + N_t.$$

3. Suppose that the function h satisfies

$$\|h\|_1 := \int_0^{+\infty} h(s)\, \mathrm{d}s < 1. \tag{5.10}$$

   Prove that $\Psi(u) = \sum_{n\geq 1} h^{n*}(u)$ is well defined. Then, show that

$$\lambda_t = \lambda_0 + \int_0^t \Psi(t-s)\, \lambda_0\, ds + \int_0^t \Psi(t-s)\, d\tilde{N}_s. \tag{5.11}$$

4. Under the condition (5.10) on the function h, prove that

$$\mathbb{E}[\lambda_t] = \lambda_0 \left(1 + \int_0^t \Psi(s)\, ds\right),$$

$$\lim_{t\to\infty} \mathbb{E}[\lambda_t] = \frac{\lambda_0}{1 - \|h\|_1}.$$

For a hint, see Chap. 14.

5.6 Subordinated Brownian Motion

In this section, we briefly describe a method in order to obtain certain types of Lévy processes. This section requires the knowledge of Brownian motion which is not covered in this text. In fact, if you do not know what Brownian motion is, we recommend that you skip this section in its entirety.

Although this is not an extension of Lévy processes but an alternative definition of some Lévy processes we include it here because of the need to know Brownian motion.

Definition 5.6.1 Let Z be a Lévy increasing process with $Z_0 = 0$ and associated Lévy measure ν. Let $B = \{B_t; t \geq 0\}$ be a Brownian motion. The process $V_t := B_{Z_t}$ is called a Brownian motion subordinated to the Lévy process Z.

Once this definition is given, various properties can be obtained.

Proposition 5.6.2 *The characteristic function of $V_t = B_{Z_t}$ is given by*

$$\begin{aligned}\mathbb{E}[e^{i\theta V_t}] &= \mathbb{E}\left[\exp\left(-\frac{\theta^2}{2}Z_t\right)\right] \\ &= \exp\left(t\int(e^{-\frac{\theta^2 s}{2}} - 1)\nu(ds)\right).\end{aligned}$$

In particular, the law of V is symmetric and the Lévy measure associated with V is $\int_0^\infty \frac{e^{-\frac{x^2}{2s}}}{\sqrt{2\pi s}}\nu(ds)dx$.

Note that the Laplace transform of Z_t, $\mathbb{E}[e^{-\theta Z_t}]$, $\theta > 0$, is finite because it is an increasing process with $Z_0 = 0$ and therefore positive.

Proof The characteristic function is computed conditioning first with respect to Z in order to obtain the result. On the other hand we also have

$$\int_{\mathbb{R}\times[0,\infty)} (e^{i\theta x} - 1 - i\theta x\,\mathbf{1}_{|x|\leq 1})\frac{e^{-\frac{x^2}{2s}}}{\sqrt{2\pi s}}dx\nu(ds) = \int_0^\infty (e^{-\frac{\theta^2 s}{2}} - 1)\nu(ds).$$

Exercise 5.6.3 Compute the generator of V in the particular case of the function $h(x) = e^{i\theta x}$. That is, compute

$$Ah(x) = \lim_{\varepsilon\to 0}\frac{\mathbb{E}[h(V_{t+\varepsilon})/V_t = x] - h(x)}{\varepsilon}.$$

Hint: This exercise requires the knowledge of the strong Markov property of Brownian motion.

Exercise 5.6.4 1. The following is an integration problem: Compute the integral of $\int_0^\infty e^{-(au+\frac{b}{u})^2} du$ for $ab < 0$.

2. Use the above result to compute the following Laplace transform: $\int_0^\infty e^{-\theta s} \frac{e^{-\frac{x^2}{2s}}}{\sqrt{2\pi s^3}} ds$.
3. Use Fubini's theorem and the gamma function in order to prove that V as in Definition 5.6.1 follows a stable law in the special case that Z has the Lévy measure given as $\nu(ds) = cs^{-\alpha}$ for $\alpha \in (0, 1)$.
4. Finally, conclude that the characteristic function is given by $\mathbb{E}[e^{i\theta Z_t}] = e^{-Ct|\theta|^{2\alpha}}$, $C > 0$. From here conclude that the random variable Z_t has a symmetric law. That is, $\mathbb{P}(Z_t > x) = \mathbb{P}(Z_t < x)$ for all $x \in \mathbb{R}$. Compute explicitly the Lévy measure associated with Z.

For a hint, see Chap. 14.

As it is a matter of multiplying the measure ν by a constant, without loss of generality, we can always assume that a stable process Z is a Lévy process with characteristic function given by $\mathbb{E}[e^{i\theta Z_t}] = e^{-t|\theta|^\alpha}$.

Chapter 6
Multi-dimensional Lévy Processes and Their Densities

We briefly present in this chapter the definition and the regularity properties of the law of general Lévy processes in many dimensions. We could have taken the same approach as in previous chapters going slowly from Poisson processes to compound Poisson processes, finite variation and then infinite variation Lévy processes in many dimensions.

Instead, we prefer to take a more traditional approach that students will find in classical textbooks. This is because we will need them in the chapters that follow. If you would rather continue in a basic setting, then we recommend to continue with Chap. 9.

Therefore, this chapter is written in the spirit of an introduction for more advanced texts. We assume that the reader has some acquaintance with Brownian motion (also called Wiener process in some texts).[1] We also start driving the discussion towards the study of densities.

Let us start by recalling the extension of Definition 3.2.1 in the multi-dimensional case.

Definition 6.0.1 (*Lévy process*) A stochastic process $\{Z_t; t \geq 0\}$ on $\mathbb{R}^d$ is a Lévy process if the following conditions are satisfied;

(i) For any choice of $n \in \mathbb{N}$ and $0 \leq t_1 < \cdots < t_n$, the random vectors Z_{t_0}, $Z_{t_1} - Z_{t_0}, \cdots, Z_{t_n} - Z_{t_{n-1}}$ are independent.
(ii) $Z_0 = 0$ a.s.
(iii) The distribution of $Z_{t+h} - Z_t$ does not depend on t.
(iv) For any $a > 0$ and $t \geq 0$, $\lim_{h\to 0} P(|Z_{t+h} - Z_t| > a) = 0$.
(v) There exists $\Omega_0 \in \mathscr{F}$ with $\mathbb{P}(\Omega_0) = 1$ such that, for every $\omega \in \Omega_0$, $Z_t(\omega)$ is right-continuous in $t \geq 0$ and has left limits in $t > 0$, i.e. $t \to Z_t(\omega)$ is càdlàg function on $[0, \infty)$.

[1]The presentation in this chapter follows closely [51] where you can find the proofs not provided here.

© Springer Nature Singapore Pte Ltd. 2019

A. Kohatsu-Higa and A. Takeuchi, *Jump SDEs and the Study of Their Densities*, Universitext, https://doi.org/10.1007/978-981-32-9741-8_6

6.1 Infinitely Divisible Processes in $\mathbb{R}^d$

We start with a brief review of convolution of measures.

Proposition 6.1.1 *Suppose that X_1 and X_2 are independent random variables on $\mathbb{R}^d$ with distribution measures μ_1 and μ_2, respectively. Then:*

(i) $X_1 + X_2$ has as distribution measure $\mu_1 * \mu_2$.
(ii) If X_1 or X_2 has a density function then $X_1 + X_2$ also has a density function.

Proof Since for $B \in \mathscr{B}(\mathbb{R}^d)$, $\mathbf{1}_{\{x+y\in B\}}$ is a bounded measurable function, we have

$$\mathbb{E}[\mathbf{1}_{\{X_1+X_2\in B\}}|\sigma(X_2)] = \mathbb{E}[\mathbf{1}_{\{X_1+y\in B\}}]|_{y=X_2} = \int_{\mathbb{R}^d} \mathbf{1}_{\{x+y\in B\}}\mu_1(dx)|_{y=X_2}$$
$$= \int_{\mathbb{R}^d} \mathbf{1}_{\{x+X_2\in B\}}\mu_1(dx).$$

Hence

$$\mathbb{P}(X_1 + X_2 \in B) = \mathbb{E}[\mathbf{1}_{\{X_1+X_2\in B\}}] = \mathbb{E}[\mathbb{E}[\mathbf{1}_{\{X_1+X_2\in B\}}|\sigma(X_2)]]$$
$$= \mathbb{E}\Big[\int_{\mathbb{R}^d} \mathbf{1}_{\{x+X_2\in B\}}\mu_1(dx)\Big] = \int_{\mathbb{R}^d}\int_{\mathbb{R}^d} \mathbf{1}_B(x+y)\mu_1(dx)\mu_2(dy).$$

If X_1 has a density function f_{X_1},

$$\mathbb{P}(X_1 + X_2 \in B) = \int_{\mathbb{R}^d}\int_{\mathbb{R}^d} \mathbf{1}_B(x+y) f_{X_1}(x)dx\mu_2(dy)$$
$$= \int_{\mathbb{R}^d} \mathbf{1}_B(z)\Big\{\int_{\mathbb{R}^d} f_{X_1}(z-y)\mu_2(dy)\Big\}dz.$$

Hence $\int_{\mathbb{R}^d} f_{X_1}(z-y)\mu_2(dy)$ is a density function of $X_1 + X_2$.

Exercise 6.1.2 Prove that if X_1 and X_2 are as in Proposition 6.1.1 and μ_1 is continuous, then $X_1 + X_2$ has a distribution measure which is continuous.

We will denote by μ^{n*} the n-fold convolution of a probability measure μ with itself, that is,

$$\mu^{n*} = \mu * \cdots * \mu.$$

Example 6.1.3 This exercise reviews some of the basic properties of the convolution.

1. Given a sequence of i.i.d. random vectors X_i, $i \in \mathbb{N}$ with law μ, prove that the law of the sum $\sum_{i=1}^n X_i$ is the convolution μ^{n*}.
2. Prove that the convolution operation is linear and commutative.

Definition 6.1.4 (*Infinitely divisible*) A probability measure μ on $\mathbb{R}^d$ is infinitely divisible if for any $n \in \mathbb{N}$, there exists a probability μ_n on $\mathbb{R}^d$ such that $\mu = \mu_n^{n*}$.

Example 6.1.5 Note that the above definition may be taken as the reverse of the procedure described in Exercise 6.1.3.1. In fact, if one thinks of the central limit theorem we can state the following simple exercise. Prove that the probability measure associated with a d-dimensional normal random variable with mean $x_0 \in \mathbb{R}^d$ and covariance matrix $\Sigma \in \mathbb{R}^{d\times d}$ is an infinite divisible distribution. This exercise is related to the study of the possible laws for the limit of sums of i.i.d. random variables.

In general, we have the following result.

Theorem 6.1.6 *Let $\{Z_t\}_{t\geq 0}$ be a Lévy process on $\mathbb{R}^d$. Then for every $t \geq 0$, the distribution of Z_t is infinitely distribution.*

Proof Let $t_k := kt/n$ and let μ be the distribution of Z_t and μ_n be the distribution of $Z_{t_k} - Z_{t_{k-1}}$. Since

$$Z_t = (Z_{t_1} - Z_{t_0}) + \cdots + (Z_{t_n} - Z_{t_{n-1}}),$$

from Proposition 6.1.1, we obtain $\mu = \mu_n^{n*}$.

Lemma 6.1.7 *If μ_1 and μ_2 are infinitely divisible then $\mu_1 * \mu_2$ is infinitely divisible.*

Proof For each $n \in \mathbb{N}$, there exist measures $\mu_{1,n}$ and $\mu_{2,n}$ such that $\mu_1 = \mu_{1,n}^{n*}$ and $\mu_2 = \mu_{2,n}^{n*}$. Hence $\mu_1 * \mu_2 = (\mu_{1,n} * \mu_{2,n})^{n*}$.

Definition 6.1.8 The characteristic function $\hat{\mu}$ of a probability measure μ on $\mathbb{R}^d$ is defined as

$$\hat{\mu}(\theta) = \int_{\mathbb{R}^d} e^{i\langle\theta,x\rangle}\mu(dx), \quad \theta \in \mathbb{R}^d.$$

Theorem 6.1.9 (Lévy Khintchine representation)

(i) If μ is an infinitely divisible distribution on $\mathbb{R}^d$, then

$$\hat{\mu}(\theta) = \exp\Big[-\frac{1}{2}\langle\theta, A\theta\rangle + i\langle\gamma,\theta\rangle + \int_{\mathbb{R}^d}\{e^{i\langle\theta,x\rangle} - 1 - i\langle\theta,x\rangle \mathbf{1}_{\{|x|\leq 1\}}(x)\}\nu(dx)\Big], \tag{6.1}$$

where A is a symmetric non-negative $d\times d$- matrix, ν is a measure on $\mathbb{R}^d$ satisfying

$$\nu\{0\} = 0 \quad and \quad \int_{\mathbb{R}^d}(|x|^2 \wedge 1)\nu(dx) < \infty, \tag{6.2}$$

and $\gamma \in \mathbb{R}^d$.

(ii) The triplet (A, ν, γ) in the representation of $\hat{\mu}$ in (i) is unique.

(iii) Conversely, if A is a symmetric non-negative $d \times d$-matrix, ν is a measure satisfying (6.2) and $\gamma \in \mathbb{R}^d$, then there exists an infinitely divisible distribution μ whose characteristic function is given by (6.1).

Definition 6.1.10 We call (A, ν, γ) in Theorem 6.1.9 the generating triplet of μ. A and the ν are called, respectively, the Gaussian covariance matrix and the Lévy measure associated with μ. When $A = 0$, μ is called purely non-Gaussian.

In fact, note that if $\nu \equiv 0$ then μ corresponds to the characteristic function of a Gaussian random vector with mean γ and covariance matrix A. On the other hand, if $A = 0$ and $\gamma = 0$ we can see the relation with the characteristic function of previously defined Lévy processes in Theorem 3.1.2, Corollary 4.1.13 and Theorem 4.1.7.

Theorem 6.1.11 (Lévy Khintchine Formula for a Lévy Process) *Let $\{Z_t\}_{t\geq 0}$ be a Lévy process. Then P_{Z_1} is an infinitely divisible distribution on $\mathbb{R}^d$. Furthermore, there exists a generating triplet (A, ν, γ) such that*

$$\int_{\mathbb{R}^d} e^{i\langle\theta,x\rangle} P_{Z_1}(dx)$$
$$= \exp\Big[-\frac{1}{2}\langle\theta, A\theta\rangle + i\langle\gamma,\theta\rangle + \int_{\mathbb{R}^d}\{e^{i\langle\theta,x\rangle} - 1 - i\langle\theta, x\rangle\mathbf{1}_{\{|x|\leq 1\}}(x)\}\nu(dx)\Big].$$

The triplet (A, ν, γ) of P_{Z_1} is called the generating triplet of Lévy process $\{Z_t\}_{t\geq 0}$ or Lévy triplet.

Definition 6.1.12 (*Characteristic exponent*) Let X be a random variable on $\mathbb{R}^d$. The function $\Psi(\theta, X) := -\log \mathbb{E}[e^{i\langle\theta,X\rangle}]$ is called the characteristic exponent of X.

Note that from Theorem 6.1.11, the characteristic exponent of Z_1 is represented by

$$\Psi(\theta, Z_1) = \frac{1}{2}\langle\theta, A\theta\rangle - i\langle\gamma,\theta\rangle - \int_{\mathbb{R}^d}\{e^{i\langle\theta,x\rangle} - 1 - i\langle\theta, x\rangle\mathbf{1}_{\{|x|\leq 1\}}(x)\}\nu(dx).$$

Example 6.1.13 The following exercise serves to show that there is no uniqueness in the above representation in the sense that one may change the indicator function $\mathbf{1}_{\{|x|\leq 1\}}(x)$ by other equivalent functions changing the value of γ. This remark is important when you read other texts as the generating triplet may change definition according to the localization function chosen for the representation.

Let $h : \mathbb{R}^d \to \mathbb{R}$ such that $\int |x||h(x) - \mathbf{1}_{\{|x|\leq 1\}}(x)|\nu(dx) < \infty$. Then there exists $\gamma' \in \mathbb{R}^d$ such that

$$\Psi(\theta, Z_1) = \frac{1}{2}\langle\theta, A\theta\rangle - i\langle\gamma',\theta\rangle - \int_{\mathbb{R}^d}\{e^{i\langle\theta,x\rangle} - 1 - i\langle\theta, x\rangle h(x)\}\nu(dx).$$

Similarly the compensation used in Chap. 5 in order to obtain the convergence of the processes $Z^{\pm}$ could be changed using the localization function h instead of $\mathbf{1}_{\{|x|\leq 1\}}(x)$ with similar results. Give the definitions of the processes $\bar{Z}_t^{(\varepsilon', \varepsilon, +)}$ that would correspond to this change of localization function.

Theorem 6.1.14 *Let $\{Z_t\}_{t\geq 0}$ be a Lévy process on $\mathbb{R}^d$. Define $\Psi_t(\theta) = \Psi(\theta, Z_t)$. Then*

$$t\Psi_1(u) = \Psi_t(u).$$

Proof Fix $t > 0$ and $t_j = \frac{jt}{n}$. From Proposition 6.1.1, we have

$$\mathbb{E}[e^{i\langle\theta, Z_t\rangle}] = \mathbb{E}[e^{i\langle\theta, \sum_{j=1}^{n}(Z_{t_j} - Z_{t_{j-1}})\rangle}] = \prod_{j=1}^{n}\mathbb{E}[e^{i\langle\theta, Z_{t_j}\rangle}] = \left\{\mathbb{E}[e^{i\langle\theta, Z_{\frac{t}{n}}\rangle}]\right\}^n.$$

So $\Psi_t(\theta) = -n\log\mathbb{E}[e^{i\langle\theta, Z_{\frac{t}{n}}\rangle}] = n\Psi_{\frac{t}{n}}(\theta)$ for any $t > 0,\ n \in \mathbb{N}$. Since for any $n, m \in \mathbb{N}$,

$$m\Psi_1(\theta) = \Psi_m(\theta) = n\Psi_{\frac{m}{n}}(\theta),$$

we have $\frac{m}{n}\Psi_1(\theta) = \Psi_{\frac{m}{n}}(\theta)$. Therefore for any $q \in \mathbb{Q}$, $q\Psi_1(\theta) = \Psi_q(\theta)$. Let q_n be a sequence of rational numbers such that $q_n \downarrow t \in \mathbb{R}$. Since the paths of Z_t are a.s. right continuous, we have by the dominated convergence theorem

$$\begin{aligned} t\Psi_1(\theta) &= \lim_{n\to\infty} q_n\Psi_1(\theta) = \lim_{n\to\infty}\left\{-\log\mathbb{E}[e^{i\langle\theta, Z_{q_n}\rangle}]\right\} \\ &= -\log\mathbb{E}[\lim_{n\to\infty} e^{i\langle\theta, Z_{q_n}\rangle}]\} = \Psi_t(\theta). \end{aligned}$$

Corollary 6.1.15 *The characteristic function of a Lévy process at time t is given by*

$$\mathbb{E}[e^{i\langle\theta, Z_t\rangle}] = \exp\left[t\left\{-\frac{1}{2}\langle\theta, A\theta\rangle + i\langle\gamma, \theta\rangle + \int_{\mathbb{R}^d}\{e^{i\langle\theta, x\rangle} - 1 - i\langle\theta, x\rangle\mathbf{1}_{\{|x|\leq 1\}}(x)\}\nu(dx)\right\}\right],$$

and it satisfies that $\mathbb{E}[e^{i\langle\theta, Z_1\rangle}]^t = \mathbb{E}[e^{i\langle\theta, Z_t\rangle}]$.

Exercise 6.1.16 In this exercise, we consider various cases of multi-dimensional Lévy processes.

1. Prove that if Z^i, $i = 1, ..., d$ are d independent one-dimensional Lévy processes with Lévy measure ν_i then the vector $(Z^1, ..., Z^d)$ is a d-dimensional Lévy process with Lévy measure given by the product Lévy measure. In particular, prove that this product measure satisfies the conditions stated in (6.2).
2. Prove that the following measure is a Lévy measure. That is, it can be used to construct a multi-dimensional Lévy process.

$$\nu(dr, d\phi) = \frac{1}{r^{1+\alpha}} dr d\phi_1 ... d\phi_{d-1},$$
$$r > 0, \phi_i \in [0, \pi), i = 1, ..., d-2, \phi_{d-1} \in [0, 2\pi), \alpha \in [0, 2).$$

Here the above measure is expressed in spherical coordinates. This example is the symmetric generalization of stable process whose coordinates are not independent. Therefore it is different from considering each Z^i to be a stable process in 1. above.[2]

Definition 6.1.17 Let $\{Z_t\}_{t\geq 0}$ be a Lévy process on $\mathbb{R}^d$ with generating triplet (A, ν, γ). It is said to be of:

(i) type A if $A = 0$, and $\nu(\mathbb{R}^d) < \infty$,
(ii) type B if $A = 0$, $\nu(\mathbb{R}^d) = \infty$, and $\int_{|x|\leq 1} |x|\nu(dx) < \infty$,
(iii) type C if $A \neq 0$ or $\int_{|x|\leq 1} |x|\nu(dx) = \infty$.

Given the study that we have done in previous chapters it should be clear that types A and B correspond to a Lévy process which has no Brownian component. Type A is a compound Poisson process with $\lambda = \nu(\mathbb{R}^d)$, type B has paths of finite variation and type C has paths of infinite variation.

Exercise 6.1.18 Check which of the conditions stated in the above definition are satisfied for each of the examples considered in Examples 3.3.10, 4.1.20, 5.1.10, 5.1.11 and 5.1.19.

Exercise 6.1.19 Prove that there is equivalence between:

- $\{Z_t\}_{t\geq 0}$ is a Lévy process of type B or C.
- $A \neq 0$ or $\nu(\mathbb{R}^d) = \infty$.

6.2 Classification of Probability Measures

When studying how to characterize the regularity of random variables generated by a stochastic process one has to first understand the different types of possibilities that the distribution of the law of the random variable may have. In this section we give a brief description of the various possibilities and what happens when one combines them.

Definition 6.2.1 Let ρ be a non-trivial measure[3] on $\mathscr{B}(\mathbb{R}^d)$.

(i) ρ is called discrete if there is a countable set C such that $\rho(\mathbb{R}^d \setminus C) = 0$,
(ii) ρ is called continuous if $\rho(\{x\}) = 0$ for every $x \in \mathbb{R}^d$,

[2]For more on this matter in the case $d = 2$, see Sect. 6.4.

[3]That is, non-zero measures.

(iii) ρ is called singular if there is $B \in \mathscr{B}(\mathbb{R}^d)$ such that $\rho(\mathbb{R}^d \setminus B) = 0$ and $Leb(B) = 0$,
(iv) ρ is called absolutely continuous if $\rho(B) = 0$ for every $B \in \mathscr{B}(\mathbb{R}^d)$ satisfying $Leb(B) = 0$,
(v) ρ is called pure if it is either discrete, absolutely continuous, or continuous singular.

Exercise 6.2.2 Let ρ be the law of a Gaussian vector with mean zero and covariance matrix $A \neq 0$ which is non-negative definite. Prove that ρ is a continuous measure. Furthermore give an example where ρ is not absolutely continuous.

For a hint, see Chap. 14.

Lemma 6.2.3 (Lebesgue decomposition) *If ρ is a σ-finite measure, then there are measures $\rho_d, \rho_{ac}, \rho_{cs}$ such that $\rho = \rho_d + \rho_{ac} + \rho_{cs}$, ρ_d is discrete, ρ_{ac} is absolutely continuous and ρ_{cs} is continuous singular. The measures $\rho_d, \rho_{ac}, \rho_{cs}$ are uniquely determined by ρ.*

For a proof see e.g., [10].

Lemma 6.2.4 *If ρ is discrete, then ρ is not continuous.*

Proof Let $C = \bigcup_{n \in \mathbb{N}} \{x_n\}$ be countable set with $\rho(\mathbb{R}^d \setminus C) = 0$. Since

$$\rho(\mathbb{R}^d) = \rho(\mathbb{R}^d \setminus C) + \rho(C) = \rho(C) = \sum_{n \in \mathbb{N}} \rho(\{x_n\}) > 0,$$

then there exists some $n \in \mathbb{N}$ such that $\rho(\{x_n\}) \neq 0$.

Lemma 6.2.5 *Let ρ_1 and ρ_2 be non-zero finite measures on $\mathbb{R}^d$, define $\rho = \rho_1 * \rho_2$. Then:*

(i) ρ is continuous if and only if ρ_1 or ρ_2 is continuous,
(ii) ρ is discrete if and only if ρ_1 and ρ_2 are discrete,
(iii) ρ is absolutely continuous if ρ_1 or ρ_2 is absolutely continuous,
(iv) ρ_1 or ρ_2 is continuous singular if ρ is continuous singular.

Proof (i), (ii). If ρ_1 is continuous, for every $x \in \mathbb{R}^d$,

$$\rho(\{x\}) = \int_{\mathbb{R}^d} \int_{\mathbb{R}^d} \mathbf{1}_{\{x\}}(z + y) \rho_1(dz) \rho_2(dy) = \int_{\mathbb{R}^d} \rho_1(\{x - y\}) \rho_2(dy) = 0.$$

Hence, ρ is continuous. From Lemma 6.2.4, if ρ is discrete, ρ_1 and ρ_2 are discrete. If ρ_1 and ρ_2 are discrete, there are countable sets C_1 and C_2 such that

$$\rho_1(\mathbb{R}^d \setminus C_1) = \rho_2(\mathbb{R}^d \setminus C_2) = 0.$$

Let $C := C_1 + C_2$.[4] Then since

$$\begin{aligned}
&\rho(\mathbb{R}^d \setminus C) \\
&= \int_{\mathbb{R}^d}\int_{\mathbb{R}^d} \mathbf{1}_{\mathbb{R}^d\setminus C}(x+y)\rho_1(dx)\rho_2(dy) \\
&= \int_{\mathbb{R}^d}\int_{\mathbb{R}^d} \left\{\mathbf{1}_{C_1}(x)\mathbf{1}_{\mathbb{R}^d\setminus C_2}(y) + \mathbf{1}_{\mathbb{R}^d\setminus C_1}(x)\mathbf{1}_{C_2}(y) + \mathbf{1}_{\mathbb{R}^d\setminus C_1}(x)\mathbf{1}_{\mathbb{R}^d\setminus C_2}(y)\right\}\rho_1(dx)\rho_2(dy) \\
&= \rho_1(C_1)\rho_2(\mathbb{R}^d \setminus C_2) + \rho_1(\mathbb{R}^d \setminus C_1)\rho_2(C_2) + \rho_1(\mathbb{R}^d \setminus C_1)\rho_2(\mathbb{R}^d \setminus C_2) = 0,
\end{aligned}$$

ρ is discrete. From Lemma 6.2.4 and (i), if ρ is continuous, ρ_1 or ρ_2 is continuous.
(iii) Suppose that ρ_1 is absolutely continuous. If $B \in \mathscr{B}(\mathbb{R}^d)$ satisfies $Leb(B) = 0$, then $Leb(B - y) = 0$, for every $y \in \mathbb{R}^d$, and

$$\rho(B) = \int_{\mathbb{R}^d}\int_{\mathbb{R}^d} \mathbf{1}_B(x+y)\rho_1(dx)\rho_2(dy) = \int_{\mathbb{R}^d} \rho_1(B-y)\rho_2(dy) = 0.$$

(iv) Suppose that neither ρ_1 nor ρ_2 is continuous singular. Then

$$(\rho_1)_d + (\rho_1)_{ac} \neq 0, \ \ (\rho_2)_d + (\rho_2)_{ac} \neq 0.$$

It follows from (ii) and (iii) that $(\rho_1)_d * (\rho_2)_d$ is discrete and $(\rho_1)_d * (\rho_2)_{ac}$, $(\rho_1)_{ac} * (\rho_2)_d$ and $(\rho_1)_{ac} * (\rho_2)_{ac}$ are absolutely continuous. Hence we have

$$\begin{aligned}
\rho_1 * \rho_2 &= \Big((\rho_1)_d + (\rho_1)_{ac}\Big) * \Big((\rho_2)_d + (\rho_2)_{ac}\Big) \\
&= (\rho_1)_d * (\rho_2)_d + (\rho_1)_d * (\rho_2)_{ac} + (\rho_1)_{ac} * (\rho_2)_d + (\rho_1)_{ac} * (\rho_2)_{ac}
\end{aligned}$$

has a discrete or absolutely continuous part.

6.3 Densities for Lévy Processes

We will turn to the study of the densities of Lévy process. The main tool is either direct calculation or deriving properties through the characteristic function.

We start first with the case of the compound Poisson process for which its law can be computed explicitly.

Proposition 6.3.1 *If $\{Z_t = \sum_{j=1}^{N_t} Y_j\}$ is a compound Poisson process with intensity λ on $\mathbb{R}^d$ where $P_{Y_j} = \nu/\lambda$, then*

$$P_{Z_t} = e^{-t\lambda} \sum_{k=0}^{\infty} \frac{t^k}{k!} \nu^{k*},$$

[4] That is, $C = \{x + y; x \in C_1, y \in C_2\}$.

where $\nu^{0*} = \delta_0$ *and* P_{Z_t} *is not continuous at* 0.

Proof For any $B \in \mathscr{B}(\mathbb{R}^d)$,

$$\begin{aligned}
P_{Z_t}(B) &= \mathbb{P}\Big(\sum_{j=1}^{N_t} Y_j \in B\Big) = \sum_{k=0}^{\infty} \mathbb{P}\Big(\sum_{j=1}^{k} Y_j \in B, N_t = k\Big) \\
&= \sum_{k=0}^{\infty} \mathbb{P}\Big(\sum_{j=1}^{k} Y_j \in B\Big)\mathbb{P}\Big(N_t = k\Big) = \sum_{k=0}^{\infty} \Big(\frac{\nu}{\lambda}\Big)^{k*}(B)\frac{(\lambda t)^k}{k!}e^{-\lambda t} = e^{-t\lambda}\sum_{k=0}^{\infty}\frac{t^k}{k!}\nu^{k*}(B).
\end{aligned}$$

Since $P_{Z_t}\{0\} = e^{-t\lambda}\sum_{k=0}^{\infty} t^k/k!\, \nu^{k*}\{0\} \geq e^{-t\lambda} > 0$, P_{Z_t} is not continuous at 0.

Exercise 6.3.2 In the set-up of Proposition 6.3.1. Suppose that $\nu(dx) = f(x)dx$, where f is an integrable positive function. Prove that for any measurable set A such that $0 \notin A$ and any continuous bounded function g we have that

$$\mathbb{E}[g(Z_t)\,\mathbf{1}_A(Z_t)] = \int_A g(y)p_t(y)dy.$$

Here p_t can be written explicitly. Also deduce that Z_t has a density at any point except zero.[5] After solving all these problems, one can also see that the law of Z_t is absolutely continuous with respect to the Lebesgue measure for sets away from zero.[6]

Let μ be a probability measure on $\mathbb{R}^d$ and $D(\mu) := \sup_{x\in\mathbb{R}^d} \mu(\{x\})$. If μ is the distribution of random variable X, we write $D(X) := D(\mu)$.

Lemma 6.3.3 *If* $\mu = \mu_1 * \mu_2$*, then* $D(\mu) \leq \min\{D(\mu_1), D(\mu_2)\}$.

Proof For any $x \in \mathbb{R}^d$,

$$\mu(\{x\}) = \int_{\mathbb{R}^d}\int_{\mathbb{R}^d} \mathbf{1}_{\{z+y=x\}}\mu_1(dz)\mu_2(dy) = \int_{\mathbb{R}^d} \mu_1(\{x-y\})\mu_2(dy) \leq D(\mu_1).$$

Theorem 6.3.4 (Continuity) *Let* $\{Z_t\}_{t\geq 0}$ *be a Lévy process on* $\mathbb{R}^d$ *with generating triplet* (A, ν, γ)*. Then the following three statements are equivalent:*

(i) P_{Z_t} *is continuous for every* $t > 0$.
(ii) P_{Z_t} *is continuous for some* $t > 0$.
(iii) $\{Z_t\}_{t\geq 0}$ *is of type B or C (that is,* $A \neq 0$ *or* $\nu(\mathbb{R}^d) = \infty$*).*

---

[5]That is, deduce that the distribution function of Z_t can be differentiated at any point except zero.

[6]After this exercise, one may think that the differentiability of the distribution function and absolute continuity are equivalent. One has to be careful about these statements as they involve sets of Lebesgue measure zero.

Proof The statement (ii) implies (iii). If $A = 0$ and $\nu(\mathbb{R}^d) < \infty$, there exists $\gamma_0 \in \mathbb{R}^d$ such that $Z_t - t\gamma_0$ is a compound Poisson process. From Lemma 6.3.1 we have

$$P_{Z_t}(\{t\gamma_0\}) = \mathbb{P}(Z_t - t\gamma_0 = 0) = P_{Z_t - t\gamma_0}(\{0\}) > 0,$$

hence P_{X_t} is not continuous at $t\gamma_0$ for every $t \geq 0$.

The statement (iii) implies (i). If $A \neq 0$, then there exist independent infinite divisible random variables W_t and J_t with characteristic functions corresponding to the generating triplets $(A, 0, 0)$ and $(0, \nu, \gamma)$, respectively. Then

$$\begin{aligned}
&\mathbb{E}[e^{i\langle\theta, W_t\rangle}]\mathbb{E}[e^{i\langle\theta, J_t\rangle}] \\
&= \exp\Big[-\frac{t}{2}\langle\theta, A\theta\rangle\Big]\exp\Big[t\Big\{i\langle\gamma, \theta\rangle + \int_{\mathbb{R}^d}\{e^{i\langle\theta,x\rangle} - 1 - i\langle\theta, x\rangle\mathbf{1}_{\{|x|\leq 1\}}(x)\}\nu(dx)\Big\}\Big] \\
&= \exp\Big[t\Big\{-\frac{1}{2}\langle\theta, A\theta\rangle + i\langle\gamma, \theta\rangle + \int_{\mathbb{R}^d}\{e^{i\langle\theta,x\rangle} - 1 - i\langle\theta, x\rangle\mathbf{1}_{\{|x|\leq 1\}}(x)\}\nu(dx)\Big\}\Big] \\
&= \mathbb{E}[e^{i\langle\theta, Z_t\rangle}].
\end{aligned}$$

Therefore $P_{Z_t} = P_{W_t + J_t}$. Note that W_t is the characteristic function of a Gaussian random vector with mean zero and covariance matrix A. From Exercises 6.2.2 and 6.1.2, we have that as μ_1 is continuous we have that P_{Z_t} is continuous for every $t > 0$. In the following, suppose that $\nu(\mathbb{R}^d) = \infty$.

Case 1. Assume that ν is discrete. Then there exists a countable set $C = \bigcup_{j=1}^{\infty}\{x_j\} \in \mathbb{R}^d$ such that $\infty = \nu(\mathbb{R}^d) = \nu(C) = \sum_{j=1}^{\infty}\nu(\{x_j\})$. Let $m_j = \nu(\{x_j\})$ and $m'_j = m_j \wedge 1$. Then we have $\sum_{j=1}^{\infty} m'_j = \infty$. Let $\nu_n := \sum_{j=1}^{n} m'_j\delta_{x_j}$. Since

$$\int_{\mathbb{R}^d}(|x|^2 \wedge 1)\nu_n(dx) = \sum_{j=1}^{n}(|x_j|^2 \wedge 1)m'_j \leq n < \infty,$$

then ν_n is a Lévy measure. Let $\gamma'_j := \int_{\mathbb{R}^d} x_j\mathbf{1}_{\{|x|\leq 1\}}(x)\nu_n(dx)$ and $\gamma' := (\gamma'_1, \cdots, \gamma'_d)$. One may construct two independent Lévy processes $\{Y_t^{(n)}\}_{t\geq 0}$ and $\{J_t^{(n)}\}_{t\geq 0}$ with generating triplet $(0, \nu_n, \gamma')$ and $(A, \nu - \nu_n, \gamma - \gamma')$, respectively. Then $\{Y_t^{(n)}\}_{t\geq 0}$ is a compound Poisson process and

$$\begin{aligned}
&\mathbb{E}[e^{i\langle\theta, Y_t^{(n)}\rangle}]\mathbb{E}[e^{i\langle\theta, J_t^{(n)}\rangle}] \\
&= \exp\Big[t\int_{\mathbb{R}^d}(e^{i\langle\theta,x\rangle} - 1)\nu_n(dx)\Big]\exp\Big[t\Big\{-\frac{1}{2}\langle\theta, A\theta\rangle + i\langle\gamma - \gamma', \theta\rangle \\
&\qquad + \int_{\mathbb{R}^d}\{e^{i\langle\theta,x\rangle} - 1 - i\langle\theta, x\rangle\mathbf{1}_{\{|x|\leq 1\}}(x)\}(\nu - \nu_n)(dx)\Big\}\Big] \\
&= \exp\Big[t\Big\{-\frac{1}{2}\langle\theta, A\theta\rangle + i\langle\gamma, \theta\rangle + \int_{\mathbb{R}^d}\{e^{i\langle\theta,x\rangle} - 1 - i\langle\theta, x\rangle\mathbf{1}_{\{|x|\leq 1\}}(x)\}\nu(dx)\Big\}\Big] \\
&= \mathbb{E}[e^{i\langle\theta, Z_t\rangle}].
\end{aligned}$$

Hence $P_{Z_t} = P_{Y_t^{(n)}} * P_{J_t^{(n)}}$ and from Lemma 6.3.3

$$D(Z_t) \le D(Y_t^{(n)}).$$

Let $c_n := \nu_n(\mathbb{R}^d)$ and $\sigma_n := \nu_n/c_n$, then

$$D(\sigma_n^{k*}) \le D(\sigma_n) = \sup_{x\in\mathbb{R}^d} \frac{\nu_n(\{x\})}{c_n} = \frac{1}{c_n} \sup_{x\in\mathbb{R}^d} \sum_{j=1}^{n} m'_j \delta_{x_j}(\{x\}) \le \frac{1}{c_n},$$

$$\begin{aligned} \mathbb{P}(Y_t^{(n)} = x) = P_{Y_t^{(n)}}(\{x\}) &= e^{-tc_n}\,\mathbf{1}\{x=0\} + e^{-tc_n} \sum_{k=1}^{\infty} \frac{(tc_n)^k}{k!} \sigma_n^{k*}(\{x\}) \\ &\le e^{-tc_n} + \frac{1}{c_n} e^{-tc_n} e^{tc_n} = e^{-tc_n} + \frac{1}{c_n}. \end{aligned}$$

Since

$$\lim_{n\to\infty} c_n = \lim_{n\to\infty} \nu_n(\mathbb{R}^d) = \lim_{n\to\infty} \sum_{j=1}^{n} m'_j \delta_{x_j}(\mathbb{R}^d) = \sum_{j=1}^{\infty} m'_j = \infty,$$

we have

$$D(Z_t) \le D(Y_t^{(n)}) \le e^{-tc_n} + \frac{1}{c_n} \to 0 \ as\ n \to \infty.$$

Therefore $D(Z_t) = 0$. That is, Z_t has a continuous distribution.

Case 2. Suppose that ν is continuous. Let $\{Y_t^{(n)}\}_{t\ge0}$ be the compound Poisson process with Lévy measure $\nu_n(\cdot) := \nu(\cdot \cap \{|x| > 1/n\})$. Again by Lemma 6.3.3, we have $D(Z_t) \le D(Y_t^{(n)})$. For $k \ge 1$, ν_n^{k*} is continuous by Lemma 6.2.5 (i). Since $\nu_n(\mathbb{R}^d) = \nu(|x| > 1/n) \to \nu(\mathbb{R}^d) = \infty$,

$$D(Z_t) \le D(Y_t^{(n)}) = \sup_{x\in\mathbb{R}^d} P_{Y_t^{(n)}}(\{x\}) = \sup_{x\in\mathbb{R}^d} e^{-t\nu_n(\mathbb{R}^d)} \sum_{k=0}^{\infty} \frac{t^k}{k!} \nu_n^{k*}(\{x\}) = 0.$$

Therefore $D(Z_t) = 0$, that is, Z_t has a continuous distribution.

Remaining case. Let ν_d and ν_c be the discrete and the continuous part of ν, respectively, then ν_d or ν_c has infinite total measure. If $\nu_d(\mathbb{R}^d) = \infty$, then let $\{Y_t\}_{t\ge0}$ be the Lévy process with generating triplet $(0, \nu_d, 0)$. By Case 1, Y_t has a continuous distribution for any $t > 0$. Hence the distribution of Z_t is continuous by Lemma 6.2.5 (i). If $\nu_c(\mathbb{R}^d) = \infty$, then, similarly use Case 2.

Corollary 6.3.5 *Let $\{Z_t\}_{t\ge0}$ be a Lévy process on $\mathbb{R}^d$ with generating triplet (A, ν, γ). Then the following three statements are equivalent:*

(i) P_{Z_t} *is discrete for every* $t > 0$.
(ii) P_{Z_t} *is discrete for some* $t > 0$.
(iii) $\{Z_t\}_{t\geq 0}$ *is of type A and* ν *is discrete.*

In this section, we have learned that the study of the properties of the law of a random variable can be very detailed and it has to be divided into many cases. In the second part of the book, we will mostly be interested in only absolute continuity and regularity of the law.

6.4 Stable Laws in Dimension Two

In the previous chapters we have used as a typical example the case of the Lévy measure given by $\nu(dx) = \frac{C}{|x|^{1+\alpha}}$. This corresponds to the so-called stable process case. In this section, we will discuss the regularity of the associated density in the two-dimensional case for a subclass of stable laws.

We define the following Lévy measure in $\mathbb{R} - \{0\}$ in polar coordinates:

$$\nu(dr, d\beta) = \frac{dr}{r^{1+\alpha}} d\beta, \ r > 0, \beta \in [0, 2\pi).$$

Lemma 6.4.1 *Let* $\alpha \in (0, 2)$. *There exists a positive constant C such that*

$$\int (e^{i\langle\theta,x\rangle} - 1 - i\langle\theta, x\rangle \mathbf{1}_{|x|\leq 1}(x))\nu(dx) = -C|\theta|^{\alpha}.$$

Proof Note that as in the one-dimensional case:

$$e^{i\langle\theta,x\rangle} - 1 = -2\sin^2\left(\frac{\langle\theta, x\rangle}{2}\right) + i\sin(\langle\theta, x\rangle).$$

Therefore due to the symmetry of the Lévy measure we have

$$\begin{aligned}\int (e^{i\langle\theta,x\rangle} - 1 - i\langle\theta, x\rangle \mathbf{1}_{|x|\leq 1}(x))\nu(dx) &= -2\int \sin^2(\frac{\langle\theta, x\rangle}{2})\nu(dx)\\ &= -2\int_0^{2\pi}\int_0^{\infty} \sin^2\left(\frac{|\theta|}{2} r\cos(\beta + \eta)\right)\frac{dr}{r^{1+\alpha}} d\beta.\end{aligned}$$

Here $\eta = \arg(\theta)$. Then letting $|\theta|r = u$ we obtain that there exists a positive constant C such that

$$\int (e^{i\langle\theta,x\rangle} - 1 - i\langle\theta, x\rangle \mathbf{1}_{|x|\leq 1}(x))\nu(dx) = -C|\theta|^{\alpha}.$$

Recall that the characteristic function of a Gaussian random variable with mean zero and variance σ^2 is $e^{-\frac{\theta^2\sigma^2}{2}}$. Therefore the above result points to the direction that stable laws are some generalization of the Gaussian laws but where the required renormalization (sometimes also called scaling property) is not related with square root but a general power law. This loose statement can be stated in different forms. Here is one such statement.

Lemma 6.4.2 *Let* $\{Z_t\}_{t\in[0,1]}$ *be a stable stochastic process. Then* $Z_t \stackrel{\mathscr{L}}{=} t^{1/\alpha} Z_1$.

Example 6.4.3 Prove the above statement using the Lévy–Khinchine formula in Corollary 6.1.15.

For the above reasons, one usually says that as the Brownian motion corresponds to the Laplacian,[7] the stable process of order α corresponds to the α-fractional power of the Laplacian.[8]

Now we can state the main property of the regularity of densities associated with stable laws.

Theorem 6.4.4 *Let Z be a stable process. Then the law of Z_t is infinitely differentiable.*

Proof It is enough to note that $\mathbb{E}[e^{i\langle\theta,Z_t\rangle}] = e^{-C|\theta|^\alpha} \leq C|\theta|^{-p}$ for any $p > 0$ and therefore due to Exercise 1.1.11, the result follows.

Exercise 6.4.5 Repeat the construction of subordinated Brownian motion in the multi-dimensional case as done in Sect. 5.6 for the one-dimensional case. In particular, obtain the corresponding results in Exercise 5.6.4, parts 4 and 5.

---

[7]In fact, the Itô formula implies that $\partial_t \mathbb{E}[f(B_t)] = \mathbb{E}[\Delta f(B_t)]$.

[8]In fact, it takes some calculations with the theory of fractional differentiation to discover that $\int f(x+y) - f(x) - \langle \nabla f(x), y\rangle \mathbf{1}_{\{|x|\leq 1\}}(x) \frac{dx}{|x|^{1+\alpha}}$ corresponds to $\Delta^\alpha f$.

Chapter 7
Flows Associated with Stochastic Differential Equations with Jumps

In this chapter, we will discuss how to obtain the flow properties for solutions of stochastic differential equations with jumps. This chapter is needed for the second part of this book and as the final goal is not to give a detailed account of the theory of stochastic differential equations driven by jump processes, we only give the main arguments, referring the reader to any specialized text on the subject. For example, see [2] (Sect. 6.6) or [48].

7.1 Stochastic Differential Equations

Through this chapter $\mathscr{N}$ is a Poisson random measure on $\mathbb{R} \times [0, T]$ with $T > 0$ and compensator given by the σ-finite measure $\widehat{\mathscr{N}}(dz, ds) = \nu(dz)ds$. Here ν is a Lévy measure satisfying $\nu(\{0\}) = 0$ and $\int(1 \wedge |z|^2)\nu(dz) < \infty$.

The way to define this measure without using the underlying stochastic process is by using compound Poisson process for each set which has a finite measure.

Exercise 7.1.1 Let $\mathbb{R} = \sum_{i=1}^{\infty} A_i$ (i.e. disjoint union) such that $\nu(A_i) < \infty$. Define the compound Poisson process associated with the renormalization of the measure $\nu(dz \cap A_i)ds$. Use independent copies of these processes to build approximations of the associated process under the condition that $\int(|z|^2 \wedge 1)\nu(dz) < \infty$. Prove that the corresponding limit exists and it corresponds to the Poisson random measure described before this exercise.

Hint: Note that in this case in order to define the associated Lévy process one will need to compensate the process in a fashion similar to Chap. 5. See also Exercise 5.4.1.

Proposition 7.1.2 (Burkholder–Davis–Gundy/Kunita's inequality) *Let* $\{a(t, z)\,;\ t \in [0, T],\ 0 < |z| \leq 1\}$ *be a* $\{\mathscr{F}_t\}_{t\geq 0}$*-predictable* $\mathbb{R}$*-valued random field*[1] *such that, for any* $p \geq 2$,

[1] A random field is a family of random variables which depend on the various parameters. In this case, it is the time variable $t \geq 0$ and $z \in \mathbb{R}$. As in Definition 3.5.1, one can extend the definition in order to include the cases of random fields.

© Springer Nature Singapore Pte Ltd. 2019

A. Kohatsu-Higa and A. Takeuchi, *Jump SDEs and the Study of Their Densities*, Universitext, https://doi.org/10.1007/978-981-32-9741-8_7

$$\mathbb{E}\left[\left(\int_{[-1,1]\times[0,t]} |a(s,z)|^2 \widehat{\mathcal{N}}(dz,ds)\right)^{p/2}\right] + \mathbb{E}\left[\int_{[-1,1]\times[0,t]} |a(s,z)|^p \widehat{\mathcal{N}}(dz,ds)\right] < +\infty.$$

Then, for any $p \geq 2$,

$$\mathbb{E}\left[\sup_{t\leq T}\left|\int_{[-1,1]\times[0,t]} a(s,z)\,\widetilde{\mathcal{N}}(dz,ds)\right|^p\right] \leq C\left\{\mathbb{E}\left[\left(\int_{[-1,1]\times[0,T]} |a(s,z)|^2 \widehat{\mathcal{N}}(dz,ds)\right)^{p/2}\right]\right.$$
$$\left. + \mathbb{E}\left[\int_{[-1,1]\times[0,T]} |a(s,z)|^p \widehat{\mathcal{N}}(dz,ds)\right]\right\}.$$

Proof See Theorems 4.4.21, 4.4.23, and Corollary 4.4.24 in [2]. The version for continuous martingales is usually called the Burkholder–Davis–Gundy inequality. In the jump case, [2] calls them Kunita's inequalities. □

Hypothesis 7.1.3 For all $p \geq 2$, the measurable functions $b : \mathbb{R} \to \mathbb{R}$ and $a : \mathbb{R} \times \mathbb{R} \to \mathbb{R}$ satisfy

$$|b(x) - b(y)|^p + \int_{|z|\leq 1} |a(x,z) - a(y,z)|^p\, \nu(\mathrm{d}z) \leq C\,|x-y|^p, \tag{7.1}$$

$$|b(x)|^p + \int_{|z|\leq 1} |a(x,z)|^p\, \nu(\mathrm{d}z) \leq C\,(1+|x|^p). \tag{7.2}$$

Let $\{Z_t\,;\, t \in [0,T]\}$ be a pure-jump Lévy process defined by

$$Z_t = \int_{[-1,1]\times[0,t]} z\,\widetilde{\mathcal{N}}(\mathrm{d}z,\mathrm{d}s) + \int_{[-1,1]^c\times[0,t]} z\,\mathcal{N}(\mathrm{d}z,\mathrm{d}s).$$

For a non-random point $x \in \mathbb{R}$, consider the $\mathbb{R}$-valued process $\{X_t\,;\, t \in [0,T]\}$ determined by the solution of the stochastic differential equation of the form:

$$X_t = x + \int_0^t b(X_s)\,\mathrm{d}s + \int_{[-1,1]\times[0,t]} a(X_{s-},z)\,\widetilde{\mathcal{N}}(\mathrm{d}z,\mathrm{d}s) + \int_{[-1,1]^c\times[0,t]} a(X_{s-},z)\,\mathcal{N}(\mathrm{d}z,\mathrm{d}s). \tag{7.3}$$

Note that the fact that $\nu(\{0\}) = 0$ implies that the value of $a(x,0)$ is irrelevant. We will therefore assume without loss of generality that $a(x,0) = 0$.

Proposition 7.1.4 (Existence and uniqueness of the solution) *Under Hypothesis 7.1.3, there exists a unique solution to the Eq.* (7.3).

Proof Our strategy, which has been explained in previous sections (see Sect. 4.2), is almost parallel to [2]-Theorem 6.2.3 (p. 367) and [30]-Theorem IV-9.1 (p. 245). Define the sequence of $\{\mathscr{F}_t\}_{t\geq 0}$-stopping times $\{\tau_n\,;\, n \in \mathbb{Z}_+\}$ by

$$\tau_n = \begin{cases} 0 & (n=0) \\ \inf\left\{t > \tau_{n-1}\,;\, \int_{[-1,1]^c\times[\tau_{n-1},t]} \mathcal{N}(\mathrm{d}z,\mathrm{d}s) \neq 0\right\} \wedge T & (n \in \mathbb{N}). \end{cases}$$

(i) Consider $\{X_t^{(k)}\,;\, t \in [0, T]\}$ ($k \in \mathbb{Z}_+$) be the $\mathbb{R}$-valued process given by

$$X_t^{(k)} = \begin{cases} x & (k=0) \\ x + \int_0^t b(X_s^{(k-1)})\,\mathrm{d}s + \int_{[-1,1]\times[0,t]} a(X_{s-}^{(k-1)}, z)\,\widetilde{\mathscr{N}}(\mathrm{d}z, \mathrm{d}s) & (k \in \mathbb{N}). \end{cases} \tag{7.4}$$

The Cauchy–Schwarz inequality and Hypothesis 7.1.3 implies that for any $t > 0$

$$\mathbb{E}\left[|X_t^{(1)} - X_t^{(0)}|^2\right] \leq C\,(1 + |x|^2).$$

Moreover, the Cauchy–Schwarz inequality, Doob's inequality and Hypothesis 7.1.3 lead us to have

$$\mathbb{E}\left[\sup_{s\leq t} |X_s^{(k+1)} - X_s^{(k)}|^2\right] \leq C \int_0^t \mathbb{E}\big[|X_s^{(k)} - X_s^{(k-1)}|^2\big]\,\mathrm{d}s.$$

Iterating such a procedure implies that for any $t > 0$

$$\begin{aligned} \mathbb{E}\left[\sup_{s\leq t} |X_s^{(k+1)} - X_s^{(k)}|^2\right] &\leq C^k \int_0^t \mathrm{d}s_1 \int_0^{s_1} \mathrm{d}s_2 \cdots \int_0^{s_{k-1}} \mathrm{d}s_k\, \mathbb{E}\left[|X_{s_k}^{(1)} - X_{s_k}^{(0)}|^2\right] \\ &\leq C\,(1 + |x|^2)\,\frac{(C\,t)^k}{k!}. \end{aligned}$$

Hence, by the Chebyshev inequality, we can get

$$\sum_{k=0}^{+\infty} \mathbb{P}\left[\sup_{t\leq T} |X_t^{(k+1)} - X_t^{(k)}| > 2^{-k}\right] \leq \sum_{k=0}^{+\infty} C\,(1 + |x|^2)\,\frac{(4\,C\,T)^k}{k!} < +\infty.$$

From the Borel Cantelli lemma, for almost all ω, the sequence $\{X_t^{(k)}\,;\, k \in \mathbb{N}\}$ converges uniformly on $[0, T]$. For each $t \in [0, T]$, write

$$X_t := \lim_{k\to+\infty} X_t^{(k)}.$$

Then, we see from the Cauchy–Schwarz inequality and Hypothesis 7.1.3, that

$$\mathbb{E}\left[\left|\int_0^t \left\{b(X_s^{(k-1)}) - b(X_s)\right\}\mathrm{d}s\right|^2\right] \leq C\,\mathbb{E}\left[\sup_{t\leq T} |X_t^{(k-1)} - X_t|^2\right],$$

$$\mathbb{E}\left[\left|\int_{[-1,1]\times[0,t]} \left\{a(X_{s-}^{(k-1)}, z) - a(X_{s-}, z)\right\}\widetilde{\mathscr{N}}(\mathrm{d}z, \mathrm{d}s)\right|^2\right] \leq C\,\mathbb{E}\left[\sup_{t\leq T} |X_t^{(k-1)} - X_t|^2\right].$$

Therefore, the process $\{X_t\,;\, t \in [0, T]\}$ satisfies

$$X_t = x + \int_0^t b(X_s)\,\mathrm{d}s + \int_{[-1,1]\times[0,t]} a(X_{s-}, z)\,\widetilde{\mathscr{N}}(\mathrm{d}z, \mathrm{d}s). \tag{7.5}$$

Now, we shall prove the uniqueness of the solutions to the Eq. (7.5). Denote by $X = \{X_t\,;\ t \in [0,T]\}$ and $\tilde{X} = \{\tilde{X}_t\,;\ t \in [0,T]\}$ the solutions to the Eq. (7.5). From Hypothesis 7.1.3, we have

$$\mathbb{E}\left[\sup_{s\le t} |X_s - \tilde{X}_s|^2\right] \le C \int_0^t \mathbb{E}\left[\sup_{s\le u} |X_s - \tilde{X}_s|^2\right] \mathrm{d}u.$$

Thus, the Gronwall inequality yields that

$$\mathbb{E}\left[\sup_{t\le T} |X_t - \tilde{X}_t|^2\right] = 0,$$

which leads us to get that $\sup_{t\le T} |X_t - \tilde{X}_t| = 0$ a.s. With this information, we know that there is a unique solution to the stochastic equation (7.3) for $t < \tau_1$.
(ii) For $t = \tau_1$, we shall define

$$X_{\tau_1} := X_{\tau_1 -} + a\big(X_{\tau_1 -}, \Delta Z_{\tau_1}\big),$$

where $\Delta Z_{\tau_1} := Z_{\tau_1} - Z_{\tau_1 -}$. Then, it can be easily checked that $\{X_t\,;\ t \in [0,\tau_1]\}$ is the unique solution to the Eq. (7.3).
(iii) Let $\tau_1 < t < \tau_2$. Replace the Eq. (7.3) by

$$X_t = X_{\tau_1} + \int_{\tau_1}^t b(X_s)\,\mathrm{d}s + \int_{[-1,1]\times[\tau_1,t]} a(X_{s-}, z)\,\widetilde{\mathscr{N}}(\mathrm{d}z, \mathrm{d}s).$$

Then, a similar study to (i) enables us to get the $\mathbb{R}^d$-valued process $\{X_t\,;\ t \in [0,\tau_2)\}$ which is a unique solution to the Eq. (7.3).
(iv) Consider the case of $t = \tau_2$. Similarly to (ii), we have only to define

$$X_{\tau_2} := X_{\tau_2 -} + a\big(X_{\tau_2 -}, \Delta Z_{\tau_2}\big),$$

which leads us to get the unique solution $\{X_t\,;\ t \in [0,\tau_2]\}$ to the Eq. (7.3).

Iterating this procedure leads us to the conclusion. □

Exercise 7.1.5 Write in detail the above iterative procedure. In particular, state in which sense existence and uniqueness of solutions has to be understood.

For a hint, see Chap. 14.

Exercise 7.1.6 Rewrite the above results with appropriate conditions on the coefficients and driving random Poisson measure as in Hypothesis 7.1.3 for the following r-dimensional stochastic differential equation driven by d-independent random point measures:

$$X_t = x + \int_0^t b(X_s)\,ds + \sum_{i=1}^{d} \int_{[-1,1]\times[0,t]} a_i(X_{s-}, z)\,\widetilde{\mathcal{N}}^i(dz, ds)$$

$$+ \sum_{i=1}^{d} \int_{[-1,1]^c\times[0,t]} a_i(X_{s-}, z)\,\mathcal{N}^i(dz, ds).$$

Exercise 7.1.7 In the spirit of Exercise 7.1.6, describe a stochastic equation driven by the symmetric stable process defined in Sect. 6.4. Prove the existence and uniqueness of solutions under appropriate conditions on the coefficients. Note that in this case, one has to be careful about the fact that a stable process does not have finite variance.

7.2 Stochastic Flows

For a proof of the following result, see [2]-Theorem 1.1.18 (p. 21), and [34]-Theorem II-2.8 (p. 53) and Problem II-2.9 (p. 55).

Lemma 7.2.1 (Kolmogorov continuity criterion) *For $\Lambda \in \mathscr{B}(\mathbb{R}^k)$, let $K = \{K_\rho\,;\ \rho \in \Lambda\}$ be the $\mathbb{R}^n$-valued random field*[2] *with the following property: there exist constants α, β such that*

$$\mathbb{E}\big[|K_{\rho_2} - K_{\rho_1}|^\alpha\big] \le C\,|\rho_2 - \rho_1|^{k+\beta} \quad (\rho_1,\ \rho_2 \in \Lambda).$$

Then, we can find a continuous version $\{\tilde{K}_\rho\,;\ \rho \in \Lambda\}$ of the random field K.

We remark that the above conclusion means that $\mathbb{P}(K_\rho = \tilde{K}_\rho)$ for all $\rho \in \Lambda$. Therefore we will work with the random field $\tilde{K}$ instead of K because its paths satisfy the additional property of continuity a.s.

Recall that $C_b^\infty(\mathbb{R}\,;\ \mathbb{R})$ denotes the family of $\mathbb{R}$-valued smooth functions on $\mathbb{R}$ such that all partial derivatives of any orders are bounded. A similar remark is also valid for $C_b^\infty(\mathbb{R}\times\mathbb{R}_0\,;\ \mathbb{R})$.

Hypothesis 7.2.2 The coefficients b and a of the Eq. (7.3) satisfy:

(i) $b \in C_b^\infty(\mathbb{R}\,;\ \mathbb{R})$.
(ii) $a \in C_b^\infty(\mathbb{R}\times\mathbb{R}_0\,;\ \mathbb{R})$.
(iii) For any $i \in \mathbb{N}_0$, $\nabla^i a(y, 0)\ \big(:= \lim_{|z|\to 0} \nabla^i a(y, z)\big) = 0$. Here $\nabla a(y, z)$ is the derivative with respect to $y \in \mathbb{R}^d$, while the derivative in $z \in \mathbb{R}^m$ will be denoted[3] by $\partial a(y, z)$.

---

[2]That is, K is a family of random variables indexed by elements of Λ.

[3]In general, we will use ∇ to denote the derivatives with respect to the main variables of the function under consideration. ∂ will be used when differentiating with respect to what may be considered as a parameter of the function.

Remark 7.2.3 Hypothesis 7.2.2 implies Hypothesis 7.1.3, which is obvious by the mean value theorem.

We shall denote by $X_{\cdot}(x)$ the solution to the Eq. (7.3) with the initial point $X_0(x) = x$.

Proposition 7.2.4 (Continuity of X in the initial point) *Assume that the coefficients b and a of the Eq. (7.3) satisfy Hypothesis 7.2.2. Then, for each $t \in [0, T]$, the mapping $\mathbb{R} \ni x \longmapsto X_t(x) \in \mathbb{R}$ admits a continuous modification.*

Proof Our argument is based upon [27].

Recall the definition of the sequence $\{\tau_n \,;\, n \in \mathbb{Z}_+\}$ of $\{\mathscr{F}_t\}_{t\geq 0}$-stopping times defined in Proposition 7.1.4. Let $p \geq 2$, and take $x,\ y \in \mathbb{R}$.
(i) Let $0 < t < \tau_1$. Since

$$X_t(x) = x + \int_0^t b(X_s(x))\,\mathrm{d}s + \int_{[-1,1]\times[0,t]} a(X_{s-}(x), z)\,\widetilde{\mathscr{N}}(\mathrm{d}z, \mathrm{d}s),$$
$$X_t(y) = y + \int_0^t b(X_s(y))\,\mathrm{d}s + \int_{[-1,1]\times[0,t]} a(X_{s-}(y), z)\,\widetilde{\mathscr{N}}(\mathrm{d}z, \mathrm{d}s),$$

we can get

$$\mathbb{E}\big[|X_t(x) - X_t(y)|^p\big] \leq C\,|x-y|^p + C\int_0^t \mathbb{E}\big[|X_s(x) - X_s(y)|^p\big]\,\mathrm{d}s$$

from Proposition 7.1.2, the Hölder inequality and Hypothesis 7.2.2. The Gronwall inequality leads us to get

$$\mathbb{E}\big[|X_t(x) - X_t(y)|^p\big] \leq C\,|x-y|^p \exp\big[C\,T\big],$$

so Proposition 7.2.1 tells us to see that for each $t \in [0, \tau_1)$, the mapping $\mathbb{R} \ni x \longmapsto X_t(x) \in \mathbb{R}$ has a continuous version.

(ii) Since

$$X_{\tau_1}(x) = X_{\tau_1-}(x) + a(X_{\tau_1-}(x), \Delta Z_{\tau_1}),$$

the conclusion follows from (i) immediately.

An iterative procedure leads us to the conclusion. □

Exercise 7.2.5 Write in detail the above iterative procedure.

Exercise 7.2.6 Suppose that $b = 0$ and $a(x, z) = a(x)z$, where a is a Lipschitz function and $\int_{[-1,1]^c} |z|^p \nu(dz) < \infty$ for some $p \geq 2$. Prove the following moment estimates:

$$\mathbb{E}\big[|X_t(x) - X_t(y)|^p\big] \leq C\,|x-y|^p \exp\big[C\,T\big].$$

Here X is the solution to (7.3).

For a hint, see Chap. 14

Proposition 7.2.7 (Differentiability of X in the initial point) *Assume that the coefficients b and a of the Eq. (7.3) satisfy Hypothesis 7.2.2.*

(a) For each $t \in [0, T]$, the mapping $\mathbb{R} \ni x \longmapsto X_t(x) \in \mathbb{R}$ admits a C^1-modification.
(b) The $\mathbb{R}$-valued process $Y = \{Y_t := \partial_x X_t(x)\,;\ t \in [0, T]\}$ satisfies a linear stochastic differential equation of the form:

$$\begin{aligned} Y_t = 1 + \int_0^t \nabla b(X_s(x))\, Y_s\, ds + \int_{[-1,1]\times[0,t]} \nabla a(X_{s-}(x), z)\, Y_{s-}\, \widetilde{\mathcal{N}}(dz, ds) \\ + \int_{[-1,1]^c\times[0,t]} \nabla a(X_{s-}(x), z)\, Y_{s-}\, \mathcal{N}(dz, ds). \end{aligned} \tag{7.6}$$

(c) For any $p > 1$, $\sup_{t\le T} \|Y_t\| \in \mathbb{L}^p(\Omega, \mathbb{P})$.

Proof Our argument is based upon [27]. Choose $\alpha \in \mathbb{R}\backslash\{0\}$, and define

$$N_t(x, \alpha) = \frac{X_t(x+\alpha) - X_t(x)}{\alpha}.$$

Then, the $\mathbb{R}$-valued random field $\{N(x, \alpha)\,;\ t \in [0, T]\}$ satisfies the equation:

$$\begin{aligned} N_t(x, \alpha) = 1 &+ \int_0^t \frac{b(X_s(x+\alpha)) - b(X_s(x))}{\alpha}\, \mathrm{d}s \\ &+ \int_{[-1,1]\times[0,t]} \frac{a(X_{s-}(x+\alpha), z) - a(X_{s-}(x), z)}{\alpha}\, \widetilde{\mathcal{N}}(\mathrm{d}z, \mathrm{d}s) \\ &+ \int_{[-1,1]^c\times[0,t]} \frac{a(X_{s-}(x+\alpha), z) - a(X_{s-}(x), z)}{\alpha}\, \mathcal{N}(\mathrm{d}z, \mathrm{d}s). \end{aligned} \tag{7.7}$$

Recall the sequence $\{\tau_n\,;\ n \in \mathbb{Z}_+\}$ of $\{\mathcal{F}_t\}_{t\ge 0}$-stopping times introduced in Proposition 7.1.4.
(i) Let $0 \le t < \tau_1$. From a similar argument to Proposition 7.2.4, we see that

$$\mathbb{E}\left[\sup_{t<T\wedge\tau_1} |N_t(x, \alpha) - N_t(y, \beta)|^p\right] \le C\left\{|x-y|^p + |\alpha-\beta|^p\right\}.$$

Then, Lemma 7.2.1 yields that for each $t \in [0, \tau_1)$, the mapping $\mathbb{R}\backslash\{0\} \ni \alpha \longmapsto N_t(x, \alpha) \in \mathbb{R}$ admits a continuous version with respect to the parameter $\alpha \in \mathbb{R}\backslash\{0\}$, and can be extended to $\alpha = 0$ continuously. Write $Y_t := \lim_{\alpha\to 0} N_t(x, \alpha) \equiv \partial X_t(x)$. Then, the $\mathbb{R}$-valued process $\{Y_t\,;\ t \in [0, \tau_1)\}$ satisfies the equation

$$Y_t = 1 + \int_0^t \nabla b(X_s(x))\, Y_s\, \mathrm{d}s + \int_{[-1,1]\times[0,t]} \nabla a(X_{s-}(x), z)\, Y_{s-}\, \widetilde{\mathcal{N}}(\mathrm{d}z, \mathrm{d}s). \tag{7.8}$$

Moreover, since ∇b and ∇a are bounded from Hypothesis 7.2.2, it is routine work to check that $\sup_{t\le u}\|Y_t\| \in \bigcap_{p>1}\mathbb{L}^p(\Omega,\mathbb{P})$ for each $u \ge 0$, via Proposition 7.1.2, the Hölder inequality and the Gronwall inequality.
(ii) Let $t = \tau_1$. By using the result in (i), it holds that the mapping

$$\mathbb{R} \ni x \longmapsto X_{\tau_1}(x) = X_{\tau_1-}(x) + a\big(X_{\tau_1-}(x), \Delta Z_{\tau_1}\big) \in \mathbb{R}$$

has a C^1-modification. Write

$$Y_{\tau_1} := \partial_x X_{\tau_1}(x) = \big\{1 + \nabla a\big(X_{\tau_1-}(x), \Delta Z_{\tau_1}\big)\big\}\, Y_{\tau_1-}.$$

Then, the $\mathbb{R}$-valued process $\{Y_t\,;\ t \in [0,\tau_1]\}$ satisfies the Eq. (7.6). Moreover, it is routine work to check that $\sup_{t\le u\wedge\tau_1}\|Y_t\| \in \bigcap_{p>1}\mathbb{L}^p(\Omega,\mathbb{P})$ for each $u \in [0,T]$, via the study in (i).

As before, iterating the procedure (see Exercise 7.2.6) stated above enables us to check the conclusion. □

Exercise 7.2.8 Note that it is important in the above result that the derivative of a with respect to x is a bounded function. Add assumptions for the case that $a(x,z) = a(x)z$ in the spirit of Exercise 7.2.6.

Hypothesis 7.2.9

$$\inf_{y\in\mathbb{R}}\inf_{z\in\mathbb{R}_0}|1 + \nabla a(y,z)| > 0. \tag{7.9}$$

□

Proposition 7.2.10 (Invertibility of Y) *Assume that the coefficients b and a of the Eq.* (7.3) *satisfy Hypotheses 7.2.2 and 7.2.9.*

(a) For each $t \in [0,T]$, the $\mathbb{R}$-valued random variable Y_t is invertible almost surely.
(b) The $\mathbb{R}$-valued process $\{\zeta_t := Y_t^{-1}\,;\ t \in [0,T]\}$ satisfies a linear stochastic differential equation of the form:

$$\begin{aligned}\zeta_t = 1 &- \int_0^t \zeta_s\, \nabla b(X_s)\, ds \\ &+ \int_{[-1,1]\times[0,t]} \zeta_{s-}\,\big\{\big(1+\nabla a(X_{s-},z)\big)^{-1} - 1\big\}\,\widetilde{\mathcal{N}}(dz,ds) \\ &+ \int_{[-1,1]^c\times[0,t]} \zeta_{s-}\,\big\{\big(1+\nabla a(X_{s-},z)\big)^{-1} - 1\big\}\,\mathcal{N}(dz,ds) \\ &+ \int_{[-1,1]\times[0,t]} \zeta_s\,\big\{\big(1+\nabla a(X_s,z)\big)^{-1} - 1 + \nabla a(X_s,z)\big\}\,\widehat{\mathcal{N}}(dz,ds).\end{aligned} \tag{7.10}$$

(c) For any $p > 1$, $\sup_{t\le T}\|\zeta_t\| \in \mathbb{L}^p(\Omega,\mathbb{P})$.

Proof Our argument is based upon [27]. Let $\{\zeta_t\ ;\ t \in [0, T]\}$ be the $\mathbb{R}$-valued process determined by the Eq. (7.10). From the Itô formula, we can easily obtain $\zeta_t\, Y_t = 1$, which implies the invertibility of the $\mathbb{R}$-valued process $\{Y_t\ ;\ t \in [0, T]\}$. The last assertion can be proved similarly to Proposition 7.2.7. □

7.3 Remark

Here, we shall give a remark on the higher-order moments for the process X.

Proposition 7.3.1 *Suppose Hypothesis 7.1.3. If* $\sup_{t\le T} Z_t \in \mathbb{L}^{p_0}(\Omega, \mathbb{P})$ *for some* $p_0 > 1$, *then* $\sup_{t\le T} |X_t| \in \mathbb{L}^p(\Omega, \mathbb{P})$ *for any* $1 < p \le p_0$.

Proof Recall the sequence $\{\tau_n\ ;\ n \in \mathbb{Z}_+\}$ of $\{\mathscr{F}_t\}_{t\ge 0}$-stopping times introduced in the proof of Proposition 7.1.4, again.
(i) Since

$$X_t = x + \int_0^t b(X_s)\, \mathrm{d}s + \int_{[-1,1]\times[0,t]} a(X_{s-}, z)\, \tilde{\mathscr{N}}(\mathrm{d}z, \mathrm{d}s)$$

for $0 \le t < \tau_1$, Proposition 7.1.2, the Hölder inequality and Hypothesis 7.1.3 lead us to see that

$$\mathbb{E}\left[\sup_{s\le t\wedge\tau_1} |X_s|^p\right] \le C\,|x|^p + C \int_0^t \mathbb{E}\left[\sup_{s\le u\wedge\tau_1} \left(1 + |X_s|^p\right)\right] \mathrm{d}u$$

for any $p \ge 2$. Then, we can get

$$\mathbb{E}\left[\sup_{s<T\wedge\tau_1} |X_s|^p\right] \le \left(1 + C\,|x|^p + C\,T\right) \exp\left[C\,T\right]$$

from the Gronwall inequality, so $\sup_{s<T\wedge\tau_1} |X_s| \in \mathbb{L}^p(\Omega, \mathbb{P})$. The estimate in the case of $1 < p < 2$ can be justified via the Jensen inequality.
(ii) Since $X_{\tau_1} = X_{\tau_1-} + a\big(X_{\tau_1-}, \Delta Z_{\tau_1}\big)$, we can get

$$\begin{aligned}
\mathbb{E}\left[\sup_{t\le\tau_1} |X_t|^p\right] &\le \mathbb{E}\left[\sup_{t<\tau_1} |X_t|^p\right] + \mathbb{E}\big[|X(\tau_1)|^p\big] \\
&\le C\,\mathbb{E}\left[\sup_{t<\tau_1} |X_t|^p\right] + C\,\mathbb{E}\Big[\big|a\big(X_{\tau_1-}, \Delta Z_{\tau_1}\big)\big|^p\Big] \\
&\le C\,\mathbb{E}\left[\sup_{t<\tau_1} |X_t|^p\right] + C \sup_{y\in\mathbb{R}} \sup_{z\in\mathbb{R}_0} \|\partial_z a(y, z)\|\, \mathbb{E}\big[|\Delta Z_{\tau_1}|^p\big]
\end{aligned}$$

from the mean value theorem. Then, the assertion for $1 < p \le p_0$ can be checked from the study in (i) and Hypothesis 7.1.3.

As usual, an iterative procedure leads us to the conclusion. □

Exercise 7.3.2 Use Proposition 7.1.2 in order to find conditions as in Hypothesis 7.1.3, which ensures that the condition $\sup_{t\leq T} Z_t \in \mathbb{L}^{p_0}(\Omega, \mathbb{P})$ for some $p_0 > 1$ is satisfied. In particular, test your result with the Lévy measures given in Examples 3.3.10, 4.1.20, 5.1.10, 5.1.11 and 5.1.19.

Part II
Densities of Jump SDEs

Chapter 8
Overview

8.1 Introduction

We learn early in any probability theory course that in order to compute any significant quantity we need to have information regarding the distribution function of random variables. This issue appears not only in applied problems where actual computation needs to be carried out but also in theoretical problems where qualitative information of the distribution function is needed.

We also learn how to obtain the probability distribution of a function of a vector of random variables using the change of variables theorem. The main topic of discussion in this second part is the existence, regularity between other properties of densities for random variables that are generated by an infinite number of independent random variables. This is the case of a functional of the solution of a stochastic differential equation.

We plan to discuss one of the possible methods to attack this problem which for us was very intuitive and simple to explain to early graduate students. We hope that this will be so for other younger generations as this was our main motivation to write this part of the book.[1]

Therefore we do not strive for a wide generality but rather to find simple examples and situations where one may explain the techniques simply. It is up to the user to devise the corresponding result that will suit their application. Also many of the techniques given here can be combined to obtain a very powerful and flexible result. We try only to explain the basic set-up of each method, leaving the possible combination or further extension of various methodologies to the user.

Therefore we do not develop so deeply a general theory to deal with random variables on an infinite-dimensional space but rather concentrate on the case of stochastic equations and try in that setting to describe the basic method and its uses. We will not explore the full power of each method and its variants. This will be

[1]Clearly there are many other ways to deal with the present problem. Notably, there are the analytical methods and most distinguishably methods developed in the partial differential equation literature.

© Springer Nature Singapore Pte Ltd. 2019

A. Kohatsu-Higa and A. Takeuchi, *Jump SDEs and the Study of Their Densities*, Universitext, https://doi.org/10.1007/978-981-32-9741-8_8

determined by each particular example, although we may give some comments in the remarks.

It should be clear from the present text that it cannot be a replacement for any of the authoritative and complete treatises on the topic: [9, 30, 46] etc. The approach, which is based on starting with finite-dimensional integration by parts formula and then extending the formula to infinite dimensions, is easily stated but it takes some work to carry it out in general. We explain it heuristically in the next section.

8.2 Explaining the Methods in Few Words

Learning the properties of the density function of the real random variable X can be done using $\mathbb{E}[G(X)]$ for a variety of functions G. For example, if $G(y) = \mathbf{1}(y \leq x)$ then we obtain that $\mathbb{E}[\mathbf{1}(X \leq x)]$ is the distribution function of X and therefore the existence of densities is related to the existence of the derivative of $\mathbb{E}[1(X \leq x)]$ with respect to x. An alternative way to obtain the same result is through the use of the so-called integration by parts (IBP) formula. To explain this method, suppose that X has finite moments.[2] Then an IBP formula for (X, Y) makes it possible to write $\mathbb{E}[G'(X)Y] = \mathbb{E}[G(X)H]$ for a function $G \in C_p^1(\mathbb{R}; \mathbb{R})$ and a suitable random variable H with finite moments which is independent of G.[3] Using this formula for $Y \equiv 1$ and using an approximation of the indicator function of an interval instead of f' one can prove the existence of the density of X.[4] In fact, one may also understand this property by taking limits, and we see that the IBP formula leads to the formula $\mathbb{E}[\mathbf{1}(X \leq x)] = \mathbb{E}[(x - X)_+ H]$, where

$$(x)_+ = \begin{cases} 0 & (x \leq 0), \\ x & (x > 0). \end{cases}$$

In fact, if we apply the IBP formula k times we may also obtain $\mathbb{E}[\mathbf{1}(X \leq x)] = \mathbb{E}\left[\frac{(x-X)_+^k}{k!} H_k\right]$ for an appropriate random variable H_k which is obtained by iteration.[5] This property will imply that the density of X is k times differentiable. In fact, one may also obtain upper bounds for the density of X and its derivatives easily using Markov-type inequalities.

With this short discussion we would like to convince the reader of the central role that is played by the IBP formula. In general, any IBP formula is based on a parallel finite-dimensional formula. It is one of the goals of this text to prove that

---

[2]That is, $\mathbb{E}[|X|^p] < \infty$ for all $p > 0$.

[3]Note that the indicator function used for the distribution function can also be written with $G(y) = \mathbf{1}(y - x \leq 0)$ and therefore $\partial_x \mathbb{E}[G(X)] = -\mathbb{E}[G'(X)]$.

[4]More exactly, one proves the absolute continuity of the probability measure induced by X. For details, see Lemma 2.1.1 in [46].

[5]Changing the values of Y being used. In fact, in the last step one uses $Y = H_{k-1}$.

this is so. Therefore the infinite-dimensional formulas will be obtained as limits of the respective finite-dimensional formulas.

In order to carry out this project, one needs first the theorem that assures that the random variable X can be expanded in terms of finite-dimensional random variables. In the set-up of Wiener functionals (that is, random variables that arise due to the Wiener process and their combinations thereof), this theorem is the Itô chaos expansion which can be interpreted as some form of Fourier series decomposition. In the set-up of jump processes, this expansion is naturally provided by the compound Poisson process approximation described in the previous chapter. This is the approach that will be explained in the next few chapters.

Clearly this is but one option of analysis. Changing the expansion used will result in a different infinite-dimensional analysis technique. As far as we know, this set-up has been pioneered by Norris [45] and later explained in [6] and applied to the Boltzmann equation in [7].

Still, in the finite-dimensional case, there are still many different variations (good and bad) that are possible which again will lead to different formulations, and these variations are what we now try to explain in heuristic terms.

The first step is to recognize that any IBP formula is a weak-type problem. That is, that the problem can be represented as an expectation. In general, any expectation is represented using functionals of random variables. Changing the random variables used for the representation and/or the functionals used will imply a new structure for the same quantity. This will in turn lead to a new representation and therefore to a different IBP formula. The main question is how to obtain different representation formulas which will lead to different IBP formulas and how to recognize the role and advantage of each representation.

In particular, studying finite-dimensional models will allow us to see more clearly the use and role of each representation and the possible advantages and disadvantages of each method.

Although there is not a unique way to obtain all possible representations, we list some of the possible methods to obtain them.

a. Write a "natural" finite-dimensional approximation and carry out a direct IBP formula with respect to any fixed random variable in the problem with smooth density. Usually these formulas are unstable if the random variable in question only affects the final result in an infinitesimal way as it is the case of diffusions.

b. Given a. with respect to a range of single infinitesimal random variables, find a way of putting them together in a stable manner. Here we describe a summation method which may stabilize the final formula.

c. Do a change of variables before carrying out the IBP. Then return the change back. Another form of the same procedure is as follows. Carry out a parametric infinitesimal change of measure in the finite-dimensional integrals and then differentiate with respect to the parameter. Return to the original model and then represent it back as an expectation. This formulation goes better with the original ideas of Bismut of using a change of measure (Girsanov theorem).

This formula is related to a "torsion"-type formula which is interesting when the change of variables involves many variables at once. In fact, we claim that the approach presented here allows the user to see the effect of such a method on the density of the basic random variable clearly.

d. Representations can be further localized by using convolutions of expressing random variables as the sum of two independent random variables, one comprising the regular part and another expressing the irregular part of the density. Then the IBP is carried with respect to the regular part (when possible).

e. Representations can also be obtained by randomization. One clear example of this is the subordination of Brownian motion to study Lévy processes.

Finally, we would not like to make a long appeal to the applicability of the results, just briefly stating that in our experience the understanding of the behavior of densities has aided in various theoretical/applied problems.

Chapter 9
Techniques to Study the Density

In the previous chapters we started studying the density of some Lévy processes using some ad-hoc techniques (see e.g. Exercises 4.1.23 and 5.1.17). In order to study densities of random variables there are many different techniques. We will briefly describe some of them in this chapter. Most of these techniques are analytic in nature and they give a different range of results. We concentrate here on the multi-dimensional results, while in Chap. 1 some basic results were discussed in the one-dimensional case.

The first result, which is the most classical for undergraduate students, is the derivative of the distribution function.

Theorem 9.0.1 *Let X be an $\mathbb{R}^d$-valued random vector with distribution function $F_X(x) := \mathbb{P}(X \leq x)$.*[1] *If $f_X := \partial_{1,\ldots,d} F_X$ exists a.e., then for any bounded continuous function g, we have*

$$\mathbb{E}[g(X)] = \int_{\mathbb{R}^d} g(x) f_X(x) dx. \tag{9.1}$$

Exercise 9.0.2 Prove the above statement using the classical approximation argument for the definition of Riemann integrals.

Sometimes there are situations when in order to characterize the density function of a random vector one tries to directly obtain Eq. (9.1), which is the reason why it is used as a definition.

Definition 9.0.3 We say that the random vector X has a density function if, for any bounded and continuous function $g : \mathbb{R}^d \to \mathbb{R}$, there exists an integrable function $f_X : \mathbb{R}^d \to [0, \infty)$ such that

$$\mathbb{E}[g(X)] = \int_{\mathbb{R}^d} g(x) f_X(x) dx.$$

---

[1]Here $\{X \leq x\} = \{X_i \leq x_i; i = 1, \ldots, d\}$.

© Springer Nature Singapore Pte Ltd. 2019

A. Kohatsu-Higa and A. Takeuchi, *Jump SDEs and the Study of Their Densities*, Universitext, https://doi.org/10.1007/978-981-32-9741-8_9

Exercise 9.0.4 Show that if the random vector X has a density function f_X then $\int_{\mathbb{R}^d} f_x(x)dx = 1$. Also show that the above definition implies that (9.1) is also valid for bounded measurable functions $g : \mathbb{R}^d \to \mathbb{R}$.

A variation of Theorem 9.0.1 is given through the concept of the Radon–Nikodým derivative. In order to continue with the discussion, we will extend the notation of distribution function F_X into the measure $F_X(dx) = P_X(dx)$, which can be obtained using the Carathéodory extension theorem. That is:

Theorem 9.0.5 *Let $F_X(x) = \mathbb{P}(X \leq x)$. Then F_X is a function of bounded variation. Furthermore, if F_X is absolutely continuous with respect to the Lebesgue measure on $\mathbb{R}^d$ then X has a density function. This density is the Radon–Nikodým derivative of F_X with respect to the Lebesgue measure.*

For basic facts and proofs of the above statements, see your favorite book on measure theory. For example, see Chap. 7 in [50].

Problem 9.0.6 This exercise gives you a method to obtain inequalities for densities. Given a fixed real random variable X. Assume that there exists a function $H \in L^1(\mathbb{R}, \mathbb{R}_+)$ such that for any bounded measurable function $f \geq 0$, one has

$$\mathbb{E}[f(X)] \leq \int f(y)H(y)dy.$$

Prove that X has a density and that its density $p_X(x)$ satisfies the inequality $p_X(x) \leq H(x)$, $a.e. - x$ with respect to the Lebesgue measure.

9.1 On an Approximation Argument

There are various ways of measuring the regularity of the density of a real random variable X. One of these ways is the integration by parts formula as described in the introduction to Chap. 8. In that case and in other situations, approximations are needed.

The first technique is to use approximations of the so-called Dirac delta Schwartz distribution functions. Rather than giving a general account of Schwartz distributions, we will give some examples. Recall that the convolution of two functions f and g is given by

$$f * g(x) := \int f(z)g(x - z)dz.$$

Define $\phi_a(z) := \frac{1}{\sqrt{2\pi a}} e^{-\frac{|z|^2}{2a}}$.

Lemma 9.1.1 *Let $f \in C_0^0(\mathbb{R}^d)$ then*

$$\lim_{a \downarrow 0} \phi_a * f(0) = \lim_{a \downarrow 0} \int \frac{1}{\sqrt{2\pi a}} e^{-\frac{|z|^2}{2a}} f(z)dz = f(0).$$

Proof The proof can be obtained if we just consider the behavior of f around zero and away from zero. In fact, for given $\varepsilon > 0$ there exists $\delta > 0$ such that for all $|z| < \delta$ then $|f(z) - f(0)| < \varepsilon$. Then we have

$$\begin{aligned}\left|\int \frac{1}{\sqrt{2\pi a}} e^{-\frac{|z|^2}{2a}} f(z)dz - f(0)\right| &\leq \left|\int_{B_\delta(0)} \frac{1}{\sqrt{2\pi a}} e^{-\frac{|z|^2}{2a}} (f(z) - f(0))dz\right| \\ &\quad + 2\|f\|_\infty \left|\int_{B_\delta(0)^c} \frac{1}{\sqrt{2\pi a}} e^{-\frac{|z|^2}{2a}} dz\right| \\ &\leq \varepsilon + 2\|f\|_\infty \int_{a^{-1/2}B_\delta(0)^c} \frac{1}{\sqrt{2\pi}} e^{-\frac{|z|^2}{2}} dz.\end{aligned}$$

If we let $a \to 0$ then the region of integration of the second integral will vanish. Furthermore, due to the integrability of the function $\frac{1}{\sqrt{2\pi}} e^{-\frac{|z|^2}{2}}$, one obtains that the second term goes to zero, which gives that

$$\lim_{a \downarrow 0} \left|\int \frac{1}{\sqrt{2\pi a}} e^{-\frac{|z|^2}{2a}} f(z)dz - f(0)\right| \leq \varepsilon,$$

for any $\varepsilon > 0$. Therefore taking $\varepsilon \to 0$ the result follows.

Exercise 9.1.2 The above result is in fact the beginning of a deep theory related to Schwartz distributions and their approximation. We will not discuss these matters fully but just give a few exercises that hint at some results that one can obtain.

For example, here we generalize the above result using a general kernel function instead of a Gaussian one. That is, let $g : \mathbb{R}^d \to \mathbb{R}_+$ be an integrable function with $\int g(z)dz = 1$. Prove that

$$\lim_{a \downarrow 0} \int a^{-1} g(za^{-1}) f(z)dz = f(0)$$

for $f \in C_0^0(\mathbb{R}^d)$.

Exercise 9.1.3 The following generalization deals with the case that f is not continuous. Suppose that $f : \mathbb{R} \to [-1, 1]$ is a càdlàg function[2] and that g satisfies besides the hypotheses in Exercise 9.1.2 that $g(z) = g(-z)$ for all $z \in \mathbb{R}$. Prove that

$$\lim_{a \downarrow 0} \int \frac{1}{a} g\left(\frac{z - x}{a}\right) f(z)dz = \frac{f(x+) + f(x-)}{2}.$$

In particular, note that in this result the value of the function f at the point x is completely irrelevant, because the Lebesgue measure of the point x is zero.

[2]That is, both side limits exist at any point.

Exercise 9.1.4 This exercise tries to show that even when the symmetry of g is not assumed, one can still compute the above limits. Let $t > 0$ and $f : \mathbb{R} \to [-1, 1]$ be a càdlàg function. Prove that

$$\lim_{z \to 0+} \int_0^t f(s) \frac{z}{s\sqrt{2\pi s}} e^{-\frac{z^2}{2s}} ds = C_1 f(0+),$$
$$\lim_{z \to 0-} \int_0^t f(s) \frac{z}{s\sqrt{2\pi s}} e^{-\frac{z^2}{2s}} ds = C_2 f(0-).$$

Find the values of C_1 and C_2.

For a hint, see Chap. 14.

Remark 9.1.5 All exercises above have generalizations in the multi-dimensional case. We leave the interested reader to try to foresee these extensions.

Now we try to point out the relation of the above results with the notion of weak convergence of probability laws.

Lemma 9.1.6 *Let $Z_a \sim N(0, a)$ be a one-dimensional Gaussian random variable with mean zero and variance a. Let X be a d-dimensional random vector with continuous density function independent of Z_a. Then one has that for $f \in \mathcal{M}_b(\mathbb{R})$*

$$\lim_{a \downarrow 0} \mathbb{E}[f(Z_a + X)] = \mathbb{E}[f(X)].$$

Proof There are two ways of proving this fact. From the analytical point of view one may rewrite the expectation as follows, using Fubini's theorem:

$$\begin{aligned}\mathbb{E}[f(Z_a + X)] &= \int_{\mathbb{R}^2} f(z + x) \frac{1}{\sqrt{2\pi a}} e^{-\frac{|z|^2}{2a}} p_X(x) dx dz \\ &= \int_{\mathbb{R}} dz \frac{1}{\sqrt{2\pi a}} e^{-\frac{|z|^2}{2a}} \int_{\mathbb{R}} dw f(w) p_X(w - z).\end{aligned}$$

Here p_X stands for the density function of the random variable X. Now note that due to the boundedness of f that $z \to \int_{\mathbb{R}} f(w) p_X(w - z) dw$ is a continuous function due to the dominated convergence theorem.[3] Therefore

$$\lim_{a \downarrow 0} \mathbb{E}[f(Z_a + X)] = \int_{\mathbb{R}} f(w) p_X(w) dw = \mathbb{E}[f(X)].$$

Exercise 9.1.7 Prove that the characteristic function of $Z_a + X$ converges to the characteristic function of X. Use this result to prove

$$\lim_{a \downarrow 0} \mathbb{E}[f(Z_a + X)] = \mathbb{E}[f(X)]$$

---

[3]Prove this. Hint: Recall the proof of Lemma 9.1.1.

for $f \in C_c^0(\mathbb{R}^d)$. Essentially this proves that the convergence of characteristic functions implies that certain expectations with respect to these laws converge.

Exercise 9.1.8 Prove that the density of $Z_a + X$ in the above lemma is given by the convolution $\phi_a * p_X$.

9.2 Using Characteristic Functions

Another method to study the densities of random vectors is through the use of characteristic functions (Fourier transforms). In this case, one needs to use the inverse Fourier transform or the characteristic function of the random variable X as $\varphi(\theta) \equiv \varphi_X(\theta) := \mathbb{E}[\exp(i\theta \cdot X)]$ for all $\theta \in \mathbb{R}^d$.

We start with a simple exercise:

Exercise 9.2.1 One learns in most basic probability courses that for a random vector $X \sim N(0, A)$ its characteristic function is given by

$$\varphi(\theta) = \exp(-\frac{\theta^* A\theta}{2}).$$

Prove this fact.

Hint: First prove it in the one-dimensional case. Then prove it when A is a diagonal matrix and finally in the general case when A is a positive definite matrix.

For a hint, see Chap. 14.

Theorem 9.2.2 (Lévy's inversion theorem) *Let* $(\Omega, \mathscr{F}, \mathbb{P})$ *be any probability space,* X *be an* $\mathbb{R}^d$*-valued random vector defined on that space and* $\varphi(\theta) = \mathbb{E}[e^{i\langle\theta, X\rangle}]$ *be its characteristic function. If* $\varphi \in L^1(\mathbb{R}^d)$*, then* f_X*, the density function of the law of* X*, exists and is continuous. Moreover, for any* x *in* $\mathbb{R}^d$*, we have*

$$f_X(x) = \frac{1}{(2\pi)^d} \int_{\mathbb{R}^d} e^{i\langle\theta, x\rangle} \varphi(\theta) d\theta.$$

The following corollary of Theorem 9.2.2 gives us a more precise criterion for the regularity of the density.

Corollary 9.2.3 *Let* X *be a random vector under the same setting as in Theorem 9.2.2 and* φ *be its characteristic function. Assume that there exists a constant* $\lambda > 0$ *such that*

$$\int_{\mathbb{R}^d} |\varphi(\theta)||\theta|^\lambda d\theta < +\infty.$$

Then the density function of the law of X *exists and it belongs to the set* C^λ.

Here we also give the proof of Theorem 9.2.2 in the particular case of the above corollary.

Proof Applying Fubini to the following integral and Exercise 9.2.1, we obtain for any $a > 0$

$$\int e^{-i\langle\theta,y\rangle}\frac{1}{(2\pi a)^{d/2}}e^{-\frac{|\theta|^2}{2a}}\varphi(\theta)d\theta = \int P_X(dx)\int e^{i\langle\theta,x-y\rangle}\frac{1}{(2\pi a)^{d/2}}e^{-\frac{|\theta|^2}{2a}}d\theta$$
$$= \int P_X(dx)e^{-\frac{|x-y|^2 a}{2}}.$$

Note that the above integral is finite due to the Gaussian term in the integral. Furthermore the last term corresponds to the density of $X + Z_a$ where $Z \sim N(0, aI)$. This result already proves that if two random vectors have the same characteristic function then their distribution functions are the same. In fact, using Lemma 9.1.6 we obtain this result if these two random vectors have a continuous density.

In order to obtain further information from the above equality we need to first obtain an alternative expression for the distribution function of X. This is obtained by integrating the above formula with respect to $y \in A(w) := \{z \in \mathbb{R}^d; w_i - h_i \leq z_i \leq w_i, i = 1,\ldots,d\}$, $w_i \in \mathbb{R}$, $h_i > 0$, $i \in \{1,\ldots d\}$. Doing so, one obtains

$$\int_{A(w)}\int e^{-i\langle\theta,y\rangle}\frac{1}{(2\pi a)^{d/2}}e^{-\frac{|\theta|^2}{2a}}\varphi(\theta)d\theta dy$$
$$= \int F_X(dx)(\frac{2\pi}{a})^{d/2}\prod_{j=1}^{d}(\Phi((w_j - x_j)a) - \Phi((w_j - h_j - x_j)a)). \tag{9.2}$$

Multiplying the above equality by $(\frac{a}{2\pi})^{d/2}$ and taking the limit as $a \uparrow \infty$, we obtain using the dominated convergence theorem

$$\int_{A(w)} F_X(dx) = \lim_{a\uparrow\infty}\int\prod_{j=1}^{d}\frac{e^{-i\theta_j(w_j-h_j)} - e^{-i\theta_j w_j}}{i\theta_j}\frac{1}{(2\pi)^d}e^{-\frac{|\theta|^2}{2a}}\varphi(\theta)d\theta.$$

From here, one obtains that if two random vectors have the same characteristic functions then their distribution functions are the same.

If $\varphi \in L^1(\mathbb{R}^d)$ then we have

$$|\int_{A(w)} dF_X(x)| \leq \frac{1}{(2\pi)^d}\prod_{j=1}^{d}h_j\int|\varphi(\theta)|d\theta.$$

Therefore the absolute continuity of the distribution function F_X is proven[4] and therefore the density of X exists. Furthermore the density can be expressed as

[4]Try to prove this in detail.

$$f_X(w) = \frac{1}{(2\pi)^d} \int e^{-i\langle\theta,w\rangle}\varphi(\theta)d\theta.$$

We leave as an exercise to prove that the above function is continuous in w. In a similar fashion, one also obtains the expression for the derivatives as

$$\partial_\alpha f_X(w) = \frac{1}{(2\pi)^d} \int (-i\theta)^\alpha e^{-i\langle\theta,w\rangle}\varphi(\theta)d\theta,$$
$$(-i\theta)^\alpha := \prod_{j=1}^{\lambda}(-i\theta_{\alpha_j}),$$

for any multi-index α with $|\alpha| = \lambda$. To discuss fractional orders $\lambda \in (0, 1)$, one considers

$$f_X(w) - f_X(w') = \frac{1}{(2\pi)^d} \int (e^{-i\langle\theta,w\rangle} - e^{-i\langle\theta,w'\rangle})\varphi(\theta)d\theta.$$

Now note that the following inequalities are satisfied[5]:

$$|e^{-i\langle\theta,w-w'\rangle} - 1| \leq |\langle\theta, w - w'\rangle| \wedge 2 \leq C_\lambda|\langle\theta, w - w'\rangle|^\lambda.$$

From here the Hölder property of f_X follows.

Exercise 9.2.4 Use the proof of the above result to prove that if X_n, $n \in \mathbb{N}$, X are random vectors such that their characteristic functions belong and converge in $L^1(\mathbb{R}^d)$ then their density functions converge pointwise.

Problem 9.2.5 Here, we consider some further properties that one can deduce from characteristic functions, as well as an important counterexample.

1. Suppose that the characteristic function of a real-valued random variable X satisfies that $|\varphi(\theta)| \leq \frac{1}{1+|\theta|^2}$. Prove that the density function of X exists and is bounded. Note that this result could also give you an extension of Corollary 9.2.3.
2. Prove that if furthermore the derivative of $\varphi(\theta)$ exists and satisfies the inequality $|\varphi'(\theta)| \leq \frac{1}{1+|\theta|^2}$ then the density function satisfies the inequality $f_X(x) \leq \frac{C}{1+|x|}$. State and prove similar results for the derivatives of f_X.
3. For $\varepsilon > 0$ and $x_0 \in \mathbb{R}$ fixed, consider the real valued random variable X such that $\mathbb{P}(X \in A) = \varepsilon\, \mathbf{1}(x_0 \in A) + (1 - \varepsilon) \int_A \frac{e^{-x^2/2}}{\sqrt{2\pi}}dx$. Prove that the characteristic function associated with X, $\varphi_X(\theta)$, does not converge to zero as $\theta \uparrow \infty$.

Exercise 9.2.6 (*Tail estimates*) Let X be a real random variable. Suppose that its characteristic function admits an analytic extension[6] and it satisfies:

---

[5]Prove this using the fact that $|\sin(x)| \leq |x|$ for values of x close to zero.

[6]This is a strong restriction. For example, note that $e^{-c\theta^2}$ does not satisfy this condition. Still a good approximation will satisfy this condition.

1. For any $R > 0$ and any $\eta \in \mathbb{R}$, $|\varphi(\pm R + i\eta)| \leq CR^{-\epsilon}$, for some $C > 0$ and $\varepsilon > 0$.
2. For any $a > 0$ and $\eta \in \mathbb{R}$,

$$\begin{aligned}|\varphi(\theta + i\eta)| \leq & F_1(\eta, a) \text{ for } |\theta| \leq a,\\ |\varphi(\theta + i\eta)| \leq & F_2(\theta, a) \text{ for } |\theta| > a.\end{aligned}$$

Prove that the density satisfies the following upper bound for any $\eta > 0$ and $a > 0$:

$$f_X(x) \leq Ce^{-\eta x}\left(2aF_1(\eta, a) + \int_{\mathbb{R}\setminus[-a,a]} F_2(\theta, a)d\theta\right), \quad x > 0.$$

Note that once this result is proven one can optimize on η and a as functions of x in order to obtain an upper bound for the density. The case $x < 0$ follows similarly.

For a hint, see Chap. 14.

Problem 9.2.7 Use the Laplace transform to find conditions under which

$$\mathbb{E}\left[\left((W^0_{Y^0_t})^2 + (W^1_{Y^1_s})^2\right)^{-1}\right] < \infty,$$

where $t, s > 0$, $W^i, i = 0, 1$ are independent Wiener processes and $Y^i, i = 0, 1$ are independent gamma processes with parameter a, that is, a Lévy process with Lévy measure $\nu(dz) = \frac{e^{-az}}{z}dz$. If you do not know much about Wiener processes consider the same problem for $W_{Y_t} \stackrel{\mathscr{L}}{=} Z\sqrt{Y_t}$, where Z is a standard Gaussian random variable. Hint: Use the identity $\int_0^\infty e^{-\lambda x}d\lambda = x^{-1}$ for $x > 0$.

An extension of the previous results which demands that the characteristic function be in $L^2(\mathbb{R}^d)$ rather than in $L^1(\mathbb{R}^d)$ is as follows.

Lemma 9.2.8 *Let X be a random vector such that its characteristic function satisfies that $\varphi_X \in L^2(\mathbb{R}^d)$. Then X has a density.*

Proof We leave the details to the reader, just mentioning the main points of the proof. Recall that in the proof of Corollary 9.2.3, we have used the random variable $X^n = X + Z_n$ where Z_n is an r.v. independent of X with mean zero and variance n^{-1} which follows a Gaussian law. Furthermore, by the Plancherel equality we have that if f_n denotes the density of X^n then $\int f_n^2(x)dx = C_d \int |\varphi_{X^n}(\theta)|^2 d\theta \leq C_d \int |\varphi_X(\theta)|^2 d\theta =: C < \infty$. Due to the weak compactness of the balls of $L^2(\mathbb{R}^d, dx)$, we may extract a subsequence n_k and find a function $f \in L^2(\mathbb{R}^d, dx)$ such that f_{n_k} converges weakly in $L^2(\mathbb{R}^d, dx)$ to f. But, on the other hand, $F_{X^n}(dx) = f_n(x)\,dx$ tends weakly (in the sense of measures) to $P_X \equiv F_X$. As a consequence, $F_X(dx)$ is nothing but $f(x)\,dx$.

Exercise 9.2.9 Another way of proving the above result is to use (1.1). That is, let $g \in L^2(\mathbb{R})$ prove that $\mathbb{E}[g(\hat{Z}_t^n)] = \int \mathcal{F}(g)(\theta)\mathbb{E}[e^{i\theta\hat{Z}_t^n}]d\theta$, where $\mathcal{F}(g)(\theta) = \frac{1}{2\pi}\int_{\mathbb{R}} g(z)\, e^{-iz\theta}\, dz$. Finally, prove that if $\mathbb{E}[e^{i\theta\hat{Z}_t^n}]$ is square integrable as a function of θ then the law of Z^n is absolutely continuous with respect to the Lebesgue measure.

According to [20] this lemma is optimal in the sense that $\varphi_X \in L^p$ with $p > 2$ does not imply that X has a density, quoting [33] for counterexamples.

Our next result explains a basic technique which uses approximations in order to prove the regularity of the density of X. This idea will resurface in Example 10.4.5.

Theorem 9.2.10 *Let $\{X_n\}_{n\in\mathbb{N}}$ be a sequence of random variables approximating X in the following sense: For some $C > 0$, $r > 0$ and $s > 0$,*

$$|\varphi_{X_n}(\theta) - \varphi_X(\theta)| \le C|\theta|^r n^{-s}.$$

Also assume that the density of X_n has the following asymptotic regularity property for $u > 0$ and $k > 0$, $|\varphi_{X_n}(\theta)| \le \frac{Cn^u}{(1+|\theta|)^k}$. Then we have that the asymptotic regularity property for X is

$$|\varphi_X(\theta)| \le \frac{C}{(1+|\theta|)^{\frac{ks-ru}{s+u}}}.$$

Therefore if $\alpha = \frac{ks-ru}{s+u} - 1 > 0$ then the density of X exists and belongs to the class $C^{\alpha-\varepsilon}(\mathbb{R},\mathbb{R})$ for any $\varepsilon > 0$.

Proof To obtain the estimate on $|\varphi_X(\theta)|$ we just need to consider the following inequality:

$$\begin{aligned}
|\varphi_X(\theta)| &\le |\varphi_{X_n}(\theta) - \varphi_X(\theta)| + |\varphi_{X_n}(\theta)| \\
&\le C|\theta|^r n^{-s} + \frac{Cn^u}{(1+|\theta|)^k}.
\end{aligned}$$

Now, one chooses n in an asymptotic optimal way as $n = \lfloor |\theta|^{r/(s+u)}(1+|\theta|)^{k/(s+u)} \rfloor$. From the resulting inequality one proves that $\varphi_X \in L^\alpha(\mathbb{R}^d)$ by performing an appropriate change of variables to cylindrical coordinates.

Exercise 9.2.11 Find a counterexample to Theorem 9.2.10. That is, define a sequence of random variables X_n with smooth density such that the sequence converges to X almost surely and the limit random variable does not admit a density.

Conversely, give an example of a sequence of random variables X_n such that they do not admit a density but in the limit have a smooth density. For this, it may be helpful to recall 3 in Exercise 9.2.5.

The following localization argument will also be of constant use.

Lemma 9.2.12 *For $\delta > 0$, let $f_\delta : \mathbb{R}_+ \mapsto [0, 1]$, be a function vanishing on $[0, \delta]$, strictly positive on (δ, ∞) and globally Lipschitz continuous (with Lipschitz constant 1). Consider the random variable X and its associated distribution function F_X and a function $\sigma : \mathbb{R} \mapsto \mathbb{R}_+$. Assume that for each $\delta > 0$, the measure $F_\delta(dx) = f_\delta(\sigma(x))F_X(dx)$ has a density. Thus, X has a density on $\{x \in \mathbb{R}, \sigma(x) > 0\}$.*

Proof We prove the absolute continuity property. Let $A \subset \mathbb{R}$ be a Borel set with Lebesgue measure 0. We have to prove that $P_X(A \cap \{\sigma > 0\}) = 0$. For each $\delta > 0$, the measures $\mathbf{1}_{\{\sigma(x)>\delta\}}F_X(dx)$ and $F_\delta(dx)$ are absolutely continuous with respect to each other. By assumption, $F_\delta(A) = 0$ for each $\delta > 0$, whence $P_X(A \cap \{\sigma > \delta\}) = 0$. Hence, $P_X(A \cap \{\sigma > 0\}) = \lim_{\delta \to 0} P_X(A \cap \{\sigma > \delta\}) = 0$.

Another criterion for proving that the density belongs to a Besov space is the criterion given by [21] (for a proof, see this reference). The spaces $B^s_{1,\infty}$, $s > 0$, can be defined as follows. Let $f : \mathbb{R}^d \to \mathbb{R}$. For $x, h \in \mathbb{R}^d$ set $(\Delta^1_h f)(x) = f(x + h) - f(x)$. Then, for any $n \in \mathbb{N}$, $n \geq 2$, let

$$(\Delta^n_h f)(x) = \left(\Delta^1_h(\Delta^{n-1}_h f)\right)(x) = \sum_{j=0}^{n}(-1)^{n-j}\binom{n}{j} f(x + jh).$$

For any $0 < s < n$, we define the norm

$$\|f\|_{B^s_{1,\infty}} = \|f\|_{L^1} + \sup_{|h|\leq 1} |h|^{-s}\|\Delta^n_h f\|_{L^1}.$$

Note that the operators Δ^n_h are higher-order difference operators and the above norms measures the functions and weak derivatives in the L^1-sense. But one allows for some explosions which are not so strong in the sense that they do not make the above integral infinite.

It can be proved that for two distinct $n, n' > s$ the norms obtained using n or n' are equivalent. Then we define $B^s_{1,\infty}$ to be the set of L^1-functions with $\|f\|_{B^s_{1,\infty}} < \infty$. We refer the reader to [56] for more details.

Exercise 9.2.13 Prove that if we denote by C^α_c the set of Hölder continuous functions of degree α with compact support, then this space is included in its corresponding Besov space ($B^\alpha_{1,\infty}$). That is, Besov spaces are a weaker version of Hölder spaces.

The next lemma establishes a criterion on the existence of densities.

Lemma 9.2.14 *Let X be a random variable. Assume that there exist $0 < \alpha \leq a < 1$, $n \in \mathbb{N}$ and a constant C_n such that for all $\phi \in C^\alpha_b$, and all $h \in \mathbb{R}$ with $|h| \leq 1$,*

$$\left|\int_{\mathbb{R}} \Delta^n_h\phi(y)F_X(dy)\right| \leq C_n\|\phi\|_{C^\alpha_b}|h|^a. \tag{9.3}$$

Then X has a density with respect to the Lebesgue measure, and this density belongs to the Besov space $B^{a-\alpha}_{1,\infty}(\mathbb{R})$.

9.3 An Application for Stable Laws and a Probabilistic Representation

We have studied in Sect. 5.6 and in particular, in Exercise 5.6.4 that the characteristic function for a stable law is given by $\varphi(\theta) = \mathbb{E}[e^{i\theta Z_t}] = e^{-C_\alpha t|\theta|^\alpha}$ for a suitable constant C_α. This example, which is different from the Gaussian case (note that the fact that the characteristic function is a one-to-one transformation follows from Lévy's inversion theorem, Theorem 9.2.2) will be a first example for which we can apply all our results of this section. In particular, $\varphi \in L^1(\mathbb{R})$ and therefore the density function of Z_t exists and due to the symmetry of the characteristic function, this density can be written as

$$f_{Z_t}(x) = \frac{1}{\pi} Re \int_0^\infty e^{i\theta x}\varphi(\theta)d\theta = \frac{1}{\pi}\int_0^\infty \cos(\theta x)\varphi(\theta)d\theta.$$

From Corollary 9.2.3 we obtain that the density function of Z_t is infinitely differentiable.

Exercise 9.3.1 Use the method described in Exercise 9.2.6 to obtain an upper bound for the density of a stable-like process. In fact, consider

$$\varphi(\theta) = \exp\left(t\int_{[-1,1]}(e^{i\theta x} - 1 - i\theta x)\frac{dx}{|x|^{1+\alpha}}\right).$$

Prove that this function has an analytic extension. Prove that its density exists and is smooth. Find an upper bound for its density using the technique explained in Exercise 9.2.6.

Without loss of generality, we assume that $x > 0$, $\alpha \in (0, 2)$. In the present case, note that

$$|\varphi(\theta)| = \exp\left(-2t\int_{[-1,1]}\sin^2\left(\frac{\theta x}{2}\right)\frac{dx}{|x|^{1+\alpha}}\right).$$

Note that the function sin has an analytic extension which still satisfies that for $|\theta|$ in any bounded set there exists a constant $C > 0$ which depends on this set such that

$$\left|\sin\left(\frac{\theta x}{2}\right)\right| \leq C\frac{|\theta x|}{2}.$$

Then we have that using the Cauchy integral theorem as in Exercise 9.2.6 that we have to estimate (we use the same notation as in the hint for Exercise 9.2.6)

$$e^{-rx}\left(\int_{B_1} e^{-t|r|^\alpha}d\theta + \int_{B_2} e^{-t|\theta|^\alpha}d\theta\right) \leq e^{-rx}\left(2ae^{-t|r|^\alpha} + Ce^{-t|a|^\alpha}|a|^{1-\alpha}\right).$$

The last estimate follows by using L'Hôpital rule for

$$\lim_{a \uparrow \infty} \frac{\int_{B_2} e^{-t|\theta|^\alpha} d\theta}{e^{-t|a|^\alpha} |a|^{1-\alpha}} = \frac{1}{t\alpha}.$$

Finally, try to optimize on the values of a and r in order to find an explicit upper bound.

Chapter 10
Basic Ideas for Integration by Parts Formulas

As explained in Sect. 8.2, the goal of this second part is to show how to obtain integration by parts (IBP) formulas for random variables which are obtained through systems based on an infinite sequence of independent random variables. In this chapter, we will show a few basic ideas of how this can be done for stochastic difference equations generated by a finite number of input noises which have smooth densities. We will revisit different methods to deal with this problem which lead to techniques used in the setting of continuously varying stochastic equations with jumps which may depend on an infinite number of random variables. This idea, which appears in various other fields in different forms and levels of difficulty, was developed by Paul Malliavin in the eighties within the context of hypoelliptic operators in stochastic analysis on the Wiener space. For this reason, it is known as Malliavin calculus.

10.1 Basic Set-Up

In this section we will consider the following simple difference equation:

$$X_{k+1} = X_k + A(X_k)Z_k, \ k = 0, ..., n-1. \tag{10.1}$$

Here, $X_0 = x \in \mathbb{R}$, $A : \mathbb{R} \to \mathbb{R}$, $Z := (Z_0, ..., Z_{n-1})$ is a random vector composed of i.i.d. real-valued random variables. Suppose that Z_0 has a density function of the form $\frac{1}{\Delta^\alpha} g(\frac{x}{\Delta^\alpha})$, where g is a one-dimensional $C_0^\infty(\mathbb{R})$ density function,[1] $\alpha \geq \frac{1}{2}$ and $\Delta = \frac{T}{n}$.

[1]This restriction is tailor-made in order to fit α^{-1} stable distributions as our main example. In particular, one may simplify the set-up to $\alpha = \frac{1}{2}$ and $g(x) = \frac{1}{\sqrt{2\pi}} \exp(-\frac{x^2}{2})$ (the density of a standard normal random variable).

© Springer Nature Singapore Pte Ltd. 2019

A. Kohatsu-Higa and A. Takeuchi, *Jump SDEs and the Study of Their Densities*, Universitext, https://doi.org/10.1007/978-981-32-9741-8_10

Exercise 10.1.1 Prove by induction that for each $k = 1, ..., n$, there exists a function $f_k : \mathbb{R}^k \to \mathbb{R}$ such that $X_k = f_k(Z_0, ..., Z_{k-1})$.[2]

Exercise 10.1.2 1. Find the conditional density of X_{k+1} given X_k.
2. Prove that regardless of the regularity of the function A, this conditional density is a smooth function if A satisfies for some positive constants c, C that $0 < c < |A(x)| < C$ for any $x \in \mathbb{R}$.

The above exercise is important to understand the goal of the following chapters. In fact, the question we will address is if these regularity properties are satisfied as $\Delta = \frac{T}{n} \to 0$. In particular, one is willing to accept some additional conditions on A. Note that in the multi-dimensional case A somehow replaces the concept of the covariance matrix in the particular case that Z_k is a standard Gaussian random variable. Therefore the requirement that $0 < c < |A(x)| < C$ states that the covariance matrix is non-degenerate.[3]

If you have some experience with difference equations you will realize that these are approximations of ordinary differential equations in the case that one considers $y_{k+1} = y_k + B(y_k)\Delta$. The limiting ordinary differential equation $y_t' = B(y_t)$ has a unique solution if B is a Lipschitz function.

One of the surprising results for younger students is that in fact the regularity of the noise term Z_k may imply that the limit of the sequence $\{X_n\}_n$ for $X_{k+1} = X_k + B(X_k)\Delta + A(X_k)Z_k, k = 0, ..., n-1$ may exist and the limiting equation may have a unique solution even in cases where B does not satisfy the Lipschitz condition.[4]

Problem 10.1.3 1. Prove that $y_n = y_0 + \sum_{i=0}^{n-1} B(y_i)\Delta$.
2. If you know the generalization of the Ascoli–Arzelà theorem for càdlàg functions[5] you may also use it to prove that the functions $f_n(t) = y_k$, $t \in [k\Delta, (k+1)\Delta)$, $k = 0, ..., n-1$ and $f_n(1) = y_n$ converge uniformly to a continuous limit if B is bounded.
3. Furthermore, suppose that B is a continuous function. Use the above result and prove that the limit of f_n satisfies the equation

$$f(t) = y_0 + \int_0^t B(f(s))ds.$$

4. Use Gronwall's lemma to prove that there is a unique solution to the above equation if we further assume that B is Lipschitz.

---

[2] In the continuous time setting this property corresponds to the adaptedness of the solution of Eq. (10.1).

[3] The concept that replaces this quantity in Malliavin Calculus is usually called the Malliavin (or stochastic) covariance matrix.

[4] This result requires much more knowledge than we give in this text and we therefore do not pursue it. This topic of research is commonly known as "regularization by noise".

[5] See for example, [11] on the space $D[0, 1]$.

10.1.1 The Unstable IBP Formula

Definition 10.1.4 Given a function $G \in C_c^1(\mathbb{R};\mathbb{R})$. If there exists an integrable random variable H (i.e., $\mathbb{E}[|H|] < \infty$) such that

$$\mathbb{E}\left[G'(X_n)\right] = -\mathbb{E}\left[G(X_n)H\right] \tag{10.2}$$

then we say that X_n satisfies an IBP formula (for G with weight H).

Exercise 10.1.5 For $n = 1$ provide enough conditions on g so that one has an IBP formula. In particular, prove that $H = g'g^{-1}(\frac{X_1 - X_0}{A(X_0)\Delta^\alpha})(A(X_0)\Delta^\alpha)^{-1}$. Apply this result in the case that g is the density of a standard Gaussian random variable. Find an IBP formula for the second derivative. That is, $\mathbb{E}[G''(X_n)] = \mathbb{E}[G(X_n)H]$.

Remark 10.1.6 As we will soon see, there is no uniqueness of the representation of H in (10.2). In fact, any random variable H' such that the conditional expectations $\mathbb{E}\left[H \middle| X_n\right] = \mathbb{E}\left[H' \middle| X_n\right]$ will satisfy this property. This (advantageous) property is related with the fact that one may enlarge the probability space so that the conditional expectation of H with respect to X_n remains unchanged. As a disadvantageous aspect of this fact, it is mathematically clear that enlarging the space will only increase the variance of H. For a precise statement see Exercise 10.1.7.

Exercise 10.1.7 Suppose that $G \in C_c^1(\mathbb{R};\mathbb{R})$ and that X_n satisfies an IBP formula for G with weight H which is a function of X_n alone. Prove that any random variable H' such that $\mathbb{E}\left[H \middle| X_n\right] = \mathbb{E}\left[H' \middle| X_n\right]$ is also a weight for the IBP formula. Also prove that $\mathrm{Var}[H] \leq \mathrm{Var}[H']$.

Problem 10.1.8 In this exercise, consider $G \in C_c^1(\mathbb{R};\mathbb{R})$. Then one can prove the following two IBP formulas, $\mathbb{E}\left[G'(X)\right] = -\mathbb{E}\left[G(X)H_i\right]$, $i = 1, 2$ for $X = \max_{0\leq s\leq 1} W_s$,[6] where W is a Brownian motion.

1. Suppose that $\mathrm{supp}(G) \subseteq (0, \infty)$, then the IBP formula is valid with $-H_1 = X$.
2. In the general case, i.e. $G \in C_c^1(\mathbb{R};\mathbb{R})$ but $G(0) \neq 0$ prove by computing $\mathbb{E}\left[G(X)H_2\right]$ that the IBP formula is satisfied with

$$\begin{aligned} -H_2 = -2{p_N}^{-1}\Bigg\{ & N(N+1)\,\mathbf{1}\left(\frac{1}{N+1} < X < \frac{1}{N}\right) \\ & -N(N-1)\,\mathbf{1}\left(\frac{1}{N} < X < \frac{1}{N-1}\right)\Bigg\} + X. \end{aligned}$$

Here N is a discrete random variable independent of W with distribution function $\{p_n;\, n \in \mathbb{N}\}$ with $p_n > 0$ for all $n \in \mathbb{N}$.

---

[6] If one prefers to put this in the framework of (10.1) then take $n = 1$, $X_0 = 0$ and $Z_0 = \max_{0\leq s\leq 1} W_s = X_1$.

Problem 10.1.9 Another way of obtaining an IBP formula in Problem 10.1.8 is using the reflection principle for Brownian motion as follows. Prove that the following IBP formula is satisfied: $\mathbb{E}[G'(|W_1|)\operatorname{sgn}(W_1)] = \mathbb{E}[G(|W_1|)W_1]$.[7]

Problem 10.1.10 Give an IBP formula for $G \in C_c^1([0,\infty);\mathbb{R})$ for $X = \varphi(W_1)$ and $G(0) = 0$ in the following cases:

1. $\varphi(x) = |x|$.
2. For $\alpha \neq 0$ and $\beta \neq 0$ consider:

$$\varphi(x) = \begin{cases} \alpha x & \text{if } x \geq 0, \\ \beta x & \text{if } x \leq 0. \end{cases}$$

3. For $\varphi(x) = x^2$ the following additional condition is needed $G(u) = o(u)$ as $u \to 0$.

Problem 10.1.11 The following result which will be assumed through this exercise is a result used for simulation of stable laws and is called the Chambers–Mallows–Stuck simulation method. For simplicity we assume that $\alpha \in (0,1) \cup (1,2)$, let W be an exponential random variable of mean 1 and U be a $(-\frac{\pi}{2}, \frac{\pi}{2})$-uniform r.v. independent of W. Then the following random variable follows an α-stable law

$$Z = \frac{\sin(\alpha U)}{\cos^{1/\alpha}(U)} \left(\frac{\cos((1-\alpha)U)}{W} \right)^{\frac{1-\alpha}{\alpha}}.$$

That is, the characteristic function of Z is given by $\mathbb{E}[e^{i\theta Z}] = e^{-|\theta|^\alpha}$. Use this result to find an IBP formula for $\mathbb{E}[G'(Z)]$ under appropriate conditions on the function G.

The above exercises show the range of possibilities when performing integration by parts formulas when one random variable is involved.

We would like to deal with the system (10.1), which is a composition of a large number of random variables $\{Z_i\}_i$. Still, we will start with the most simple cases, as in Exercise 10.1.5, in order to try to develop a general theory. For this, we will need the concept of stochastic derivatives.

Definition 10.1.12 Let $Y = U(Z)$ where $U : \mathbb{R}^n \to \mathbb{R}^k$ is a C^1 function. We define the stochastic derivative as the random vector $(\partial_0 Y, \ldots, \partial_{n-1} Y)$, where $\partial_k Y := \partial_k U(Z)$. In this sense, we will say that the random variable Y is a C^1 random variable. Similarly, we may say that a random variable is C_b^1, C_p^k, etc.

We say that the random variable Y is uniformly elliptic with respect to Z_k if there exists a positive constant c_0 such that $\partial_k Y \geq c_0$ a.s.

Remark 10.1.13 (i) The uniform ellipticity condition just states that the function U transfers all the variability of the random variable Z_k if all other variables are left

[7] This is a way of hiding the boundary terms that appeared in the previous exercise.

fixed. In particular, if Z_k has a density then U will "transfer" this density to the density of Y. As a typical example of non-uniformly elliptic random variable, consider the case that U is constant in some part of the support of Z.

(ii) We remark that the above definition of differentiability depends on the basic random variable(s) Z used in the representation of the random variable Y. It may be true that a representation of the random variable Y may not be differentiable with respect to Z but differentiable with respect to a different representation using a different random variable Z_1. It is not difficult to think of an example.

Exercise 10.1.14 Let Z_1, Z_2, Z_3 three independent standard normal random variables. Define $X := 1(Z_1 \in A)Z_2 + 1(Z_1 \in A^c)Z_3$ for any measurable set A. Prove that X is not differentiable with respect to Z_1. Still, the law of X corresponds to a standard Gaussian law and therefore it has a smooth density.

Exercise 10.1.15 Recall the basic set-up (10.1). Prove that $\partial_k X_n = A(X_k) \prod_{j=k+1}^{n-1} (1 + A'(X_j)Z_j)$. In particular, note that $\partial_k X_j = 0$ if $j \leq k$.[8]

Theorem 10.1.16 *Suppose that $A \in C_c^2(\mathbb{R}; \mathbb{R}_+)$ satisfying that $A(x) \geq c_0 > 0$ and $|A'(x)Z_k| \leq c_1 < 1$ for all $x \in \mathbb{R}$. Furthermore suppose that $(\log g)'(\frac{Z_k}{\Delta^\alpha})$ is integrable. Then X_n satisfies the IBP formula with weight*

$$H = \partial_k(\partial_k X_n)^{-1} + (\partial_k X_n)^{-1}(\log g)'\left(\frac{Z_k}{\Delta^\alpha}\right)\frac{1}{\Delta^\alpha}. \tag{10.3}$$

Remark 10.1.17 The above formula is called unstable IBP because H usually does not converge (say, in law) as $\Delta \to 0$.[9] The concept of non-degeneracy appears here as one needs that $(\partial_k X_n)^{-1}$ has to be well defined.

Proof of Theorem 10.1.16 The idea of the proof is straightforward. Just write the expectation in (10.2) as an integral and carry out an IBP with respect to the density of the random variable Z_k. We just give the main idea here, leaving the details to the reader. First, it is clear by induction that X_n is a $C^2(\mathbb{R}^n; \mathbb{R})$ function of Z. Furthermore, we have for $k < n$ that X_n is uniformly elliptic with respect to Z_k. In fact,

$$\partial_k X_n = A(X_k) \prod_{j=k+1}^{n-1} (1 + A'(X_j)Z_j) \geq c_0(1 - c_1)^{n-k-1} > 0. \tag{10.4}$$

Here, we interpret the product as 1 if $k = n - 1$. Similarly, $\partial_k X_n \leq C_0(1 + c_1)^{n-k-1}$ with $C_0 := \sup_x |A(x)|$. Then we compute the IBP, using a conditional expectation

---

[8]Note that this property is related to the so-called adaptedness condition of X. In short, as X_j only depends on the values of $Z_0, ..., Z_{j-1}$, then $\partial_k X_j = 0$ if $j \leq k$.

[9]Exercise: Think of an example. Actually it is much harder to think of an example where this formula converges.

with respect to Z_k and the independence of the random variables Z_j, $j = 1, ..., n$ as follows:

$$\begin{aligned}\mathbb{E}\left[G'(X_n)\right] &= \mathbb{E}\left[G'(X_n)\partial_k X_n(\partial_k X_n)^{-1}\right] \\ &= \mathbb{E}\left[\partial_k\{G(X_n)\}(\partial_k X_n)^{-1}\right] \\ &= \frac{1}{\Delta^\alpha}\int_{\mathbb{R}}\mathbb{E}\left[\partial_k\{G(X_n)\}(\partial_k X_n)^{-1}\big|_{Z_k=z}\right] g\left(\frac{z}{\Delta^\alpha}\right)dz \\ &= \frac{1}{\Delta^\alpha}\int_{\mathbb{R}}\mathbb{E}\left[\left(G(X_n)|_{Z_k=\cdot}\right)'(z)\ (\partial_k X_n)^{-1}\big|_{Z_k=z}\right] g\left(\frac{z}{\Delta^\alpha}\right)dz \\ &= -\mathbb{E}\left[G(X_n)H\right].\end{aligned} \tag{10.5}$$

In order to obtain the last step one needs to carry out the classical IBP formula taking into account that the boundary terms will cancel as $G \in C_c^1$, $\lim_{z\to\infty} g\left(\frac{z}{\Delta^\alpha}\right) = 0$ and the assumed hypotheses on A. Finally the formula is obtained by converting the resulting expression into an expectation.

Remark 10.1.18 The interpretation of (10.3) is clear. The second term corresponds to the derivative of the density g in the integration by parts formula. The first term is a compensation term which appears because $f'(X_n)$ does not correspond exactly to a stochastic derivative.

Exercise 10.1.19 Prove that in the particular case that $k = n - 1$ in Theorem 10.1.16 then the restriction $|A'(x)Z_k| \le c_1 < 1$ is not needed. Therefore in this particular case the result applies in the case that g is the standard Gaussian density.

Exercise 10.1.20 As the derivative A' is bounded in Theorem 10.1.16, then the restriction $|A'(x)Z_k| \le c_1 < 1$ in Theorem 10.1.16 implies that the random variable Z_k should take values on a bounded set. This is a restrictive condition.

Using an approximation argument, give the alternative condition to the condition $|A'(x)Z_k| \le c_1 < 1$ so that under moment conditions on $1 + A'(X_k)Z_k$ the IBP formula for X_n (10.3) is valid. Still, the basic example of Z_k distributed according to the standard Gaussian distribution cannot be handled in general.

Exercise 10.1.21 Obtain the same result as in Theorem 10.1.16, using instead the following idea: Define $H^\xi : \mathbb{R} - \{0\} \to \mathbb{R}$ for any parameter $\xi \ge 0$ as $H^\xi(z) = z\exp(\xi z^{-1})$. This is a sequence of smooth space transformations satisfying the following conditions:

$$H^\xi(z)\big|_{\xi=0} = z, \quad \partial_z H^\xi(z)\big|_{\xi=0} = 1, \quad \partial_\xi H^\xi(z)\big|_{\xi=0} = 1.$$

Consider $\mathbb{E}[G(X_n)F(Z)]$ for an appropriate functional F and use the above space transformation H^ξ for values of ξ close to zero to carry out an integration by parts in order to prove Theorem 10.1.16.[10] This idea will be used in Sect. 10.3.

[10] Hint: After carrying out the appropriate change of variables, differentiate the formula with respect to ξ and evaluate it for $\xi = 0$.

Exercise 10.1.22 In this exercise, we will analyze an example with a two-dimensional noise. Here, Z_k is a two-dimensional random vector such that it has density given by $\Delta^{-1}g(\frac{x}{\sqrt{\Delta}}, \frac{y}{\sqrt{\Delta}})$, $x, y \in \mathbb{R}$ for $g \in C_0^\infty$. Furthermore, the probabilistic model is given by $X_{k+1} = X_k + \langle A(X_k), Z_k \rangle$ where $A : \mathbb{R} \to \mathbb{R}^2$ with $A_1(x) \neq 0$ and $A_2(x) \neq 0$ for all $x \in \mathbb{R}$. The goal is to find various integration by parts formulas for $\mathbb{E}[G'(X_{k+1})]$ for $G \in C_c^1$.

1. Use integration by parts formulas in one dimension in order to obtain two integration by parts formulas based on each component of g.
2. Use the divergence theorem[11] in order to obtain an integration by parts formula using the two variables in g.
3. Propose a way to do this integration by parts weakening the condition that $A_1(x) \neq 0$ and $A_2(x) \neq 0$ for all $x \in \mathbb{R}$.

For a hint, see Chap. 14.

Exercise 10.1.23 Obtain the integration by parts formula in the following two-dimensional case: $Y = U(Z^1, Z^2)$ where the random vector (Z^1, Z^2) has a joint smooth density $g(x_1, x_2)$ on a compact domain D with smooth boundary with g vanishing at ∂D. Obtain the integration by parts formula using each coordinate separately and then using both coordinates. This is of course related to Green's/Stokes/Divergence theorem as in Exercise 10.1.22.

Exercise 10.1.24 In this exercise, we will try to state a way to bring all problems of integration by parts to the Gaussian law.

1. Let U be a random variable with density function $g \in C^\infty$ with support equal to $\mathbb{R}$. Find the density of $F(U)$ where F is a strictly increasing differentiable function so that $F(\mathbb{R}) = \mathbb{R}$.
2. Find an explicit example so that $F(U)$ is a standard Gaussian random variable.

Using this idea with the proof of Theorem 10.1.16 allows the use of Gaussian random variables for Z_k.

For a hint, see Chap. 14.

Exercise 10.1.25 Note that the previous set-up is modeled after the case when the Z_0 is standard normal random variable and therefore $\alpha = \frac{1}{2}$. Obtain a result similar to Theorem 10.1.16 in the case that the density of Z_0 is of the type $\frac{1}{\Delta^\alpha}g(\frac{x}{\Delta^\alpha})$ for some $\alpha \geq \frac{1}{2}$ and $g \in C_c^\infty$.[12] Recall that an extension argument as in Exercise 10.1.20 will have to be used.

---

[11] Also known as Gauss's theorem or Ostrogradsky's theorem.

[12] This example corresponds to α^{-1}-stable process. For example, $\alpha = 1$ corresponds to the Cauchy process for which $g(x) = \frac{1}{\pi(1+x^2)}$, $x \in \mathbb{R}$.

10.2 Stability by Summation

From the result in Theorem 10.1.16 we see that we have $n-1$ IBP formulas for each $k = 1, \ldots, n-1$. All these formulas will be unstable (that is, H will blow up) in the case that we want to take limits.

In order to stabilize these formulas we need to collect these $n-1$ formulas and somehow add them (although see Remark 10.2.5 for other possibilities). Intuitively we would like to obtain formulas such that in the eventual case that we want to take limits as $n \to \infty$ (or equivalently $\Delta \to 0$) the IBP formula will remain stable. The stability of the formula can be obtained in a number of ways. One of them uses a summation argument as we explain in the following theorem.

Theorem 10.2.1 *Suppose that $A \in C_c^2(\mathbb{R}; \mathbb{R}_+)$ satisfying that $A(x) \geq c_0 > 0$ and $|A'(x)Z_k| \leq c_1 < 1$ for all $x \in \mathbb{R}$ and any $k = 1, .., n$. For any auxiliary sequence of C_b^1 bounded random variables a_k, $k = 1, \ldots, n-1$ such that $a_k \geq c_0$ and under the same conditions as in Theorem 10.1.16 we have that an IBP for X_n is valid with*

$$\begin{aligned} H = \sum_{k=0}^{n-1} \Bigg\{ \partial_k \left(a_k \left(\sum_{j=0}^{n-1} a_j \partial_j X_n \right)^{-1} \right) \\ + \left(a_k \left(\sum_{j=0}^{n-1} a_j \partial_j X_n \right)^{-1} \right) (\log g)' \left(\frac{Z_k}{\Delta^\alpha} \right) \frac{1}{\Delta^\alpha} \Bigg\}. \end{aligned} \tag{10.6}$$

Proof The proof follows the same lines as in the proof of Theorem 10.1.16 with one clear difference. Note in particular that H is well defined as

$$\sum_{j=0}^{n-1} a_j \partial_j X_n \geq c_0^2 c_1^{-1} (1 - (1 - c_1)^n).$$

In step (10.5), we carry out the calculation as follows:

$$\begin{aligned} \mathbb{E}\left[G'(X_n)\right] &= \mathbb{E}\left[G'(X_n) \frac{\sum_{k=0}^{n-1} a_k \partial_k X_n}{\sum_{j=0}^{n-1} a_j \partial_j X_n} \right] \\ &= \sum_{k=0}^{n-1} \mathbb{E}\left[G'(X_n) \frac{a_k \partial_k X_n}{\sum_{j=0}^{n-1} a_j \partial_j X_n} \right] \\ &= \sum_{k=0}^{n-1} \frac{1}{\Delta^\alpha} \int_{\mathbb{R}} \mathbb{E}\left[\partial_k \{G(X_n)\} a_k \left(\sum_{j=0}^{n-1} \partial_j a_j X_n \right)^{-1} \Bigg|_{Z_k = z} \right] g\left(\frac{z}{\Delta^\alpha}\right) dz \\ &= \sum_{k=0}^{n-1} \frac{1}{\Delta^\alpha} \int_{\mathbb{R}} \mathbb{E}\left[\left(G(X_n)|_{Z_k=\cdot}\right)'(z)\, a_k \left(\sum_{j=0}^{n-1} a_j \partial_j X_n \right)^{-1} \Bigg|_{Z_k = z} \right] g\left(\frac{z}{\Delta^\alpha}\right) dz \end{aligned} \tag{10.7}$$

$$= -\mathbb{E}[G(X_n)H].$$

Remark 10.2.2 Why do we call the above formula a stable formula? In fact, if $a_k = O(\Delta)$ then most sums in (10.6) can be interpreted as Riemann sums (and therefore they are approximations to Lebesgue integrals) with the exception of the last sum. For this last term we hope that[13]

$$a_k (\log g)' \left(\frac{Z_k}{\Delta^\alpha} \right) \frac{1}{\Delta^\alpha} = \begin{cases} O(Z_k) & (\alpha = \frac{1}{2}), \\ O(\frac{\Delta}{Z_k}) & (\alpha \neq \frac{1}{2}). \end{cases}$$

This will lead to the idea that the last term should converge to a stochastic integral (or the so-called Skorohod integral). In fact, as an exercise, verify that if $g(x) = \frac{1}{\sqrt{2\pi}} \exp(-\frac{x^2}{2})$ and $a_k = \Delta$ then the above property is satisfied.

Exercise 10.2.3 In the spirit of Example 10.1.25 (try first, for example Cauchy or gamma increments) and find a_k so that the integration by parts formula is stable by summation.

Exercise 10.2.4 Suppose that $g(x) = 2^{-1}\, \mathbf{1}_{(-1,1)}(x)$. Clearly this function does not satisfy the basic hypothesis $g \in C_0^\infty(\mathbb{R})$ stated right after (10.1). In order to carry out the integration by parts formula, one can use the following localization argument.

Repeat the arguments in Theorem 10.2.1 using instead $\sum_{j=0}^{n-1} a_j \partial_j X_n$, the function $\sum_{j=0}^{n-1} a_j \partial_j X_n \psi(Z_j)$ for $\psi \in C^\infty(\mathbb{R}; [0, 1])$ such that $\psi(x) = 0$ for x in a neighborhood of $x = -1, 1$. State appropriate conditions for the calculations to be satisfied.

Using this argument one can obtain a stable integration by parts formula without boundary terms.

10.2.1 Example: Diffusion Case

Here we will not touch various delicate issues such as the necessary weak convergence arguments and we will only give an idea of how the application of Theorem 10.2.1 leads to the usual IBP formula. In this case one lets $\Omega = C([0, T])$ and on this space we have a family of random variables $Z_k \sim N(0, \Delta)$ with $\Delta = \frac{T}{n}$. Then $X_n \to Y_T$ in $L^p(\Omega)$ for any $p \in \mathbb{N}$, where Y is the solution of the one-dimensional stochastic differential equation (SDE)

$$Y_t = x + \int_0^t A(Y_s) dW_s.$$

[13] How to resolve the issue that Z_k appears in the denominator will be further discussed in Example 10.3.4.

In this case, $g(x) = \frac{1}{\sqrt{2\pi}} \exp\left(-\frac{x^2}{2}\right)$. Therefore $(\log g)'(x) = -x$ and taking $a_k = \Delta$ we have as promised that

$$a_k (\log g)' \left(\frac{Z_k}{\Delta^{\frac{1}{2}}}\right) \frac{1}{\Delta^{\frac{1}{2}}} = -Z_k.$$

Furthermore, for any sequence $k \equiv k_n$ such that $t \in [\frac{Tk}{n}, \frac{T(k+1)}{n})$, we have that

$$\begin{aligned}\partial_k X_n &= A(X_k) \prod_{j=k+1}^{n-1} (1 + A'(X_j) Z_j) \\ &\to A(Y_t) \exp\left(\int_t^T A'(Y_s) dW_s - \frac{1}{2} \int_t^T A'(Y_s)^2 ds\right) =: \partial_t Y_T.\end{aligned}$$

Therefore the limit formula for H (where we have slightly abused the notation understanding that ∂_s is a linear operator) is

$$H = \int_0^T \partial_s \left(\left(\int_0^T \partial_t Y(T) dt\right)^{-1}\right) ds + \left(\int_0^T \partial_t Y(T) dt\right)^{-1} W_T.$$

In fact, ∂_t above is the well-known stochastic derivative (a.k.a. Skorohod derivative) of Malliavin calculus. The random variable H is sometimes called the Malliavin weight. For more details, see any standard reference on the topic.

Remark 10.2.5 Note that this is not the only way of performing the IBP; we remark that there is a way of obtaining stability using an argument based on products. The idea is based on the parametrix method of partial differential equations. As this method is complicated to introduce here we just mention it in passing, giving as Ref. [42] or [58].

10.3 The Change of Variables Method

Another method to obtain an IBP formula which is different from the previous one is obtained through a change of variables. This is somewhat the discrete equivalent version of the Bismut method for the IBP which is based on Girsanov's formula. As we will see later, this point of view is interesting as the conditions under which each version of the IBP formula will be valid are different as we only require that the finite-dimensional formula converges.

Theorem 10.3.1 *Let $H^{\xi}:\mathbb{R}\to\mathbb{R}$ for $\xi\geq 0$ be a sequence of C_b^1-smooth space transformations*[14] *satisfying the following conditions:*

$$H^{\xi}(z)\big|_{\xi=0}=z,\quad \partial_z H^{\xi}(z)\big|_{\xi=0}=1.$$

Define $h(z):=\partial_{\xi}H^{\xi}(z)\big|_{\xi=0}$. For a sequence of real numbers $\{a_k\}_{k=1}^n$ we have that the following IBP for X_n is valid with either

$$\begin{aligned}H=&\,\partial_k\left\{(\partial_k X_n)^{-1}h(Z_k)^{-1}\right\}h(Z_k)\\&+(\partial_k X_n)^{-1}h(Z_k)^{-1}\left\{(\log g)'\left(\frac{Z_k}{\Delta^{\alpha}}\right)\frac{h(Z_k)}{\Delta^{\alpha}}+h'(Z_k)\right\},\end{aligned}\tag{10.8}$$

or

$$\begin{aligned}H=&\sum_{k=0}^{n-1}\left\{\partial_k\left(a_k\left(\sum_{j=0}^{n-1}a_j\partial_j X_n h(Z_j)\right)^{-1}\right)h(Z_k)\right.\\&\left.+\left(a_k\left(\sum_{j=0}^{n-1}a_j\partial_j X_n h(Z_j)\right)^{-1}\right)\left\{(\log g)'\left(\frac{Z_k}{\Delta^{\alpha}}\right)\frac{h(Z_k)}{\Delta^{\alpha}}+h'(Z_k)\right\}\right\}.\end{aligned}\tag{10.9}$$

Here we assume that the random variables above have finite expectation.

Proof We proceed as follows: we consider a general expectation $\mathbb{E}[G(X_n)F(Z)]$, we rewrite it as an integral with respect to Z_k, we perform the change of variables $z=H^{\xi}(u)$, and then we differentiate with respect to $\xi=0$ in order to obtain an IBP. This gives

$$\mathbb{E}[G(X_n)F(Z)]=\int_{\mathbb{R}}\mathbb{E}\left[G(X_n)F(Z)|_{Z_k=H^{\xi}(u)}\right]g\left(\frac{H^{\xi}(u)}{\Delta^{\alpha}}\right)\frac{\partial_u H^{\xi}(u)}{\Delta^{\alpha}}du.$$

Differentiating the above expression with respect to ξ and taking $\xi=0$, we obtain

$$\begin{aligned}&\mathbb{E}\left[\partial_k\{G(X_n)F(Z)\}h(Z_k)\right]\\&=-\mathbb{E}\left[G(X_n)F(Z)\left\{(\log g)'\left(\frac{Z_k}{\Delta^{\alpha}}\right)\frac{h(Z_k)}{\Delta^{\alpha}}+h'(Z_k)\right\}\right].\end{aligned}\tag{10.10}$$

Therefore if we choose $F(Z)=(\partial_k X_n)^{-1}h(Z_k)^{-1}$ we obtain the unstable IBP.

Now we remark that the above argument also applies if

[14] In particular, our definition of "space transformation" includes the property that the transformation is one-to-one and that the inverse is differentiable.

$$F(Z) = F_k(Z) = a_k \left(\sum_{j=0}^{n-1} a_j \partial_j X_n h(Z_j) \right)^{-1}$$

and also if we sum (10.10) with respect to k. Therefore we have that due to the property $\sum_{k=0}^{n-1} \partial_k X_n h(Z_k) F_k(Z) = 1$ we obtain the corresponding stable formula (10.9).

$$\begin{aligned}
\mathbb{E}\left[G'(X_n)\right] &= \mathbb{E}\left[G'(X_n) \frac{\sum_{k=0}^{n-1} a_k \partial_k X_n h(Z_k)}{\sum_{j=0}^{n-1} a_j \partial_j X_n h(Z_j)} \right] \\
&= \sum_{k=0}^{n-1} a_k \mathbb{E}\left[\partial_k G(X_n) F_k(Z) h(Z_k)\right] \qquad (10.11) \\
&= \sum_{k=0}^{n-1} a_k \left\{ \mathbb{E}\left[\partial_k (G(X_n) F_k(Z)) h(Z_k)\right] - \mathbb{E}\left[(G(X_n) \partial_k F_k(Z)) h(Z_k)\right] \right\}.
\end{aligned}$$

From here the result follows.

Remark 10.3.2 Some remarks about the change of variables method are in order.

(i) The interpretation of the restrictions should be clear. Note that the conditions $H^{\xi}(z)\big|_{\xi=0} = z$ and $\partial_z H^{\xi}(z)\big|_{\xi=0} = 1$ are needed so that after differentiation we return to the original model. After careful consideration of a slightly modified model one can see that the condition $\partial_z H^{\xi}(z)\big|_{\xi=0} = 1$ can be dispensed with (this is an exercise for the reader).

(ii) Note the differences between the formulas (10.3) and (10.8). In fact, $h \equiv 1$ gives (10.3). The appearance of the last term is due to the space twist deformation. This reveals the geometric nature of the Girsanov formula.

(iii) The fact that the term $(\log g)'\left(\frac{Z_k}{\Delta^{\alpha}}\right)\frac{h(Z_k)}{\Delta^{\alpha}} + h'(Z_k)$ appears in (10.9) will be very important later. In fact, this way of writing the formula allows us to control degeneration behavior of the first term by choosing an appropriate h. Such analysis is not easy to see in the general formulations of Malliavin calculus for jump processes.

(iv) Using a function H^{ξ} which may depend on all the coordinates of Z (or even two of them) leads to interesting IBP formulas. Again the present form of the formula allows the understanding of the role of the change of variables.

(v) If one compares the statement of Theorem 10.2.1 and Theorem 10.3.1, one realizes that the fact that H is integrable is related with the ellipticity of A. In particular, using the notation introduced in the proof of Theorem 10.3.1, the conditions in Theorem 10.2.1 are related to F_k^{-1}.

Exercise 10.3.3 Prove that the formula (10.8) can also be obtained by a simpler argument using the classical change of variables and integration by parts formula.

Example 10.3.4 Consider the density function $g(x) = \frac{1}{2}x^{-1/2}$ for $x \in (0, 1]$.[15] Suppose that we can apply the result (10.6). Then in the weight H, the term $(\log g)'\left(\frac{Z_k}{\Delta^\alpha}\right) = -\frac{\Delta^\alpha}{2Z_k}$ which leads to instabilities of the IBP formula in the form of integrability problems. In fact, it is clear that $\mathbb{E}[(Z_k)^{-1}] = \infty$. Therefore the result in (10.6) cannot be satisfied in general.

Instead, consider $H^\xi(z) = z\exp(\xi z)$, $z \in [0, 1]$. Then $h(x) = x^2$ and the conditions of Theorem 10.3.1 are satisfied. Then, in this case

$$(\log g)'\left(\frac{Z_k}{\Delta^\alpha}\right)\frac{h(Z_k)}{\Delta^\alpha} + h'(Z_k) = \frac{3}{2}Z_k.$$

Therefore this will lead to a stable IBP formula in the sense that the previous integrability problem has been resolved. Furthermore, the sum appearing in (10.9) may now lead to a stochastic integral when limits are taken. This kind of problem typically appears in stable-like densities. Note that the above function H^ξ is not the only one that will achieve the same goal. In fact, $H^\xi(z) = z(1+\xi)$ will also achieve the same goal. In general, one may consider the following differential equation for a sufficiently regular function $v \equiv v_{\alpha,\Delta}$ and $C \equiv C_{\alpha,\Delta}$:

$$C\frac{h(x)}{x} + h'(x) = v(x). \tag{10.12}$$

Then the general solution is of the form

$$h(x) = x^{-C}\int x^C v(x)dx. \tag{10.13}$$

The exact initial conditions and the function G are to be chosen by the user in order to establish the desired regularity properties. In particular, if one desires that $h(0) = h(1) = 0$. For example, if $v(x) = a + bx$ then the parameters a and b should satisfy $\frac{a}{C+1} + \frac{b}{C+2} = 0$.

In fact, most of the hypotheses in Theorem 10.3.1 are local conditions around $\xi = 0$.

This example shows the flexibility of the change of variables method. We will see more of this in Example 10.4.5. In fact, it is this flexibility in the choice of h which will allow different goals in different problems. We will use a particular choice of h in order to analyze the Boltzmann equation in Chap. 13 later on.

Example 10.3.5 In this example, we discuss an application of the above change of variables in a two-dimensional setting. For this discussion, one needs to reconsider the generalization of the IBP formula to two dimensions, as it has been carried out in Exercises 10.1.22 and 10.1.23.

---

[15]Note that this example does not satisfy our hypothesis of smoothness of g in $x = 0, 1$, stated after (10.1), therefore some modifications will be needed. More discussion appears in Sect. 9.1.

For example, consider the set-up of Exercise 10.1.23. Carry out the change of variables $(x_1, x_2) = h(z_1, z_2)$. Find the corresponding IBP formula as in Theorem 10.3.1. Discuss the possible optimality considerations for h.

As the set-up may be rather vague for the reader, we develop this example further. For this, let $X = U(Z^1, Z^2)$ with (Z^1, Z^2) a two-dimensional random vector with smooth density $g : \mathbb{R}^2 \to \mathbb{R}_+$. Again let us assume the existence of the sequence of smooth space transformations $H^\xi(z_1, z_2)$. Then as in the proof of Theorem 10.3.1 we have

$$\begin{aligned}\mathbb{E}[G(X)F(Z)] &= \int_{\mathbb{R}^2} G(U(x_1, x_2))F(x_1, x_2)g(x_1, x_2)dx_1dx_2\\ &= \int_{\mathbb{R}^2} G(U(H^\xi(z_1, z_2)))F(H^\xi(z_1, z_2))g(H^\xi(z_1, z_2))J_{H^\xi}(z_1, z_2)dz_1dz_2.\end{aligned}$$

Here J_{H^ξ} stands for the Jacobian associated with H^ξ. Differentiating with respect to ξ and taking $\xi = 0$, we obtain

$$\mathbb{E}[\partial_k\{G(X)F(Z)\}] = -\mathbb{E}\left[G(X)F(Z)\left\{\nabla(\log g)(Z)\,h(Z) + J_h(Z)\right\}\right].$$

The main advantage of the above formula is the fact that now there are two functional parameters $h = (h_1, h_2)$ in the problem that could be chosen depending on the objective of the problem. In particular, one can combine the correlation effects between the components of Z.

Note that from the above IBP formulas, as well as in order to obtain an IBP formula for $\mathbb{E}[G^{(n)}(X)]$ in inductive form it is much better to provide a general integration by parts formula for $\mathbb{E}[G'(X)F(X)]$ for a function F with enough regularity conditions. There is another method to obtain a change of variables formula which also implies the change of probabilistic model which introduces also the motivation of the next section.

10.4 IBP with Change of Weak Representation Formula

In probability theory an expectation can be expressed in different forms using different random variables. Each expression will bring a different integration by parts formula. This technique is ad-hoc at the present moment and therefore we present it through some examples.

Specifically, in this section we want to discuss how the IBP formula depends on the representation formula used for X_n as a function of other random variables. A very simple example of this is given in Problem 10.1.11. Furthermore, the formula described in Theorem 10.3.1 is one formula of this type. The change of variables used corresponds to writing X_n as a function of $(H^\xi)^{-1}(Z_k)$, $k = 0, ..., n-1$.

- An alternative probabilistic representation can be obtained by rewriting $Z_k = Z_k^1 + Z_k^2$, where $(Z_k^j)_{k=0,\dots,n}^{j=1,2}$ is a sequence of independent random variables. Then we can obtain an IBP formula based on $(Z_0^1, \dots, Z_{n-1}^1)$. This method is sometimes called partial Malliavin calculus. Here we prefer to call it "alternative representation by convolution", the reason being that in the analytical sense the sum of random variables is represented by a convolution.
- Another representation can be obtained by randomization. That is, let $(\varepsilon_k^j)_{k=0,\dots,n}^{j=1,2}$ be a sequence of i.i.d. Bernoulli random variables such that $Z_k = \varepsilon_k^1 Z_k^1 + \varepsilon_k^2 Z_k^2$. We call this "alternative representation by sum".

10.4.1 The Plain Change of Variables Formula

This is the most straightforward version of the formula, although there is a change in the representation of the model.

Theorem 10.4.1 *Let h be a strictly increasing C_b^1-function. Define the random variables $U_k = h(Z_k)$. Suppose the same conditions as in Theorem 10.1.16, where the conditions on the density g are replaced by the corresponding density of U_k given by $\hat{g}(x) = \frac{1}{\Delta^\alpha} g\left(\frac{h^{-1}(x)}{\Delta^\alpha}\right)(h'(h^{-1}(x)))^{-1}$. Then we have the IBP formula*

$$H = \left(\hat{\partial}_k(\hat{\partial}_k X_n)^{-1} + (\hat{\partial}_k X_n)^{-1}(\log \hat{g})'(U_k)\right). \tag{10.14}$$

Here $\hat{\partial}_k X$ denotes the stochastic derivative of the random variable X when written as a functional of the i.i.d. sequence of random variables $U := \{U_k; k = 0, \dots, n-1\}$.

Proof As the proof is similar to the proof of Theorem 10.1.16 we only give a sketch. We perform the change $x = h(z)$, defining $U_k := h(Z_k)$:

$$\begin{aligned}
\mathbb{E}\left[G'(X_n)\right] &= \int_{\mathbb{R}} \mathbb{E}\left[\hat{\partial}_k\{G(X_n)\}(\hat{\partial}_k X_n)^{-1}\Big|_{U_k=x}\right]\hat{g}(x)dx \\
&= \int_{\mathbb{R}} \mathbb{E}\left[\left(G(X_n)|_{U_k=\cdot}\right)'(x)\,(\hat{\partial}_k X_n)^{-1}\Big|_{U_k=x}\right]\hat{g}(x)\,dx \\
&= -\mathbb{E}\left[G(X_n)H\right].
\end{aligned}$$

Remark 10.4.2 (i) We remark that writing the term $\hat{\partial}_k(\hat{\partial}_k X_n)^{-1}$ implies the assumption that $\hat{\partial}_k X_n$ can be written as a non-zero functional of U which is differentiable and integrable.

(ii) The formula (10.14) is not so different from (10.3). It is just that the basic random variables are different. This introduces a change in the expression for H that has to be clearly understood. In particular, the behavior of the logarithmic derivative of $\hat{g}$ will become important.

We finish this section with the corresponding stabilization formula.

Theorem 10.4.3 *For any auxiliary sequence of $\wedge$-uniformly elliptic C_b^1-bounded random variables a_k, $k = 1, ..., n-1$*[16] *and under the same conditions as in Theorem 10.4.1 we have that the IBP formula is valid with*

$$H = \sum_{k=0}^{n-1} \left\{ \hat{\partial}_k \left(a_k \left(\sum_{j=0}^{n-1} a_j \hat{\partial}_j X_n \right)^{-1} \right) + \left(a_k \left(\sum_{j=0}^{n-1} a_j \hat{\partial}_j X_n \right)^{-1} \right) (\log \hat{g})' (U_k) \right\}. \tag{10.15}$$

Exercise 10.4.4 Here, we introduce a second idea in order to deal with boundary effects in the density g. Assume the same setting as in Exercise 10.2.4. First define the change of variables $z = \tan(\frac{x\pi}{2})$ so that the uniform density on $[-1, 1]$ is now replaced by a Cauchy density which has support on $\mathbb{R}$. Prove that the model defined in (10.1) with $Z_k \sim U[-1, 1]$ is now equivalent in law to the model where Z_k are i.i.d. r.v.s with Cauchy distribution. In general, for this method to be successful, it requires that the random variable X_n has a particular behavior when $U = -1, 1$.

10.4.2 The Change of Representation Argument by Sum and Convolution

In this section we will put some techniques together in order to describe how to carry out an integration by parts formula for stable-like random variables Z_i. The main feature in this section is that the integration by parts will be carried out with respect to the laws of the small jumps. This will introduce further problems for which the solution is briefly described here. Most of these techniques will be used later on in Chap. 13.

Example 10.4.5 First, we will represent a stable law as $Z_k = \varepsilon_k Z_k^1 + (1 - \varepsilon_k) Z_k^2$ such that Z_k^1 will represent the jumps in the interval $(\zeta, 1)$ and Z_k^2 will represent the rest of the other jumps. This example has a deep relation with the Boltzmann equation to be treated later in Chap. 13 and it was introduced in [7]. In that case the density function of Z_k^1 is given by $f_\zeta(x) = C_\zeta^{-1} x^{-(1+\alpha)} \mathbf{1}(\zeta < x < 1)$ for $\zeta < (1+\alpha)^{-\frac{1}{\alpha}}$, $\alpha \in (0, 2)$ and $C_\zeta := \frac{\zeta^{-\alpha}-1}{\alpha} > 1$. The desired change of variables is given by the function $h^{-1}(z) = (\zeta^{-\alpha} - \alpha z)^{-\frac{1}{\alpha}}$. In this case $h(x) = \frac{\zeta^{-\alpha} - x^{-\alpha}}{\alpha}$, for $x \in (\zeta, 1)$, $\alpha \in (0, 2)$. Now, we let $\{(\varepsilon_k, Z_k^1); k = 0, ..., n-1\}$ be a sequence of i.i.d. vectors such

[16] $\wedge$ means that the requirement is related to the representation of X_n using the new random variables U.

that ε_k is a Bernoulli r.v. with parameter $p \equiv p(\zeta, \Delta) = C_\zeta \Delta < 1$[17] independent of Z_k^1. Z_k^1 has as density the function f_ζ.[18] The goal here is to find an IBP formula for $X_n(\zeta)$ that will be stable as $\zeta \to 0$ and $\Delta \to 0$. In order to explain things slowly, we give first the change of probabilistic representation:

$$\mathbb{E}\left[G'(X_n)\right] = C_\zeta^{-1}\Delta \int_\zeta^1 \mathbb{E}\left[G'(X_n)\middle|\, Z_k^1 = x, \varepsilon_k = 1\right] x^{-(1+\alpha)}dx \\ +\mathbb{E}\left[G'(X_n)\middle|\, \varepsilon_k = 0\right](1 - C_\zeta \Delta).$$

Now using the change of variables $x = h^{-1}(z) = (\zeta^{-\alpha} - \alpha z)^{-\frac{1}{\alpha}}$, for $z \in [0, C_\zeta]$. Then we obtain that as $f_\zeta(h^{-1}(z))h'(h^{-1}(z))^{-1} = C_\zeta^{-1}$,[19]

$$\begin{aligned} &C_\zeta^{-1} \int_\zeta^1 \mathbb{E}\left[G'(X_n)\middle|\, Z_k^1 = x, \varepsilon_k = 1\right] x^{-(1+\alpha)}dx \\ &= C_\zeta^{-1} \int_0^{C_\zeta} \mathbb{E}\left[G'(X_n)\middle|\, h(Z_k^1) = z, \varepsilon_k = 1\right] dz. \end{aligned} \tag{10.16}$$

Therefore we can use instead of the random variables Z_k^1 a sequence of i.i.d. r.v. s $U_k := h(Z_k^1)$ which are uniformly distributed on $[0, C_\zeta]$. Finally, the probabilistic model representation can be changed accordingly. We leave the rest of the probabilistic description to the reader as a non-trivial exercise. This idea in order to obtain an integration by parts formula is interesting as the density f_ζ around zero blows up with a higher degree each time a derivative is taken for $\zeta \approx 0$. This new representation allows the exchange of this blow-up behavior with a stretching of the region of integration.[20]

Now another problem appears: the density of U does not vanish at the endpoints of the support. This can be solved with a well-known localization technique which is just an alternative representation by sum.[21] Or one may improve the approximation technique as follows: Let $\varphi_{\zeta,1}(x)$ be a C_c^∞ function which satisfies

---

[17] This is a condition on Δ.

[18] Note that the set-up is slightly different from the one given in Sect. 10.1 with respect to the form of the density of Z_k.

[19] We see here the differential equation related to the change of variable quoted in Remark 10.3.2(iii). That is, the last term in (10.14) vanishes. Otherwise if such a change is not used, one would obtain a term of the order -1. Compare this with Example 10.3.4.

[20] That is, $C_\zeta \to \infty$ as $\zeta \to 0$. Clearly this can be done only in the case that $\varepsilon_k = 1$. But this can be done often if we choose ζ as an appropriate function of Δ so that $\mathbb{P}(\varepsilon_k = 1)$ is close to one.

[21] See Exercise 10.4.7.

$$\varphi_{\zeta,1}(x) = \begin{cases} 0 & (x \le \zeta), \\ \text{increasing} & (\zeta < x < 2\zeta), \\ 1 & (2\zeta \le x \le 1-\zeta), \\ \text{decreasing} & (1-\zeta < x < 1), \\ 0 & 1 \le x. \end{cases}$$

Explicit functions of this type can be obtained using the function $\exp(-\frac{1}{x})$. One can now modify (10.16) as follows:

$$C_\zeta^{-1} \int_\zeta^1 \mathbb{E}\left[G'(X_n) \middle| Z_k^1 = x, \varepsilon_k = 1 \right] x^{-(1+\alpha)} \varphi_{\zeta,1}(x) dx$$
$$= C_\zeta^{-1} \int_{\mathbb{R}} \mathbb{E}\left[G'(X_n) \middle| h(Z_k^1) = z, \varepsilon_k = 1 \right] \varphi_{\zeta,1}(h(z)) dz.$$

The final argument will depend on estimates of the error as ζ becomes closer to zero. Therefore the above argument is an approximation argument which could lead to an exact IBP by using techniques similar to Problem 10.1.8. Otherwise one may also use the approximation result in Theorem 9.1.10.

An important remark on this method: We do not need to express the first expectation on the same original variables as the one used in the first set-up problem. Also we may use the function $\varphi_{\zeta,1}(h(z))$ to construct a new random variable with smooth density and bounded support.

In some way, these ideas have a deep link with the stratification method described in [19]. Some of these ideas will be put to work in the case of the Boltzmann equation in Chap. 13, so if you want some hints for the solutions to the proposed exercises just read that chapter.

Exercise 10.4.6 (*Continuation of Example* 10.4.5) Propose an alternative to the method proposed previously in the light of Example 10.3.4. In fact, let us for the sake of argument assume that $\Delta = 1$. Then as stated previously, any acceptable change of variables will lead to the formula (10.12) and its corresponding general solution (10.13). Propose some alternative change of variables.

Exercise 10.4.7 This exercise shows how to perform an IBP formula when the random variables are defined on a close interval, and is another variant for Problem 10.1.8. Find an alternative density function $g(x)$ in order to replace f_ζ such that it satisfies the following conditions

- $g(x)\,\mathbf{1}(x \notin (\zeta, 1)) = 0$.
- $g(x) = Cx^{-(1+\alpha)}$ for x in a interval on length of at least $(1-\zeta)/2$.
- g is a smooth function with bounded derivatives.

This new density can be used instead of f_ζ in order to obtain similar results. The difference is that the above density function will not generate boundary terms when performing integration by parts.

For a hint, see Chap. 14.

10.5 The Case of Jump Processes

So far, our goal has been to obtain an IBP formula of the type $\mathbb{E}\left[G'(X)\right] = -\mathbb{E}\left[G(X)H\right]$. In the case that X is a random variable which depends on jump processes the goal of obtaining the IBP may be too difficult sometimes, and therefore one may reduce expectations and obtain the IBP formula for a particular type of functions G. For example, if $G(x) = e^{i\theta x}$, then an IBP formula could help in order to determine the existence and regularity of the density of the random variable X, as explained in Chap. 9.

In this section, we will introduce some simple methods that will be later exploited for studying the density of jump-driven stochastic differential equations.

To start, we have to simply recognize that in comparison with the Gaussian situation there may be cases where one may not deal with the case of Poisson processes in the same fashion as in Theorem 10.1.16 and its stable extension Theorem 10.2.1. The first reason for this is that as we know the Poisson process does not have a density. In fact, recall from Proposition 2.1.16 that if $\{N_t; t \geq 0\}$ be a Poisson process of parameter $\lambda > 0$, then we have that N_t is a discrete random variable and its law is given by

$$\mathbb{P}(N_t = k) = \frac{e^{-\lambda t}(\lambda t)^k}{k!}, \quad k \in \mathbb{N}^*.$$

On the other hand, we have also seen that in Proposition 6.3.1 and Exercise 6.3.2, that in the case of a compound Poisson process with jumps, densities may exist away from zero if the Lévy measure is absolutely continuous and finite.

Following Picard's approach [47], we will try to concentrate first on a very particular integration by parts formula that essentially applies only to the case of the characteristic function.

Exercise 10.5.1 Using the explicit characteristic function of a Gaussian random variable, prove that the Wiener process at time $t > 0$, denoted by W_t, satisfies the IBP formula for the function $e^{i\theta x}$ with weight $H = t^{-1}W_t$,

$$t\mathbb{E}[i\theta e^{i\theta W_t}] = \mathbb{E}[e^{i\theta W_t} W_t].$$

On the other hand, use the explicit form of the characteristic function of the compound/simple Poisson process to show that the above equality is not satisfied in general if we replace W by a simple or compound Poisson process. For more information, see Theorem 10.5.2.

For a hint, see Chap. 14.

In fact, a different version of the above formula which may still be used as an IBP formula is the following partial result.

Theorem 10.5.2 *Let $Z_t = \sum_{i=1}^{N_t} Y_i$ be a compound Poisson process where Y_i are i.i.d. random variables with a $C_0^1(\mathbb{R})$ density function g. Define $h(z) = \partial_z \log(g(z))$ and assume that $\int_0^t \int h(z)\mathcal{N}(dz, ds) = \sum_{i=0}^{N_t} h(Y_i) \in L^1(\Omega)$. Then the following* *"modified" IBP formula for $e^{i\theta z}$ is satisfied:*

$$\mathbb{E}[i\theta e^{i\theta Z_t} N_t \mathbf{1}(N_t > 0)] = -\mathbb{E}[e^{i\theta Z_t} \int_0^t \int h(z)\mathcal{N}(dz, ds)\mathbf{1}(N_t > 0)].$$

Here we recall the reader that $\mathcal{N}(dz, ds)$ stands for the Poisson random measure associated to Z.

Remark 10.5.3 1. The reason why the above is called a modified IBP formula is because one has $e^{i\theta \cdot}$ on the right-hand side of the above equation and its derivative $i\theta e^{i\theta \cdot}$ on the left-hand side. It is a modified formula because it is not satisfied on the whole probability space but on the set $\{N_t > 0\}$ with some special weights (r.v.s) multiplying the particular exponential functions.
Note that in the case of $N_t = 0$ there is no density and therefore the IBP formula cannot be expected to hold in the form above. This remark is also related to the fact that the reasonable way to make out of this a fully fledged IBP formula is to take a sequence of Poisson processes with parameter λ converging to infinity so that $\mathbb{P}(N_T = 0) = e^{-\lambda T} \to 0$.
2. From the above formula one also understands that the factor that appears in the IBP formula is due to the logarithmic derivative of the density of the jumps. Therefore one should understand why in exercise 10.5.1 the IBP formulas for the Wiener and compound Poisson process do not have the same form. Similarly, we will also see that in many cases the logarithmic derivative of the density of jumps is not integrable. In that case we may use the technique explained in Theorem 10.3.1.
3. There are two reasons why the exponential functions are used in this section. The first is that obtaining IBP formulas for exponential functions may indicate that these formulas may be valid in general if one understands the series development of Fourier expansions. The other reason is that the study of the density of a random variable is done through the study of its characteristic function.

Proof We start by computing

$$\begin{aligned}
&\mathbb{E}[e^{i\theta Z_t} \int_0^t \int h(z)\mathcal{N}(dz, ds)\mathbf{1}(N_t > 0)] \\
&= \sum_{k=1}^{\infty} \mathbb{P}(N_t = k) \sum_{l=1}^{k} \mathbb{E}[\prod_{j=1}^{k} e^{i\theta Y_j} h(Y_l)] \\
&= \sum_{k=1}^{\infty} \mathbb{P}(N_t = k) \sum_{l=1}^{k} \left(\prod_{\substack{j=1 \\ j \neq l}}^{k} \mathbb{E}[e^{i\theta Y_j}] \right) \mathbb{E}[e^{i\theta Y_l} h(Y_l)] \\
&= -i\theta \mathbb{E}[e^{i\theta Z_t} N_t \mathbf{1}(N_t > 0)].
\end{aligned}$$

Note that in the last step, we have used that for $h(z) = \partial_z \log(g(z))$, we obtain using the IBP formula on the distribution of the jump size that $\mathbb{E}[e^{i\theta Y}h(Y)] = \int e^{i\theta y}h(y)g(y)dy = -i\theta\mathbb{E}[e^{i\theta Y}]$. This is the reason for the last equality if one uses the independence of the jump sizes $\{Y_i; i \in \mathbb{N}\}$.

Exercise 10.5.4 Find conditions on h so that $\int h(z)\nu(dz) = 0$. Therefore the formula in Theorem 10.5.2 can now be written in the form

$$\mathbb{E}[i\theta e^{i\theta Z_t} N_t \mathbf{1}(N_t > 0)] = -\mathbb{E}[e^{i\theta Z_t} \int_0^t \int h(z)\widetilde{\mathcal{N}}(dz, ds)\mathbf{1}(N_t > 0)].$$

In general, this will imply that conditions of the function h can be weakened.

Exercise 10.5.5 Propose the same result for kth-order derivatives. That is, find the proper function h_k such that one can prove the equality

$$\mathbb{E}[(i\theta)^k e^{i\theta Z_t} N_t \mathbf{1}(N_t > 0)] = (-1)^k\mathbb{E}[e^{i\theta Z_t} \int_0^t \int h_k(z)\mathcal{N}(dz, ds)\mathbf{1}(N_t > 0)].$$

Analyze the proper conditions that should be required for g in order for the formula to be valid.[22]

For a hint, see Chap. 14.

Exercise 10.5.6 Here, we present a different way of obtaining a similar result.

Prove that in the case of second-order derivatives, there exists h such that the following formula is satisfied:

$$\begin{aligned}
&\mathbb{E}[(i\theta)^2 e^{i\theta Z_t} N_t(N_t - 1)\mathbf{1}(N_t > 0)] \\
=& - \mathbb{E}[e^{i\theta Z_t} \int_0^t \int \int_0^s \int h(z)\mathcal{N}(dz, du)h(w)\mathcal{N}(dw, ds)\mathbf{1}(N_t > 0)].
\end{aligned}$$

Consider if the above formulation can be generalized for kth-order derivatives.

10.5.1 The Picard Method

In this section we want to treat briefly the example of a Lévy process which has a smooth law, although the corresponding Lévy measure is not absolutely continuous with respect to the Lebesgue measure. For this, recall Exercises 4.1.23, 5.1.17 and 5.1.19.

We will not go deeply into the full details and power of this method and we refer the reader to the associated literature in [47] or [31].

The first important point in this section is that we will give up on the idea of obtaining a clean and clear IBP formula like the ones stated in the previous section.

[22] Hint: Do not forget the boundary conditions for the integrals of h_k at $\pm\infty$.

On the other hand, this section is modeled through Exercise 5.1.19 where a sequence of point masses in a discrete-type Lévy measure will generate an infinite differentiable density. As we have always started with Lévy measures with densities, we will take a sequence of approximations to a discrete measure.

In order to describe an IBP formula for the particular case of characteristic functions, we will first generalize Theorem 10.5.2 to the case of sums of compound Poisson processes. That is, let $\bar{Z}_t^n = \sum_{i=-n}^n Z_t^i$, where $\{Z^i; i = -n, ..., n\}$ is a sequence of independent compound Poisson processes with parameters $\lambda_i, g_i, i = -n, ..., n$, where g_i are the corresponding Lévy densities which have disjoint supports. N^i denotes the corresponding jump counting process for Z^i, $\bar{N}_t^n := \sum_{i=1}^n N_t^i$. For simplicity, we assume that $\lambda_0 = 0$ so that there are only an even number of compound Poisson processes being considered. Then in this case we also have the following result. Its proof is a direct application of Theorem 10.5.2 once one knows that the Lévy measure associated with $\bar{Z}^n$ is $\sum_{i=-n}^n \lambda_i g_i(z)dz$.

Theorem 10.5.7 *Let $h(z) = \partial_z \log(\sum_{i=-n}^n \lambda_i g_i(z))$ be such that $\int_0^t \int h(z)\mathcal{N}(dz, ds)$ is integrable. Then the modified integration by parts formula for $e^{i\theta z}$ is satisfied:*

$$\mathbb{E}[i\theta e^{i\theta \bar{Z}_t^n} \bar{N}_t^n \mathbf{1}(\bar{N}_t^n > 0)] = -\mathbb{E}[e^{i\theta \bar{Z}_t^n} \int_0^t \int h(z)\mathcal{N}(dz, ds)\mathbf{1}(\bar{N}_t^n > 0)].$$

Exercise 10.5.8 State a condition that assures that $\int_0^t \int h(z)\mathcal{N}(dz, ds)$ is integrable. As stated in Exercise 10.5.4, find formulas like the one above which require conditions on the integrability of $\int_0^t \int h(z)\widetilde{\mathcal{N}}(dz, ds)$.

For a hint, see Chap. 14.

Exercise 10.5.9 Think of a situation in Theorem 10.5.7 related to Exercise 5.1.19 where $\int_0^t \int h(z)\mathcal{N}(dz, ds)$ does not converge weakly as $n \to \infty$.

From this exercise, one understands that the previous formula, which is a multiplicative-type IBP formula (this is the reason for the appearance of the logarithm in the definition of h) cannot be extended to a general situation. In exchange, one can add formulas of the above type. This is the direction we will take in what follows. We have to note that the complication in the current setting is that we will give up on the idea of a perfect IBP formula and just obtain an approximative formula which is based not on the jump size as in Theorem 10.5.7 but on the structure of the underlying Poisson processes.

Lemma 10.5.10 *The Poisson process N satisfies the following approximative integration by parts formula for $e^{i\theta x}$ is valid for any $\theta \in \mathbb{R}$ and $t > 0$:*

$$(e^{i\theta} - 1)\mathbb{E}[e^{i\theta N_t}]\lambda t = \mathbb{E}[e^{i\theta N_t}\tilde{N}_t].$$

Proof The proof is a simple calculation which is based on the following equality: $k\mathbb{P}(N_t = k) = \mathbb{P}(N_t = k - 1)\lambda t$.

Note that if one interprets that $e^{i\theta} - 1 \simeq i\theta$ then the above formula is some type of IBP formula. In the next result, we will "mount" the jump structure onto the above approximative IBP formula.

Theorem 10.5.11 *Let $w : \mathbb{R} \to \mathbb{R}$ be a measurable function so that*

$$\int_0^t \int |w(z)| \widehat{\mathcal{N}}(dz, ds) < \infty.$$

As before assume that $\bar{Z}_t^n = \sum_{i=-n}^n Z_t^i$, then the following "modified" IBP formula for $e^{i\theta z}$ is satisfied:

$$\int_0^t \int (e^{i\theta z} - 1) w(z) \widehat{\mathcal{N}}(dz, ds) \mathbb{E}[e^{i\theta \bar{Z}_t^n}] = \mathbb{E}[e^{i\theta \bar{Z}_t^n} \int_0^t \int w(z) \widetilde{\mathcal{N}}(dz, ds)]. \tag{10.17}$$

Remark 10.5.12 The reason why the above formula is a modified IBP formula should be clear. On the left side, one does not obtain $i\theta$ but $e^{i\theta z} - 1$, which is an approximation of $i\theta z$ but it is further integrated with respect to the compensator of $\mathcal{N}$. Still, many things remain unclear with this formula at this point.

For example, does it really provide an IBP formula, in order to obtain the necessary decreasing property of the characteristic function as $|\theta| \to \infty$?[23] Before discussing this more in detail, let us provide the proof of the theorem.

Proof In fact, it is just a matter of direct calculation. We will do the proof in the particular case of $Z_t^1 = \sum_{k=1}^{N_T^1} Y_i^1$ and leave the rest of the details of the proof in the general case for the reader.

$$\begin{aligned}
\mathbb{E}[e^{i\theta Z_t^1} \int_0^t \int w(z) \mathcal{N}^1(dz, ds)] =& \sum_{k=1}^{\infty} \mathbb{P}(N_T^1 = k) (\mathbb{E}[e^{i\theta Y^1}])^{k-1} \mathbb{E}[e^{i\theta Y^1} w(Y^1)] k \\
=& \sum_{k=0}^{\infty} \mathbb{P}(N_T^1 = k) \lambda^1 t (\mathbb{E}[e^{i\theta Y^1}])^k \mathbb{E}[e^{i\theta Y^1} w(Y^1)] \\
=& \mathbb{E}\left[e^{i\theta Z_t^1} \int_0^t \int e^{i\theta z} w(z) \widehat{\mathcal{N}}^1(dz, ds) \right].
\end{aligned}$$

Note that in the second equality we have used the identity $\mathbb{P}(N_T^1 = k) = \frac{\lambda^1 t}{k} \mathbb{P}(N_T^1 = k - 1)$ together with a change of variables in the sum.

To finish the proof one just needs to subtract $\mathbb{E}\left[e^{i\theta Z_t^1}\right] \int_0^t \int w(z) \widehat{\mathcal{N}}(dz, ds) < \infty$ from both sides of the above equality.

Note that in the above result we have a compensated Poisson random measure integral on the right-hand side of (10.17) and on the left-hand side the compensator. For this

[23] Recall the result in Exercise 1.1.11.

reason, some people say that the IBP formula (10.17) is based on a duality formula for Poisson processes. But note that this feature is linked strongly with the fact that one uses the exponential function. It is also a formula that only uses the structure of the Poisson process and not of the jumps as stated in the next exercise.

Exercise 10.5.13 Prove the full statement of Theorem 10.5.11. Note that the proof of this result does not depend on the particular form of the Lévy measure $\sum_{i=-n}^{n} \lambda_i g_i(z)dz$. This form will be used in the next Example 10.5.16.

Exercise 10.5.14 State and prove an extension of Theorem 10.5.11 for functions $w(s, z)$ which are continuous with respect to the time variable s. Hint: First prove an statement similar to the one in Theorem 10.5.11 for simple functions with respect to the time variable and then take limits.

Exercise 10.5.15 Use the following martingale-type argument in order to prove Lemma 10.5.10. Define $u(t-s, x) := \mathbb{E}[e^{i\theta N_t} \mid N_s = x]$. Use Itô's formula for $u(t-r, N_r), r \in [0, t]$ in order to obtain the result. Note that the introduction of the function u is crucial for the argument to work. If one applies directly the Itô formula to $e^{i\theta N_t}\tilde{N}_t$ the time variable on $e^{i\theta N_t}$ will start to change dynamically in the interval $[0, t]$ which will not allow us to recover the same function as stated in Lemma 10.5.10.

We will now move forward towards proving a result similar to the one in Exercise 4.1.23 and 5.1.19 using the technique presented in Theorem 10.5.11.

Rather than dealing with a general situation, we will show the strategy in a simpler example with a weaker result.

Example 10.5.16 Consider the following particular example: We define for $\lambda_j \equiv 1$, $j = 0, ..., n$ the sequence of functions for $x \in (-1, 1)$ and $\alpha \in (0, 2)$,

$$g_j(x) := \mathbf{1}_{(2^{-(j+1)}, 2^{-j})}(x)|x|^{-(1+\alpha)},$$
$$g_{-j}(x) := \mathbf{1}_{(-2^{-j}, -2^{-(j+1)})}(x)|x|^{-(1+\alpha)}.$$

For $\alpha \in (0, 2)$ and $\varepsilon \in (0, \alpha/2)$, prove that there exists a constant C independent of n such that for all $\theta \in \mathbb{R}$

$$|\mathbb{E}[e^{i\theta Z_t}]| \leq \frac{C}{1 + |\theta|^{\alpha/2-\varepsilon}}. \tag{10.18}$$

Here Z stands for the Lévy process with Lévy measure given by $\nu(dx) = \mathbf{1}_{(-1,1)}\frac{dx}{|x|^{1+\alpha}}$ whose construction was given in Chap. 5 as the compensated limit of $\bar{Z}^n$. Using Lemma 9.1.8, prove that the above inequality implies that for $\alpha > 1$ a density fpr Z_t exists.

Proof We apply Theorem 10.5.11 for the function $w(z) = |z|^r \mathbf{1}_{(-1,1)}(z)$, $r > -1$. We leave for the reader to check that the conditions in Theorem 10.5.11 are satisfied.

First, we will find a lower bound for

$$\int_0^t \int_{-1}^1 (e^{i\theta z} - 1)|z|^r \widehat{\mathcal{N}}(dz, ds).$$

Due to the symmetry of the sequence g_j and the fact that the sine function is an odd function, we have as in Exercise 4.1.23,

$$I_n := \left| \int_0^t \int (e^{i\theta z} - 1)|z|^r \widehat{\mathcal{N}}(dz, ds) \right| = 2t \left| \int_{2^{-(n+1)}}^1 \frac{|\sin^2(\frac{\theta z}{2})|}{|z|^{\alpha+1-r}} dz \right|$$

$$= 2^{r-\alpha} |\theta|^{\alpha-r} t \int_{2^{-(n+2)}\theta}^{2^{-1}\theta} \frac{|\sin^2(u)|}{|u|^{\alpha+1-r}} du.$$

Now for any fixed $|\theta| > 1$, we have

$$\lim_{n\to\infty} I_n \geq 2^{r-\alpha} |\theta|^{\alpha-r} t \int_0^{\theta/2} \frac{|\sin^2(u)|}{|u|^{\alpha+1-r}} du.$$

Note that the above integral is finite in the neighborhood of zero if $r > \alpha - 2$ because $|\sin(u)| \leq |u|$. It will also be integrable at ∞ if $\alpha > r$ because the sin function is bounded.

Next, we need to find an upper bound for the right-hand side of (10.17). To do this, is enough to use Proposition 7.1.2 (the Burkholder–Davis–Gundy inequality) in order to obtain that for $r > \alpha/2$

$$\mathbb{E}\left[\left| \int_0^t \int w(z) \widetilde{\mathcal{N}}(dz, ds) \right| \right] \leq ct \left(\int_{[-1,1]} |z|^{2r-1-\alpha} dz \right)^{1/2} < \infty.$$

Of all three restrictions for r the most restrictive is the last one, which leads to the inequality

$$|\mathbb{E}[e^{i\theta \bar{Z}_t^n}]| \leq \frac{C}{1 + |\theta|^{\alpha/2-\varepsilon}}.$$

The inequality (10.18) follows by noting that $\sum_{j=-n}^{j=n} \int_{-1}^1 x g_j(x) dx = 0$ and that as shown in Chap. 5, the process $\tilde{Z}_t^n = \bar{Z}_t^n - t \sum_{j=-n}^{j=n} \int_{-1}^1 x g_j(x) dx$ converges *a.s.* to the Lévy process Z. The existence of the density then follows by Lemma 9.1.8 as the characteristic function of Z is square integrable in $\theta \in \mathbb{R}$.

Remark 10.5.17 (i) Note that in order to obtain the argument in other cases, one may need to further compensate the integrals appearing in Theorem 10.5.11.

(ii) Note also that the argument is somewhat imprecise when one compares the above with the exact calculation of the characteristic function. This is inherent to the method.

(iii) In order to obtain further regularity properties of the density of Example 10.5.16 as they were obtained in Exercise 4.1.23, we will need to repeat the procedure in Theorem 10.5.11 as needed and the above arguments as much as needed and then apply Corollary 9.1.3. The advantage of the present method in comparison to the one used in Example 4.1.23 is that the argument presented here can be generalized to other situations such as stochastic equations.

Problem 10.5.18 Prove again the result in Exercise 5.1.19 by using the fact that the proof of Theorem 10.5.11 can also be stated in the setting of Exercise 5.1.19. In particular, prove the existence of the density of the process defined in Exercise 5.1.19.

Hint: Note the parallel with the argument of Example 10.5.16.

10.5.1.1 The Difference Operator Approach

The plan for future chapters is to give an explanation for Theorem 10.5.7 from the setting of differential operators on probability spaces.

We will try to briefly explain the result in Theorem 10.5.11 from the point of view of IBP using difference operators instead of derivative operators. We start with a trivial exercise about the discrete IBP formula.

Exercise 10.5.19 Let $\{a_n\}_{n\in\mathbb{N}}$ and $\{b_n\}_{n\in\mathbb{N}}$ be two sequence of real numbers. Prove the following discrete IBP formula:

$$\sum_{k=1}^{m}(a_k - a_{k-1})b_k = -\sum_{k=1}^{m}(b_k - b_{k-1})a_{k-1} + a_m b_m - a_0 b_0. \tag{10.19}$$

Give conditions so that the above formula is also valid for $m \to \infty$. Use (10.19) to prove the classical IBP formula: $\int_a^b f'(x)g(x)dx = -\int_a^b g'(x)f(x)dx + f(b)g(b) - f(a)g(a)$ for continuous differentiable real-valued functions f and g and $a < b$.

The idea of the following result is to modify the proof of Theorem 10.5.2 using instead the above discrete IBP formula on the law of the Poisson process.

Theorem 10.5.20 *Let $\{a(k)\}_{k\geq -1}$ be a sequence of real numbers; then we have the following two IBP formulas for all $\theta \in \mathbb{R}$ and $t > 0$ assuming $\mathbb{E}[|a(N_t)|N_t] < \infty$ for the first formula and $\mathbb{E}[\frac{|a(N_t)|}{N_t}] < \infty$ for the second formula:*

$$\begin{aligned}
(e^{i\theta} - 1)\lambda t\mathbb{E}[e^{i\theta N_t}a(N_t)] =& \mathbb{E}[e^{i\theta N_t}b(N_t)]\\
b(k) :=& a(k-1)k - a(k)\lambda t,\\
(e^{i\theta} - 1)\mathbb{E}[e^{i\theta N_t}a(N_t)] =& \mathbb{E}[e^{i\theta(N_t+1)}c(N_t+1)] - a(0)e^{-\lambda t}\\
c(k) :=& a(k-1) - \frac{a(k)}{k}\lambda t.
\end{aligned}$$

Proof We will do the first leaving the second for the reader. The proof is just a rearranging of the expectations expressed as sums as follows:

$$(e^{i\theta}-1)\mathbb{E}[e^{i\theta N_t}a(N_t)] = \sum_{k=1}^{\infty} e^{i\theta k}a(k-1)\mathbb{P}(N_t=k-1) - \sum_{k=0}^{\infty} e^{i\theta k}a(k)\mathbb{P}(N_t=k).$$

Then one uses that $\mathbb{P}(N_t = k-1) = \frac{k}{\lambda t}\mathbb{P}(N_t = k)$ to obtain the result.

Again, we see in the above formula an approximative IBP formula. We leave as an exercise for the reader to check that Lemma 10.5.10 follows from the above result.

Exercise 10.5.21 The difference between the results in Theorems 10.5.2/10.5.7 and 10.5.11 is that in the latter one can take limits with respect to n. In fact, one has to understand that both methodologies are different in nature. In Theorem 10.5.7, one uses a multiplicative format and an IBP. In the one explained above, which is additive, one uses a discrete version of the IBP formula. In fact, one just has to recall that the discrete IBP formula in Exercise 10.5.19 gives that

$$\sum_{k=1}^{\infty}(e^{i\theta k}-e^{i\theta(k-1)})k\mathbb{P}(N_t=k) = \sum_{k=1}^{\infty} e^{i\theta(k-1)}\{(k-1)\mathbb{P}(N_t=k-1)-k\mathbb{P}(N_t=k)\}.$$

From here, prove that

$$\mathbb{E}[e^{i\theta N_t}(e^{i\theta}-1)\lambda t] = \mathbb{E}[e^{i\theta N_t}(N_t-\lambda t)].$$

Find the relation between the above result and the statement of Theorem 10.5.11.

Exercise 10.5.22 Use the general form of Theorem 10.5.20 in order to find an approximative IBP formula of order two for $e^{i\theta x}$, that is, a formula of the type

$$(e^{i\theta}-1)^2\mathbb{E}[e^{i\theta N_t}] = \mathbb{E}[e^{i\theta N_t}H],$$

for an appropriate random variable H. Clearly, in order to be able to carry out all these formulas for processes like Z further non-trivial arguments have to be implemented.

For a hint, see Chap. 14.

Exercise 10.5.23 Use the generalization of Exercise 10.5.15 to prove the following approximative IBP formula for $e^{i\theta x}$:

$$(e^{i\theta}-1)\mathbb{E}[e^{i\theta N_t}\int_0^t\int w(s,z)\widehat{\mathcal{N}}(dz,ds)] = \mathbb{E}[e^{i\theta N_t}\int_0^t\int w(s,z)\widetilde{\mathcal{N}}(dz,ds)].$$

Here w is a bounded predictable random process and $\mathcal{N}$ is the Poisson random measure associated with $N_s, s\in[0,T]$. Apply this result for a particular type of process w, which is a stochastic integral, in order to obtain the result of Exercise 10.5.22.

For a hint, see Chap. 14.

From the above relation we also see that a stochastic integral appears on the right side and a Lebesgue integral appears on the left together with the approximative derivative term $e^{i\theta} - 1$. This structure resembles the structure of the classical IBP formula in the following way.

If the derivative operator is defined for functions $f \in C_c(\mathbb{R})$ as $Df(x) = f'(x)$, then the IBP formula can be rewritten as

$$\begin{aligned}\langle Df, g\rangle &= -\langle f, Dg\rangle,\\ \langle f, g\rangle &:= \int f(x)g(x)dx.\end{aligned}$$

Therefore in the language of operator theory it means that $D^* = -D$, where D^* stands for the adjoint operator. In the same way, one may reinterpret heuristically the result in Exercise 10.5.23 in similar terms as follows:

$$\begin{aligned}D(e^{i\theta N_t}) &:=(e^{i\theta} - 1)e^{i\theta N_t}\\ D^*(w) &:= \int_0^t \int w(s,z)\widetilde{\mathcal{N}}(dz, ds)\\ \langle F, g\rangle &:=\mathbb{E}\left[\int_0^t F \int g(z,s)\widehat{\mathcal{N}}(dz, ds)\right].\end{aligned}$$

In the case of Theorem 10.5.2, we have

$$\begin{aligned}D(e^{i\theta N_t}) &=i\theta e^{i\theta N_t}\\ D^*(1) &:= \int_0^t \int h(z)\mathcal{N}(dz, ds)\\ \langle F, g\rangle &:=\mathbb{E}\left[F\int_0^t \int g(z,s)\mathcal{N}(dz, ds)\, \mathbf{1}(N_t > 0)\right].\end{aligned}$$

In the above sense, IBP formulas are linked to the dual operator of a derivative (or difference) operator. This issues will appear again and may become more clear in Chap. 12.

Some conclusions: In this chapter, we tried to briefly give different techniques that may be used in various problems when one wants to perform an IBP. It is our purpose to show that very basic techniques can solve a wide range of complicated problems. This is not by far a complete list and we encourage the reader to keep adding items to this list. The main characteristics, requirements and goals that can be achieved should be clear for each method before proceeding. We also hope that it is clear that approximation procedures are always needed such as in the case that we are considering difference equations as an approximation for continuous time models. Also that the blow-up of Lévy measures near zero has forced us in the last example to approximate the jump model using cutoff functions.

As we will see in further chapters the above techniques will allow us obtain IBP formulas for complex models. Still, the method will fall short in providing upper/lower bounds for densities between other properties. This remains an open problem from the point of view of the previously described methods. In some cases an analytical method will provide these bounds.

Chapter 11
Sensitivity Formulas

In many applied problems, one needs to compute expectations of a function of a random variable which are obtained through a certain theoretical development. This is the case of $\mathbb{E}[G(Z_t)]$, where Z is a Lévy process with Lévy measure ν which may depend on various parameters. Similarly, G is a real-valued bounded measurable function which may also depend on some parameters and is not necessarily smooth. For many stability reasons one may be interested in having explicit expressions for the partial derivatives of the previous expectation with respect to the parameters in the model. These quantities are called "Greeks" in finance but they may have different names in other fields.

In this chapter, we provide some examples of how to compute these quantities in the case that G is not necessarily differentiable. This problem is clearly related to the smoothness of the density of the random variable Z_t because $\mathbb{E}[G(Z_t)] = \int G(z)\mu(dz)$, where μ denotes the law of Z_t which depends on the Lévy measure. In particular, we are interested in formulas of the type $\partial\mathbb{E}[G(Z_t)] = \mathbb{E}[G(Z_t)H]$ for some appropriate random variable H and where the derivative is taken with respect to some parameter in Z or in the function G. At the beginning we will consider formulas of the type $\mathbb{E}[G'(Z_t)Y] = \mathbb{E}[G(Y_t)H]$ for some appropriate random variable Y. This will show how to carry out the first step of the procedure which involves the integration by parts (IBP) formula. In a second step, we will show how to obtain formulas for $\mathbb{E}[G'(Z_t)]$ by using some sort of inversion procedure with respect to Y. In that step we will need to prove that the inverse moments of Y are finite. For this, it is useful to recall the calculations in (10.5). The final formulas which may be amenable to computation through Monte Carlo simulation are obtained using the IBP formula. The ideas introduced here will serve the reader as a guide for the method to be introduced in the next chapter. These are based on ideas previously explained in Exercise 10.1.21 and in Sects. 10.3 and 10.4.

© Springer Nature Singapore Pte Ltd. 2019

A. Kohatsu-Higa and A. Takeuchi, *Jump SDEs and the Study of Their Densities*, Universitext, https://doi.org/10.1007/978-981-32-9741-8_11

11.1 Models Driven by Gamma Processes

Let T, a and b be positive constants, and $Z = \{Z_t\,;\; t \in [0, T]\}$ the gamma process with the parameters (a, b) whose Lévy measure is given in the form:

$$\nu(\mathrm{d}z) = f(z)dz := \frac{a}{z}\, e^{-bz}\, \mathbf{1}_{(z>0)}\, \mathrm{d}z.$$

Denote by $\mathscr{N}(\mathrm{d}s, \mathrm{d}z)$ the Poisson random measure on $[0, T] \times (0, \infty)$ with the intensity measure $\hat{\mathscr{N}}(\mathrm{d}s, \mathrm{d}z) := \mathrm{d}s\, \nu(\mathrm{d}z)$. Then, for each $t \in [0, T]$, the random variable Z_t has the gamma distribution with the characteristic function

$$\mathbb{E}\big[\exp(i\,\theta\, Z_t)\big] = \exp\left[t \int_0^\infty \left(e^{i\,\theta\, z} - 1\right) \nu(\mathrm{d}z)\right] = \left(1 - \frac{i\,\theta}{b}\right)^{-at}, \tag{11.1}$$

and the density of Z_t is given in closed form:

$$p_t^Z(y) = \frac{b^{at}}{\Gamma(at)}\, y^{at-1}\, e^{-by}\, \mathbf{1}_{(y>0)}. \tag{11.2}$$

Problem 11.1.1 Prove (11.1). Using the Taylor series expansion for the complex exponential.

The Lévy–Itô decomposition (i.e. Theorem 4.1.9) leads us to see that Z_t can be expressed as

$$Z_t = \int_{(0,\infty)\times[0,t]} z\, \mathscr{N}(\mathrm{d}s, \mathrm{d}z).$$

That is, Z is an increasing process of bounded variation in compact intervals.

Let $S_0 > 0$, $\rho \in \mathbb{R}$ and $c \in \mathbb{R}$ be constants. We shall introduce the simple asset price models $S = \{S_t\,;\; t \in [0, T]\}$ given by

$$S_t = S_0\, \exp(\rho\, Z_t + c\, t).$$

We are not going to give here all the motivations for this financial model. We just note that S is positive, it always jumps on the same direction (e.g. if $\rho > 0$ then it always increases when it jumps) and it has a continuous part which is always monotone depending on the sign of c. Then we have the following IBP formula.

Theorem 11.1.2 *For $G \in C_c^1(\mathbb{R})$, it holds that*

$$\mathbb{E}\big[G'(S_t)\, S_t\, \rho\, Z_t\big] = \mathbb{E}\big[G(S_t)\, (b\, Z_t - at)\big]. \tag{11.3}$$

Proof For $0 < \varepsilon < 1$, we shall introduce the new process $Z^\varepsilon = \left\{Z_t^\varepsilon\,;\; t \in [0, T]\right\}$ defined by

$$Z_t^\varepsilon = \int_{(0,\infty)\times[0,t]} z\,\mathcal{N}^{(\varepsilon,+)}(\mathrm{d}s,\mathrm{d}z).$$

Moreover, the marginal Z_t^ε can be also expressed as

$$Z_t^\varepsilon = \begin{cases} 0 & (N_t^\varepsilon = 0), \\ \displaystyle\sum_{k=1}^{N_t^\varepsilon} Y_k^\varepsilon & (N_t^\varepsilon \geq 1), \end{cases}$$

where $N^\varepsilon = \{N_t^\varepsilon\,;\, t \in [0,T]\}$ is the Poisson process with the parameter $\lambda_\varepsilon := \nu\big(\{z \in (0,\infty)\,;\, z > \varepsilon\}\big)$, and $\{Y_k^\varepsilon\,;\, k \in \mathbb{N}\}$ are i.i.d.r.v.s, independent of N^ε, with the common law

$$\mathbb{P}\big[Y_k^\varepsilon \in \mathrm{d}z\big] = \frac{\mathbf{1}_{(z>\varepsilon)}}{\lambda_\varepsilon}\,\nu(\mathrm{d}z)\,(=: \nu_\varepsilon(\mathrm{d}z))\,,$$

which is independent of the process N^ε. For $\xi < b$, define the function $H^\xi : (0,\infty) \to (0,\infty)$ by

$$H^\xi(z) := \frac{b\,z}{b-\xi}. \tag{11.4}$$

Define the new processes $Z^{\varepsilon,\xi} = \big\{Z_t^{\varepsilon,\xi}\,;\, t \in [0,T]\big\}$ and $S^{\varepsilon,\xi} = \big\{S_t^{\varepsilon,\xi}\,;\, t \in [0,T]\big\}$ by

$$\begin{aligned} Z_t^{\varepsilon,\xi} &:= \int_{(0,\infty)\times[0,t]} H^\xi(z)\,\mathcal{N}^{(\varepsilon,+)}(\mathrm{d}s,\mathrm{d}z), \\ S_t^{\varepsilon,\xi} &:= S_0 \exp\big(\rho\, Z_t^{\varepsilon,\xi} + c\,t\big). \end{aligned}$$

Let $G \in C_c^1(\mathbb{R})$. Note that then $\mathbb{E}[|G(S_t^\varepsilon)|^p] < \infty$ for all $p > 1$ and $\varepsilon \geq 0$. Define the derivative operator D as

$$D\big(G(S_t^\varepsilon)\big) \equiv \partial_\xi \big(G(S_t^{\varepsilon,\xi})\big)\Big|_{\xi=0} = \frac{1}{b}\,G'(S_t^\varepsilon)\,S_t^\varepsilon\,\rho\,Z_t^\varepsilon.$$

It should be clear here that $Z^{\varepsilon,\xi}$ is an ad-hoc perturbation of the jump structure of the process Z^ε. Therefore the above definition is a directional derivative. From the definition of the derivative, we have

$$\begin{aligned} &\mathbb{E}\big[D\big(G(S_t^\varepsilon)\big)\big] = \frac{1}{b}\,\mathbb{E}\big[G'(S_t^\varepsilon)\,S_t^\varepsilon\,\rho\,Z_t^\varepsilon\,\mathbf{1}_{(N_t^\varepsilon\geq 1)}\big] \\ &= \frac{1}{b}\mathbb{E}\left[G'\left(S_0 \exp\left\{\rho\sum_{j=1}^{N_t^\varepsilon} Y_j^\varepsilon + c\,t\right\}\right) S_0 \exp\left\{\rho\sum_{j=1}^{N_t^\varepsilon} Y_j^\varepsilon + c\,t\right\}\,\rho\sum_{k=1}^{N_t^\varepsilon} Y_k^\varepsilon\,\mathbf{1}_{(N_t^\varepsilon\geq 1)}\right] \end{aligned}$$

$$
\begin{aligned}
&= \sum_{N \geq 1} \frac{\mathbb{P}[N_t^\varepsilon = N]}{b} \sum_{k=1}^{N} \int_0^{+\infty} \nu_\varepsilon(\mathrm{d}z_k) \\
&\quad \times \mathbb{E}\left[G'\left(S_0 \exp\left\{ \rho \sum_{j \neq k} Y_j^\varepsilon + \rho\, z_k + c\, t \right\} \right) S_0 \exp\left\{ \rho \sum_{j \neq k} Y_j^\varepsilon + \rho\, z_k + c\, t \right\} \rho\, z_k \right] \\
&=: \sum_{N \geq 1} \frac{\mathbb{P}[N_t^\varepsilon = N]}{b} \sum_{k=1}^{N} I_{k,N}^\varepsilon .
\end{aligned}
$$

An application of the divergence formula (or simply "the IBP formula") gives

$$
\begin{aligned}
I_{k,N}^\varepsilon &= \int_\varepsilon^\infty \frac{z_k\, f(z_k)}{\lambda_\varepsilon}\, \mathrm{d}z_k\, \mathbb{E}\left[\partial_{z_k} \left\{ G\left(S_0 \exp\left\{ \rho \sum_{j \neq k} Y_j^\varepsilon + \rho\, z_k + c\, t \right\} \right) \right\} \right] \\
&= \left[\mathbb{E}\left[G\left(S_0 \exp\left\{ \rho \sum_{j \neq k} Y_j^\varepsilon + \rho\, z_k + c\, t \right\} \right) \right] \frac{z_k\, f(z_k)}{\lambda_\varepsilon} \right]_\varepsilon^{+\infty} \\
&\quad - \int_\varepsilon^\infty \mathbb{E}\left[G\left(S_0 \exp\left\{ \rho \sum_{j \neq k} Y_j^\varepsilon + \rho\, z_k + c\, t \right\} \right) \right] \frac{\partial_{z_k}\big(z_k\, f(z_k) \big)}{\lambda_\varepsilon}\, \mathrm{d}z_k \\
&=: I_{k,N,1}^\varepsilon + I_{k,N,2}^\varepsilon .
\end{aligned}
$$

By differentiation of $z_k\, f(z_k)$, we have

$$
\begin{aligned}
I_{k,N,2}^\varepsilon &= \int_0^\infty \mathbb{E}\left[G\left(S_0 \exp\left\{ \rho \sum_{j \neq k} Y_j^\varepsilon + \rho\, z_k + c\, t \right\} \right) b\, z_k \right] \nu_\varepsilon(\mathrm{d}z_k) \\
&= \mathbb{E}\left[G\left(S_0 \exp\left\{ \rho \sum_{j=1}^{N} Y_j^\varepsilon + c\, t \right\} \right) b\, Y_k^\varepsilon \right].
\end{aligned}
$$

As for $I_{k,N,1}^\varepsilon$, we have

$$
I_{k,N,1}^\varepsilon = \begin{cases} -G\left(S_0 \exp\{ \rho\, \varepsilon + c\, t \} \right) \dfrac{a\, e^{-b\,\varepsilon}}{\lambda_\varepsilon} & (N = 1), \\[2ex] -\mathbb{E}\left[G\left(S_0 \exp\left\{ \rho \sum_{j \neq k} Y_j^\varepsilon + \rho\, \varepsilon + c\, t \right\} \right) \dfrac{a\, e^{-b\,\varepsilon}}{\lambda_\varepsilon} \right] & (N \geq 2). \end{cases}
$$

Thus, we obtain

$$
\frac{1}{b}\, \mathbb{E}\big[G'(S_t^\varepsilon)\, S_t^\varepsilon\, \rho\, Z_t^\varepsilon \big]
$$

$$
\begin{aligned}
&= -\frac{1}{b}\mathbb{P}\big[N_t^\varepsilon = 1\big]\, G\left(S_0 \exp\{\rho\,\varepsilon + c\,t\}\right)\, \frac{a\, e^{-b\,\varepsilon}}{\lambda_\varepsilon} \\
&\quad - \frac{1}{b}\sum_{N\ge 2}\mathbb{P}\big[N_t^\varepsilon = N\big]\sum_{k=1}^{N}\mathbb{E}\left[G\left(S_0 \exp\left\{\rho\sum_{j\neq k} Y_j^\varepsilon + \rho\,\varepsilon + c\,t\right\}\right)\frac{a\, e^{-b\,\varepsilon}}{\lambda_\varepsilon}\right] \\
&\quad + \frac{1}{b}\sum_{N\ge 1}\mathbb{P}\big[N_t^\varepsilon = N\big]\sum_{k=1}^{N}\mathbb{E}\left[G\left(S_0 \exp\left\{\rho\sum_{j=1}^{N} Y_j^\varepsilon + c\,t\right\}\right) b\, Y_k^\varepsilon\right] \\
&= -\frac{a\, e^{-b\varepsilon}}{b\,\lambda_\varepsilon}\,\mathbb{P}\big[N_t^\varepsilon = 1\big]\, G\left(S_0 \exp\{\rho\,\varepsilon + c\,t\}\right) \\
&\quad - \frac{a\, e^{-b\varepsilon}}{b\,\lambda_\varepsilon}\sum_{N\ge 2} N\,\mathbb{P}\big[N_t^\varepsilon = N\big]\,\mathbb{E}\left[G\left(S_0 \exp\left\{\rho\sum_{j=1}^{N-1} Y_j^\varepsilon + \rho\,\varepsilon + c\,t\right\}\right)\right] \\
&\quad + \mathbb{E}\left[G\big(S_t^\varepsilon\big)\, Z_t^\varepsilon\right] \\
&= -\frac{at}{b}\, e^{-b\varepsilon}\,\mathbb{P}\big[N_t^\varepsilon = 0\big]\, G\left(S_0 \exp\{\rho\,\varepsilon + c\,t\}\right) \\
&\quad - \frac{at}{b}\, e^{-b\varepsilon}\sum_{N\ge 1}\mathbb{P}\big[N_t^\varepsilon = N\big]\,\mathbb{E}\left[G\left(S_0 \exp\left\{\rho\sum_{j=1}^{N} Y_j^\varepsilon + \rho\,\varepsilon + c\,t\right\}\right)\right] \\
&\quad + \mathbb{E}\left[G\big(S_t^\varepsilon\big)\, Z_t^\varepsilon\right] \\
&= -\frac{at}{b}\, e^{-b\varepsilon}\,\mathbb{E}\left[G(S_t^\varepsilon\, e^{\rho\,\varepsilon})\right] + \mathbb{E}\left[G\big(S_t^\varepsilon\big)\, Z_t^\varepsilon\right].
\end{aligned}
$$

Taking the limit as $\varepsilon \downarrow 0$ leads us to obtain that

$$
\frac{1}{b}\,\mathbb{E}\big[G'(S_t)\, S_t\, \rho\, Z_t\big] = -\frac{at}{b}\,\mathbb{E}\left[G(S_t^\varepsilon)\right] + \mathbb{E}\left[G(S_t)\, Z_t\right].
$$

The proof is complete. □

Remark 11.1.3 (i) Note that there is an important step in the penultimate last step of the above proof where one uses a regenerative property of the Poisson distribution. That is, $\lambda_\varepsilon^{-1} N\,\mathbb{P}\big[N_t^\varepsilon = N\big] = \mathbb{P}\big[N_t^\varepsilon = N-1\big]$.

(ii) In the IBP formula (11.3), the term $-at\mathbb{E}\big[G(S_t)\big]$ appears due to the fact that the jumps are strictly positive. That is, this term corresponds to boundary terms in the IBP formula. On the other hand, the expression $b\,\mathbb{E}\big[G(S_t)\, Z_t\big]$ corresponds to integral term in the classical IBP formula.

Problem 11.1.4 1. Prove the formula (11.3) directly, by using the closed form (11.2) of the density, and the divergence formula.[1]

2. Find an IBP formula for $\mathbb{E}\big[G'(S_t)\big]$ if $at > 1$ and $G \in C_c^1$.

[1] Still, as we will find out later the above methodology is far more general as in many cases explicit densities cannot be obtained.

3. Find a formula for $\partial_\rho \mathbb{E}\big[G(S_t)\big]$ for a function G which is measurable and bounded.

Problem 11.1.5 In this exercise, we will perform the IBP with the same transformation as in Theorem 11.1.2 but in an order that is closer to the argument in Exercise 10.1.21. Consider the transformation in (11.4). Rather than performing the IBP on the noise appearing in S^ε perform a similar calculation using the process $S^{\varepsilon,\xi}$. Once the IBP is obtained take $\xi = 0$. Compare the formula obtained with the one in (11.3). Note that the reason why both results coincide is because $\partial_z H^\xi(z)$ is a constant fo any $\xi > 0$.

Problem 11.1.6 Using the function $H^\xi(z) := ze^{-\xi z^r}$ obtain a result similar to (11.3). Note that in that formula the process Z is now replaced by the power process which is the limit as $\varepsilon \to 0$ of

$$V_t^\varepsilon = \int_{(0,\infty)\times[0,t]} z^{r+1}\, \mathcal{N}^{(\varepsilon,+)}(\mathrm{d}s, \mathrm{d}z).$$

The idea presented in this section of performing IBP with respect to the law of the jump size will be the main idea in the next chapter.

11.2 Stable Processes

In this section, we will give an IBP formula for general stable processes based on the structure of gamma processes described in the previous section. We start by giving a general definition for stable processes that includes the ones considered in previous chapters. Particular cases of stable process have already been discussed in Chaps. 4 and 5.

Let $0 < \alpha < 2$, $-1 \le \beta \le 1$, $\sigma > 0$ and $\gamma \in \mathbb{R}$. We say that the Lévy process $Z = \{Z_t\,;\, t \ge 0\}$ is a $S_\alpha(\sigma, \beta, \gamma)$-stable process if its characteristic function is of the form:

$$\frac{1}{t}\ln \mathbb{E}\Big[\exp\big(i\,\theta\, Z_t\big)\Big] = \begin{cases} -\sigma^\alpha |\theta|^\alpha \left(1 - i\,\mathrm{sgn}(\theta)\,\beta\,\tan\dfrac{\pi\,\alpha}{2}\right) + i\,\gamma\,\theta & (\alpha \neq 1), \\ -\sigma|\theta|\left(1 + i\,\mathrm{sgn}(\theta)\,\beta\,\dfrac{2}{\pi}\,\ln|\theta|\right) + i\,\gamma\,\theta & (\alpha = 1). \end{cases} \tag{11.5}$$

The parameters (γ, σ, β) represent the “mean”, “variability" and “asymmetry” parameters. Notice that as stated previously, the theoretical mean or variance of Z_t may not exist for stable random variables. In particular, for the case $\beta = \gamma = 0$ gives the stable process introduced in Chaps. 4 (for $\alpha \in (0, 1)$) and 5.

On the other hand, the characteristic function of the process Z can be parametrized in an alternative form, based upon the analytic viewpoint, as explained in [61]. For this, define

$$K(\alpha) = \begin{cases} \alpha & (0 < \alpha < 1), \\ \alpha - 2 & (1 < \alpha < 2), \end{cases}$$

and define

$$\beta_2 = \begin{cases} \dfrac{2}{\pi\, K(\alpha)} \arctan\left(\beta \tan \dfrac{\pi\,\alpha}{2}\right) & (\alpha \neq 1), \\ \beta & (\alpha = 1), \end{cases} \tag{11.6}$$

$$\sigma_2 = \begin{cases} \sigma \left(1 + \beta^2 \tan^2 \dfrac{\pi\,\alpha}{2}\right)^{1/2\alpha} & (\alpha \neq 1), \\ \dfrac{2\,\sigma}{\pi} & (\alpha = 1). \end{cases} \tag{11.7}$$

Then, we can get

$$\frac{1}{t} \ln \mathbb{E}\Big[\exp\big(i\,\theta\, Z_t\big)\Big] = \begin{cases} -\sigma_2^\alpha |\theta|^\alpha \exp\left(-i \operatorname{sgn}(\theta)\, \beta_2 \dfrac{\pi\, K(\alpha)}{2}\right) + i\,\gamma\,\theta & (\alpha \neq 1), \\ -\sigma_2 |\theta| \left(\dfrac{\pi}{2} + i \operatorname{sgn}(\theta)\, \beta_2 \ln|\theta|\right) + i\,\gamma\,\theta & (\alpha = 1). \end{cases} \tag{11.8}$$

We remark that expressions (11.5) and (11.8) are equal, although the parametrization is different.

11.2.1 Alternative Probabilistic Representation for Stable Random Variables

Now, we shall describe an alternative probabilistic representation for $S_\alpha(\sigma, \beta, \gamma)$-stable processes. This representation will help us provide an integration by parts formula as explained in Exercise 10.1.11. In this section we will use an associated jump structure using gamma processes. The main reason for this choice is that, as explained previously, the mean or variance of Z_t may not exist and therefore using Z_t in the integration by parts formula may not be possible.

Let V be a uniformly random variable on $\big(-\pi/2, \pi/2\big)$, and E a standard exponential random variable (i.e. $\mathbb{E}[E] = 1$) independent of V.

Theorem 11.2.1 *The following representations are valid for fixed $t > 0$:*

$$Z_t \overset{\mathcal{L}}{=} \begin{cases} t^{1/\alpha}\, \sigma \left(1 + \beta^2 \tan^2\left(\dfrac{\pi\,\alpha}{2}\right)\right)^{1/2\alpha} U\, E^{-(1-\alpha)/\alpha} + \gamma\, t & (\alpha \neq 1), \\ t\,\sigma\, \dfrac{2}{\pi} \big(U - \beta \ln E\big) + \gamma\, t & (\alpha = 1), \end{cases} \tag{11.9}$$

where $\rho_2 := \beta_2\, \pi\, K(\alpha)/(2\alpha)$ for $\alpha \neq 1$ and

$$U := \begin{cases} \dfrac{\sin\big(\alpha(V+\rho_2)\big)}{\big(\cos V\big)^{1/\alpha}} \left\{\cos\big((1-\alpha)\,V - \alpha\,\rho_2\big)\right\}^{(1-\alpha)/\alpha} & (\alpha \neq 1), \\ \left(\dfrac{\pi}{2} + \beta_2\,V\right)\tan V - \beta_2 \ln\left(\dfrac{\cos V}{\frac{\pi}{2} + \beta_2\,V}\right) & (\alpha = 1). \end{cases}$$

For a proof see (3.2) in Theorem 3.1 in [59], where the case of $\sigma_2 = 1$ and $\gamma = 0$ is discussed. Heuristically speaking, the above representation is obtained using an alternative integral representation of the Lévy characteristic exponent in the spirit of Exercise 4.1.22.

Using the above probabilistic representation, we will introduce an alternative jump structure representation that will produce an IBP formula. Before doing this, for comparison purposes, we recall the notation for the standard construction of stable processes used in Chap. 5.

Let $\alpha \in (0,1) \cup (1,2)$.[2] We shall consider two independent Lévy processes $Z_- = \{Z_{-,t}\,;\, t \geq 0\}$ and $Z_+ = \{Z_{+,t}\,;\, t \geq 0\}$ with the characteristic functions

$$\frac{1}{t}\ln\mathbb{E}\Big[\exp\big(i\,\theta\,Z_{-,t}\big)\Big] = -\sigma_-^\alpha\,|\theta|^\alpha\,\left(1 + i\,\mathrm{sgn}(\theta)\,\tan\frac{\pi\,\alpha}{2}\right), \tag{11.10}$$

$$\frac{1}{t}\ln\mathbb{E}\Big[\exp\big(i\,\theta\,Z_{+,t}\big)\Big] = -\sigma_+^\alpha\,|\theta|^\alpha\,\left(1 - i\,\mathrm{sgn}(\theta)\,\tan\frac{\pi\,\alpha}{2}\right), \tag{11.11}$$

where $C_-,\ C_+ \geq 0$ are constants, and

$$\sigma_- := \left(-C_-\,\Gamma(-\alpha)\,\cos\frac{\pi\,\alpha}{2}\right)^{1/\alpha},$$

$$\sigma_+ := \left(-C_+\,\Gamma(-\alpha)\,\cos\frac{\pi\,\alpha}{2}\right)^{1/\alpha}.$$

The processes $Z_\pm$ are $S_\alpha(\sigma_\pm, \pm 1, 0)$-stable processes without a Gaussian component, respectively. Note that they correspond to the positive and negative jumps of the process Z. Moreover, their corresponding Lévy measures $\nu_-(dz)$ and $\nu_+(dz)$ are given by

$$\nu_-(dz) = \frac{C_-}{|z|^{1+\alpha}}\,\mathbf{1}_{(-\infty,0)}(z)\,dz,$$

$$\nu_+(dz) = \frac{C_+}{z^{1+\alpha}}\,\mathbf{1}_{(0,+\infty)}(z)\,dz.$$

Write

$$\sigma := \left(\sigma_-^\alpha + \sigma_+^\alpha\right)^{1/\alpha} = \left\{-(C_- + C_+)\,\Gamma(-\alpha)\,\cos\frac{\pi\,\alpha}{2}\right\}^{1/\alpha},$$

---

[2] As is usually the case. The case $\alpha = 1$ has to be dealt with separately due to the difference in representations.

$$\beta := \frac{C_+ - C_-}{C_+ + C_-}.$$

Then, the process $Z = \{Z_t := Z_{+,t} + Z_{-,t}\ ;\ t \geq 0\}$ is also an $S_\alpha(\sigma, \beta, 0)$-stable process such that the characteristic function of the marginal Z_t is

$$\begin{aligned} \frac{1}{t} \ln \mathbb{E}\Big[\exp\big(i\,\theta\, Z_t\big)\Big] &= \frac{1}{t} \ln \mathbb{E}\Big[\exp\big(i\,\theta\, Z_{-,t}\big)\Big] + \frac{1}{t} \ln \mathbb{E}\Big[\exp\big(i\,\theta\, Z_{+,t}\big)\Big] \\ &= C_-\, \Gamma(-\alpha) \left(\cos \frac{\pi\,\alpha}{2}\right) |\theta|^\alpha \left(1 + i\,\mathrm{sgn}(\theta)\, \tan \frac{\pi\,\alpha}{2}\right) \\ &\quad + C_+\, \Gamma(-\alpha) \left(\cos \frac{\pi\,\alpha}{2}\right) |\theta|^\alpha \left(1 - i\,\mathrm{sgn}(\theta)\, \tan \frac{\pi\,\alpha}{2}\right) \\ &= -\sigma^\alpha\, |\theta|^\alpha \left(1 - i\,\mathrm{sgn}(\theta)\, \beta\, \tan \frac{\pi\,\alpha}{2}\right). \end{aligned} \tag{11.12}$$

Now we consider the jump structure which arises from the exponential random variables in the representation of Theorem 11.2.1. Define

$$U_- := \frac{\sin\big(\alpha(V_- + \rho_-)\big)}{\big(\cos V_-\big)^{1/\alpha}} \Big\{\cos\big((1-\alpha)\, V_- - \alpha\, \rho_-\big)\Big\}^{(1-\alpha)/\alpha},$$

$$U_+ := \frac{\sin\big(\alpha(V_+ + \rho_+)\big)}{\big(\cos V_+\big)^{1/\alpha}} \Big\{\cos\big((1-\alpha)\, V_+ - \alpha\, \rho_+\big)\Big\}^{(1-\alpha)/\alpha},$$

where V_-, V_+ are independent uniformly random variables on $\big(-\pi/2, \pi/2\big)$, E_-, E_+ are independent standard exponential random variables independent of V_-, V_+, and

$$\rho_- := \arctan\left(-\tan \frac{\pi\,\alpha}{2}\right)\Big/\alpha, \quad \rho_+ := \arctan\left(\tan \frac{\pi\,\alpha}{2}\right)\Big/\alpha.$$

Consider the processes $\tilde{Z}_\pm = \{\tilde{Z}_{\pm,t}\ ;\ t \geq 0\}$ given by

$$\tilde{Z}_{-,t} := t^{1/\alpha}\, \sigma_- \left(1 + \tan^2\left(\frac{\pi\,\alpha}{2}\right)\right)^{1/2\alpha} U_-\, E_-^{-(1-\alpha)/\alpha}, \tag{11.13}$$

$$\tilde{Z}_{+,t} := t^{1/\alpha}\, \sigma_+ \left(1 + \tan^2\left(\frac{\pi\,\alpha}{2}\right)\right)^{1/2\alpha} U_+\, E_+^{-(1-\alpha)/\alpha}, \tag{11.14}$$

It is clear that the processes $\tilde{Z}_\pm$ are independent. As seen in (11.9), we see that for every $t > 0$

$$Z_{-,t} \stackrel{\mathcal{L}}{=} \tilde{Z}_{-,t}, \quad Z_{+,t} \stackrel{\mathcal{L}}{=} \tilde{Z}_{+,t}.$$

Then, we can conclude that $\tilde{Z}_t := \tilde{Z}_{+,t} - \tilde{Z}_{-,t}$ has the same law of Z_t for each $t > 0$. Note that only the marginal laws (that is for fixed t) of Z and $\tilde{Z}$ are equal. In fact, it is easy to see that $\tilde{Z}_\pm$ is always differentiable with respect to $t > 0$. This construction of $\tilde{Z}$ will be used when computing sensibilities.

11.2.2 Sensitivity Analysis on Stable Processes

Now, we shall study the sensitivity analysis for the $S_\alpha(\sigma, \pm 1, 0)$-stable process introduced in the previous subsection. As stated previously, due to issues of existence of moments we will consider the case $1 < \alpha < 2$.

In order to avoid lengthy expressions, we shall write

$$\tilde{\sigma}_{\pm,t} := t^{1/\alpha}\,\sigma_\pm \left(1 + \tan^2\left(\frac{\pi\,\alpha}{2}\right)\right)^{1/2\alpha} U_\pm,$$
$$\tilde{Z}^y_{\pm,t} := \tilde{\sigma}_{\pm,t}\, y^{(\alpha-1)/\alpha}, \quad \tilde{Z}^{y,z}_t := \tilde{Z}^z_{+,t} + \tilde{Z}^y_{-,t}.$$

Denote by $\mathcal{N}^\pm$ two independent Poisson random measures associated with gamma processes on $\pm(0,\infty)$ with parameter values $a = b = 1$. Define

$$E_- = -\int_{(-\infty,0)\times[0,1]} z\,\mathcal{N}^-(dz, ds), \quad E_+ = \int_{(0,\infty)\times[0,1]} z\,\mathcal{N}^+(dz, ds),$$
$$\tilde{Z}^{E_\pm}_{\pm,t} = \tilde{\sigma}_{\pm,t}\,(E_\pm)^{(\alpha-1)/\alpha} = \tilde{Z}_{\pm,t}, \quad \tilde{Z}^{E_-,E_+}_t = \tilde{Z}_{-,t} + \tilde{Z}_{+,t} = \tilde{Z}_t.$$

For $0 < \varepsilon < 1$, recall that $\mathcal{N}^{(\varepsilon,\pm)}$ denotes the Poisson random measure associated with the jumps that are larger/smaller than $\pm\varepsilon$:

$$E^\varepsilon_- := -\int_{(-\infty,0)\times[0,1]} z\,\mathcal{N}^{(\varepsilon,-)}(dz, ds), \quad E^\varepsilon_+ := \int_{(0,\infty)\times[0,1]} z\,\mathcal{N}^{(\varepsilon,+)}(dz, ds),$$
$$\tilde{Z}^\varepsilon_{\pm,t} := \tilde{\sigma}_{\pm,t}\,(E^\varepsilon_\pm)^{(\alpha-1)/\alpha}\big(= \tilde{Z}^{E^\varepsilon_\pm}_{\pm,t}\big), \quad \tilde{Z}^\varepsilon_t := \tilde{Z}^\varepsilon_{-,t} + \tilde{Z}^\varepsilon_{+,t}\big(= \tilde{Z}^{E^\varepsilon_-,E^\varepsilon_+}_t\big).$$

We remark that the random variables $E^\varepsilon_\pm$ can be expressed as

$$E^\varepsilon_\pm = \sum_{k=1}^{N^\varepsilon_\pm} Z^\varepsilon_{\pm,k},$$

where $N^\varepsilon_\pm$ are the Poisson random variables with the parameter $\lambda^\varepsilon_\pm$, and $\{Z^\varepsilon_{\pm,k}\,;\,k \in \mathbb{N}\}$ are i.i.d.r.v.s, which are also independent of $N^\varepsilon_\pm$, with the common law

$$\mathbb{P}\big[Z^\varepsilon_{-,k} \in dz\big] = \frac{\mathbf{1}_{(z\le-\varepsilon)}}{\lambda^\varepsilon_-}\,f(z)\,dz, \quad \mathbb{P}\big[Z^\varepsilon_{+,k} \in dz\big] = \frac{\mathbf{1}_{(z\ge\varepsilon)}}{\lambda^\varepsilon_+}\,f(z)\,dz,$$

where

$$f(z) := e^{-|z|} \quad (z \in (-\infty, 0) \cup (0, \infty)),$$
$$\lambda_-^\varepsilon := \int_{-\infty}^{-\varepsilon} f(z)\,\mathrm{d}z\big(= e^{-\varepsilon}\big), \quad \lambda_+^\varepsilon := \int_\varepsilon^\infty f(z)\,\mathrm{d}z\big(= e^{-\varepsilon}\big).$$

Theorem 11.2.2 *For $G \in C_c^1(\mathbb{R})$, it holds that*

$$\mathbb{E}\big[G'(Z_t)\,Z_t\big] = \mathbb{E}\left[\big\{G\big(\tilde{Z}_t\big) - G(0)\big\}\,\frac{\alpha\big(- E_- + E_+ - 2\big)}{\alpha - 1}\right]. \tag{11.15}$$

Proof Note that as $\alpha \in (1, 2)$ we have that $\mathbb{E}[|G(\tilde{Z}_t)|\,|E_- + E_+|] < \infty$.[3] Write $\delta = (\alpha - 1)/\alpha$. For $\xi \in \mathbb{R}$ with $|\xi| \le 1$, we shall define $H^\xi(z) = z/(1-\xi)$, as seen in the proof of Theorem 11.1.2. Define by

$$\begin{aligned}
E_-^{\varepsilon,\xi} &:= -\int_{(-\infty,0)\times[0,1]} H^\xi(z)\,\mathcal{N}^{(\varepsilon,-)}(\mathrm{d}z, \mathrm{d}s),\\
E_+^{\varepsilon,\xi} &:= \int_{(0,\infty)\times[0,1]} H^\xi(z)\,\mathcal{N}^{(\varepsilon,+)}(\mathrm{d}z, \mathrm{d}s),\\
\tilde{Z}_t^{\varepsilon,\xi} &:= \tilde{Z}_{-,t}^{E_-^{\varepsilon,\xi}} + \tilde{Z}_{+,t}^{E_+^{\varepsilon,\xi}} \big(= \tilde{Z}_t^{E_-^{\varepsilon,\xi}, E_+^{\varepsilon,\xi}} \big).
\end{aligned}$$

Since $\partial_\xi E_\pm^{\varepsilon,\xi}\big|_{\xi=0} = E_\pm^\varepsilon$ and

$$D\big(G(\tilde{Z}_t^\varepsilon)\big) = \partial_\xi\big(G(\tilde{Z}_t^{\varepsilon,\xi})\big)\big|_{\xi=0} = G'(\tilde{Z}_t^\varepsilon)\,\partial_\xi \tilde{Z}_t^{\varepsilon,\xi}\big|_{\xi=0} = \delta\, G'(\tilde{Z}_t^\varepsilon)\,\tilde{Z}_t^\varepsilon,$$

we see that

$$\begin{aligned}
\mathbb{E}\big[D\big(G(\tilde{Z}_t^\varepsilon)\big)\big] &= \delta\,\mathbb{E}\big[G'(\tilde{Z}_t^\varepsilon)\,\tilde{Z}_{-,t}^\varepsilon\,\mathbf{1}_{(N_-^\varepsilon \ge 1)}\big] + \delta\,\mathbb{E}\big[G'(\tilde{Z}_t^\varepsilon)\,\tilde{Z}_{+,t}^\varepsilon\,\mathbf{1}_{(N_+^\varepsilon \ge 1)}\big]\\
&=: I_1^\varepsilon + I_2^\varepsilon.
\end{aligned}$$

Now, we shall focus on the study of I_2^ε only, because I_1^ε can be computed similarly. We remark that

$$\begin{aligned}
I_2^\varepsilon &= \delta\,\mathbb{E}\Big[G'\big(\tilde{Z}_{-,t}^\varepsilon + \tilde{\sigma}_{+,t}\,(E_+^\varepsilon)^\delta\big)\,\tilde{\sigma}_{+,t}\,(E_+^\varepsilon)^\delta\,\mathbf{1}_{(N_+^\varepsilon \ge 1)}\Big]\\
&= \delta \sum_{N=1}^\infty \mathbb{P}\big[N_+^\varepsilon = N\big]\,\mathbb{E}\Big[G'\Big(\tilde{Z}_{-,t}^\varepsilon + \tilde{\sigma}_{+,t}\,\big(\textstyle\sum_{k=1}^N Z_{+,k}^\varepsilon\big)^\delta\Big)\,\tilde{\sigma}_{+,t}\,\big(\textstyle\sum_{j=1}^N Z_{+,j}^\varepsilon\big)^\delta\Big]\\
&= \sum_{N=1}^\infty \mathbb{P}\big[N_+^\varepsilon = N\big]
\end{aligned}$$

---

[3]In the case that $G \in C_b^1$ recall, for example, Exercise 5.1.13. Since $\mathbb{E}\big[E_+^p\big] = \int_0^\infty \mathbb{P}[E_+ > \lambda^{1/p}]\,d\lambda$, all we have to check is $\mathbb{P}[E_+ > \lambda^{1/p}]$ as $\lambda \to \infty$, which can be seen in [36]. We can also obtain the pth moment on E_-.

$$\times \sum_{j=1}^{N} \mathbb{E}\left[\frac{\partial}{\partial z_j} \left\{ G\left(\tilde{Z}^{\varepsilon}_{-,t} + \tilde{\sigma}_{+,t} \left(\sum_{k \neq j} Z^{\varepsilon}_{+,k} + z_j \right)^{\delta} \right) - G(0) \right\} \Big|_{z_j = Z^{\varepsilon}_{+,j}} Z^{\varepsilon}_{+,j} \right]$$

$$=: \sum_{N=1}^{\infty} \mathbb{P}[N^{\varepsilon}_{+} = N] \sum_{j=1}^{N} I^{\varepsilon}_{j,N}.$$

Since $G \in C^1_c(\mathbb{R})$ and $zf(z) \to 0$ as $|z| \to \infty$, the divergence formula tells us to see that

$$\begin{aligned}
I^{\varepsilon}_{j,N} &= \frac{1}{\lambda^{\varepsilon}_{+}} \int_{\varepsilon}^{\infty} \mathbb{E}\left[\frac{\partial}{\partial z_j} \left\{ G\left(\tilde{Z}^{\varepsilon}_{-,t} + \tilde{\sigma}_{+,t} \left(\sum_{k \neq j} Z^{\varepsilon}_{+,k} + z_j \right)^{\delta} \right) - G(0) \right\} z_j \, f(z_j) \right] \mathrm{d}z_j \\
&= \frac{1}{\lambda^{\varepsilon}_{+}} \left[\mathbb{E}\left[\left\{ G\left(\tilde{Z}^{\varepsilon}_{-,t} + \tilde{\sigma}_{+,t} \left(\sum_{k \neq j} Z^{\varepsilon}_{+,k} + z_j \right)^{\delta} \right) - G(0) \right\} z_j \, f(z_j) \right] \right]_{z_j = \varepsilon}^{z_j = \infty} \\
&\quad - \frac{1}{\lambda^{\varepsilon}_{+}} \int_{\varepsilon}^{\infty} \mathbb{E}\left[\left\{ G\left(\tilde{Z}^{\varepsilon}_{-,t} + \tilde{\sigma}_{+,t} \left(\sum_{k \neq j} Z^{\varepsilon}_{+,k} + z_j \right)^{\delta} \right) - G(0) \right\} \left(z_j \, f(z_j) \right)' \right] \mathrm{d}z_j \\
&= -\frac{1}{\lambda^{\varepsilon}_{+}} \mathbb{E}\left[\left\{ G\left(\tilde{Z}^{\varepsilon}_{-,t} + \tilde{\sigma}_{+,t} \left(\sum_{k \neq j} Z^{\varepsilon}_{+,k} + \varepsilon \right)^{\delta} \right) - G(0) \right\} \varepsilon \, f(\varepsilon) \right] \\
&\quad - \mathbb{E}\left[\left\{ G\left(\tilde{Z}^{\varepsilon}_{-,t} + \tilde{\sigma}_{+,t} \left(\sum_{1 \le k \le N} Z^{\varepsilon}_{+,k} \right)^{\delta} \right) - G(0) \right\} \frac{\left(z_j \, f(z_j) \right)' \big|_{z_j = Z^{\varepsilon}_{+,j}}}{f(Z^{\varepsilon}_{+,j})} \right].
\end{aligned}$$

Moreover, since $\left(z_j \, f(z_j) \right)' \big|_{z_j = Z^{\varepsilon}_{+,j}} = (1 - Z^{\varepsilon}_{+,j}) \, f(Z^{\varepsilon}_{+,j})$, we have

$$\begin{aligned}
I^{\varepsilon}_{2} &= -\varepsilon \sum_{N=1}^{\infty} \mathbb{P}[N^{\varepsilon}_{+} = N] \\
&\qquad \times \sum_{j=1}^{N} \mathbb{E}\left[\left\{ G\left(\tilde{Z}^{\varepsilon}_{-,t} + \tilde{\sigma}_{+,t} \left(\sum_{k \neq j} Z^{\varepsilon}_{+,k} + \varepsilon \right)^{\delta} \right) - G(0) \right\} \right] \\
&\quad - \sum_{N=1}^{\infty} \mathbb{P}[N^{\varepsilon}_{+} = N] \\
&\qquad \times \sum_{j=1}^{N} \mathbb{E}\left[\left\{ G\left(\tilde{Z}^{\varepsilon}_{-,t} + \tilde{\sigma}_{+,t} \left(\sum_{1 \le k \le N} Z^{\varepsilon}_{+,k} \right)^{\delta} \right) - G(0) \right\} (1 - Z^{\varepsilon}_{+,j}) \right] \\
&= -\varepsilon \sum_{N=1}^{\infty} N \, \mathbb{P}[N^{\varepsilon}_{+} = N] \, \mathbb{E}\left[\left\{ G\left(\tilde{Z}^{\varepsilon}_{-,t} + \tilde{\sigma}_{+,t} \left(\sum_{k=1}^{N-1} Z^{\varepsilon}_{+,k} + \varepsilon \right)^{\delta} \right) - G(0) \right\} \right] \\
&\quad - \mathbb{E}\left[\left\{ G\left(\tilde{Z}^{\varepsilon}_{-,t} + \tilde{Z}^{\varepsilon}_{+,t} \right) - G(0) \right\} \left(1 - E^{\varepsilon}_{+} \right) \right].
\end{aligned}$$

Taking the limit as $\varepsilon \downarrow 0$ leads us to get

$$\lim_{\varepsilon \downarrow 0} I_2^\varepsilon = -\mathbb{E}\Big[\big\{G(\tilde{Z}_t^\varepsilon) - G(0)\big\}\,\big(1 - E_+^\varepsilon\big)\Big].$$

Similarly, we can also derive

$$\lim_{\varepsilon \downarrow 0} I_1^\varepsilon = -\mathbb{E}\Big[\big\{G(\tilde{Z}_t^\varepsilon) - G(0)\big\}\,\big(1 + E_-^\varepsilon\big)\Big].$$

Therefore, we have

$$\begin{aligned}\mathbb{E}\big[G'(Z_t)\,Z_t\big] &= \mathbb{E}\big[G'(\tilde{Z}_t)\,\tilde{Z}_t\big] \\ &= \frac{1}{\delta}\lim_{\varepsilon \downarrow 0}\big(I_1^\varepsilon + I_2^\varepsilon\big) \\ &= \mathbb{E}\left[\big\{G(\tilde{Z}_t) - G(0)\big\}\,\frac{-E_- + E_+ - 2}{\delta}\right].\end{aligned}$$

The proof is complete.

Problem 11.2.3 Prove the formula (11.15) in the above theorem, by using the densities of exponential random variables $E_\pm$, via the usual IBP.

Easily, one can extend the previous result to the case of stable random variables with non-zero mean. That is, as explained in (11.9), we know that $Z_t \stackrel{\mathscr{L}}{=} \tilde{\sigma}_t\,E^\delta + \gamma\,t$, where $\tilde{\sigma}_t := t^{1/\alpha}\,\sigma_2\,U$ and σ_2 is the constant given by (11.7). Write $\tilde{Z}_t^y := \tilde{\sigma}_t\,y^\delta + \gamma\,t$. Then, we have $\tilde{Z}_t^E \stackrel{\mathscr{L}}{=} Z_t$. Similarly to Theorem 11.2.2, we can also get

Corollary 11.2.4 *For $G \in C_c^1(\mathbb{R})$, it holds that*

$$\mathbb{E}\big[G'(Z_t)\,(Z_t - \gamma\,t)\big] = \mathbb{E}\left[\big\{G(\tilde{Z}_t) - G(\gamma\,t)\big\}\,\frac{\alpha(E-1)}{\alpha - 1}\right]. \tag{11.16}$$

Proof. For $0 < \varepsilon < 1$, let N_ε be the Poisson random variables with the parameter λ_ε, and $\big\{Y_k^\varepsilon\,;\,k \geq 1\big\}$ the i.i.d.r.v.s, which are also independent of N_ε, with the common law

$$\mathbb{P}\big[Z_k^\varepsilon \in \mathrm{d}z\big] = \frac{\mathbf{1}_{(z \geq \varepsilon)}}{\lambda_\varepsilon}\,f(z)\,\mathrm{d}z,$$

where $f(z) := e^{-z}\ \ (z > 0)$, $\lambda_\varepsilon := \int_\varepsilon^\infty f(z)\,\mathrm{d}z\big(= e^{-\varepsilon}\big)$. Write $E_\varepsilon := \sum_{k=1}^{N_\varepsilon} Z_k^\varepsilon$ and $\tilde{Z}_t^\varepsilon := \tilde{Z}_t^{E_\varepsilon}$.

Write $\delta = (\alpha - 1)/\alpha$. In the proof of Theorem 11.2.2, we have already studied that, for $\Phi \in C_c^1(\mathbb{R})$,

$$\mathbb{E}\big[\Phi'(\tilde{\sigma}_{+,t}\,(E_+^\varepsilon)^\delta)\,\tilde{\sigma}_{+,t}\,(E_+^\varepsilon)^\delta\big] = \mathbb{E}\left[\big\{\Phi(\tilde{\sigma}_{+,t}\,(E_+^\varepsilon)^\delta) - \Phi(0)\big\}\,\frac{E_+^\varepsilon - 1}{\delta}\right]. \tag{11.17}$$

In particular, we shall choose $\Phi(x) = G(x + \gamma\, t)$, and replace $\tilde{\sigma}_{+,t}$ and E_+^ε by $\tilde{\sigma}_t$ and E_ε, in the equation (11.17). Since $\tilde{\sigma}_t\, E_\varepsilon^\delta = \tilde{Z}_t^\varepsilon - \gamma\, t$, we can get

$$\mathbb{E}\big[G'(\tilde{Z}_t^\varepsilon)\,(\tilde{Z}_t^\varepsilon - \gamma\, t)\big] = \mathbb{E}\left[\big\{G(\tilde{Z}_t^\varepsilon) - G(\gamma\, t)\big\}\,\frac{E_\varepsilon - 1}{\delta}\right].$$

Therefore, we have

$$\begin{aligned}\mathbb{E}\big[G'(Z_t)\,(Z_t - \gamma\, t)\big] &= \mathbb{E}\big[G'(\tilde{Z}_t)\,(\tilde{Z}_t - \gamma\, t)\big]\\ &= \lim_{\varepsilon\downarrow 0}\mathbb{E}\big[G'(\tilde{Z}_t^\varepsilon)\,(\tilde{Z}_t^\varepsilon - \gamma\, t)\big]\\ &= \lim_{\varepsilon\downarrow 0}\mathbb{E}\left[\big\{G(\tilde{Z}_t^\varepsilon) - G(\gamma\, t)\big\}\,\frac{E_\varepsilon - 1}{\delta}\right]\\ &= \mathbb{E}\left[\big\{G(\tilde{Z}_t) - G(\gamma\, t)\big\}\,\frac{E - 1}{\delta}\right].\end{aligned}$$

The proof is complete. □

Problem 11.2.5 Prove the formula (11.16), by using the densities of the exponential random variable E, via the usual IBP.

A similar idea using subordinated Brownian motions (recall Sect. 5.6) was used in [35].

11.3 Sensitivity Analysis for Truncated Stable Processes

Let $\kappa > 0$ and $0 < \alpha < 1$. The truncated stable process $Z = \big\{Z_t\,;\, t \in [0, T]\big\}$ is the pure-jump Lévy process with the Lévy measure

$$\nu(\mathrm{d}z) = f(z)\,\mathrm{d}z, \quad f(z) = \kappa\, z^{-1-\alpha}\,\mathbf{1}_{(0<z\leq 1)}.$$

Then, the Lévy–Itô decomposition theorem leads us to see that Z_t can be expressed as

$$Z_t = \int_{(0,\infty)\times[0,t]} z\,\mathcal{N}^+(\mathrm{d}z, \mathrm{d}s).$$

Let $0 < \varepsilon < 1$, and $\xi \in \mathbb{R}$ with $|\xi| \leq 1$. Write

$$\lambda_\varepsilon := \int_\varepsilon^\infty \nu(\mathrm{d}z)\left(= \frac{\kappa}{\alpha}(\varepsilon^{-\alpha} - 1)\right), \quad \nu_\varepsilon(\mathrm{d}z) := \frac{\mathbf{1}_{(z>\varepsilon)}}{\lambda_\varepsilon}\,\nu(\mathrm{d}z).$$

Let

$$Z_t^\varepsilon := \int_{(0,\infty)\times[0,t]} z\, \mathcal{N}^{(\varepsilon,+)}(\mathrm{d}z, \mathrm{d}s),$$
$$Z_t^{\varepsilon,\xi} := \int_{(0,\infty)\times[0,t]} H^\xi(z)\, \mathcal{N}^{(\varepsilon,+)}(\mathrm{d}z, \mathrm{d}s),$$

where $\mathcal{N}^{(\varepsilon,+)}(\mathrm{d}z, \mathrm{d}s)$ is the Poisson random measure with the intensity $\hat{\mathcal{N}}^{(\varepsilon,+)}(\mathrm{d}z, \mathrm{d}s) := \nu_\varepsilon(\mathrm{d}z)\,\mathrm{d}s$, $\tilde{\mathcal{N}}^{(\varepsilon,+)}(\mathrm{d}z, \mathrm{d}s) := \mathcal{N}^{(\varepsilon,+)}(\mathrm{d}z, \mathrm{d}s) - \hat{\mathcal{N}}^{(\varepsilon,+)}(\mathrm{d}z, \mathrm{d}s)$. Here, the transformation function is defined as $H^\xi(z) := z\exp(\xi\,\varphi(z))$ with $\varphi \in C^1((0,\infty))$ such that

$$|\varphi(z)| \le C\,(|z| \wedge 1). \tag{11.18}$$

We remark that the processes $Z^\varepsilon = \{Z_t^\varepsilon\,;\, t \ge 0\}$ and $Z^{\varepsilon,\xi} = \{Z_t^{\varepsilon,\xi}\,;\, t \ge 0\}$ can be expressed as

$$Z_t^\varepsilon = \sum_{k=1}^{N_t^\varepsilon} Y_k^\varepsilon,$$
$$Z_t^{\varepsilon,\xi} = \sum_{k=1}^{N_t^\varepsilon} H^\xi(Y_k^\varepsilon),$$

where $N^\varepsilon = \{N_t^\varepsilon\,;\, t \ge 0\}$ is the Poisson process with the parameter λ_ε, and $\{Y_k^\varepsilon\,;\, k \in \mathbb{N}\}$ are the i.i.d.r.v.s, independent of N^ε, with the common law $\mathbb{P}[Y_k^\varepsilon \in \mathrm{d}z] = \nu_\varepsilon(\mathrm{d}z)$.

Theorem 11.3.1 *Suppose that $\varphi(\varepsilon) = o(\varepsilon^\alpha)$ as $\varepsilon \to 0$. Then, for $G \in C_c^1(\mathbb{R})$, it holds that*

$$\begin{aligned}
&\mathbb{E}\left[G'(Z_t)\int_{(0,\infty)\times[0,t]} z\,\varphi(z)\,\mathcal{N}^+(dz, ds)\right] \\
&= -\mathbb{E}\left[G(Z_t)\int_{(0,\infty)\times[0,t]} \big(\varphi'(z)\,z - \alpha\varphi(z)\big)\,\mathcal{N}^+(dz, ds)\right] \\
&\quad + \kappa\, t\, \varphi(1)\,\mathbb{E}\big[G(Z_t + 1)\big].
\end{aligned} \tag{11.19}$$

Proof First, note that in this case, $\mathbb{E}[|G(Z_t)|^p] < \infty$, for any $p > 0$. Furthermore, since $\partial_\xi (H^\xi(z))\big|_{\xi=0} = z\,\varphi(z)$, we see that

$$D\big(G(Z_t^\varepsilon)\big) = \partial_\xi\big(G(Z_t^{\varepsilon,\xi})\big)\big|_{\xi=0} = G'(Z_t^\varepsilon)\sum_{k=1}^{N_t^\varepsilon} Y_k^\varepsilon\,\varphi(Y_k^\varepsilon).$$

Hence, we have evaluating the expectation with respect to N_t^ε and conditioning with respect to all jump sizes except Y_k^ε,

$$
\begin{aligned}
\mathbb{E}\big[D\big(G(Z_t^\varepsilon)\big)\big] &= \mathbb{E}\left[G'(Z_t^\varepsilon) \sum_{k=1}^{N_t^\varepsilon} Y_k^\varepsilon \, \varphi(Y_k^\varepsilon) \, \mathbf{1}_{(N_t^\varepsilon \geq 1)} \right] \\
&= \sum_{N=1}^{\infty} \mathbb{P}\big[N_t^\varepsilon = N\big] \, \mathbb{E}\left[G'\big(\textstyle\sum_{j=1}^N Y_j^\varepsilon\big) \displaystyle\sum_{k=1}^{N} Y_k^\varepsilon \, \varphi(Y_k^\varepsilon) \right] \\
&= \sum_{N=1}^{\infty} \mathbb{P}\big[N_t^\varepsilon = N\big] \sum_{k=1}^{N} \mathbb{E}\big[\partial_{z_k} \big\{G\big(\textstyle\sum_{j \neq k} Y_j^\varepsilon + z_k\big)\big\}\big|_{z_k = Y_k^\varepsilon} Y_k^\varepsilon \, \varphi(Y_k^\varepsilon)\big] \\
&= \sum_{N=1}^{\infty} \mathbb{P}\big[N_t^\varepsilon = N\big] \sum_{k=1}^{N} \int_0^\infty \mathbb{E}\big[\partial_{z_k} \big\{G\big(\textstyle\sum_{j \neq k} Y_j^\varepsilon + z_k\big)\big\} \, z_k \, \varphi(z_k)\big] \, \nu_\varepsilon(dz_k) \\
&=: \sum_{N=1}^{\infty} \mathbb{P}\big[N_t^\varepsilon = N\big] \sum_{k=1}^{N} I_k^\varepsilon .
\end{aligned}
$$

The divergence formula in the usual sense leads us to get that

$$
\begin{aligned}
I_k^\varepsilon &= \int_0^\infty \mathbb{E}\big[\partial_{z_k} \big\{G\big(\textstyle\sum_{j \neq k} Y_j^\varepsilon + z_k\big)\big\} \, z_k \, \varphi(z_k)\big] \, \nu_\varepsilon(dz_k) \\
&= \int_\varepsilon^1 \mathbb{E}\big[\partial_{z_k} \big\{G\big(\textstyle\sum_{j \neq k} Y_j^\varepsilon + z_k\big)\big\}\big] \, \dfrac{z_k \, \varphi(z_k) \, f(z_k)}{\lambda_\varepsilon} \, dz_k \\
&= \left[\mathbb{E}\big[G\big(\textstyle\sum_{j \neq k} Y_j^\varepsilon + z_k\big)\big] \, \dfrac{z_k \, \varphi(z_k) \, f(z_k)}{\lambda_\varepsilon} \right]_{z_k = \varepsilon}^{z_k = 1} \\
&\quad - \int_\varepsilon^1 \mathbb{E}\big[G\big(\textstyle\sum_{j \neq k} Y_j^\varepsilon + z_k\big)\big] \, \dfrac{\big(z_k \, \varphi(z_k) \, f(z_k)\big)'}{\lambda_\varepsilon} \, dz_k \\
&= \mathbb{E}\big[G\big(\textstyle\sum_{j \neq k} Y_j^\varepsilon + 1\big)\big] \, \dfrac{\varphi(1) \, f(1)}{\lambda_\varepsilon} - \mathbb{E}\big[G\big(\textstyle\sum_{j \neq k} Y_j^\varepsilon + \varepsilon\big)\big] \, \dfrac{\varepsilon \, \varphi(\varepsilon) \, f(\varepsilon)}{\lambda_\varepsilon} \\
&\quad - \mathbb{E}\left[G\big(\textstyle\sum_{1 \leq j \leq N} Y_j^\varepsilon\big) \, \dfrac{\big(z_k \, \varphi(z_k) \, f(z_k)\big)'\big|_{z_k = Y_k^\varepsilon}}{f(Y_k^\varepsilon)} \right].
\end{aligned}
$$

We remark that reordering the random variables Y_j^ε and using the relation $N \, \mathbb{P}\big[N_t^\varepsilon = N\big] = \lambda_\varepsilon t \, \mathbb{P}\big[N_t^\varepsilon = N - 1\big]$,

$$
\begin{aligned}
&\sum_{N=1}^{\infty} \mathbb{P}\big[N_t^\varepsilon = N\big] \\
&\quad \times \sum_{k=1}^{N} \left\{ \mathbb{E}\big[G\big(\textstyle\sum_{j \neq k} Y_j^\varepsilon + 1\big)\big] \, \dfrac{\varphi(1) \, f(1)}{\lambda_\varepsilon} - \mathbb{E}\big[G\big(\textstyle\sum_{j \neq k} Y_j^\varepsilon + \varepsilon\big)\big] \, \dfrac{\varepsilon \, \varphi(\varepsilon) \, f(\varepsilon)}{\lambda_\varepsilon} \right\} \\
&= \mathbb{P}\big[N_t^\varepsilon = 1\big] \left\{ G(1) \, \dfrac{\varphi(1) \, \kappa}{\lambda_\varepsilon} - G(\varepsilon) \, \dfrac{\varphi(\varepsilon) \, \kappa}{\lambda_\varepsilon \, \varepsilon^\alpha} \right\}
\end{aligned}
$$

$$
\begin{aligned}
&+ \sum_{N=2}^{\infty} \mathbb{P}[N_t^\varepsilon = N] \\
&\quad \times \sum_{k=1}^{N} \left\{ \mathbb{E}\big[G(\textstyle\sum_{j\neq k} Y_j^\varepsilon + 1)\big] \dfrac{\varphi(1)\, f(1)}{\lambda_\varepsilon} - \mathbb{E}\big[G(\textstyle\sum_{j\neq k} Y_j^\varepsilon + \varepsilon)\big] \dfrac{\varepsilon\, \varphi(\varepsilon)\, f(\varepsilon)}{\lambda_\varepsilon} \right\} \\
&= \mathbb{P}[N_t^\varepsilon = 1] \left\{ G(1)\, \frac{\varphi(1)\, \kappa}{\lambda_\varepsilon} - G(\varepsilon)\, \frac{\varphi(\varepsilon)\, \kappa}{\lambda_\varepsilon\, \varepsilon^\alpha} \right\} \\
&+ \sum_{N=2}^{\infty} N\, \mathbb{P}[N_t^\varepsilon = N] \\
&\quad \times \left\{ \mathbb{E}\big[G(\textstyle\sum_{1\leq j\leq N-1} Y_j^\varepsilon + 1)\big] \dfrac{\varphi(1)\, \kappa}{\lambda_\varepsilon} - \mathbb{E}\big[G(\textstyle\sum_{1\leq j\leq N-1} Y_j^\varepsilon + \varepsilon)\big] \dfrac{\varphi(\varepsilon)\, \kappa}{\lambda_\varepsilon\, \varepsilon^\alpha} \right\} \\
&= t\, \mathbb{P}[N_t^\varepsilon = 0] \left\{ G(1)\, \varphi(1)\, \kappa - G(\varepsilon)\, \frac{\varphi(\varepsilon)\, \kappa}{\varepsilon^\alpha} \right\} \\
&+ t \sum_{N=1}^{\infty} \mathbb{P}[N_t^\varepsilon = N] \\
&\quad \times \left\{ \mathbb{E}\big[G(\textstyle\sum_{1\leq j\leq N} Y_j^\varepsilon + 1)\big]\, \varphi(1)\, \kappa - \mathbb{E}\big[G(\textstyle\sum_{1\leq j\leq N} Y_j^\varepsilon + \varepsilon)\big] \dfrac{\varphi(\varepsilon)\, \kappa}{\varepsilon^\alpha} \right\} \\
&= t \left\{ \mathbb{E}\big[G(\textstyle\sum_{1\leq j\leq N_t^\varepsilon} Y_j^\varepsilon + 1)\big]\, \varphi(1)\, \kappa - \mathbb{E}\big[G(\textstyle\sum_{1\leq j\leq N_t^\varepsilon} Y_j^\varepsilon + \varepsilon)\big] \dfrac{\varphi(\varepsilon)\, \kappa}{\varepsilon^\alpha} \right\} \\
&= t \left\{ \mathbb{E}\big[G(Z_t^\varepsilon + 1)\big]\, \varphi(1)\, \kappa - \mathbb{E}\big[G(Z_t^\varepsilon + \varepsilon)\big] \frac{\varphi(\varepsilon)\, \kappa}{\varepsilon^\alpha} \right\} \\
&\to t\, \mathbb{E}\big[G(Z_t^\varepsilon + 1)\big]\, \varphi(1)\, \kappa
\end{aligned}
$$

as $\varepsilon \downarrow 0$ due to the condition $\varphi(\varepsilon) = o(\varepsilon^\alpha)$ as $\varepsilon \to 0$. On the other hand, it holds that

$$
\begin{aligned}
&\sum_{N=1}^{\infty} \mathbb{P}[N_t^\varepsilon = N]\, \mathbb{E}\left[G(\textstyle\sum_{1\leq j\leq N} Y_j^\varepsilon) \displaystyle\sum_{k=1}^{N} \frac{\big(z_k\, \varphi(z_k)\, f(z_k)\big)'\big|_{z_k = Y_k^\varepsilon}}{f(Y_k^\varepsilon)} \right] \\
&= \mathbb{E}\left[G(\textstyle\sum_{1\leq j\leq N_t^\varepsilon} Y_j^\varepsilon) \displaystyle\sum_{k=1}^{N_t^\varepsilon} \frac{\big(z_k\, \varphi(z_k)\, f(z_k)\big)'\big|_{z_k = Y_k^\varepsilon}}{f(Y_k^\varepsilon)} \right] \\
&= \mathbb{E}\left[G(Z_t^\varepsilon) \int_{(0,\infty)\times[0,t]} \frac{\big(z\, \varphi(z)\, f(z)\big)'}{f(z)}\, \mathscr{N}^{(\varepsilon,+)}(\mathrm{d}z, \mathrm{d}s) \right] \\
&\to \mathbb{E}\left[G(Z_t) \int_{(0,\infty)\times[0,t]} \frac{\big(z\, \varphi(z)\, f(z)\big)'}{f(z)}\, \mathscr{N}^{+}(\mathrm{d}z, \mathrm{d}s) \right]
\end{aligned}
$$

as $\varepsilon \downarrow 0$. Note that the limit processes are well defined due to the condition that $\varphi \in C^1((0,\infty))$. The proof is complete.

Exercise 11.3.2 Find an example of the function φ satisfying the assumptions stated above.

The result and the proof in the case of $1 \le \alpha < 2$ can be done similarly. In fact, one just needs to change the definitions in the same fashion as in Chaps. 4 and 5. This is done in the next exercise.

Problem 11.3.3 In the case $1 \le \alpha < 2$, we have the following definitions and properties:

$$Z_t = \int_{(0,\infty)\times[0,t]} z\,\tilde{\mathscr{N}}^{+}(\mathrm{d}z, \mathrm{d}s).$$

$$Z_t^{\varepsilon} := \int_{(0,\infty)\times[0,t]} z\,\tilde{\mathscr{N}}^{(\varepsilon,+)}(\mathrm{d}z, \mathrm{d}s),$$

$$Z_t^{\varepsilon,\xi} := \int_{(0,\infty)\times[0,t]} H^{\xi}(z)\,\tilde{\mathscr{N}}^{(\varepsilon,+)}(\mathrm{d}z, \mathrm{d}s).$$

$$Z_t^{\varepsilon} = \sum_{k=1}^{N_t^{\varepsilon}} Y_k^{\varepsilon} - \int_{(0,\infty)\times[0,t]} z\,\hat{\mathscr{N}}^{(\varepsilon,+)}(\mathrm{d}z, \mathrm{d}s),$$

$$Z_t^{\varepsilon,\xi} = \sum_{k=1}^{N_t^{\varepsilon}} H^{\xi}(Y_k^{\varepsilon}) - \int_{(0,\infty)\times[0,t]} z\,\hat{\mathscr{N}}^{(\varepsilon,+)}(\mathrm{d}z, \mathrm{d}s),$$

where $N^{\varepsilon} = \{N_t^{\varepsilon}\,;\, t \ge 0\}$ is the Poisson process with the parameter λ_{ε}, and $\{Y_k^{\varepsilon}\,;\, k \in \mathbb{N}\}$ are the i.i.d.r.v.s, independent of N^{ε}, with the common law $\mathbb{P}[Y_k^{\varepsilon} \in \mathrm{d}z] = \nu_{\varepsilon}(\mathrm{d}z)$. With the above definitions prove an analogous result as in Theorem 11.3.1.

11.4 Sensitivity Analysis for Tempered Stable Processes

Let a and b be positive constants, and $0 < \alpha < 2$. The tempered stable process $Z = \{Z_t\,;\, t \in [0,T]\}$ is the pure-jump Lévy process with the Lévy measure

$$\nu(\mathrm{d}z) = f(z)\,\mathrm{d}z, \quad f(z) = a\,z^{-1-\alpha}\,e^{-bz}\,\mathbf{1}_{(z>0)}.$$

In particular, the case of $\alpha = 1/2$ is called *the inverse Gaussian process*. Furthermore, for each $t \in [0,T]$, the marginal Z_t has the explicit form of the characteristic function of

$$\mathbb{E}\big[\exp(i\,y\,Z_t)\big] = \exp\big[a\,\Gamma(-\alpha)\,t\,\big\{(b-iy)^{\alpha} - b^{\alpha} - \alpha\,i\,y\,b^{\alpha-1}\,\mathbf{1}_{(1<\alpha<2)}\big\}\big].$$

The Lévy–Itô decomposition theorem leads us to see that Z_t can be expressed as

$$Z_t = \int_{(0,\infty)\times[0,t]} z\,\widetilde{\mathcal{N}^+}(dz, ds) + \int_{(0,\infty)\times[0,t]} z\,\widehat{\mathcal{N}^+}(dz, ds)\,\mathbf{1}_{(0<\alpha<1)}.$$

In this section, we will deduce the sensibility formulas for both cases $0 < \alpha < 1$ and $1 \le \alpha < 2$ using the above compact notation with indicator functions. Let $0 < \varepsilon < 1$, and $\xi \in \mathbb{R}$ with $|\xi| \le 1$. Write

$$\lambda_\varepsilon := \int_\varepsilon^\infty \nu(dz), \quad \nu_\varepsilon(dz) := \frac{\mathbf{1}_{(z>\varepsilon)}}{\lambda_\varepsilon}\,\nu(dz).$$

Let

$$\begin{aligned}
Z_t^\varepsilon &:= \int_{(0,\infty)\times[0,t]} z\,\tilde{\mathcal{N}}^{(\varepsilon,+)}(dz, ds) + \int_{(0,\infty)\times[0,t]} z\,\hat{\mathcal{N}}^{(\varepsilon,+)}(dz, ds)\,\mathbf{1}_{(0<\alpha<1)}\\
&= \int_{(0,\infty)\times[0,t]} z\,\mathcal{N}^{(\varepsilon,+)}(dz, ds) - \int_{(0,\infty)\times[0,t]} z\,\hat{\mathcal{N}}^{(\varepsilon,+)}(dz, ds)\,\mathbf{1}_{(1\le\alpha<2)},\\
Z_t^{\varepsilon,\xi} &:= \int_{(0,\infty)\times[0,t]} H^\xi(z)\,\mathcal{N}^{(\varepsilon,+)}(dz, ds) - \int_{(0,\infty)\times[0,t]} z\,\hat{\mathcal{N}}^{(\varepsilon,+)}(dz, ds)\,\mathbf{1}_{(1\le\alpha<2)},
\end{aligned}$$

where $\mathcal{N}^{(\varepsilon,+)}(dz, ds)$ is the Poisson random measure with the intensity

$$\begin{aligned}
&\hat{\mathcal{N}}^{(\varepsilon,+)}(dz, ds) := \nu_\varepsilon(dz)\,ds,\\
&\tilde{\mathcal{N}}^{(\varepsilon,+)}(dz, ds) := \mathcal{N}^{(\varepsilon,+)}(dz, ds) - \hat{\mathcal{N}}^{(\varepsilon,+)}(dz, ds),
\end{aligned}$$

and $\varphi \in C^1((0, \infty))$ such that

$$|\varphi(z)| \le C\,(|z| \wedge 1), \tag{11.20}$$

and $H^\xi(z) := z\exp(\xi\,\varphi(z))$. We remark that the processes $Z^\varepsilon = \{Z_t^\varepsilon\,;\, t \ge 0\}$ and $Z^{\varepsilon,\xi} = \{Z_t^{\varepsilon,\xi}\,;\, t \ge 0\}$ can be expressed as

$$\begin{aligned}
Z_t^\varepsilon &= \sum_{k=1}^{N_t^\varepsilon} Y_k^\varepsilon - \int_{(0,\infty)\times[0,t]} z\,\hat{\mathcal{N}}^{(\varepsilon,+)}(dz, ds)\,\mathbf{1}_{(1\le\alpha<2)},\\
Z_t^{\varepsilon,\xi} &= \sum_{k=1}^{N_t^\varepsilon} H^\xi(Y_k^\varepsilon) - \int_{(0,\infty)\times[0,t]} z\,\hat{\mathcal{N}}^{(\varepsilon,+)}(dz, ds)\,\mathbf{1}_{(1\le\alpha<2)},
\end{aligned}$$

where $N^\varepsilon = \{N_t^\varepsilon\,;\, t \ge 0\}$ is the Poisson process with the parameter λ_ε, and $\{Y_k^\varepsilon\,;\, k \in \mathbb{N}\}$ are the i.i.d.r.v.s, independent of N^ε, with the common law $\mathbb{P}\big[Y_k^\varepsilon \in dz\big] = \nu_\varepsilon(dz)$.

Theorem 11.4.1 *Suppose that $\varphi(\varepsilon) = o(\varepsilon^\alpha)$ and $\varphi'(\varepsilon) = O(\varepsilon^{(\alpha-1)^+})$ as $\varepsilon \to 0$. Then, for $G \in C_c^1(\mathbb{R})$, it holds that*

$$\begin{aligned}&\mathbb{E}\left[G'(Z_t)\int_{(0,\infty)\times[0,t]} z\,\varphi(z)\,\mathcal{N}^+(dz,ds)\right]\\&=\mathbb{E}\left[G(Z_t)\int_{(0,\infty)\times[0,t]}\big(\varphi'(z)\,z-(\alpha+bz)\,\varphi(z)\big)\,\mathcal{N}^+(dz,ds)\right].\end{aligned}\tag{11.21}$$

Proof The strategy to prove the assertion is almost parallel to Theorem 11.3.1. Also recall Theorem 11.3.1. We shall focus on the case of $0<\alpha<1$, only. The proof for $1\le\alpha<2$ can be done similarly. Since

$$D\big(G(Z_t^\varepsilon)\big)=\partial_\xi\big(G(Z_t^{\varepsilon,\xi})\big)\big|_{\xi=0}=G'(Z_t^\varepsilon)\sum_{k=1}^{N_t^\varepsilon}Y_k^\varepsilon\,\varphi(Y_k^\varepsilon),$$

we have

$$\begin{aligned}\mathbb{E}\big[D\big(G(Z_t^\varepsilon)\big)\big]&=\mathbb{E}\left[G'(Z_t^\varepsilon)\sum_{k=1}^{N_t^\varepsilon}Y_k^\varepsilon\,\varphi(Y_k^\varepsilon)\,\mathbf{1}_{(N_t^\varepsilon\ge1)}\right]\\&=\sum_{N=1}^{\infty}\mathbb{P}[N_t^\varepsilon=N]\sum_{k=1}^{N}\int_0^\infty\mathbb{E}\big[\partial_{z_k}\big\{G(\textstyle\sum_{j\neq k}Y_j^\varepsilon+z_k)\big\}\,z_k\,\varphi(z_k)\big]\,\nu_\varepsilon(\mathrm{d}z_k)\\&=:\sum_{N=1}^{\infty}\mathbb{P}[N_t^\varepsilon=N]\sum_{k=1}^{N}I_k^\varepsilon.\end{aligned}$$

The divergence formula in the usual sense leads us to get that

$$\begin{aligned}I_k^\varepsilon&=\int_\varepsilon^\infty\mathbb{E}\big[\partial_{z_k}\big\{G(\textstyle\sum_{j\neq k}Y_j^\varepsilon+z_k)\big\}\big]\,\dfrac{z_k\,\varphi(z_k)\,f(z_k)}{\lambda_\varepsilon}\,\mathrm{d}z_k\\&=\left[\mathbb{E}\big[G(\textstyle\sum_{j\neq k}Y_j^\varepsilon+z_k)\big]\,\dfrac{z_k\,\varphi(z_k)\,f(z_k)}{\lambda_\varepsilon}\right]_{z_k=\varepsilon}^{z_k=\infty}\\&\quad-\int_\varepsilon^\infty\mathbb{E}\big[G(\textstyle\sum_{j\neq k}Y_j^\varepsilon+z_k)\big]\,\dfrac{\big(z_k\,\varphi(z_k)\,f(z_k)\big)'}{\lambda_\varepsilon}\,\mathrm{d}z_k\\&=\lim_{M\to\infty}\mathbb{E}\big[G(\textstyle\sum_{j\neq k}Y_j^\varepsilon+M)\big]\,\dfrac{M\,\varphi(M)\,f(M)}{\lambda_\varepsilon}-\mathbb{E}\big[G(\textstyle\sum_{j\neq k}Y_j^\varepsilon+\varepsilon)\big]\,\dfrac{\varepsilon\,\varphi(\varepsilon)\,f(\varepsilon)}{\lambda_\varepsilon}\\&\quad-\mathbb{E}\left[G(\textstyle\sum_{1\le j\le N}Y_j^\varepsilon)\,\dfrac{\big(z_k\,\varphi(z_k)\,f(z_k)\big)'\big|_{z_k=Y_k^\varepsilon}}{f(Y_k^\varepsilon)}\right].\end{aligned}$$

Since the function G is bounded and φ grows linearly, and

$$\frac{M\,\varphi(M)\,f(M)}{\lambda_\varepsilon}=\frac{a\,\varphi(M)}{\lambda_\varepsilon M^\alpha}\,e^{-bM}\to0$$

as $M\to\infty$, we see that

$$\sum_{N=1}^{\infty} \mathbb{P}\big[N_t^\varepsilon = N\big] \sum_{k=1}^{N} \lim_{M\to\infty} \mathbb{E}\big[G(\textstyle\sum_{j\neq k} Y_j^\varepsilon + M)\big] \dfrac{M\,\varphi(M)\,f(M)}{\lambda_\varepsilon} = 0.$$

On the other hand, we remark that due to $\varphi(\varepsilon) = o(\varepsilon^\alpha)$

$$\begin{aligned}
&\sum_{N=1}^{\infty} \mathbb{P}\big[N_t^\varepsilon = N\big] \sum_{k=1}^{N} \mathbb{E}\big[G(\textstyle\sum_{j\neq k} Y_j^\varepsilon + \varepsilon)\big] \dfrac{\varepsilon\,\varphi(\varepsilon)\,f(\varepsilon)}{\lambda_\varepsilon} \\
&= \mathbb{P}\big[N_t^\varepsilon = 1\big]\, G(\varepsilon)\, \frac{a\,\varphi(\varepsilon)\,e^{-b\varepsilon}}{\lambda_\varepsilon\,\varepsilon^\alpha} \\
&\quad + \sum_{N=2}^{\infty} \mathbb{P}\big[N_t^\varepsilon = N\big] \sum_{k=1}^{N} \mathbb{E}\big[G(\textstyle\sum_{j\neq k} Y_j^\varepsilon + \varepsilon)\big] \dfrac{a\,\varphi(\varepsilon)\,e^{-b\varepsilon}}{\lambda_\varepsilon\,\varepsilon^\alpha} \\
&= \mathbb{P}\big[N_t^\varepsilon = 1\big]\, G(\varepsilon)\, \frac{a\,\varphi(\varepsilon)\,e^{-b\varepsilon}}{\lambda_\varepsilon\,\varepsilon^\alpha} \\
&\quad + \sum_{N=2}^{\infty} N\,\mathbb{P}\big[N_t^\varepsilon = N\big]\, \mathbb{E}\big[G(\textstyle\sum_{1\le j\le N-1} Y_j^\varepsilon + \varepsilon)\big] \dfrac{a\,\varphi(\varepsilon)\,e^{-b\varepsilon}}{\lambda_\varepsilon\,\varepsilon^\alpha} \\
&= t\,\mathbb{P}\big[N_t^\varepsilon = 0\big]\, G(\varepsilon)\, \frac{a\,\varphi(\varepsilon)\,e^{-b\varepsilon}}{\varepsilon^\alpha} \\
&\quad + t \sum_{N=1}^{\infty} \mathbb{P}\big[N_t^\varepsilon = N\big]\, \mathbb{E}\big[G(\textstyle\sum_{1\le j\le N} Y_j^\varepsilon + \varepsilon)\big] \dfrac{a\,\varphi(\varepsilon)\,e^{-b\varepsilon}}{\varepsilon^\alpha} \\
&= t\,\mathbb{E}\big[G(\textstyle\sum_{1\le j\le N_t^\varepsilon} Y_j^\varepsilon + \varepsilon)\big] \dfrac{a\,\varphi(\varepsilon)\,e^{-b\varepsilon}}{\varepsilon^\alpha} \\
&= t\,\mathbb{E}\big[G(Z_t^\varepsilon + \varepsilon)\big] \frac{a\,\varphi(\varepsilon)\,e^{-b\varepsilon}}{\varepsilon^\alpha} \\
&\to 0
\end{aligned}$$

as $\varepsilon \downarrow 0$. Moreover, using that $\varphi'(\varepsilon) = O(\varepsilon^{(\alpha-1)^+})$ as $\varepsilon \to 0$, it holds that

$$\begin{aligned}
&\sum_{N=1}^{\infty} \mathbb{P}\big[N_t^\varepsilon = N\big] \mathbb{E}\left[G(\textstyle\sum_{1\le j\le N} Y_j^\varepsilon) \displaystyle\sum_{k=1}^{N} \frac{\big(z_k\,\varphi(z_k)\,f(z_k)\big)'\big|_{z_k=Y_k^\varepsilon}}{f(Y_k^\varepsilon)} \right] \\
&= \mathbb{E}\left[G(\textstyle\sum_{1\le j\le N_t^\varepsilon} Y_j^\varepsilon) \displaystyle\sum_{k=1}^{N_t^\varepsilon} \frac{\big(z_k\,\varphi(z_k)\,f(z_k)\big)'\big|_{z_k=Y_k^\varepsilon}}{f(Y_k^\varepsilon)} \right] \\
&= \mathbb{E}\left[G(Z_t^\varepsilon) \int_{(0,\infty)\times[0,t]} \frac{\big(z\,\varphi(z)\,f(z)\big)'}{f(z)}\, \mathscr{N}^{(\varepsilon,+)}(\mathrm{d}z, \mathrm{d}s) \right] \\
&\to \mathbb{E}\left[G(Z_t) \int_{(0,\infty)\times[0,t]} \frac{\big(z\,\varphi(z)\,f(z)\big)'}{f(z)}\, \mathscr{N}^{+}(\mathrm{d}z, \mathrm{d}s) \right]
\end{aligned}$$

as $\varepsilon \downarrow 0$. Since

$$\frac{\left(z\,\varphi(z)\,f(z)\right)'}{f(z)} = \varphi'(z)\,z - (\alpha + bz)\,\varphi(z),$$

the proof is complete. □

Exercise 11.4.2 Find an example of the function φ satisfying the assumptions stated above.

For a hint, see Chap. 14.

Exercise 11.4.3 Compute and prove an IBP formula for stable processes with Lévy measure $\nu(dx) = \frac{dx}{x^{1+\alpha}}\mathbf{1}_{\{x>0\}}$, $\alpha \in (0, 2)$. That is, use a transformation function $H^{\xi}(z) := z\exp(\xi\,\varphi(z))$ and find appropriate conditions in order to obtain an IBP formula.

11.5 Inverting the Stochastic Covariance Matrix

In order to obtain a flexible IBP formula or a proper sensibility formula, we need to show how to modify all the previous results in order to obtain a formula of the type $\mathbb{E}[G'(Z_t)Y] = \mathbb{E}[G(Z_t)H]$ for a large class of integrable random variable Y and where H has to be a random variable so that the right-hand side is well defined.

Take as an example the result in Theorem 11.1.2. If we wish to find a sensibility formula for $\partial_c\mathbb{E}[G(S_t)]$, where the function G is not necessarily differentiable, we will need to perform the following steps:

1. Approximate G using functions $G_n \in C_c^1$ such that $G_n \to G$ in an appropriate topology.
2. Compute $\partial_c\mathbb{E}[G_n(S_t)]$. This will probably give $\partial_c\mathbb{E}[G_n(S_t)] = \mathbb{E}[G_n'(S_t)S_t]t$.
3. The problem now is how to apply formula (11.3) here. One idea is to find an auxiliary function $g_n \in C_c^1$ such that $g_n'(S_t)S_tZ_t = G_n'(S_t)S_t$. For the moment, let us suppose that such a function exists.[4] In order to perform this step one can intuitively see that we will need to prove that Z_t^{-1} is properly defined in L^p. This is usually called the inverse moment problem or non-degeneracy condition, among other names in the literature.
4. Now (11.3) can be applied. Finally, we need to prove using a limit procedure that the formulas obtained converge to $\partial_c\mathbb{E}[G(S_t)]$.

Exercise 11.5.1 Let $Z = \{Z_t\}_{t\geq 0}$ be the gamma process described in Sect. 11.1. Find conditions on the parameters (a, b, t) so that $\mathbb{E}[Z_t^{-p}] < \infty$ for some $p \geq 1$, using the following two methods.

(i) Compute directly via the explicit density function of Z_t.

[4] The existence of such a function can be assured, if one assumes enough conditions on the function G_n.

(ii) Prove the assertion via the Laplace transform of Z_t.

Hint: Prove and use the following useful identity for appropriate positive random variables X: $\mathbb{E}[X^{-p}] = \frac{1}{\Gamma(p)} \int_0^\infty \theta^{p-1} \mathbb{E}[e^{-\theta X}] d\theta$.

For a hint, see Chap. 14.

Exercise 11.5.2 Let $Z = \{Z_t\}_{t \geq 0}$ be the stable process with index $0 < \alpha < 2$, described in Sect. 11.2. Prove that $\mathbb{E}[Z_t^{-p}] < \infty$ for any $p \geq 1$, via the Laplace transform of Z_t.

For a hint, see Chap. 14.

Exercise 11.5.3 Let $Z = \{Z_t\}_{t \geq 0}$ be the truncated stable process with the index $0 < \alpha < 2$, described in Sect. 11.3. Assume that $\varphi(z) = c\, z$ for $|z| \leq 1$. Define $\Gamma_t := \int_{(0,\infty)\times[0,t]} z\, \varphi(z)\, \mathcal{N}^+(dz, ds)$. Prove that $\mathbb{E}[\Gamma_t^{-p}] < \infty$ for any $p \geq 1$, via the Laplace transform of Z_t.

For a hint, see Chap. 14.

Exercise 11.5.4 Let $Z = \{Z_t\}_{t \geq 0}$ be the tempered stable process with the index $0 < \alpha < 2$, described in Sect. 11.4. Assume that $\varphi(z) = cz$ for $|z| \leq 1$. Define $\Gamma_t := \int_{(0,\infty)\times[0,t]} z\, \varphi(z)\, \mathcal{N}^+(dz, ds)$. Prove that $\mathbb{E}[\Gamma_t^{-p}] < \infty$ for some $p \geq 1$, via the Laplace transform of Z_t.

For a hint, see Chap. 14.

Let $\{X_t \equiv X_t^c\}_{t\in[0,T]}$ be a process depending on the parameter $c \in \mathbb{R}$, which we have seen in the previous sections, e.g. the geometric gamma process in Sect. 11.1, the stable process in Sect. 11.2, the truncated stable process in Sect. 11.3, and the tempered stable process in Sect. 11.4. Suppose that, for each $t > 0$, the law of X_t has a density $p_{X_t}(c, x)$ with respect to the Lebesgue measure on $\mathbb{R}$, and the random variable X_t is differentiable with respect to c. As seen in Theorems 11.1.2, 11.2.2, 11.3.1 and 11.4.1 and Corollary 11.2.4, we have already obtained that in some cases the following result is valid:

$$\mathbb{E}\big[G'(X_t)\, Y_t\, H_t\big] = \mathbb{E}\big[G(X_t)\, \Theta(Y_t, H_t)\big] \tag{11.22}$$

for $G \in C_b^1(\mathbb{R})$, where $\{Y_t\}_{t\in[0,T]}$ is the process associated with the process X such that the inverse of Y_t has a higher-order moment, $\{H_t\}_{t\in[0,T]}$ is an auxiliary process such that H_t has a higher-order moment, and $\{\Theta(Y_t, H_t)\}_{t\in[0,T]}$ is a random variable depending on Y_t and H_t such that the right-hand side is well defined.

Theorem 11.5.5 *Assume that* (11.22) *is valid for* $H_t = \partial_c X_t Y_t^{-1}$. *For* $G \in C_b^1(\mathbb{R})$, *it holds that*

$$\mathbb{E}[G'(X_t)\, \partial_c X_t] = \mathbb{E}\Big[G(X_t)\, \Theta\Big(Y_t, \frac{\partial_c X_t}{Y_t}\Big)\Big]. \tag{11.23}$$

Proof From the equality (11.22), we have

$$\mathbb{E}\big[G'(X_t)\,\partial_c X_t\big] = \mathbb{E}\left[G'(X_t)\,Y_t\,\frac{\partial_c X_t}{Y_t}\right]$$
$$= \mathbb{E}\left[G(X_t)\,\Theta\Big(Y_t,\,\frac{\partial_c X_t}{Y_t}\Big)\right].$$

□

Remark 11.5.6 (*Negative-order moments*) Consider the case of the geometric Lévy process of the form:

$$X_t = X_0\,\exp(\rho\,Z_t + \tilde{c}\,t),$$

where $X_0 > 0$, $\rho > 0$, and $\tilde{c} \in \mathbb{R}$. For $G \in C_b^1(\mathbb{R})$, we have already obtained the following formulas:

(i) If Z is the gamma process with the parameters (a, b), then it holds that

$$\mathbb{E}\big[G'(X_t)\,X_t\,\rho\,Z_t\big] = \mathbb{E}[G(X_t)\,(b\,Z_t - a\,t)]$$

in Theorem 11.1.2 of Sect. 11.1.

(ii) If Z is the stable process with the index $1 < \alpha < 2$, then it holds that

$$\mathbb{E}\big[G'(X_t)\,X_t\,\rho\,Z_t\big] = \mathbb{E}\left[\big\{G(\tilde{X}_t) - G(X_0)\big\}\,\frac{\alpha\,(-E_- + E_+ - 2)}{\alpha - 1}\right]$$

in Theorem 11.2.2 of Sect. 11.2, where $\tilde{X}_t = X_0\,\exp(\rho\,\tilde{Z}_t + \tilde{c}\,t)$.

(iii) If Z is the truncated stable process with the index $0 < \alpha < 2$, then it holds that

$$\mathbb{E}\big[G'(X_t)\,X_t\,\rho\,\Gamma_t\big] = -\mathbb{E}\left[G(X_t)\int_{(0,\infty)\times[0,t]}\big(\varphi'(z)\,z - \varphi(z)\,z^2\big)\,\mathscr{N}^+(dz, ds)\right]$$
$$+\,\kappa\,t\,\varphi(1)\,\mathbb{E}\big[G(X_0\,\exp(\rho\,(Z_t + 1) + \tilde{c}\,t))\big]$$

in Theorem 11.3.1 of Sect. 11.3, where

$$\Gamma_t = \int_{(0,\infty)\times[0,t]} z\,\varphi(z)\,\mathscr{N}^+(dz, ds).$$

Here, we consider

$$Z_t = \int_{(0,\infty)\times[0,t]} z\,\tilde{\mathscr{N}}^+(dz, ds) + \int_{(0,\infty)\times[0,t]} z\,\hat{\mathscr{N}}^+(dz, ds)\,\mathbb{I}_{(0<\alpha<1)}.$$

(iv) If Z is the tempered stable process with the index $0 < \alpha < 2$, then it holds that

$$\mathbb{E}\left[G'(X_t)\, X_t\, \rho\, \Gamma_t\right]$$
$$= \mathbb{E}\left[G(X_t) \int_{(0,\infty)\times[0,t]} \left(\varphi'(z)\, z - (\alpha + b\, z)\, \varphi(z)\right) \mathcal{N}^+(dz, ds)\right]$$

in Theorem 11.4.1 of Sect. 11.4, where

$$\Gamma_t = \int_{(0,\infty)\times[0,t]} z\, \varphi(z)\, \mathcal{N}^+(dz, ds).$$

Here, we consider

$$Z_t = \int_{(0,\infty)\times[0,t]} z\, \tilde{\mathcal{N}}^+(dz, ds) + \int_{(0,\infty)\times[0,t]} z\, \hat{\mathcal{N}}^+(dz, ds)\, \mathbb{I}_{(0<\alpha<1)}.$$

In the formula (11.22), we shall substitute for

$$(Y_t, H_t) = \begin{cases} \left(\partial_c X_t\, Z_t,\, 1/Z_t\right) & \text{in the case of (i) and (ii),} \\ \left(\partial_c X_t\, \Gamma_t,\, 1/\Gamma_t\right) & \text{in the case of (iii) and (iv),} \end{cases}$$

and the parameter c is either X_0 or ρ.

In order to obtain such a formula we need to perform again similar procedures as in previous theorems but with specific test functions g. Just to consider one specific example, consider the case of a tempered stable process and the sensibility $\partial_c \mathbb{E}[G(cZ_t)] = \mathbb{E}[G'(cZ_t)Z_t]$. In order to deal with this problem we will need to enlarge the dimension of the process and consider $\partial_c \mathbb{E}[g(cX_t, \Gamma_t)]$ for $g(x, y) = \rho^{-1} G(x) y^{-1}$. The procedure is parallel to the one carried in the proof of Theorem 11.4.1, just longer because of the amount of terms. Clearly when doing this, one will need an argument to prove that $\mathbb{E}[\Gamma_t^{-p}]$ for integer values of $p > 1$.

Then, our purpose is to study the higher-order moments of $1/Z_t$ and $1/\Gamma_t$. In the case of Z_t, this has been studied in Exercises 11.5.1–11.5.4. The study of the inverse moments for Γ_t requires a careful choice of the deformation function $H^\xi(z) = z\exp(\xi\varphi(z))$. Again, we remark that this is linked to the decrease of the Laplace transform of Γ_t.

Exercise 11.5.7 Find proper conditions on φ so that the inverse moments of $\Gamma_t = \int_{(0,\infty)\times[0,t]} z\, \varphi(z)\, \mathcal{N}^+(dz, ds)$, are finite.

For a hint, see Chap. 14.

11.6 Sensitivity for Non-smooth Test Functions

Now, we shall study a sensitivity formula of $\mathbb{E}[G(X_t^c)]$ with respect to the parameter c for a function G which is not necessarily differentiable. Instead of giving a general theory we consider the particular case of an indicator function. In what follows, we may simplify the notation as $X_t \equiv X_t^c$ if the meaning is clear.

Theorem 11.6.1 *Define the following class of test functions:*

$$\mathfrak{F}(\mathbb{R}) = \left\{ G = \sum_{k=1}^{m} \alpha_k \, G_k \, \mathbb{I}_{A_k} \, ; \; m \in \mathbb{N}, \; \alpha_k \in \mathbb{R}, \; G_k \in C_c(\mathbb{R}), \; A_k\text{: an interval of } \mathbb{R} \right\}.$$

Suppose that for any compact set K,

$$\sup_{c \in K} \mathbb{E}\Big[\Big| \Theta_t\Big(Y, \frac{\partial_c X_t}{Y}\Big) \Big| \Big] < \infty.$$

Then, for $G \in \mathfrak{F}(\mathbb{R})$, it holds that

$$\partial_c \big(\mathbb{E}[G(X_t)] \big) = \mathbb{E}\Big[G(X_t) \, \Theta_t\Big(Y, \frac{\partial_c X_t}{Y}\Big) \Big]. \tag{11.24}$$

Proof For $G \in C_c^1(\mathbb{R})$, we can check the sensitivity formula (11.24) directly, because

$$\begin{aligned} \partial_c \big(\mathbb{E}[G(X_t)] \big) &= \mathbb{E}\big[G'(X_t) \, \partial_c X_t \big] \\ &= \mathbb{E}\Big[G(X_t) \, \Theta_t\Big(Y, \frac{\partial_c X_t}{Y}\Big) \Big]. \end{aligned}$$

The strategy to remove the regularity conditions on G and to extend to the class $\mathfrak{F}(\mathbb{R})$ can be found in e.g. [54].

(i) For $G \in C_c(\mathbb{R})$, we can find the sequence $\{G_n \, ; \, n \in \mathbb{N}\}$ in $C_c^1(\mathbb{R})$ such that $\|G_n - G\|_\infty \to 0$ as $n \to \infty$, where

$$\|\varphi\|_\infty = \sup_{t \in [0,T]} |\varphi(t)|.$$

Then, for each compact set $K \subset \mathbb{R}$, we have

$$\big| \mathbb{E}[G_n(X_t)] - \mathbb{E}[G(X_t)] \big| \le \|G_n - G\|_\infty,$$

$$\sup_{c \in K} \Big| \partial_c \big(\mathbb{E}[G_n(X_t)] \big) - \mathbb{E}\Big[G(X_t) \, \Theta_t\Big(Y, \frac{\partial_c X_t}{Y}\Big) \Big] \Big| \le \sup_{c \in K} \mathbb{E}\Big[\Big| \Theta_t\Big(Y, \frac{\partial_c X_t}{Y}\Big) \Big| \Big] \, \|G_n - G\|_\infty,$$

which tend to 0 as $n \to \infty$. Hence, we can get the formula (11.24) for $G \in C_c(\mathbb{R})$.

(ii) Let $G \in C_b(\mathbb{R})$[5] and fix $0 < \varepsilon < 1$. Then, there exists the sequence $\{G_n \, ; \, n \in \mathbb{N}\}$ of continuous functions defined by

$$G_n(x) = \begin{cases} G(x) & (x \in [-n + \varepsilon, n - \varepsilon]), \\ 0 & (x \in (-\infty, -n - \varepsilon] \cup [n + \varepsilon, \infty)), \end{cases}$$

[5] If Z_t has a second-order moment for each $t \in [0, T]$, the function $G \in C_b(\mathbb{R})$ can be extended to the continuous function with linear growth order.

and $G_n(x) \in [0, G(x)]$ for each $x \in (-n-\varepsilon, -n+\varepsilon) \cup (n-\varepsilon, n+\varepsilon]$. Here, $[0, -1]$ should be understood as $[-1, 0]$. Then, it is clear that

$$G_n \in C_c(\mathbb{R}), \quad \sup_{n \in \mathbb{N}} \|G_n\|_\infty = \|G\|_\infty.$$

The dominated convergence theorem enables us to see that

$$\left|\mathbb{E}[G(X_t)] - \mathbb{E}[G_n(X_t)]\right| \to 0$$

as $n \to \infty$. Moreover, since

$$\begin{aligned}\mathbb{E}\big[|G_n(X_t) - G(X_t)|^2\big] &\le \mathbb{E}\big[|G_n(X_t)|^2\, \mathbb{I}_{(|X_t|>n-\varepsilon)}\big] \\ &\le \|G\|_\infty^2\, \mathbb{P}[|Xt| > n-\varepsilon] \to 0\end{aligned}$$

as $n \to \infty$, we have, for each compact set $K \subset \mathbb{R}$, that

$$\begin{aligned}&\sup_{c \in K} \left|\partial_c\big(\mathbb{E}[G_n(X_t)]\big) - \mathbb{E}\Big[G(X_t)\,\Theta_t\Big(Y, \frac{\partial_c X_t}{Y}\Big)\Big]\right| \\ &\le \sup_{c \in K} \mathbb{E}\Big[\Big|\Theta_t\Big(Y, \frac{\partial_c X_t}{Y}\Big)\Big|^2\Big]^{1/2} \sup_{c \in K} \mathbb{E}\big[|G_n(X_t) - G(X_t)|^2\big]^{1/2}\end{aligned}$$

by the Cauchy–Schwarz inequality, which tends to 0 as $n \to \infty$. Therefore, we can get the formula (11.24) for $G \in C_b(\mathbb{R})$.

Finally, we compute the derivative

$$\begin{aligned}&\left|\frac{\mathbb{E}[G(X_t^{c+h})] - \mathbb{E}[G(X_t^c)]}{h} - \mathbb{E}\Big[G(X_t)\,\Theta_t\Big(Y, \frac{\partial_c X_t}{Y}\Big)\Big]\right| \\ &\le 2h^{-1}\|G_n - G\|_\infty + \left|\frac{\mathbb{E}[G_n(X_t^{c+h})] - \mathbb{E}[G_n(X_t^c)]}{h} - \mathbb{E}\Big[G(X_t)\,\Theta_t\Big(Y, \frac{\partial_c X_t}{Y}\Big)\Big]\right|.\end{aligned}$$

From here, one chooses n large enough in order to achieve the proof of convergence.

(iii) Let $-\infty \le a < b \le \infty$, and write

$$a_n^{\pm} = a \pm \frac{1}{n}, \quad b_n^{\pm} = b \pm \frac{1}{n}$$

for $n \in \mathbb{N}$. Now, we shall consider the case of $G = \mathbb{I}_{(a,b]}$.[6] Then, we can find a sequence $\{G_n\,;\, n \in \mathbb{N}\}$ of continuous functions such that

$$G_n(x) = \begin{cases} G(x) & (x \in (a_n^+, b_n^-)), \\ 0 & (x \in (-\infty, a_n^-] \cup [b_n^+, \infty)), \end{cases}$$

[6] We can discuss the case of $[a, b)$, (a, b) and $[a, b]$ similarly.

and $G_n(x) \in [0, G(x)]$ for $x \in (a_n^-, a_n^+) \cup (b_n^-, b_n^+)$. It is trivial that

$$G_n \in C_b(\mathbb{R}), \quad \sup_{n\in\mathbb{N}} \|G_n\|_\infty \le 1.$$

The dominated convergence theorem implies that

$$\left|\mathbb{E}[G_n(X_t)] - \mathbb{E}[G(X_t)]\right| \to 0$$

as $n \to \infty$. Moreover, for each compact set $K \in \mathbb{R}$, we see that

$$\begin{aligned}\sup_{c\in K} \mathbb{E}\big[|G_n(X_t) - G(X_t)|^2\big] &= \sup_{c\in K} \mathbb{E}\big[|G_n(X_t) - G(X_t)|^2; X_t \in (a_n^-, a_n^+) \cup (b_n^-, b_n^+)\big] \\ &\le 4 \sup_{c\in K} \mathbb{P}\big[X_t \in (a_n^-, a_n^+)\big] + 4 \sup_{c\in K} \mathbb{P}\big[X_t \in (b_n^-, b_n^+)\big] \\ &= 4 \int_{a_n^-}^{a_n^+} \sup_{c\in K} p_{X_t}(c, x)\, dx + \int_{b_n^-}^{b_n^+} \sup_{c\in K} p_{X_t}(c, x)\, dx,\end{aligned}$$

which tends to 0 as $n \to \infty$, because the Lebesgue measures of the intervals (a_n^-, a_n^+) and (b_n^-, b_n^+) converges to 0 as $n \to \infty$. Hence, we have

$$\begin{aligned}&\sup_{c\in K} \left|\partial_c\big(\mathbb{E}[G_n(X_t)]\big) - \mathbb{E}\Big[G(X_t)\,\Theta_t\Big(Y, \frac{\partial_c X_t}{Y}\Big)\Big]\right| \\ &\le \sup_{c\in K} \mathbb{E}\left[\left|\Theta_t\Big(Y, \frac{\partial_c X_t}{Y}\Big)\right|^2\right]^{1/2} \sup_{c\in K} \mathbb{E}\big[|G_n(X_t) - G(X_t)|^2\big]^{1/2}\end{aligned}$$

from the Cauchy–Schwarz inequality, which tends to 0 as $n \to \infty$.

As in (ii), we need to compute

$$\begin{aligned}&\left|\frac{\mathbb{E}[G(X_t^{c+h})] - \mathbb{E}[G(X_t^c)]}{h} - \mathbb{E}\Big[G(X_t)\,\Theta_t\Big(Y, \frac{\partial_c X_t}{Y}\Big)\Big]\right| \\ &\le 2h^{-1} \sup_{c\in K} \mathbb{E}\big[|G_n(X_t^c) - G(X_t^c)|^2\big]^{1/2} \\ &\quad + \left|\frac{\mathbb{E}[G_n(X_t^{c+h})] - \mathbb{E}[G_n(X_t^c)]}{h} - \mathbb{E}\Big[G(X_t)\,\Theta_t\Big(Y, \frac{\partial_c X_t}{Y}\Big)\Big]\right|.\end{aligned}$$

Therefore, we can get the formula (11.24) for $G = \mathbb{I}_{(a,b]}$.

(iv) Finally, from the studies in (ii) and (iii) stated above, we can get the formula (11.24) for $G \in \mathfrak{F}(\mathbb{R})$.

Chapter 12
Integration by Parts: Norris Method

In this chapter, we extend the method of analysis introduced in Chap. 11 to a general framework. This method was essentially introduced by Norris to obtain an integration by parts (IBP) formula for jump-driven stochastic differential equations. We focus our study on the directional derivative of the jump measure which respect to the direction of the Girsanov transformation. We first generalize the method in order to consider random variables on Poisson spaces and then show in various examples how the right choice of direction of integration is an important element of this formula.

12.1 Introduction

In the recent past, various efforts to develop IBP formulas on Poisson spaces have appeared. Among them, one may mention [9, 12–14, 38, 47] between others. Most of these methods are designed to obtain the regularity of the laws of solutions of stochastic differential equations driven by jump processes and at the same time they try to strive for as much generality as possible. This generalization feature, which may be very appealing from a mathematical point of view, has hampered the development of various other applications which may use differential techniques on the Poisson space. In particular, taking into consideration that most models where this differential theory may be applied have a specific compensator, one may believe that further developments on studies of upper and lower bounds for densities, studies on the support of the law and various other expansion techniques for functionals on Poisson space should have mushroomed after the discovery of these methods.

Still, this is not the case. Most probably this is due to the lack of adaptability of the method to explicit models. In this chapter, we reconsider the approach introduced in [45] which was designed with explicit compensators in mind. In particular, Norris considers directional derivatives with respect to the underlying Lévy measure, which provides an IBP formula where the interaction of the direction and the measure is

© Springer Nature Singapore Pte Ltd. 2019
A. Kohatsu-Higa and A. Takeuchi, *Jump SDEs and the Study of Their Densities*, Universitext, https://doi.org/10.1007/978-981-32-9741-8_12

extremely clear. This allows for a better choice of direction when considering explicit problems. We reintroduce the methodology of Norris in a general set-up, compare it with the Girsanov transform method of Bismut and then apply it in various applied problems which indicate clearly that this explicit IBP formula may be helpful in order to obtain various properties of the density of a random variable on Poisson spaces.

Let us explain our approach in a simpler example. Let X be a one-dimensional random variable with density function g and let $G \in C_b^1(\mathbb{R})$. Using a change of variable one has that

$$\mathbb{E}\left[G(X)\right] = \mathbb{E}\left[G(H^{\xi}(X))g(H^{\xi}(X))\partial_y H^{\xi}(X)g^{-1}(X)\right]. \tag{12.1}$$

Here we suppose for simplicity that $g(x) > 0$ for all $x \in \mathbb{R}$ and that H^{ξ} is a one-to-one differentiable transformation. Now if we differentiate the right-hand side with respect to ξ, evaluating this derivative with respect to $\xi = 0$ and under the hypothesis that $H^{\xi}(x) = x$ we obtain the following IBP formula:

$$\mathbb{E}\left[G'(X)\partial_{\xi} H^{\xi}(X)\Big|_{\xi=0}\right] = \mathbb{E}\left[G(X)\left\{\partial_x \ln(g(X)) + \partial_{\xi} \ln(\partial_y H^{\xi}(X))\Big|_{\xi=0}\right\}\right].$$

Some comments are in order at this point:

(a) In order to consider such a proper IBP formula, we will have to assume that the boundary terms in the IBP formula do not contribute to the final result. We further have to modify the argument in order to obtain $\mathbb{E}\left[G'(X)\Theta\right]$ on the left side of the above equation, where Θ is any random variable which is a function of X. We also have to reproduce this argument in infinite dimensions.

(b) In the general setting the "density function" g becomes the Radon–Nikodým derivative of the Lévy measure and therefore is not a probability density function anymore. For this reason, we take the approach of approximating the Lévy process using compound Poisson processes.

(c) In the above argument, one clearly sees the change of measure in the term $g(H^{\xi}(X))\partial_x H^{\xi}(X)g^{-1}(X)$ and the change of the size of the jump. The differentiation of the change of measure term gives rise to the dual operator. Therefore it is natural to discuss how to obtain this result using infinite-dimensional change of measure results. We will then see that the conditions required for the transformation function H^{ξ} will slightly differ if one considers a different approach. This is the case with the use of Girsanov's theorem/Escher transform. Still when applying the Girsanov method, one does the change of measure so as to recover a copy of the original process. In our case, this may not be so as the jump size is being changed in (12.1). Furthermore, we have preferred the approximation approach because this point of view provides a general definition of derivative and adjoint operator that allows a general IBP formula.

(d) One may consider that our approach is a space–time deformation approach which is not the same as Picard's approach (see [47]) or the lent particle method of

Bouleau–Denis (see [14] or [13]). It may have a link with the approach in [9], but in that case it seems that there is no space deformation like the one proposed here (this needs to be reviewed again!). In our approach the deformation is not performed on path space but on the jump laws themselves. In that sense one may consider the approach presented here a weak form of infinite-dimensional IBP with deformation in jump size.

(e) The main advantage of the current approach is its simplicity, adaptability to various situations and that it is easy to interpret. On the other hand, one may claim that this approach is restrictive to absolutely continuous Lévy measures but we believe that it can be enlarged to situations similar to Exercise 5.1.19.

The spirit of this chapter is somewhat independent from the first part of this book. In fact, we will assume some generalizations which had not been carried out here but which we hope the reader can foresee easily.

In this chapter we use the following standard notation. Set $\mathbb{Z}_+ = \mathbb{N} \cup \{0\}$. Norms of vectors are usually denoted by $|\cdot|$ and norms of matrices by $\|\cdot\|$. All vectors are considered as line vectors and the derivative of a function $f : \mathbb{R}^a \to \mathbb{R}^b$ is considered as a $b \times a$-valued matrix. For a given vector $x = (x_1, ..., x_n)$, we write by $x_{j\to n} = (x_j, ..., x_n)$.[1]

12.2 Set-Up

In this section, we shall introduce the notation, the framework, and some well-known facts, which will be used throughout the chapter. Let $(\Omega, \mathscr{F}, \mathbb{P})$ be our underlying probability space. Let $\mathbb{E} \equiv \mathbb{E}_{\mathbb{P}}$ denote the associated expectation and $L^p(\Omega\,;\,\mathbb{R})$-lim will denote the limit in p-norm.

Let $\nu(dz)$ be a measure on $\mathbb{R}^m_0 := \mathbb{R}^m\backslash\{0\}$ which satisfies

$$\int_{\mathbb{R}^m_0} \left(|z|^2 \wedge 1\right)\, \nu(dz) < +\infty.$$

Let $Z = \{Z_t\,;\, t \geq 0\}$ be the m-dimensional Lévy process with the associated Lévy–Khintchine representation of the form:

$$\begin{aligned} &\frac{1}{t} \ln \mathbb{E}_{\mathbb{P}}\big[\exp(i\,\theta \cdot Z_t)\big] \\ &= i\,\theta\cdot\gamma - \frac{|\sigma^*\theta|^2}{2} + \int_{\mathbb{R}^m_0} \big\{\exp(i\,\theta\cdot z) - 1 - i\,\theta\cdot z\,\mathbf{1}_{\bar{B}_1}(z)\big\}\,\nu(dz) \end{aligned} \tag{12.2}$$

[1] We hope that the context will not bring any confusion with the matrix notation or the dependence of constants with respect to certain parameters.

for $\theta \in \mathbb{R}^m$, where $\gamma \in \mathbb{R}^m$ and $\sigma \in \mathbb{R}^m \otimes \mathbb{R}^m$ such that $\sigma\,\sigma^*$ is non-negative definite. Then Z is the Lévy process associated with the triplet (γ, σ, ν) and ν is called its associated Lévy measure.

The Lévy–Itô decomposition theorem tells us that Z also has the following representation:

$$Z_t = \gamma\, t + \sigma\, W_t + \int_{\bar{B}_1 \times [0,t]} z\, \widetilde{\mathcal{N}}(dz, ds) + \int_{\bar{B}_1^c \times [0,t]} z\, \mathcal{N}(dz, ds), \tag{12.3}$$

where $W = \{W_t\,;\, t \geq 0\}$ is the m-dimensional Brownian motion with $W_0 = 0$, $\mathcal{N}(dz, dz) := \sharp\{t \in ds\,;\, 0 \neq Z_t - Z_{t-} \in dz\}^2$ is the Poisson random measure on $\mathbb{R}_+ \times \mathbb{R}_0^m$ with the intensity measure $\widehat{\mathcal{N}}(dz, ds) := ds\, \nu(dz)$, and $\widetilde{\mathcal{N}}(dz, ds) := \mathcal{N}(dz, ds) - \widehat{\mathcal{N}}(dz, ds)$ is the compensated Poisson random measure (cf. [51]). Here $\sharp(A)$ denotes the cardinality of the set A. Let $\mathcal{F}_t = \bigcap_{\varepsilon > 0} \sigma(Z_s\,;\, s \leq t + \varepsilon)$ be the right-continuous completed filtration generated by Z. In this chapter, we will consider the following simplifying situation:

Hypothesis 12.2.1 $\sigma = 0$, $\gamma = 1$ and the Lévy measure ν has a density. That is, $\nu(dz) = f(z)\, dz$ for some function $f : \mathbb{R}_0^m \to \mathbb{R}_+$ satisfying that $f(0) = 0$ and

$$\int_{\mathbb{R}_0^m} (|z|^2 \wedge 1) f(z)\, dz < \infty, \qquad \int_{\mathbb{R}_0^m} f(z)\, dz = \infty.$$

Let $b \in C_b^\infty(\mathbb{R}^d\,;\, \mathbb{R}^d)$ and $a \in C_b^\infty(\mathbb{R}^d \times \mathbb{R}_0^m\,;\, \mathbb{R}^d)$ with *the invertibility condition:*

$$\inf_{y \in \mathbb{R}^d} \inf_{z \in \mathbb{R}_0^m} \left\| I_d + \nabla a(y, z) \right\| > 0. \tag{12.4}$$

Here, $I_d \in \mathbb{R}^d \otimes \mathbb{R}^d$ is the identity matrix, the norm $\|\cdot\|$ is[3] in the matrix sense, and $\nabla a(y, z)$ is the derivative with respect to $y \in \mathbb{R}^d$, while the derivative in $z \in \mathbb{R}_0^m$ is denoted by $\partial a(y, z)$. For $x \in \mathbb{R}^d$, we shall introduce the $\mathbb{R}^d$-valued process $X = \{X_t\,;\, t \geq 0\}$ determined by the stochastic differential equation of the form:

$$\begin{aligned} X_t = x + \int_0^t b(X_s)\, ds &+ \int_{\bar{B}_1 \times [0,t]} a\big(X_{s-}, z\big)\, \widetilde{\mathcal{N}}(dz, ds) \\ &+ \int_{\bar{B}_1^c \times [0,t]} a\big(X_{s-}, z\big)\, \mathcal{N}(dz, ds). \end{aligned} \tag{12.5}$$

In this section, we will let Hypothesis 7.2.2 stand without any further mention.[4]

[2]Clearly, this definition is formal. Try to write the exact definition.

[3]Recall that as all matrix norms are equivalent, you can pick the one you like best.

[4]Note that the current set-up is not totally covered by the results of Part I of this book. If you prefer you may think instead of the above general setting that the Lévy measure corresponds to the tempered Lévy measure introduced in Sect. 11.4.

Exercise 12.2.2 Give an example to show that under the condition that $\int_{\mathbb{R}_0^m} f(z)\,dz < \infty$ then the law of X_t does not have a density.[5]

For a hint, see Chap. 14.

Under our situation, there exists a unique solution such that the mapping $\mathbb{R}^d \ni x \longmapsto X_t(x) \in \mathbb{R}^d$ has a C^1-modification (cf. [2] or Chap. 7). Moreover, we see that the first-order derivative process $Y = \{Y_t \equiv Y_t(x) := \partial_x X_t(x)\,;\, t \geq 0\}$ valued in $\mathbb{R}^d \otimes \mathbb{R}^d$ satisfies the linear stochastic differential equation of the form:

$$\begin{aligned} Y_t = I_d &+ \int_0^t \nabla b(X_s)\, Y_s\, ds \\ &+ \int_{\bar{B}_1 \times [0,t]} \nabla a\big(X_{s-}, z\big)\, Y_{s-}\, \widetilde{\mathcal{N}}(dz, ds) \\ &+ \int_{\bar{B}_1^c \times [0,t]} \nabla a(X_{s-}, z)\, Y_{s-}\, \mathcal{N}(dz, ds). \end{aligned} \tag{12.6}$$

Recalling Proposition 3.6.8 may be of help.

Problem 12.2.3 Prove that in the one-dimensional case, one can write the solution of the above equation for Y explicitly.

From condition (12.4), the process $Y = \big\{Y_t\,;\, t \geq 0\big\}$ is invertible a.s. in the sense that for each $t \geq 0$, the matrix inverse of Y_t exists. Moreover, the inverse process $Y^{-1} = \big\{Y_t^{-1}\,;\, t \geq 0\big\}$ also satisfies the linear stochastic differential equation of the form:

$$\begin{aligned} Y_t^{-1} = I_d &- \int_0^t Y_s^{-1}\, \nabla b(X_s)\, ds \\ &+ \int_{\bar{B}_1 \times [0,t]} Y_{s-}^{-1} \Big\{\big(I_d + \nabla a\big(X_{s-}, z\big)\big)^{-1} - I_d\Big\}\, \widetilde{\mathcal{N}}(dz, ds) \\ &+ \int_{\bar{B}_1^c \times [0,t]} Y_{s-}^{-1} \Big\{\big(I_d + \nabla a\big(X_{s-}, z\big)\big)^{-1} - I_d\Big\}\, \mathcal{N}(dz, ds) \\ &+ \int_{\bar{B}_1 \times [0,t]} Y_s^{-1} \Big\{\big(I_d + \nabla a(X_s, z)\big)^{-1} - I_d + \nabla a(X_s, z)\Big\}\, \widehat{\mathcal{N}}(dz, ds). \end{aligned} \tag{12.7}$$

Recall that the main idea to obtain an IBP formula is to approximate the solution of (12.5), using compound Poisson processes. Therefore, we shall define the corresponding approximation sequence of the $\mathbb{R}^d$-valued process $X = \{X_t\,;\, t \geq 0\}$ determined by Eq. (12.5). Let $\varepsilon > 0$ be sufficiently small, and $T > 0$ be fixed. Recall that

$$\lambda_\varepsilon = \nu\big(\{z \in \mathbb{R}_0^m\,;\, |z| > \varepsilon\}\big), \quad f_\varepsilon(z) = \frac{f(z)}{\lambda_\varepsilon}\, \mathbf{1}_{(|z|>\varepsilon)}, \quad \nu_\varepsilon(dz) = f_\varepsilon(z)\, dz. \tag{12.8}$$

---

[5] It "almost" does have a density.

We remark that $\nu_\varepsilon(dz)$ is the probability measure on $\mathbb{R}_0^m$. Define the approximation process

$$Z_t^\varepsilon = \int_{(\bar{B}_1 \cap \bar{B}_\varepsilon^c)\times[0,t]} z\,\widetilde{\mathcal{N}}(dz, ds) + \int_{\bar{B}_1^c\times[0,t]} z\,\mathcal{N}(dz, ds).$$

Define the associated sequence of $(\mathscr{F}_t)$-stopping times $\{\tau_k^\varepsilon\,;\, k \in \mathbb{Z}_+\}$ by $\tau_0^\varepsilon = 0$, and

$$\tau_k^\varepsilon = \inf\left\{t > \tau_{k-1}^\varepsilon\,;\, \int_{\bar{B}_\varepsilon^c\times(\tau_{k-1}^\varepsilon,t]} \mathcal{N}(dz, ds) \neq 0\right\} \wedge T$$

for $k \in \mathbb{N}$. These stopping times are the times where the process Z^ϵ jumps and their successive differences form a sequence of independent λ_ε-exponentially distributed random variables. We denote by $N^\varepsilon = \{N_t^\varepsilon\,;\, t \in [0, T]\}$, the Poisson process with intensity λ_ε defined using this sequence of jump times.

We denote the set of jump sizes associated with the process Z^ε in $[0, T]$ as $Y^\varepsilon = \{Y_k^\varepsilon\,;\, k \in \mathbb{N}\}$,[6] which is a sequence of independent and identically distributed $\mathbb{R}_0^m$-valued random variables with the law $\nu_\varepsilon(dz)$, which are also independent of the process N^ε. Therefore we can rewrite the $\mathbb{R}^m$-valued process $Z^\varepsilon = \{Z_t^\varepsilon\,;\, t \in [0, T]\}$ as

$$Z_t^\varepsilon = \begin{cases} 0 & (N_t^\varepsilon = 0), \\ \displaystyle\sum_{k=1}^{N_t^\varepsilon}(Y_k^\varepsilon - \mathbb{E}[Y_k^\varepsilon \mathbf{1}_{(|Y_k^\varepsilon|\le 1)}]) & (N_t^\varepsilon > 0), \end{cases}$$

which is a compound Poisson process. Now using arguments similar to Chap. 5, one obtains the following result.

Lemma 12.2.4 *For each $0 < t \le T$, the law of Z_t^ε converges weakly to the one of Z_t as $\varepsilon \downarrow 0$.*

Proof The assertion is obvious, because one can explicitly compute the characteristic function of Z_t^ε and then prove the convergence of its characteristic function towards the corresponding one for Z_t.

We shall introduce the $\mathbb{R}^d$-valued process $X^\varepsilon = \{X_t^\varepsilon\,;\, t \in [0, T]\}$ defined by

$$X_t^\varepsilon = \begin{cases} x + \displaystyle\int_0^t \tilde{a}_0(X_s^\varepsilon)\, ds & (0 \le t < \tau_1^\varepsilon), \\ X_{\tau_k^\varepsilon -}^\varepsilon + a\big(X_{\tau_k^\varepsilon -}^\varepsilon, Y_k^\varepsilon\big) & (t = \tau_k^\varepsilon,\ k \in \mathbb{N}), \\ X_{\tau_k^\varepsilon}^\varepsilon + \displaystyle\int_{\tau_k^\varepsilon}^t \tilde{a}_0(X_s^\varepsilon)\, ds & (\tau_k^\varepsilon < t < \tau_{k+1}^\varepsilon), \end{cases}$$

---

[6] We hope that this notation for the jump size random variables which has been used from the start will not confuse the reader with the derivative of the flow. The context should make clear which object we are referring to in each case.

where $\tilde{a}_0(x) = b(x) - \int_{\varepsilon<|z|\le 1} a(x,z)\,\nu(dz)$. Define

$$K^z(x) = x + a(x,z), \quad F_{s,t}(x) = x + \int_s^t \tilde{a}_0\big(F_{s,u}(x)\big)\,du.$$

Then, we have

$$\begin{aligned} X_t^\varepsilon &= F_{\tau_N^\varepsilon,t} \circ K^{Y_N^\varepsilon} \circ F_{\tau_{N-1}^\varepsilon,\tau_N^\varepsilon-} \circ K^{Y_{N-1}^\varepsilon} \circ F_{\tau_{N-2}^\varepsilon,\tau_{N-1}^\varepsilon-} \circ \cdots \circ K^{Y_1^\varepsilon} \circ F_{0,\tau_1^\varepsilon-}(x) \\ &=: \Phi_{\varepsilon,N}\big(x\,;\, Y_{1\to N}^\varepsilon\,;\, \tau_{1\to N}^\varepsilon\big) \end{aligned} \tag{12.9}$$

for $\tau_N^\varepsilon \le t < \tau_{N+1}^\varepsilon$, where as explained previously, we use the notation $Y_{1\to N}^\varepsilon = \big(Y_1^\varepsilon,\ldots,Y_N^\varepsilon\big)$ and $\tau_{1\to N}^\varepsilon = \big(\tau_1^\varepsilon,\ldots,\tau_N^\varepsilon\big)$.

From here, we see that the approximating stochastic process X^ε is generated by an iterative functional which depends on the jump sizes and the jump times associated with the process Z^ϵ. Therefore a general theory of IBP can be developed using this approximation as the main axis. This is done in Sect. 12.4.

We also remark that the above construction has the characteristics of a strong convergence construction (i.e. almost sure convergence). One can easily change the set-up to a weak convergence structure (i.e. convergence in law) and then the IBP formula to follow in that case would also be some type of weak IBP formula.

Exercise 12.2.5 Prove that the explicit stochastic process X^ε defined above is the unique solution of the stochastic differential equation

$$\begin{aligned} X_t^\varepsilon = x &+ \int_0^t b(X_s^\varepsilon)\,ds + \int_{(\bar{B}_1\cap\bar{B}_\varepsilon^c)\times[0,t]} a\big(X_{s-}^\varepsilon,z\big)\,\widetilde{\mathcal{N}}^{(\varepsilon)}(dz,ds) \\ &+ \int_{\bar{B}_1^c\times[0,t]} a\big(X_{s-}^\varepsilon,z\big)\,\mathcal{N}^{(\varepsilon)}(dz,ds). \end{aligned}$$

12.3 Hypotheses

In this section, we introduce our hypotheses in order to be able to carry the IBP. We recall that we always assume Hypothesis 12.2.1.

Hypothesis 12.3.1 $\int_{\bar{B}_1^c} |z|^p\, f(z)\,dz < \infty$ for any $p \ge 2$.

Clearly this hypothesis is needed in order to have all necessary moment conditions. The following hypothesis determines the direction of derivation in the IBP formula.

Hypothesis 12.3.2 Let $\xi \in \mathbb{R}^u$ be in a sufficiently small neighborhood around the origin (without loss of generality, we may assume that $|\xi| \le 1$ or $0 \le \xi < 1$), and $H^\xi(\cdot) : \mathbb{R}_0^m \to \mathbb{R}_0^m$ be a mapping with $H^\xi(z)\big|_{\xi=0} = z$ and $\partial_z H^\xi(z)\big|_{\xi=0} = I_m$ for $z \in$

Supp[ν]. Define $h : \Omega' \times \mathbb{R}_0^m \to \mathbb{R}_0^m \otimes \mathbb{R}^u$ by $h(z) = (\partial_\xi)_0 H^\xi(z) := \partial_\xi H^\xi(z)\big|_{\xi=0}$, $z \in \text{Supp}[\nu]$.

We will call the random function H^ξ, the deformation/transformation function, We have chosen this name as this function deforms/transforms r.v.s through the function, $H^\xi(Y_k^\varepsilon)$ and therefore the differential calculus is carried through these new r.v.s instead of Y_k^ε. Asymptotically, as $\xi \to 0$ they are equivalent in law but the IBP in each case will lead to different formulas. The following hypothesis will also be essential for the integrability of the duality formula.

Hypothesis 12.3.3 Assume that

$$\int_{\mathbb{R}_0^m} |h(z)|\, \nu(dz) < +\infty, \quad \int_{\mathbb{R}_0^m} \left| \frac{\text{div}\{h(z) f(z)\}}{f(z)} \right| \nu(dz) < +\infty.$$

Remark 12.3.4 The condition

$$|h(z)| + \left| \frac{\text{div}\{h(z) f(z)\}}{f(z)} \right| \leq C(|z|^2 \wedge 1)$$

is a sufficient condition for Hypothesis 12.3.3 to hold.

We will introduce some examples satisfying the the above conditions in Sect. 12.6.3. Unless explicitly stated, we assume the above hypotheses throughout the rest of the chapter.

The idea of the method is to change the jumps of the driving process Z from being of size z into size $H^\xi(z)$. Once this change is done one performs IBP with respect to the new jump size and take limits as $\xi \to 0$. Therefore this method introduces a new parameter represented by $h(z)$ which can be chosen depending on the model. Another interpretation of this method is that the change of jump size implies that the law of the process Z is being changed according to a change of jump law. This method can be studied using Girsanov's theorem, which is done in Sect. 12.6.1. Still, the intuitive understanding given here can help devise a proper method to develop an IBP formula in particular situations.

12.4 Variational Calculus on the Poisson Space

In this section, we shall introduce the general framework of the stochastic calculus on the Poisson space. Recall that for simplicity, we will assume that the Lévy process Z does not have a Brownian component. The general case follows in a similar fashion by replacing all expectations with conditional expectations with respect to the Brownian path.

Let $0 < \varepsilon \leq 1$ be sufficiently small, and $T > 0$ fixed. Let $U_\varepsilon = \bar{B}_\varepsilon^c = \{y \in \mathbb{R}_0^m ;\ |y| > \varepsilon\}$, and let ∂U_ε be its boundary, $n(y)$ the outer normal unit vector at

$y \in \partial U_\varepsilon$ and $\sigma(dy)$ the surface measure on ∂U_ε. As seen before, we shall recall the notation

$$\lambda_\varepsilon = \nu(U_\varepsilon), \quad \nu_\varepsilon(dz) = \frac{f(z)}{\lambda_\varepsilon} \mathbf{1}_{U_\varepsilon}(z)\, dz.$$

Denote by $\tau^\varepsilon = \{\tau_k^\varepsilon\,;\, k \in \mathbb{N}\}$ a sequence of $\mathbb{R}_+$-valued random variables such that their successive differences $\{\tau_k^\varepsilon - \tau_{k-1}^\varepsilon\,;\, k \in \mathbb{N}\}$ are independent λ_ε-exponential random variables, where $\tau_0^\varepsilon = 0$. Let $Y^\varepsilon = \{Y_k^\varepsilon\,;\, k \in \mathbb{N}\}$ be a sequence of independent and identically distributed $\mathbb{R}_0^m$-valued random variables with the law $\nu_\varepsilon(dz)$. Write

$$N^\varepsilon = \sum_{k\in\mathbb{N}} \mathbf{1}_{[0,T]}(\tau_k^\varepsilon),$$

that is, N^ε is the Poisson random variable with intensity $\lambda_\varepsilon T$. Recall the mapping $H^\xi : \mathbb{R}_0^m \to \mathbb{R}_0^m$ as introduced in Sect. 12.3 for $\xi \in \mathbb{R}^u$ near the origin, and write $H^\xi(Y^\varepsilon) = \{H^\xi(Y_k^\varepsilon)\,;\, k \in \mathbb{N}\}$. We shall use the notation

$$\begin{aligned}
Y_{1\to N^\varepsilon}^\varepsilon &= \left(Y_1^\varepsilon, \ldots, Y_{N^\varepsilon}^\varepsilon, 0, 0, \ldots\right), \\
Y_{1\to N^\varepsilon}^{\varepsilon,(j)} &= \left(Y_1^\varepsilon, \ldots, Y_{j-1}^\varepsilon, z_j, Y_{j+1}^\varepsilon, \ldots, Y_{N^\varepsilon}^\varepsilon, 0, 0, \ldots\right), \quad 1 \le j \le N^\varepsilon, \\
\tau_{1\to N^\varepsilon}^\varepsilon &= \left(\tau_1^\varepsilon, \ldots, \tau_{N^\varepsilon}^\varepsilon, 0, 0, \ldots\right).
\end{aligned}$$

We remark that the random variable $Y_{1\to N^\varepsilon}^\varepsilon$ is $(\mathbb{R}_0^m)^\infty \equiv \cup_{j=1}^\infty (\mathbb{R}_0^m)^j$-valued, while $\tau_{1\to N^\varepsilon}^\varepsilon$ is the $\mathbb{R}_+^\infty \equiv \bigcup_{j=1}^\infty (\mathbb{R}_+)^j$-valued one.

Rather than giving a topology to these infinite-dimensional spaces we will only restrict their properties to each corresponding finite-dimensional space.

In fact, we define the space $C_b^\infty\left((\mathbb{R}_0^m)^\infty\,;\, \mathbb{R}^d\right)$ as the space of all functions $f : (\mathbb{R}_0^m)^\infty \to \mathbb{R}^d$ such that this function restricted to each $(\mathbb{R}_0^m)^j \equiv (\mathbb{R}_0^m)^j \times \prod_{k=j+1}^\infty \{0\}$, denoted by $f\big|_{(\mathbb{R}_0^m)^j}$, belongs to the space $C_b^\infty\left((\mathbb{R}_0^m)^j\,;\, \mathbb{R}^d\right)$ and for any multi-index $\alpha = (\alpha_1, \ldots, \alpha_m)$, $m \in \mathbb{N}$ with $\alpha^* = \max_{i=1,\ldots,m} \alpha_i$,

$$\sum_{j=\alpha^*}^\infty \frac{\|\partial_\alpha f\big|_{\mathbb{R}^j}\|_\infty}{j!} < \infty. \tag{12.10}$$

Definition 12.4.1 Denote by $\mathscr{S}_\varepsilon(\mathbb{R}^d)$ the set of $\mathbb{R}^d$-valued random variables X which can be written as smooth functions of $(Y^\varepsilon\,;\, \tau^\varepsilon)$. That is,

$$X = \Phi\left(Y_{1\to N^\varepsilon}^\varepsilon; \tau_{1\to N^\varepsilon}^\varepsilon\right),$$

where $\Phi : (\mathbb{R}_0^m)^\infty \times \mathbb{R}_+^\infty \to \mathbb{R}^d$ is bounded such that $\Phi(\cdot\,;\, \mathbf{t}) \in C_b^\infty\left((\mathbb{R}_0^m)^\infty\,;\, \mathbb{R}^d\right)$ uniformly for $\mathbf{t} \in [0, T]^\infty$.

Furthermore they satisfy the following property:

$$\Phi\left(Y_{1\to N^\varepsilon+1}^\varepsilon; \mathbf{t}_{1\to N^\varepsilon+1}\right) \mathbf{1}_{(Y_i^\varepsilon=0, i\le N^\varepsilon+1)} = \Phi\left(Y_{1\to N^\varepsilon}^\varepsilon; \mathbf{t}_{1\to N^\varepsilon}^i\right).$$

Here, $\mathbf{t}^i_{1\to N^\varepsilon}$ has the same components as $\mathbf{t}_{1\to N^\varepsilon+1}$ except for the component i which is deleted from the vector $\mathbf{t}_{1\to N^\varepsilon+1}$.

Notice in particular that the function Φ introduced above when restricted to $(\mathbb{R}^m_0)^j$, $j \in \mathbb{N}$, is a smooth function with at most linear growth order at infinity. This space characterizes all random variables which can be written as smooth functions of the random variables associated with the jump size and jump times of the process Z^ε.

The second property states that given the function representation of X, one jump size is replaced by zero; it is as if that jump did not happen. Sometimes in order to stress the dependence of X on ε, we will prefer to use the notation $X^\varepsilon \in \mathscr{S}_\varepsilon(\mathbb{R}^d)$.

Exercise 12.4.2 Prove that for $0 < \varepsilon_0 < \varepsilon_1$ then $\mathscr{S}_{\varepsilon_1}(\mathbb{R}^d) \subset \mathscr{S}_{\varepsilon_0}(\mathbb{R}^d)$ and that X^ε_t defined in Exercise 12.2.5 is an element of $\mathscr{S}_\varepsilon(\mathbb{R}^d)$.

Proposition 12.4.3 *Let $p \geq 2$, $\varepsilon > 0$ and $N \in \mathbb{Z}_+$. Then, it holds that*

$$\mathscr{S}_\varepsilon(\mathbb{R}^d) \subset L^p(\Omega\,;\,\mathbb{R}^d) \subset L^2(\Omega\,;\,\mathbb{R}^d).$$

Proof The set of inclusions is clear; because of Hypothesis 12.3.1 and the definition of Lévy measure, we have that:

$$\int_{\bar{B}_1} |z|^2\,\nu(dz) + \int_{\bar{B}_1^c} |z|^p\,\nu(dz) < +\infty.$$

In fact, under the above conditions, we can check easily that for fixed $\varepsilon > 0$ and $p \geq 2$

$$\mathbb{E}\big[|Y^\varepsilon_k|^p\big] = \frac{1}{\lambda_\varepsilon}\int_{U_\varepsilon} |z|^p\,\nu(dz) < +\infty,$$

$$\mathbb{E}\big[(\tau^\varepsilon_k)^p\big] = \int_0^{+\infty} s^{p+k-1}\,\lambda^k_\varepsilon\,\frac{e^{-s\lambda_\varepsilon}}{(k-1)!}\,ds < +\infty.$$

The proof is complete.

Definition 12.4.4 For $\mathbf{z} = \big\{z_j\,;\, j \in \mathbb{N}\big\} \in (\mathbb{R}^m_0)^\infty$ and $X^\varepsilon \in \mathscr{S}_\varepsilon(\mathbb{R}^d)$, define the stochastic derivative operator $D_{\mathbf{z}}X^\varepsilon = \big\{D_{z_j}X^\varepsilon\,;\, j \in \mathbb{N}\big\}$ by

$$D_{z_j}X^\varepsilon = \begin{cases} \big(\partial_\xi\big)_0\Big\{\Phi\big(Y^{\varepsilon,(j,\xi)}_{1\to N^\varepsilon}\,;\,\tau^\varepsilon_{1\to N^\varepsilon}\big)\Big\} & (1 \leq j \leq N^\varepsilon),\\ 0 & (j \geq N^\varepsilon + 1),\end{cases}$$

where $Y^{\varepsilon,(j,\xi)}_{1\to N^\varepsilon} = \big(Y^\varepsilon_1,\,\ldots,\,Y^\varepsilon_{j-1},\,H^\xi(z_j),\,Y^\varepsilon_{j+1},\,\ldots,\,Y^\varepsilon_{N^\varepsilon}, 0,\,0,\ldots\big)$. When $N^\varepsilon = 0$, the above sum is defined as zero. Moreover, define $D_j = D_{Y^\varepsilon_j}$.

We remark that $D_{z_j}X^\varepsilon$ is the $\mathbb{R}^{d\times u}$-valued random variable. The meaning of the above definition of the derivative is as follows. One introduces a deformation function

H^{ξ} which will tangentially ($\xi = 0$) determine the derivative of X^{ε}. The derivative operator is well defined and the chain rule for it are straightforward properties. In fact, for $1 \le j \le N^{\varepsilon}$,

$$\begin{aligned} D_{z_j}(G(X^{\varepsilon})) &= (\partial_{\xi})_0 \Big\{ G\Big(\Phi\big(Y^{\varepsilon,(j,\xi)}_{1\to N^{\varepsilon}}\,;\ \tau^{\varepsilon}_{1\to N^{\varepsilon}}\big)\Big)\Big\} \\ &= (\nabla G)\Big(\Phi\big(Y^{\varepsilon,(j)}_{1\to N^{\varepsilon}}\,;\ \tau^{\varepsilon}_{1\to N^{\varepsilon}}\big)\Big)\,(\partial_{\xi})_0 \Big\{ \Phi\big(Y^{\varepsilon,(j,\xi)}_{1\to N^{\varepsilon}}\,;\ \tau^{\varepsilon}_{1\to N^{\varepsilon}}\big)\Big\} \\ &= (\nabla G)(X^{\varepsilon,(j)})\, D_{z_j} X^{\varepsilon} \end{aligned} \tag{12.11}$$

for $G \in C^1(\mathbb{R}^d\,;\ \mathbb{R})$, where $X^{\varepsilon,(j)} = \Phi\big(Y^{\varepsilon,(j)}_{1\to N^{\varepsilon}};\ \tau^{\varepsilon}_{1\to N^{\varepsilon}}\big)$ and $Y^{\varepsilon,(j)}_{1\to N^{\varepsilon}} \equiv Y^{\varepsilon,(j,0)}_{1\to N^{\varepsilon}}$.

Exercise 12.4.5 Rewrite $\sum_{j=1}^{N^{\varepsilon}} D_j X^{\varepsilon}$ as an integral with respect to the Poisson random measure. In particular, in the case of the solution to the equation in Exercise 12.2.5, find the linear equation satisfied by $D_{z_j} X^{\varepsilon}_t$ ($j \in \mathbb{N}, t \in [0, T]$).

For a hint, see Chap. 14.

Definition 12.4.6 Let $\Theta^{\varepsilon} \in \mathscr{S}_{\varepsilon}(\mathbb{R}^u)$ be expressed as $\Theta^{\varepsilon} = \Lambda\big(Y^{\varepsilon}_{1\to N^{\varepsilon}}\,;\ \tau^{\varepsilon}_{1\to N^{\varepsilon}}\big)$, where $\Lambda : (\mathbb{R}^m_0)^{\infty} \times \mathbb{R}^{\infty}_+ \to \mathbb{R}^u$ is bounded such that $\Phi(\cdot\,;\ \mathbf{t}) \in C^{\infty}_b\big((\mathbb{R}^m_0)^{\infty}\,;\ \mathbb{R}^u\big)$ for given $\mathbf{t} \in \mathbb{R}^{\infty}_+$. Define the operator δ by

$$\begin{aligned} \delta(\Theta^{\varepsilon}) &= -\sum_{j=1}^{N^{\varepsilon}} \frac{\operatorname{div}\big\{h(z_j)\, f(z_j)\,\Theta^{\varepsilon}(z_j)\big\}}{f(z_j)}\Bigg|_{z_j = Y^{\varepsilon}_j} \\ &= -\left(\sum_{j=1}^{N^{\varepsilon}} \frac{\operatorname{div}\big\{h(z_j)\, f(z_j)\big\}}{f(z_j)}\Bigg|_{z_j = Y^{\varepsilon}_j}\right)\Theta^{\varepsilon} - \sum_{j=1}^{N^{\varepsilon}} \operatorname{tr}\big(D_j\Theta^{\varepsilon}\big), \end{aligned}$$

where $\Theta^{\varepsilon}(z_j) := \Lambda\big(Y^{\varepsilon,(j,z_j)}_{1\to N^{\varepsilon}};\ \tau^{\varepsilon}_{1\to N^{\varepsilon}}\big)$, $j \in \mathbb{N}$. When $f(Y^{\varepsilon}_j) = 0$, the first term within the above sum is defined as zero. Moreover, if $N^{\varepsilon} = 0$, then the above two sums in the right-hand side are also defined as zero.[7]

We remark that $\delta(\Theta^{\varepsilon})$ is a $\mathbb{R}$-valued random variable.

Exercise 12.4.7 Rewrite the first sum in the definition $\delta(\Theta^{\varepsilon})$ using an integral with respect to the corresponding Poisson random measure.

For a hint, see Chap. 14.

Exercise 12.4.8 Once Exercise 12.4.7 is understood, the proof of the following lemma (recall Chap. 4 and Hypothesis 12.3.3) is straightforward.

Lemma 12.4.9 *The following sequence of random variables:*

$$\int_{U_{\varepsilon}\times[0,T]} \frac{\operatorname{div}\{h(z)\, f(z)\}}{f(z)} \mathscr{N}(dz, dt), \quad \varepsilon \in (0, 1]$$

---

[7] This sum convention is used from now on.

is uniformly integrable, and converges in $L^1(\Omega)$ *and its limit as* $\varepsilon \to 0$ *has as average* $T \int_{\mathbb{R}_0^m} \frac{\operatorname{div}\{h(z) f(z)\}}{f(z)} \nu(dz)$.

Note that the definitions of D and δ depend on the deformation function H^{ξ}. In order to find the relation between the operators D and δ, we shall introduce some notation. Let $X \in \mathscr{S}_{\varepsilon}(\mathbb{R}^d)$ and $\Theta \in \mathscr{S}_{\varepsilon}(\mathbb{R}^u)$. Then, the random variables X and Θ can be rewritten as:

$$X^{\varepsilon} = \Phi\left(Y^{\varepsilon}_{1\to N^{\varepsilon}}\,;\, \tau^{\varepsilon}_{1\to N^{\varepsilon}}\right), \quad \Theta^{\varepsilon} = \Lambda\left(Y^{\varepsilon}_{1\to N^{\varepsilon}}\,;\, \tau^{\varepsilon}_{1\to N^{\varepsilon}}\right),$$

where $\Phi : (\mathbb{R}_0^m)^{\infty} \times \mathbb{R}_+^{\infty} \to \mathbb{R}^d$ and $\Lambda : (\mathbb{R}_0^m)^{\infty} \times \mathbb{R}_+^{\infty} \to \mathbb{R}^u$ are bounded such that $\Phi(\cdot\,;\, \mathbf{t}) \in C_b^{\infty}\left((\mathbb{R}_0^m)^{\infty}\,;\, \mathbb{R}^d\right)$ and $\Phi(\cdot\,;\, \mathbf{t}) \in C_b^{\infty}\left((\mathbb{R}_0^m)^{\infty}\,;\, \mathbb{R}^u\right)$ for given $\mathbf{t} \in \mathbb{R}_+^{\infty}$. Then, we have:

Lemma 12.4.10 *Let* $0 < \varepsilon \leq 1$, $X^{\varepsilon} \in \mathscr{S}_{\varepsilon}(\mathbb{R}^d)$ *and* $\Theta^{\varepsilon} \in \mathscr{S}_{\varepsilon}(\mathbb{R}^u)$. *For* $G \in C_c^1(\mathbb{R}^d\,;\, \mathbb{R})$, *it holds that*

$$\mathbb{E}\left[\sum_{j=1}^{N^{\varepsilon}} D_j\left(G(X^{\varepsilon})\right) \Theta^{\varepsilon}(Y_j^{\varepsilon})\right] = \mathbb{E}\left[G(X^{\varepsilon})\,\delta(\Theta^{\varepsilon})\right] + \mathbb{E}\left[\mathscr{R}_{\varepsilon}\left(G(X^{\varepsilon}), \Theta^{\varepsilon}\right)\right], \tag{12.12}$$

where

$$\mathscr{R}_{\varepsilon}\left(G(X^{\varepsilon})\,;\, \Theta^{\varepsilon}\right) = \sum_{j=1}^{N^{\varepsilon}} \int_{\partial U_{\varepsilon}} n(z_j)^* h(z_j) \frac{f(z_j)}{\lambda_{\varepsilon}} G\left(\Phi\left(Y^{\varepsilon,(j)}_{1\to N^{\varepsilon}}\,;\, \tau^{\varepsilon}_{1\to N^{\varepsilon}}\right)\right)$$
$$\times \Lambda\left(Y^{\varepsilon,(j)}_{1\to N^{\varepsilon}}\,;\, \tau^{\varepsilon}_{1\to N^{\varepsilon}}\right) \sigma(dz_j).$$

That is, D and δ are approximately the adjoint[8] of each other if we interpret $\mathscr{R}_{\varepsilon}$ as a residue and use the following inner product definition:

$$\langle D(G(X^{\varepsilon})), \Theta^{\varepsilon}\rangle = \sum_{j=1}^{N^{\varepsilon}} D_j\left(G(X^{\varepsilon})\right) \Theta^{\varepsilon}(Y_j^{\varepsilon}).$$

We remark that $D_j X^{\varepsilon} = 0$ for $j > N^{\varepsilon}$. Putting this together with the chain rule in (12.11), we have that the above statement is an approximate IBP formula.

Proof For $\mathbf{z} = \left\{z_j\,;\, j \in \mathbb{N}\right\} \in (\mathbb{R}_0^m)^{\infty}$, define

$$D_{z_j} X^{\varepsilon} \equiv (\partial_{\xi})_0 \left\{\Phi\left(Y^{\varepsilon,(j,\xi)}_{1\to N^{\varepsilon}}\,;\, \tau^{\varepsilon}_{1\to N^{\varepsilon}}\right)\right\} =: \Psi_j\left(Y^{\varepsilon,(j)}_{1\to N^{\varepsilon}}\,;\, \tau^{\varepsilon}_{1\to N^{\varepsilon}}\right).$$

The chain rule tells us that for $\Theta^{\varepsilon} = \Lambda\left(Y^{\varepsilon}_{1\to N^{\varepsilon}}\,;\, \tau^{\varepsilon}_{1\to N^{\varepsilon}}\right)$, one has

[8] It may be helpful to recall the discussion at the end of Chap. 10.

$$\begin{aligned}
&\mathbb{E}\left[\sum_{j=1}^{N^\varepsilon} D_j\,(G(X^\varepsilon))\ \Theta^\varepsilon(Y_j^\varepsilon)\right] \\
&= \mathbb{E}\left[\sum_{j=1}^{N^\varepsilon} D_j\,(G(X^\varepsilon))\ \Theta^\varepsilon(Y_j^\varepsilon)\ \mathbf{1}_{(N^\varepsilon\geq 1)}\right] \\
&= \mathbb{E}\left[\sum_{j=1}^{N^\varepsilon} (\nabla G)(X^\varepsilon)\, D_j X^\varepsilon\ \Theta^\varepsilon(Y_j^\varepsilon)\ \mathbf{1}_{(N^\varepsilon\geq 1)}\right] \\
&= \mathbb{E}\left[\sum_{j=1}^{N^\varepsilon} \big[(\nabla G\circ\Phi)\ \Psi_j\ \Lambda\big](Y^\varepsilon_{1\to N^\varepsilon}\,;\ \tau^\varepsilon_{1\to N^\varepsilon})\ \mathbf{1}_{(N^\varepsilon\geq 1)}\right] \\
&= \sum_{N\geq 1}\sum_{j=1}^{N} \mathbb{E}\Big[\big[(\nabla G\circ\Phi)\ \Psi_j\ \Lambda\big](Y^\varepsilon_{1\to N}\,;\ \tau^\varepsilon_{1\to N})\ \mathbf{1}_{(N^\varepsilon=N)}\Big] \\
&= \sum_{N\geq 1}\sum_{j=1}^{N} \int \nu_\varepsilon(dz_j)\,\mathbb{E}\Big[\big[(\nabla G\circ\Phi)\ \Psi_j\ \Lambda\big]\big(Y^{\varepsilon,(j)}_{1\to N}\,;\ \tau^\varepsilon_{1\to N}\big)\ \mathbf{1}_{(N^\varepsilon=N)}\Big] \\
&=: \sum_{N\geq 1} I_{\varepsilon,N}.
\end{aligned}$$

We remark that the above integrals and the ones to follow from now on are all well defined due to Hypothesis 12.3.3. In fact, recall that

$$\begin{aligned}
(\partial_\xi)_0\Big\{\Phi\big(Y^{\varepsilon,(j,\xi)}_{1\to N}\,;\ \tau^\varepsilon_{1\to N}\big)\Big\} &= \partial_{z_j}\Phi\big(Y^{\varepsilon,(j)}_{1\to N}\,;\ \tau^\varepsilon_{1\to N}\big)\,h(z_j) \\
&= \Psi_j\big(Y^{\varepsilon,(j)}_{1\to N}\,;\ \tau^\varepsilon_{1\to N}\big).
\end{aligned}$$

Then, using Fubini's theorem and the definition of ν_ε, we have

$$\begin{aligned}
I_{\varepsilon,N} &= \sum_{j=1}^{N}\int \nu_\varepsilon(dz_j)\,\mathbb{E}\Big[\big(\nabla G\circ\Phi\big)\big(Y^{\varepsilon,(j)}_{1\to N}\,;\ \tau^\varepsilon_{1\to N}\big)\,\partial_{z_j}\Phi\big(Y^{\varepsilon,(j)}_{1\to N}\,;\ \tau^\varepsilon_{1\to N}\big)\,h(z_j) \\
&\qquad\times \Lambda\big(Y^{\varepsilon,(j)}_{1\to N}\,;\ \tau^\varepsilon_{1\to N}\big)\ \mathbf{1}_{(N^\varepsilon=N)}\Big] \\
&= \sum_{j=1}^{N}\mathbb{E}\Big[\int \nu_\varepsilon(dz_j)\partial_{z_j}\Big\{G\Big(\Phi\big(Y^{\varepsilon,(j)}_{1\to N}\,;\ \tau^\varepsilon_{1\to N}\big)\Big)\Big\}\,h(z_j) \\
&\qquad\times \Lambda\big(Y^{\varepsilon,(j)}_{1\to N}\,;\ \tau^\varepsilon_{1\to N}\big)\ \mathbf{1}_{(N^\varepsilon=N)}\Big] \\
&= \sum_{j=1}^{N}\mathbb{E}\Big[\int_{U_\varepsilon} dz_j\ \mathrm{div}_{z_j}\Big\{G\Big(\Phi\big(Y^{\varepsilon,(j)}_{1\to N}\,;\ \tau^\varepsilon_{1\to N}\big)\Big)\,h(z_j)\,\frac{f(z_j)}{\lambda_\varepsilon}
\end{aligned}$$

$$
\begin{aligned}
&\qquad\qquad \times \Lambda\big(Y^{\varepsilon,(j)}_{1\to N}\,;\,\tau^{\varepsilon}_{1\to N}\big)\Big\}\ \mathbf{1}_{(N^{\varepsilon}=N)}\Big] \\
&- \sum_{j=1}^{N} \mathbb{E}\Big[\int_{U_\varepsilon} dz_j\, G\Big(\Phi\big(Y^{\varepsilon,(j)}_{1\to N}\,;\,\tau^{\varepsilon}_{1\to N}\big)\Big) \\
&\qquad\qquad \times \operatorname{div}_{z_j}\Big\{h(z_j)\,\frac{f(z_j)}{\lambda_\varepsilon}\,\Lambda\big(Y^{\varepsilon,(j)}_{1\to N}\,;\,\tau^{\varepsilon}_{1\to N}\big)\Big\}\ \mathbf{1}_{(N^{\varepsilon}=N)}\Big] \\
&=: I_{\varepsilon,N,1} + I_{\varepsilon,N,2}.
\end{aligned}
$$

In the fourth equality, we have used the product rule for the divergence operator. That is,

$$
\begin{aligned}
&\operatorname{div}_{z_j}\Big\{G\Big(\Phi\big(Y^{\varepsilon,(j)}_{1\to N}\,;\,\tau^{\varepsilon}_{1\to N}\big)\Big)\,h(z_j)\,\frac{f(z_j)}{\lambda_\varepsilon}\,\Lambda\big(Y^{\varepsilon,(j)}_{1\to N}\,;\,\tau^{\varepsilon}_{1\to N}\big)\Big\} \\
&= \partial_{z_j}\Big\{G\Big(\Phi\big(Y^{\varepsilon,(j)}_{1\to N}\,;\,\tau^{\varepsilon}_{1\to N}\big)\Big)\Big\}\,h(z_j)\,\frac{f(z_j)}{\lambda_\varepsilon}\,\Lambda\big(Y^{\varepsilon,(j)}_{1\to N}\,;\,\tau^{\varepsilon}_{1\to N}\big) \\
&\quad + G\Big(\Phi\big(Y^{\varepsilon,(j)}_{1\to N}\,;\,\tau^{\varepsilon}_{1\to N}\big)\Big)\operatorname{div}_{z_j}\Big\{h(z_j)\,\frac{f(z_j)}{\lambda_\varepsilon}\,\Lambda\big(Y^{\varepsilon,(j)}_{1\to N}\,;\,\tau^{\varepsilon}_{1\to N}\big)\Big\}.
\end{aligned}
$$

Furthermore, we can get

$$
\begin{aligned}
I_{\varepsilon,N,2} &= -\sum_{j=1}^{N}\int \nu_\varepsilon(dz_j)\,\mathbb{E}\Big[G\Big(\Phi\big(Y^{\varepsilon,(j)}_{1\to N}\,;\,\tau^{\varepsilon}_{1\to N}\big)\Big) \\
&\qquad\qquad \times \frac{\operatorname{div}_{z_j}\Big\{h(z_j)\,f(z_j)\,\Lambda\big(Y^{\varepsilon,(j)}_{1\to N}\,;\,\tau^{\varepsilon}_{1\to N}\big)\Big\}}{f(z_j)}\ \mathbf{1}_{(N^{\varepsilon}=N)}\Big] \\
&= -\mathbb{E}\Big[G\Big(\Phi\big(Y^{\varepsilon}_{1\to N}\,;\,\tau^{\varepsilon}_{1\to N}\big)\Big) \\
&\qquad\qquad \times \sum_{j=1}^{N}\frac{\operatorname{div}_{z_j}\Big\{h(z_j)\,f(z_j)\,\Lambda\big(Y^{\varepsilon,(j)}_{1\to N}\,;\,\tau^{\varepsilon}_{1\to N}\big)\Big\}}{f(z_j)}\bigg|_{z_j=Y^{\varepsilon}_j}\ \mathbf{1}_{(N^{\varepsilon}=N)}\Big].
\end{aligned}
$$

On the other hand, the divergence formula leads us to see that

$$
\begin{aligned}
I_{\varepsilon,N,1} &\equiv \sum_{j=1}^{N}\mathbb{E}\Big[\int_{U_\varepsilon} dz_j\,\operatorname{div}_{z_j}\Big\{G\Big(\Phi\big(Y^{\varepsilon,(j)}_{1\to N}\,;\,\tau^{\varepsilon}_{1\to N}\big)\Big) \\
&\qquad\qquad \times h(z_j)\,\frac{f(z_j)}{\lambda_\varepsilon}\,\Lambda\big(Y^{\varepsilon,(j)}_{1\to N}\,;\,\tau^{\varepsilon}_{1\to N}\big)\Big\}\ \mathbf{1}_{(N^{\varepsilon}=N)}\Big] \\
&= \sum_{j=1}^{N}\mathbb{E}\Big[\int_{\partial U_\varepsilon} n(z_j)^{*}\,G\Big(\Phi\big(Y^{\varepsilon,(j)}_{1\to N}\,;\,\tau^{\varepsilon}_{1\to N}\big)\Big)
\end{aligned}
$$

$$\times\, h(z_j)\, \frac{f(z_j)}{\lambda_\varepsilon}\, \Lambda\big(Y^{\varepsilon,(j)}_{1\to N}\,;\ \tau^\varepsilon_{1\to N}\big)\, \sigma(dz_j)\ \mathbf{1}_{(N^\varepsilon=N)}\Bigg].$$

Therefore, we have

$$\begin{aligned}
&\mathbb{E}\left[\sum_{j=1}^{N^\varepsilon} D_j\,(G(X^\varepsilon))\ \Theta^\varepsilon(Y^\varepsilon_j)\right]\\
&= \mathbb{E}\Bigg[\sum_{j=1}^{N^\varepsilon} \int_{\partial U_\varepsilon} n(z_j)^*\, G\Big(\Phi\big(Y^{\varepsilon,(j)}_{1\to N^\varepsilon}\,;\ \tau^\varepsilon_{1\to N^\varepsilon}\big)\Big)\\
&\qquad\qquad \times h(z_j)\, \frac{f(z_j)}{\lambda_\varepsilon}\, \Lambda\big(Y^{\varepsilon,(j)}_{1\to N^\varepsilon}\,;\ \tau^\varepsilon_{1\to N^\varepsilon}\big)\, \sigma(dz_j)\ \mathbf{1}_{(N^\varepsilon\geq 1)}\Bigg]\\
&\quad - \mathbb{E}\Bigg[G\Big(\Phi\big(Y^\varepsilon_{1\to N^\varepsilon}\,;\ \tau^\varepsilon_{1\to N^\varepsilon}\big)\Big)\\
&\qquad\qquad \times \sum_{j=1}^{N^\varepsilon} \frac{\operatorname{div}\Big\{h(z_j)\, f(z_j)\, \Lambda\big(Y^{\varepsilon,(j)}_{1\to N^\varepsilon}\,;\ \tau^\varepsilon_{1\to N^\varepsilon}\big)\Big\}}{f(z_j)}\Bigg|_{z_j=Y^\varepsilon_j}\ \mathbf{1}_{(N^\varepsilon\geq 1)}\Bigg]\\
&= \mathbb{E}\big[\mathscr{R}_\varepsilon\big(G(X^\varepsilon)\,;\ \Theta^\varepsilon\big)\big] + \mathbb{E}\big[G(X^\varepsilon)\,\delta(\Theta^\varepsilon)\big].
\end{aligned}$$

The proof is complete.

Lemma 12.4.11 *Under the same hypotheses as in Lemma 12.4.10, it holds that*

$$\mathbb{E}\big[\mathscr{R}_\varepsilon\big(G(X^\varepsilon)\,;\ \Theta^\varepsilon\big)\big] = T\int_{U_\varepsilon} \frac{\operatorname{div}\{h(z)\, f(z)\}}{f(z)}\, \nu(dz)\, \mathbb{E}\big[G(X^\varepsilon)\,\Theta^\varepsilon\big] + o(1) \tag{12.13}$$

as $\varepsilon \to 0$.

Proof Since $\mathbb{P}\big[N^\varepsilon = N\big] = \big(\lambda_\varepsilon\, T\big)^N\, e^{-\lambda_\varepsilon T}/N!$ for $N \geq 0$, we have

$$\begin{aligned}
&\mathbb{E}\big[\mathscr{R}_\varepsilon\big(G(X^\varepsilon)\,;\ \Theta^\varepsilon\big)\big]\\
&= \sum_{N\geq 1} \mathbb{P}\big[N^\varepsilon = N\big]\\
&\times \mathbb{E}\left[\sum_{j=1}^{N} \int_{\partial U_\varepsilon} n(z_j)^*\, h(z_j)\, \frac{f(z_j)}{\lambda_\varepsilon}\, \big[(G\circ\Phi)\,\Lambda\big]\big(Y^{\varepsilon,(j)}_{1\to N}\,;\ \tau^\varepsilon_{1\to N}\big)\, \sigma(dz_j)\middle| N^\varepsilon = N\right]\\
&=: \sum_{N\geq 1} \frac{(\lambda_\varepsilon\, T)^{N-1}}{(N-1)!}\, e^{-\lambda_\varepsilon T}\, J_{\varepsilon,N}.
\end{aligned}$$

Recalling the result in Proposition 2.1.21, each term can be computed as follows:

$$J_{\varepsilon,N}=\frac{T}{N}\sum_{j=1}^{N}\int_{t_1<.<t_j<.<t_N<T}\mathbb{E}\Big[\int_{\partial U_\varepsilon} n(z_j)^* h(z_j)\, f(z_j)$$
$$\times \big[(G\circ\Phi)\,\Lambda\big]\big(Y^{\varepsilon,(j)}_{1\to N}\,;\,\mathbf{t}_{1\to N}\big)\,\sigma(dz_j)\Big]\frac{N!}{T^N}d\mathbf{t}_{1\to N}.$$

We will replace in the above expression $\big[(G\circ\Phi)\,\Lambda\big]\big(Y^{\varepsilon,(j)}_{1\to N}\,;\,\mathbf{t}_{1\to N}\big)$ with $\big[(G\circ \Phi)\,\Lambda\big]\big(Y^{\varepsilon,(j)}_{1\to N}\,;\,\mathbf{t}_{1\to N}\big)\big|_{z_j=0}$. If we denote the new expression obtained after this substitution by $J^0_{\varepsilon,N}$, we have by the definition of the class $\mathscr{S}_\varepsilon(\mathbb{R}^d)$

$$\begin{aligned}J^0_{\varepsilon,N}&=\frac{T}{N}\int_{\partial U_\varepsilon} n(z)^* h(z)\, f(z)\sigma(dz)\\&\quad\times\sum_{j=1}^{N}\int_{t_1<.<t_j<.<t_N<T}\mathbb{E}\Big[\big[(G\circ\Phi)\,\Lambda\big]\big(Y^{\varepsilon}_{1\to N-1}\,;\,\mathbf{t}_{1\to N}\big)\Big]\frac{N!}{T^N}d\mathbf{t}_{1\to N}\\&=\int_{\partial U_\varepsilon} n(z)^* h(z)\, f(z)\sigma(dz)\\&\quad\times\int_{t_1<\ldots<t_{N-1}<T}\mathbb{E}\Big[\big[(G\circ\Phi)\,\Lambda\big]\big(Y^{\varepsilon}_{1\to N-1}\,;\,\mathbf{t}_{1\to N-1}\big)\Big]\frac{(N-1)!}{T^{N-2}}d\mathbf{t}_{1\to N-1}.\end{aligned}$$

Hence, we can conclude the result once we prove that the following difference converges to zero:

$$\begin{aligned}A:=\mathbb{E}\Big[&\int_{\partial U_\varepsilon} n(z_j)^* h(z_j)\, f(z_j)\\&\times\Big(\big[(G\circ\Phi)\,\Lambda\big]\big(Y^{\varepsilon,(j)}_{1\to N};\,\mathbf{t}_{1\to N}\big)-\big[(G\circ\Phi)\,\Lambda\big]\big(Y^{\varepsilon,(j)}_{1\to N}\big|_{z_j=0};\,\mathbf{t}_{1\to N}\big)\Big)\ \sigma(dz_j)\Big].\end{aligned}$$

This term can be bounded using the mean value theorem as follows:

$$\begin{aligned}A&=\mathbb{E}\Big[\int_{\partial U_\varepsilon} n(z_j)^* h(z_j)\, f(z_j)\,\big(\partial_{z_j}\big)_{\theta_j z_j}\big[(G\circ\Phi)\,\Lambda\big]\big(Y^{\varepsilon,(j)}_{1\to N}\,;\,\mathbf{t}_{1\to N}\big)\,z_j\sigma(dz_j)\Big]\\&=o(1)\end{aligned}$$

as $\varepsilon\to 0$, where the random variable $\theta_j\in[0,1]$. The reasons for the convergence are because $G\in C^1_c(\mathbb{R}^d\,;\,\mathbb{R})$, $\Phi(\cdot\,;\,\mathbf{t})\in C^\infty_b\big((\mathbb{R}^m_0)^\infty\,;\,\mathbb{R}^d\big)$, $\Lambda(\cdot\,;\,\mathbf{t})\in C^\infty_b\big((\mathbb{R}^m_0)^\infty\,;\,\mathbb{R}^u\big)$ which satisfy (12.10) uniformly for $\mathbf{t}\in[0,T]^\infty$. Therefore, we obtain that

$$\mathbb{E}\big[\mathscr{R}_\varepsilon\big(G(X^\varepsilon)\,;\,\Theta^\varepsilon\big)\big]=T\int_{U_\varepsilon}\frac{\operatorname{div}\{h(z)\, f(z)\}}{f(z)}\,\nu(dz)\,\mathbb{E}\big[G(X^\varepsilon)\,\Theta^\varepsilon\big]+o(1)$$

as $\varepsilon\to 0$. The proof is complete.

Corollary 12.4.12 *Under the same hypotheses as in Lemma 12.4.10, it holds that*

$$\mathbb{E}\left[\sum_{j=1}^{N^\varepsilon} D_j\left(G(X^\varepsilon)\right)\,\Theta^\varepsilon(Y_j^\varepsilon)\right] = \mathbb{E}\Big[G(X^\varepsilon)\,\tilde{\delta}(\Theta^\varepsilon)\Big] + o(1) \tag{12.14}$$

as $\varepsilon \to 0$, *where*

$$\begin{aligned}\tilde{\delta}(\Theta^\varepsilon) = &-\int_{U_\varepsilon\times[0,T]} \frac{\operatorname{div}\{h(z)\,f(z)\}}{f(z)}\,\widetilde{\mathcal{N}}(dz,dt)\,\Theta^\varepsilon \\ &-\int_{U_\varepsilon\times[0,T]} \operatorname{tr}\big(D_z\Theta^\varepsilon\big)\,\mathcal{N}(dz,dt).\end{aligned} \tag{12.15}$$

Next, we move towards the generalization of the stochastic derivative to its possible domain by a limit procedure.

Definition 12.4.13 Let $p \geq 1$ and $l \in \mathbb{Z}_+$. For an $\mathbb{R}^d$-valued random variable $X \in \bigcup_{0<\varepsilon\leq 1}\mathscr{S}_\varepsilon(\mathbb{R}^d)$, define

$$\|X\|_{p,l} = \begin{cases} \|X\|_{L^p(\Omega\,;\,\mathbb{R}^d)} & (l=0), \\ \left(\|X\|^p_{L^p(\Omega\,;\,\mathbb{R}^d)} + \displaystyle\sum_{k=1}^{l} \|D^k X\|^p_{L^p(\Omega\,;\,(\mathbb{R}^d\otimes(\mathbb{R}^u)^{\otimes k})}\right)^{1/p} & (l\in\mathbb{N}),\end{cases}$$

where $DX = \sum_{j=1}^{N^\varepsilon} D_j X$ is given in Definition 12.4.4 as the $\mathbb{R}^d \otimes \mathbb{R}^u$-valued random variable, and $D^k X = \sum_{j=1}^{N^\varepsilon} D_j(D^{k-1}X)$ is defined inductively. Denote by $\mathbb{D}_{p,l}(\mathbb{R}^d)$ the completion of $\bigcup_{0<\varepsilon\leq 1}\mathscr{S}_\varepsilon(\mathbb{R}^d)$ with respect to the norm $\|\cdot\|_{p,l}$.

Exercise 12.4.14 Let $X \in \mathbb{D}_{p,l}(\mathbb{R}^d)$. For any given sequence $\varepsilon_n \to 0$ in $(0,1]$, prove that there exists a sequence of random variables $X_n \in \mathscr{S}_{\varepsilon_n}(\mathbb{R}^d)$ such that $\lim_{n\to\infty}\|X_n - X\|_{p,l} = 0$. In the case that X is a bounded random variable, one can choose the sequence of r.v.s in such a way that they are uniformly bounded.

For a hint, see Chap. 14.

Definition 12.4.15 For $q \geq 1$, $l \in \mathbb{Z}_+$ and $\Theta \in \mathbb{D}_{q,l}(\mathbb{R}^u)$, define the $\mathbb{R}$-valued random variables $\delta(\Theta)$ and $\tilde{\delta}(\Theta)$ by

$$\begin{aligned}\delta(\Theta) := &-\int_{\mathbb{R}_0^m\times[0,T]} \frac{\operatorname{div}\{h(z)\,f(z)\}}{f(z)}\,\mathcal{N}(dz,dt)\,\Theta \\ &-\int_{\mathbb{R}_0^m\times[0,T]} \operatorname{tr}\big(D_z\Theta\big)\,\mathcal{N}(dz,dt)\end{aligned}$$

$$\tilde{\delta}(\Theta) = -\int_{\mathbb{R}_0^m\times[0,T]} \frac{\operatorname{div}\{h(z)\,f(z)\}}{f(z)}\,\widetilde{\mathcal{N}}(dz,dt)\,\Theta$$

$$- \int_{\mathbb{R}_0^m \times [0,T]} \mathrm{tr}\big(D_z \Theta\big) \mathcal{N}(dz, dt).$$

We remark that as explained in Lemma 12.4.9, the right-hand side with stochastic integral in $\delta(\Theta)$ exists as an $L^1(\Omega, \mathbb{R})$-random variable, due to Hypothesis 12.3.3. Also $D_z\Theta$ denotes the stochastic derivative defined in Definition 12.4.4. Therefore the above definition includes the fact that the following $L^1(\Omega, \mathbb{R})$ limit exists: $\lim_{\varepsilon \to 0} \int_{U_\varepsilon \times [0,T]} \mathrm{tr}\big(D_z \Theta^\varepsilon\big) \mathcal{N}(dz, dt) =: \int_{\mathbb{R}_0^m \times [0,T]} \mathrm{tr}\big(D_z \Theta\big) \mathcal{N}(dz, dt)$.

Exercise 12.4.16 Prove that the above definitions extend the previous definitions of D in Definition 12.4.1, δ in Definition 12.4.6, and $\tilde{\delta}$ in Corollary 12.4.12.

Exercise 12.4.17 Assume that Θ is a bounded random variable. Furthermore, assume that the uniformly bounded sequence of r.v.s $\big\{\Theta_n\,;\, n \in \mathbb{N}\big\}$ satisfies that $\|\Theta_n - \Theta\|_{1,1} \to 0$ as $n \to \infty$. Prove that the sequence $\big\{\delta(\Theta_n)\,;\, n \in \mathbb{N}\big\}$ converges in $L^1(\Omega; \mathbb{R})$ to

$$\begin{aligned} \delta(\Theta) := &- \int_{\mathbb{R}_0^m \times [0,T]} \frac{\mathrm{div}\big\{h(z)\, f(z)\big\}}{f(z)} \mathcal{N}(dz, dt)\, \Theta \\ &- \int_{\mathbb{R}_0^m \times [0,T]} \mathrm{tr}\big(D_z \Theta\big) \mathcal{N}(dz, dt). \end{aligned}$$

Exercise 12.4.18 Assume that Θ is a bounded random variable. Prove that if we further assume that $\int_{\mathbb{R}_0^m} \left|\frac{\mathrm{div}\{h(z)\, f(z)\}}{f(z)}\right|^2 \nu(dz) \leq C$ and $\Theta \in \mathbb{D}_{2,1}(\mathbb{R}^u)$, then we have that $\delta(\Theta) \in L^2(\Omega; \mathbb{R})$ and

$$\|\delta(\Theta)\|_{L^2(\Omega;\mathbb{R})} \leq \tilde{C}\left(1 + \sqrt{C}\right)\|\Theta\|_{2,1}.$$

Furthermore prove the stability of the definition in the following sense. Let $\big\{\Theta_n\,;\, n \in \mathbb{N}\big\} \subset \bigcup_{0<\varepsilon\leq 1} \mathcal{S}_\varepsilon(\mathbb{R}^u)$ with $\|\Theta_n - \Theta\|_{2,1} \to 0$ as $n \to \infty$. Then $\delta(\Theta) = L^2(\Omega\,;\,\mathbb{R})$-$\lim_{n\to\infty} \delta(\Theta_n)$. Give the proper assumptions in order to extend this result to convergence in the space $L^q(\Omega; \mathbb{R})$, $q > 2$. Also give a similar result for $\tilde{\delta}$.

Theorem 12.4.19 *Let $p,\, q \geq 2$ such that $1/p + 1/q = 1$. Let $X \in \mathbb{D}_{p,1}(\mathbb{R}^d)$, and $\Theta \in \mathbb{D}_{q,1}(\mathbb{R}^u)$ be bounded. Then, for $G \in C_c^1(\mathbb{R}^d\,;\,\mathbb{R})$, it holds that*

$$\begin{aligned} &D\,(G(X)) = \nabla G(X)\, DX, \\ &\mathbb{E}\big[D\big(G(X)\big)\,\Theta\big] = \mathbb{E}\Big[G(X)\,\tilde{\delta}(\Theta)\Big]. \end{aligned} \tag{12.16}$$

Taking into consideration Exercises 12.4.7, 12.4.17 and (12.16), one may sometimes state, loosely speaking, that the dual operator associated with the derivative operator is the stochastic integral.

Proof The first assertion is clear via the limiting procedure in the definition of $\mathbb{D}_{p,1}(\mathbb{R}^d)$. Now, we shall prove the second assertion. From Proposition 12.4.3, we can find the sequence $\{X_n\,;\ n\in\mathbb{N}\}$ in $\bigcup_{0<\varepsilon\leq 1}\mathscr{S}_\varepsilon(\mathbb{R}^d)$ such that $\|X_n - X\|_{p,1}\to 0$ as $n\to\infty$. Since $X_n\in\bigcup_{0<\varepsilon\leq 1}\mathscr{S}_\varepsilon(\mathbb{R}^d)$ for each $n\in\mathbb{N}$, there exists $0<\varepsilon_n\leq 1$ such that $X_n\in\mathscr{S}_{\varepsilon_n}(\mathbb{R}^d)$. For sufficiently large $n\in\mathbb{N}$, we can assume without loss of generality that $0<\varepsilon_n\leq 1$ and $\varepsilon_n\to 0$ as $n\to\infty$. On the other hand, due to Exercise 12.4.14, as $\Theta\in\mathbb{D}_{q,1}(\mathbb{R}^u)$, we can find a uniformly bounded sequence $\{\Theta_n\,;\ n\in\mathbb{N}\}\subset\mathscr{S}_{\varepsilon_n}(\mathbb{R}^u)$ such that $\|\Theta_n-\Theta\|_{q,1}\to 0$ as $n\to\infty$.

From Corollary 12.4.12, one obtains that

$$\mathbb{E}\big[D\big(G(X_n)\big)\,\Theta_n\big]=\mathbb{E}\Big[G(X_n)\,\tilde{\delta}(\Theta_n)\Big]+o(1).$$

Using the uniform boundedness of Θ_n and taking the limit as $n\to\infty$ (recall Exercise 12.4.17) enable us to see that

$$\begin{aligned}\mathbb{E}\big[D\big(G(X)\big)\,\Theta\big]&=\lim_{n\to\infty}\mathbb{E}\big[D\big(G(X_n)\big)\,\Theta_n\big]\\&=\lim_{n\to\infty}\Big\{\mathbb{E}\Big[G(X_n)\,\tilde{\delta}(\Theta_n)\Big]+o(1)\Big\}\\&=\mathbb{E}\Big[G(X)\,\tilde{\delta}(\Theta)\Big].\end{aligned}$$

The proof is complete.

The following remarks point towards possible extensions of the above construction which uses approximation techniques in order to obtain IBP formulas.

Remark 12.4.20 (i) The case of unbounded Θ can be also studied under stronger conditions. See Exercise 12.4.21.

(ii) The construction we have given here is a pathwise type of construction towards the IBP formula. It should be clear that there is the option of introducing a weak type formulation using weak convergence, instead of $L^p(\Omega;\mathbb{R}^d)$-convergence.

(iii) Similarly, the fact that the Lévy measure ν is absolutely continuous with respect to the Lebesgue measure in Hypothesis 12.2.1 can be weakened by using proper approximating structures.

(iv) The appearance of the term $\frac{\mathrm{div}\{h(y)\,f(y)\}}{f(y)}$ in $\delta(\Theta)$ and $\tilde{\delta}(\Theta)$ will be extremely important in the discussion that follows. In particular, we will observe that a good definition of h will allow the control of g as can be seen from Exercise 12.4.23.

(v) Clearly the duality formula in (12.16) leads to an IBP formula due to the chain rule (12.11). See Exercise 12.4.22.

(vi) In the light of the results in Sect. 11.5, the requirement on the inverse moments of the so-called stochastic covariance matrix is currently hidden in the condition that $\Theta\in\mathbb{D}_{q,1}(\mathbb{R})$.

(vii) Note that the use of $\tilde{\delta}$ instead of δ can help change Hypothesis 12.3.3, $\int_{\mathbb{R}_0^m}\left|\frac{\mathrm{div}\{h(z)\,f(z)\}}{f(z)}\right|^i\nu(dz)<+\infty$ from $i=1$ into $i=2$.

Exercise 12.4.21 Assume the condition in Exercise 12.4.18. Prove that in that case one can prove the result in Theorem 12.4.19 without the assumption of boundedness of Θ.

Exercise 12.4.22 The following is a first simple application of the IBP formula in the one-dimensional case ($d = m = u = 1$). Consider any Lévy measure satisfying the hypotheses stated in Sect. 12.3 with the deformation function $H^{\xi}(z) = ze^{-\xi z}$. The goal of the following calculations is to provide an IBP formula for $X = \int_{(0,1]\times(0,1]} z\widetilde{\mathcal{N}}(dz, ds)$.

1. Use the duality formula in (12.16) to prove the following IBP formula, assuming necessary conditions on X, Θ and G so that the formula is valid:

$$\mathbb{E}\big[G'(X)\,\Theta\big] = \mathbb{E}\big[G(X)\,\tilde{\delta}\big((DX)^{-1}\,\Theta\big)\big]$$

2. Use this result in order to prove the following IBP for the characteristic function of X:

$$i\theta\,\mathbb{E}\big[e^{i\theta X}\big] = \mathbb{E}\big[e^{i\theta X}\,\tilde{\delta}\big((DX)^{-1}\big)\big].$$

For a hint, see Chap. 14.

Exercise 12.4.23 Recall Theorem 10.5.11. Compare the formulation with the one in this section in the one-dimensional case. In particular, provide an example of the Lévy measure which is supported on the whole real line so that the deformation function $H^{\xi}(z) = z\,e^{-\xi z}$ satisfies the hypotheses in Sect. 12.3. Provide an IBP formula in the particular case of characteristic functions. Think of other possible deformation functions which may achieve the same goal for other Lévy measures. For example, try $H^{\xi}(z) = z\,e^{-\xi\psi(z)}$ and state some minimal conditions on the function ψ. Also see Sect. 12.6.3 later.

Exercise 12.4.24 Reconsider the above proofs from the beginning of Sect. 12.4 changing (12.8) under the following set-up:

$$f_{\varepsilon}(z) = \frac{\varphi_{\varepsilon}(z) f(z)}{\lambda_{\varepsilon}}, \quad \nu_{\varepsilon}(dz) = f_{\varepsilon}(z)\,dz.$$

Here, $\varphi_{\varepsilon} \in C_b^{\infty}(\mathbb{R}^m)$ is a positive function such that

$$\varphi_{\varepsilon}(z) = \begin{cases} 0 & (|z| \leq \varepsilon), \\ 1 & (|z| \geq 2\varepsilon). \end{cases}$$

Furthermore, λ_{ε} is determined so that f_{ε} is a density function. In particular, prove that through the above approximation, one can also obtain a result similar to Theorem 12.4.19.

Now, we shall study the higher-order IBP formula in our framework. Let $G \in C_c^\infty(\mathbb{R}^d\,;\,\mathbb{R})$ and $X \in \mathbb{D}_{p,2}(\mathbb{R})$. Then, as seen in Theorem 12.4.19, we have

$$D\big(G(X)\big) = \nabla G(X)\, DX.$$

Assume that $d = u$ so that DX takes values in $\mathbb{R}^d \otimes \mathbb{R}^d$. Suppose that $(DX)^{-1}$ is well defined. For each $k = 1, \dots, d$, denote by $\boldsymbol{e}_k = \big(0, \dots, 0, 1, 0, \dots, 0\big)^* \in \mathbb{R}^d$ the k-th unit vector. For $Z \in \mathbb{D}_{q,1}(\mathbb{R})$, Theorem 12.4.19 yields that under the assumption that $\tilde{\delta}\big((DX)^{-1}\boldsymbol{e}_k\, Z\big) \in L^1(\Omega)$,

$$\begin{aligned}\mathbb{E}\big[\partial_k G(X)\, Z\big] &= \mathbb{E}\big[D\big(G(X)\big)\,(DX)^{-1}\boldsymbol{e}_k\, Z\big] \\ &= \mathbb{E}\Big[G(X)\,\tilde{\delta}\big((DX)^{-1}\boldsymbol{e}_k\, Z\big)\Big].\end{aligned}$$

For $n \in \mathbb{N}$, let $i_1, \dots, i_n \in \mathbb{N}$ and define $\Theta_{i_1,\dots,i_n}(Z)$ by

$$\Theta_{i_1,\dots,i_n}(Z) = \begin{cases}\tilde{\delta}\big(V^{-1}\boldsymbol{e}_{i_1}\, Z\big) & (n = 1), \\ \tilde{\delta}\big(V^{-1}\boldsymbol{e}_{i_{n-1}}\, \Theta_{i_1,\dots,i_{n-1}}(Z)\big) & (n \geq 2).\end{cases}$$

Then, iterative applications of Theorem 12.4.19 enable us to see that under a finite moment hypothesis on $\Theta_{i_1,\dots,i_k}(Z)$ for all finite indices $(i_1, \dots, i_k)$,

$$\begin{aligned}\mathbb{E}\big[\partial_{i_1}\cdots\partial_{i_n} G(X)\, Z\big] &= \mathbb{E}\big[\partial_{i_2}\cdots\partial_{i_n} G(X)\, \Theta_{i_1}(Z)\big] \\ &= \mathbb{E}\big[\partial_{i_3}\cdots\partial_{i_n} G(X)\, \Theta_{i_1,i_2}(Z)\big] \\ &= \cdots \\ &= \mathbb{E}\big[\partial_{i_n} G(X)\, \Theta_{i_1,\dots,i_{n-1}}(Z)\big] \\ &= \mathbb{E}\big[G(X)\, \Theta_{i_1,\dots,i_n}(Z)\big].\end{aligned}$$

Therefore we obtain the following result.

Theorem 12.4.25 *Let $d = u$, and $p, q \geq 2$ such that $1/p + 1/q = 1$. Let $X \in \mathbb{D}_{p,n+1}(\mathbb{R}^d)$ and $Z \in \mathbb{D}_{q,n}(\mathbb{R})$ such that $\Theta_{i_1,\dots,i_n}(Z) \in L^1(\Omega)$. Then for $G \in C_c^\infty(\mathbb{R}^d\,;\,\mathbb{R})$, it holds that*

$$\mathbb{E}\big[\partial_{i_1}\cdots\partial_{i_n} G(X)\, Z\big] = \mathbb{E}\big[G(X)\, \Theta_{i_1,\dots,i_n}(Z)\big]. \tag{12.17}$$

Recall that the issue of the finiteness of the moments of $\Theta_{i_1,\dots,i_n}(Z)$ is related to properties of $(DX)^{-1}$, or in other words, on the fact that $\det(DX)^{-1} \in \bigcap_{p>1} L^p(\Omega)$. This has already been discussed in Sect. 11.5.

Exercise 12.4.26 Write the details of the proof of the above theorem. In particular, prove that $(DX)^{-1}_{ij} \in \bigcap_{p>1} L^p(\Omega)$ for any $i, j = 1, \dots, d$ if one assumes that $\det(DX)^{-1} \in \bigcap_{p>1} L^p(\Omega)$.

12.5 Approximation of SDEs

In this section, we apply the above methodology to obtain an IBP in the classical situation of stochastic differential equations with jump components. We will use many ideas exposed already in Chap. 7. Although those ideas were carried out in one dimension, the multi-dimensional extension is straightforward.[9] Therefore, we will always assume Hypothesis 7.2.2 which implies Hypothesis 7.1.3. Furthermore, adding to the hypotheses in Sect. 12.3, we will also assume:

Hypothesis 12.5.1 $\sup_{\xi \in F} \left\{ \int_{\bar{B}_1} \left|H^{\xi}(z) - z\right|^2 \nu(dz) + \int_{\bar{B}_1^c} \left|H^{\xi}(z)\right|^p \nu(dz) \right\} < \infty$ for some open set $F \subset \mathbb{R}^u$ including the origin and any $p \geq 2$.

This hypothesis will be used in order to define the approximations to the solutions of stochastic differential equations and their stochastic derivatives.

Remark 12.5.2 Let $F \subset \mathbb{R}^u$ be an open set including the origin. The condition

$$\sup_{\xi \in F} \left|H^{\xi}(z) - z\right| \leq C\left(|z| \wedge 1\right)$$

is a sufficient condition for Hypothesis 12.5.1 to be satisfied, because of Hypothesis 12.3.1.

Hypothesis 12.5.3 Let $F \subset \mathbb{R}^u$ be an open set including the origin. Assume that $H^{\xi}(z)$ is once differentiable with respect to $\xi \in \mathbb{R}^u$ for all $z \in \mathbb{R}_0^m$, and that we have, for any $a,\ b \in F$,

$$\int_{\bar{B}_1} \left|\left(\partial_{\xi}^k\right)_a H^{\xi}(z) - \left(\partial_{\xi}^k\right)_b H^{\xi}(z)\right|^2 \nu(dz) + \int_{\bar{B}_1^c} \left|\left(\partial_{\xi}^k\right)_a H^{\xi}(z) - \left(\partial_{\xi}^k\right)_b H^{\xi}(z)\right| \nu(dz)$$
$$\leq C\,|a-b|$$

for $k = 0, 1$.

This hypothesis will be used in Lemma 12.5.6 in order to obtain the necessary continuity properties of the derivative flow. Recall that we are assuming Hypothesis 7.2.2 and that the stochastic differential equation of interest is

$$X_t = x + \int_0^t b(X_s)\, ds + \int_{\bar{B}_1 \times [0,t]} a\left(X_{s-}, z\right) \widetilde{\mathcal{N}}(dz, ds)$$
$$+ \int_{\bar{B}_1^c \times [0,t]} a\left(X_{s-}, z\right) \mathcal{N}(dz, ds).$$

[9] We assume that the reader can extrapolate the conditions in Chap. 7 to the multi-dimensional case, otherwise they may also assume that the current chapter applies for the case $d = m = u = 1$.

Also note that due to Hypothesis 12.3.1 and the definition of Lévy measure, we have that for any $p \geq 2$,

$$\eta_p := \int_{\bar{B}_1} |z|^2 \nu(dz) + \int_{\bar{B}_1^c} |z|^p \nu(dz) < \infty.$$

Define

$$Z_t^\varepsilon = t + \int_{(\bar{B}_1 \cap \bar{B}_\varepsilon^c) \times [0,t]} z \widetilde{\mathcal{N}}(dz, ds) + \int_{\bar{B}_1^c \times [0,t]} z \mathcal{N}(dz, ds),$$

and let X^ε be the solution associated with the Poisson random measure generated by Z^ε (see Exercise 12.2.5).

Lemma 12.5.4 *Under Hypothesis 7.2.2, we have for* $p \geq 2$

$$\mathbb{E}\Big[\sup_{0 \leq t \leq T} \big|X_t^\varepsilon - X_t\big|^p\Big] \to 0$$

as $\varepsilon \to 0$.

Proof For the proof, one applies strategies which have already been used previously. Using Hypothesis 7.1.3, together with the Burkholder–Davis–Gundy inequality (see Proposition 7.1.2) and Gronwall-type estimates, one obtains the following uniformly estimate in $L^p(\Omega)$ for the process X^ε:

$$\sup_{0 \leq \varepsilon < 1} \mathbb{E}\Big[\sup_{0 \leq s \leq T} \big|X_s^\varepsilon\big|^p\Big] \leq C(x, T, \eta_p). \tag{12.18}$$

Similarly, if one considers the difference between the processes X^ε and X, the difference satisfies

$$\begin{aligned} X_t - X_t^\varepsilon = & \int_0^t \big\{b(X_s) - b(X_s^\varepsilon)\big\}\, ds \\ & + \int_{(\bar{B}_1 \cap \bar{B}_\varepsilon^c) \times [0,t]} \big\{a(X_{s-}, z) - a(X_{s-}^\varepsilon, z)\big\} \widetilde{\mathcal{N}}(dz, ds) \\ & + \int_{\bar{B}_\varepsilon \times [0,t]} a(X_{s-}, z) \widetilde{\mathcal{N}}(dz, ds) \\ & + \int_{\bar{B}_1^c \times [0,t]} \big\{a(X_{s-}, z) - a(X_{s-}^\varepsilon, z)\big\} \mathcal{N}(dz, ds). \end{aligned}$$

A similar argument to (12.18), leads to the estimate

$$\mathbb{E}\Big[\sup_{0 \leq s \leq T} \big|X_s^\varepsilon - X_s\big|^p\Big] \leq C(x, T, \eta_p) \int_{\bar{B}_\varepsilon} |z|^p \, \nu(dz).$$

Therefore the conclusion follows.

In a complete analogous way, one defines the transformed/deformed approximation processes. For sufficiently small $\xi \in \mathbb{R}^d$, such that Hypothesis 12.5.1 is satisfied, let $X^\xi = \{X_t^\xi \,;\, t \geq 0\}$ be the $\mathbb{R}^d$-valued process determined by the stochastic differential equation of the form:

$$X_t^\xi = x + \int_0^t b(X_s^\xi)\, ds + \int_{\bar{B}_1\times[0,t]} a(X_{s-}^\xi, H^\xi(z))\, \widetilde{\mathcal{N}}(dz, ds) + \int_{\bar{B}_1^c\times[0,t]} a(X_{s-}^\xi, H^\xi(z))\, \mathcal{N}(dz, ds).$$

As in the proof of Lemma 12.5.4, we can then define the process $X^{\varepsilon,\xi}$ which neglects all jumps less than ε. Then, as before we define

$$Z_t^{\varepsilon,\xi} = t + \int_{(\bar{B}_1\cap\bar{B}_\varepsilon^c)\times[0,t]} H^\xi(z)\, \widetilde{\mathcal{N}}(dz, ds) + \int_{\bar{B}_1^c\times[0,t]} H^\xi(z)\, \mathcal{N}(dz, ds).$$

Therefore, the proof of the convergence of the sequence of processes $X^{\varepsilon,\xi}$ towards X^ξ as $\varepsilon \to 0$ is done as in Lemma 12.5.4, because due to Hypothesis 12.5.1, we have for $p \geq 2$

$$\eta_p^\xi := \int_{\bar{B}_1} |H^\xi(z)|^2\, \nu(dz) + \int_{\bar{B}_1^c} |H^\xi(z)|^p\, \nu(dz) < \infty.$$

Lemma 12.5.5 *Under Hypotheses 7.2.2 and 12.5.1, we have for* $p \geq 2$

$$\mathbb{E}\Big[\sup_{0\leq t\leq T} |X_t^{\varepsilon,\xi} - X_t^\xi|^p\Big] \to 0$$

as $\varepsilon \to 0$. *Furthermore, we have* $X^\xi\big|_{\xi=0} = X$.

Now, we shall discuss the differentiability of the process X^ξ with respect to the parameter ξ.

Lemma 12.5.6 *Under Hypotheses 7.22, 12.5.1 and 12.5.3, the* $\mathbb{R}^d$*-valued process* $X^\xi = \{X_t^\xi \,;\, t \in [0, T]\}$ *is differentiable in* ξ *a.s., and the* $\mathbb{R}^{d\times u}$*-valued derivative process* $\{V_t^\xi := \partial_\xi X_t^\xi \,;\, t \in [0, T]\}$ *satisfies the equation:*

$$\begin{aligned} V_t^\xi = &\int_{\bar{B}_1\times[0,t]} \partial a(X_{s-}^\xi, H^\xi(z))\, \partial_\xi H^\xi(z)\, \widetilde{\mathcal{N}}(dz, ds) \\ &+ \int_{\bar{B}_1^c\times[0,t]} \partial a(X_{s-}^\xi, H^\xi(z))\, \partial_\xi H^\xi(z)\, \mathcal{N}(dz, ds) \\ &+ \int_0^t \nabla b(X_s^\xi)\, V_s^\xi\, ds + \int_{\bar{B}_1\times[0,t]} \nabla a(X_{s-}^\xi, H^\xi(z))\, V_{s-}^\xi\, \widetilde{\mathcal{N}}(dz, ds) \\ &+ \int_{\bar{B}_1^c\times[0,t]} \nabla a(X_{s-}^\xi, H^\xi(z))\, V_{s-}^\xi\, \mathcal{N}(dz, ds). \end{aligned} \tag{12.19}$$

Proof This result is a direct consequence of the Kolmogorov continuity criterion for random fields. One can also adapt the set-up in Theorem 2.3 of [27] by setting the first component of the stochastic differential equation as the constant ξ and the Lévy process as a process concentrated in the functions of the type $a\big(x, H^\xi(z)\big)$ with measure $\nu(dz)$.

Similarly, we can obtain the differentiability of the process $X^{\xi,\varepsilon}$ with respect to the parameter ξ.

Lemma 12.5.7 *Under Hypotheses 7.2.2, 12.5.1 and 12.5.3, the $\mathbb{R}^d$-valued process $X^{\xi,\varepsilon} = \{X_t^{\xi,\varepsilon} ;\ t \in [0, T]\}$ is differentiable in ξ a.s., and the $\mathbb{R}^{d\times u}$-valued derivative process $\{V_t^{\xi,\varepsilon} := \partial_\xi X_t^{\xi,\varepsilon} ;\ t \in [0, T]\}$ satisfies the equation:*

$$\begin{aligned} V_t^{\xi,\varepsilon} = &\int_{(\bar{B}_1\cap\bar{B}_\varepsilon^c)\times[0,t]} \partial a\big(X_{s-}^{\xi,\varepsilon}, H^\xi(z)\big)\, \partial_\xi H^\xi(z)\, \widetilde{\mathcal{N}}(dz, ds) \\ &+ \int_{\bar{B}_1^c\times[0,t]} \partial a\big(X_{s-}^{\xi,\varepsilon}, H^\xi(z)\big)\, \partial_\xi H^\xi(z)\, \mathcal{N}(dz, ds) \\ &+ \int_0^t \nabla b(X_s^{\xi,\varepsilon})\, V_s^{\xi,\varepsilon}\, ds \\ &+ \int_{(\bar{B}_1\cap\bar{B}_\varepsilon^c)\times[0,t]} \nabla a\big(X_{s-}^{\xi,\varepsilon}, H^\xi(z)\big)\, V_{s-}^{\xi,\varepsilon}\, \widetilde{\mathcal{N}}(dz, ds) \\ &+ \int_{\bar{B}_1^c\times[0,t]} \nabla a\big(X_{s-}^{\xi,\varepsilon}, H^\xi(z)\big)\, V_{s-}^{\xi,\varepsilon}\, \mathcal{N}(dz, ds). \end{aligned} \tag{12.20}$$

As in Lemmas 12.5.4 and 12.5.5, we have:

Lemma 12.5.8 *Under Hypotheses 7.2.2, 12.5.1 and 12.5.3, we have for $p \geq 2$*

$$\mathbb{E}\Big[\sup_{0\leq t\leq T} \big|V_t^{\varepsilon,\xi} - V_t^\xi\big|^p\Big] \to 0, \quad \text{as } \varepsilon \to 0.$$

Let $C_1 < C_2$ be two positive constants, and $\Xi \in C_b^\infty(\mathbb{R}^d \otimes \mathbb{R}^d ;\ [0, 1])$ such that

$$\Xi(V) = \begin{cases} 1 & (|\det V| \in (C_1, C_2)) \\ 0 & (|\det V| \notin (C_1/2, 2C_2). \end{cases}$$

Then, we can get a similar assertion to Theorem 12.4.19. Recall that we are assuming the hypotheses in Sect. 12.3, as well as Hypotheses 7.2.2, 12.5.1 and 12.5.3.

Theorem 12.5.9 *Let $d = u$ and $G \in C_c^1(\mathbb{R}^d ;\ \mathbb{R})$. Assume that*

$$\int_{\mathbb{R}_0^m} \left|\frac{\operatorname{div}\{h(z)\, f(z)\}}{f(z)}\right| \nu(dz) + \int_{\mathbb{R}_0^m} \left|\frac{\operatorname{div}\{h(z)\, f(z)\}}{f(z)}\right|^2 \nu(dz) < \infty,$$

and $\Xi(V_T)\,V_T^{-1} \in \mathbb{D}_{2,1}(\mathbb{R}^{d\times d})$. *Then, it holds that*

$$\mathbb{E}\big[\nabla G(X_T)\,\Xi(V_T)\big] = \mathbb{E}\Big[G(X_T)\,\tilde{\delta}\big(\Xi(V_T)\,V_T^{-1}\big)\Big]. \tag{12.21}$$

Exercise 12.5.10 Prove Theorem 12.5.9.

Clearly, one may write other versions of this result. In this case we have localized the IBP formula on the set where the inverse of the matrix V_T is well defined.

12.6 Relation with Bismut's Approach

In this section, we compare the previous approach with a method based on the Girsanov theorem, which is a change of measure approach used in the most classical forms of the IBP formula. This approach is also known as Bismut's approach in the literature. We have discussed the Girsanov change of measure in the simple setting of Poisson processes on Exercise 2.1.40.

12.6.1 Girsanov Transformations

Now, we discuss the IBP formula in the light of infinite-dimensional measure transformations, better known as the Girsanov theorems. The statements which follow in this subsection are based upon [40]. See also Theorems 33.1 and 33.2 in [51]. The following lemma is a form of the martingale representation theorem. That is, we state that a martingale can be expressed using stochastic integrals.

Lemma 12.6.1 (cf. [39]-Theorem 2.1, [40]-Theorem 6.1) *Let* $\{\alpha_t\,;\,t\in[0,T]\}$ *be a positive local martingale with* $\alpha_0 = 1$. *Then, there exists a unique* $\mathbb{R}$*-valued predictable process* $\{g(t,z)\,;\,t\in[0,T],\,z\in\mathbb{R}^m_0\}$ *satisfying*

$$\int_{\bar{B}_1\times[0,T]} |g(t,z)|^2\,\widehat{\mathcal{N}}(dz,dt) < \infty, \quad \text{a.s.,}$$
$$\int_{\bar{B}_1^c\times[0,T]} \big(\exp\big[g(t,z)\big]-1\big)\,\widehat{\mathcal{N}}(dz,dt) < \infty, \quad \text{a.s.,}$$

such that if we define

$$X_t := \exp\Bigg[\int_{\bar{B}_1\times[0,t]} g(s,z)\,\widetilde{\mathcal{N}}(dz,ds) + \int_{\bar{B}_1^c\times[0,t]} g(s,z)\,\mathcal{N}(dz,ds) - \int_{\mathbb{R}^m_0\times[0,t]} \big(\exp\big[g(s,z)\big]-1-g(s,z)\,\mathbf{1}_{\bar{B}_1}(z)\big)\,\widehat{\mathcal{N}}(dz,ds)\Bigg],$$

then we have that $\alpha_t = X_t$ *for all* $t \geq 0$.

From the above lemma, we may use the notation $\alpha_t \equiv \alpha_t(g)$.

Exercise 12.6.2 Prove that the process on the right-hand side (in the definition of X_t) introduced in Lemma 12.6.1 is a local martingale.

For a hint, see Chap. 14.

Exercise 12.6.3 This exercise is a preparation for the next lemma. Let α_T be an $\mathscr{F}_T$-measurable, positive random variable such that $\mathbb{E}[\alpha_T] = 1$. Define $\mathbb{Q}(A) := \mathbb{E}[1_A \alpha_T]$ for any $A \in \mathscr{F}_T$.

1. Prove that $\mathbb{Q}$ is a measure on $(\Omega, \mathscr{F}_T)$.
2. Prove that $\mathbb{Q}$ is absolutely continuous with respect to $\mathbb{P}$.
3. Suppose that α_T^{-1} is integrable. Then, prove that $\mathbb{P}$ is absolutely continuous with respect to $\mathbb{Q}$.
4. Under all the above conditions, prove the following conditional formula:

$$\mathbb{E}_{\mathbb{Q}}\big[Y \big| \mathscr{F}_s\big] = \mathbb{E}\big[Y \alpha_T \big| \mathscr{F}_s\big] \alpha_T^{-1},$$

which is usually called the Bayes formula.

For a hint, see Chap. 14.

Lemma 12.6.4 (cf. [39]-Theorem 2.3) *Define a new probability measure $\mathbb{Q}$ on $(\Omega, \mathscr{F}_T)$ by*

$$\left.\frac{d\mathbb{Q}}{d\mathbb{P}}\right|_{\mathscr{F}_T} = \alpha_T,$$

where the coefficient g satisfies the conditions stated in Lemma 12.6.1. Then, we have that:

(i) The random measure $\widetilde{\mathcal{N}}^g(dz, ds)$ given by

$$\widetilde{\mathcal{N}}^g(dz, ds) = \mathcal{N}(dz, ds) - \exp\big[g(s, z)\big] \widehat{\mathcal{N}}(dz, ds)$$

is a martingale measure with respect to $\mathbb{Q}$. Moreover, the stochastic integral with respect to the measure $\widetilde{\mathcal{N}}^g(dz, ds)$ is well defined as a local martingale under $\mathbb{Q}$.

(ii) Suppose that g is a deterministic function. Then, the random measure $\mathcal{N}$ is still a Poisson random measure under $\mathbb{Q}$ with intensity measure $\exp\big[g(t, z)\big] \widehat{\mathcal{N}}(dz, dt)$.

Proof (i) We remark that

$$d\alpha_t = \int_{\mathbb{R}_0^m} \alpha_{t-} \big\{\exp[g(t, z)] - 1\big\} \widetilde{\mathcal{N}}(dz, dt).$$

Define

$$\Gamma_t = \int_{\mathbb{R}_0^m \times [0,t]} \varphi(s,z)\, \widetilde{\mathcal{N}}^g(dz, ds),$$

where $\{\varphi(s,z)\,;\ s \in [0,T],\ z \in \mathbb{R}_0^m\}$ is a predictable process with

$$\int_{\mathbb{R}_0^m \times [0,t]} \varphi(s,z)^2 \exp\big[g(t,z)\big]\, \widehat{\mathcal{N}}(dz, ds) < \infty, \text{ a.s.}$$

Using the Itô formula (Theorem 5.3.17), one has

$$d\,(\alpha_t\,\Gamma_t) = \int_{\mathbb{R}_0^m} \alpha_{t-}\ \Big[\Gamma_{t-}\big\{\exp\big[g(t,z)\big] - 1\big\} + \exp\big[g(t,z)\big]\,\varphi(t,z)\Big]\ \widetilde{\mathcal{N}}(dz, dt).$$

Then, the process $\Gamma = \{\Gamma_t\,;\ t \in [0,T]\}$ is a local martingale under the measure $\mathbb{Q}$ due to Bayes formula.

(ii) Suppose that the function g is deterministic. Let $l \in \mathbb{N}$, $y_1, \ldots, y_l > 0$ and $U_1, \ldots, U_l \in \mathscr{B}(\mathbb{R}_0^m)$ be disjoint such that $\widehat{\mathcal{N}}\big(U_k \times [0,t]\big) < \infty$. Define

$$\Lambda_t = \exp\left[-\int_{\mathbb{R}_0^m \times [0,t]} \sum_{k=1}^{l} y_k\, \mathbf{1}_{U_k}(z)\, \mathcal{N}(dz, ds)\right].$$

Then, the Itô formula (Theorem 5.3.17) gives that

$$d\Lambda_t = \int_{\mathbb{R}_0^m} \Lambda_{t-}\ \left\{\exp\left[-\sum_{k=1}^{l} y_k\, \mathbf{1}_{U_k}(z)\right] - 1\right\}\ \mathcal{N}(dz, dt).$$

Again, the Itô formula (Theorem 5.3.17) yields that

$$\begin{aligned} d\big(\alpha_t\,\Lambda_t\big) &= \int_{\mathbb{R}_0^m} \alpha_{t-}\,\Lambda_{t-}\left\{\exp\left[g(t,z) - \sum_{k=1}^{l} y_k\, \mathbf{1}_{U_k}(z)\right] - 1\right\}\ \widetilde{\mathcal{N}}(dz, dt) \\ &\quad + \int_{\mathbb{R}_0^m} \alpha_t\,\Lambda_t\left\{\exp\left[-\sum_{k=1}^{l} y_k\, \mathbf{1}_{U_k}(z)\right] - 1\right\} \exp\big[g(t,z)\big]\,\widehat{\mathcal{N}}(dz, dt). \end{aligned}$$

Thus, we can obtain the following linear equation:

$$\begin{aligned} \mathbb{E}\left[\frac{\alpha_t\,\Lambda_t}{\alpha_s\,\Lambda_s}\middle|\mathscr{F}_s\right] &= 1 + \int_{\mathbb{R}_0^m \times (s,t]} \mathbb{E}\left[\frac{\alpha_u\,\Lambda_u}{\alpha_s\,\Lambda_s}\middle|\mathscr{F}_s\right] \\ &\quad \times \left\{\exp\left[-\sum_{k=1}^{l} y_k\, \mathbf{1}_{U_k}(z)\right] - 1\right\} \exp\big[g(u,z)\big]\,\widehat{\mathcal{N}}(dz, du), \end{aligned}$$

which can be solved explicitly as follows:

$$\mathbb{E}\left[\frac{\alpha_t \Lambda_t}{\alpha_s \Lambda_s}\middle|\mathscr{F}_s\right] = \exp\left[\int_{\mathbb{R}_0^m\times(s,t]}\left\{\exp\left[-\sum_{k=1}^{l} y_k \mathbf{1}_{U_k}(z)\right]-1\right\}\right.$$
$$\left.\times \exp\left[g(u,z)\right]\widehat{\mathcal{N}}(dz,du)\right],$$
$$= \exp\left[\sum_{k=1}^{l}\left(e^{-y_k}-1\right)\int_{U_k\times(s,t]} \exp\left[g(u,z)\right]\widehat{\mathcal{N}}(dz,du)\right].$$

In the light of Lemma 12.6.4, we can see, by comparing with (12.1), that we have only accomplished the change of the compensating measure or frequency of jumps. In fact, the new process does not necessarily have the same laws as the original one. We will now change the size of the jump. This will be done in the next subsection.

Exercise 12.6.5 Under the setting of Lemma 12.6.4, find the characteristic function associated with

$$Z_t = \int_{\bar{B}_1\times[0,t]} z\,\widetilde{\mathcal{N}}(dz,ds) + \int_{\bar{B}_1^c\times[0,t]} z\,\mathcal{N}(dz,ds)$$

under the measure $\mathbb{Q}$. Prove that Z is a Lévy process and find its associated Lévy measure. This result is also known as the Esscher transform.

For a hint, see Chap. 14.

12.6.2 *Changing the Jump Size*

In this subsection, we achieve the invariance principle that appears in (12.1) in order to obtain the integration by parts formula. In the remainder of this subsection, we suppose that:

Hypothesis 12.6.6 The mapping H^ξ satisfies the following conditions:

(i) $H^\xi \in C^1(\mathrm{Supp}[\nu];\mathbb{R}^m)$ is a one-to-one differentiable mapping for $|\xi|\le 1$ with $H^0(z)=z,\ \nu$-a.e.

(ii) We assume that $\det \partial_z H^\xi(z) > 0$, ν-a.e. Define the random measure $\mathcal{N}^\xi(dz,ds)$ by

$$\int_{\mathbb{R}_0^m\times[0,t]} \varphi(s,z)\,\mathcal{N}^\xi(dz,ds) = \int_{\mathbb{R}_0^m\times[0,t]} \varphi\big(s,H^\xi(z)\big)\,\mathcal{N}(dz,ds),$$

and write for $z\in\{y\in\mathbb{R}_0^m;\ f(y)>0\}\subseteq\{y\in\mathbb{R}_0^m;\ f(H^\xi(y))>0\}$,

$$k^\xi(z) = \frac{f\big(H^\xi(z)\big)}{f(z)}\det \partial_z H^\xi(z).$$

Note that the assumption stated above implies that $k^\xi(z)>0$, ν-a.e.

(iii) There exists a positive constant C independent of ξ such that $|\ln k^{\xi}(z)| \leq C\,(|z| \wedge 1)$ and $|k^{\xi}(z) - 1 - \ln k^{\xi}(z)\,\mathbf{1}_{(|z|\leq 1)}| \leq C\,(|z|^2 \wedge 1)$, ν-a.e.

Then, under the assumption (iii) in Hypothesis 12.6.6, we can define the new probability measure $\mathbb{Q}_{\xi}$, via the Radon–Nikodým derivative, by

$$\begin{aligned}\left.\frac{d\mathbb{Q}_{\xi}}{d\mathbb{P}}\right|_{\mathscr{F}_T} &\equiv \alpha_T(\ln k^{\xi}) \\ &= \exp\Bigg[\bigg[\int_{\bar{B}_1\times[0,T]} \ln k^{\xi}(z)\,\widetilde{\mathscr{N}}(dz,ds) + \int_{\bar{B}_1^c\times[0,T]} \ln k^{\xi}(z)\,\mathscr{N}(dz,ds) \\ &\qquad - \int_{\mathbb{R}_0^m\times[0,T]} \left\{k^{\xi}(z) - 1 - \ln k^{\xi}(z)\,\mathbf{1}_{\bar{B}_1}(z)\right\}\widehat{\mathscr{N}}(dz,ds)\Bigg].\end{aligned}$$

Then, we have the following result.

Lemma 12.6.7 *Assume that $g = \ln k^{\xi}$ satisfies the conditions stated in Lemma 12.6.1. Then, the process $\left\{M_t^{\xi} := \alpha_t(\ln k^{\xi})\,;\ t \in [0,T]\right\}$ is an $\{\mathscr{F}_t\}_{t\in[0,T]}$-square integrable martingale under $\mathbb{P}$. Furthermore, $\mathscr{N}^{\xi}(dz,ds)$ is a Poisson random measure with intensity $\nu(dz)\,ds$ under $\mathbb{Q}_{\xi}$.*

Proof The first assertion, as in the proof of Lemma 12.6.4, is a direct consequence of the Itô formula (Theorem 5.3.17). In fact, the Itô formula (Theorem 5.3.17) leads us to

$$\begin{aligned}dM_t^{\xi} &= \int_{\bar{B}_1} M_{t-}^{\xi}\,\left(k^{\xi}(z) - 1\right)\widetilde{\mathscr{N}}(dz,dt) \\ &\quad + \int_{\bar{B}_1^c} M_{t-}^{\xi}\,\left(k^{\xi}(z) - 1\right)\left\{\mathscr{N}(dz,dt) - \widehat{\mathscr{N}}(dz,dt)\right\},\end{aligned}$$

which tells us that the process M^{ξ} is an $\{\mathscr{F}_t\}_{t\in[0,T]}$-square integrable martingale under $\mathbb{P}$.

Let $y_1,\ldots,y_l > 0$, and $U_1,\ldots,U_l \in \mathscr{B}\big(\mathbb{R}_0^m\big)$ be disjoint. Define

$$\mathscr{E}_t^{\xi} = \exp\left[-\int_{\mathbb{R}_0^m\times[0,t]} \sum_{k=1}^{l} y_k\,\mathbf{1}_{U_k}\big(H^{\xi}(z)\big)\,\mathscr{N}(dz,ds)\right].$$

As in the proof of Lemma 12.6.4 (ii), we have

$$\begin{aligned}&\mathbb{E}\left[\left.\frac{M_t^{\xi}\,\mathscr{E}_t^{\xi}}{M_s^{\xi}\,\mathscr{E}_s^{\xi}}\right|\mathscr{F}_s\right] \\ &= 1 + \int_s^t \mathbb{E}\Bigg[\frac{M_u^{\xi}\,\mathscr{E}_u^{\xi}}{M_s^{\xi}\,\mathscr{E}_s^{\xi}} \int_{\mathbb{R}_0^m} k^{\xi}(z)\end{aligned}$$

$$
\begin{aligned}
&\times \left\{ \exp\left[-\sum_{k=1}^{l} y_k \, \mathbf{1}_{U_k}\big(H^\xi(z)\big) \right] - 1 \right\} \nu(dz) \Big| \mathscr{F}_s \Bigg] du \\
&= 1 + \int_s^t \mathbb{E}\left[\frac{M_u^\xi \mathscr{E}_u^\xi}{M_s^\xi \mathscr{E}_s^\xi} \int_{\mathbb{R}_0^m} f\big(H^\xi(z)\big) \right. \\
&\quad \times \left. \left\{ \exp\left[-\sum_{k=1}^{l} y_k \, \mathbf{1}_{U_k}\big(H^\xi(z)\big) \right] - 1 \right\} dH^\xi(z) \Big| \mathscr{F}_s \right] du \\
&= 1 + \left(\int_{\mathbb{R}_0^m} \left\{ \exp\left[-\sum_{k=1}^{l} y_k \, \mathbf{1}_{U_k}(z) \right] - 1 \right\} \nu(dz) \right) \int_s^t \mathbb{E}\left[\frac{M_u^\xi \mathscr{E}_u^\xi}{M_s^\xi \mathscr{E}_s^\xi} \Big| \mathscr{F}_s \right] du \\
&= 1 + \left(\sum_{k=1}^{l} \left(e^{-y_k} - 1 \right) \nu(U_k) \right) \int_s^t \mathbb{E}\left[\frac{M_u^\xi \mathscr{E}_u^\xi}{M_s^\xi \mathscr{E}_s^\xi} \Big| \mathscr{F}_s \right] du,
\end{aligned}
$$

which can be solved explicitly as

$$
\mathbb{E}\left[\frac{M_t^\xi \mathscr{E}_t^\xi}{M_s^\xi \mathscr{E}_s^\xi} \Big| \mathscr{F}_s \right] = \exp\left[(t-s) \sum_{k=1}^{l} \left(e^{-y_k} - 1 \right) \nu(U_k) \right].
$$

Therefore, $\mathscr{N}^\xi(dz, ds)$ is the Poisson random measure with the intensity $\nu(dz)\, ds$, with respect to the measure $\mathbb{Q}_\xi$. The proof is complete.

Theorem 12.6.8 *Assume that $d = u$ and that there exist positive constants C and δ such that, for all $|\xi| < \delta$, k^ξ is differentiable in ξ with*

$$
|\partial_\xi \ln k^\xi(z)| \le C\,(|z| \wedge 1), \quad |\partial_\xi k^\xi(z) - \partial_\xi \ln k^\xi(z)\, \mathbf{1}_{\bar{B}_1}(z)| \le C\,(|z|^2 \wedge 1).
$$

Moreover, suppose that $\left\{ \partial_\xi \left(\Xi(V_T^\xi)\,(V_T^\xi)^{-1} \right);\ |\xi| < \delta \right\}$ is a uniformly integrable family of r.v.s. Then, for $G \in C_c^1(\mathbb{R}^d\,;\,\mathbb{R})$, it holds that

$$
\begin{aligned}
&\mathbb{E}\big[\nabla G(X_T)\, \Xi(V_T) \big] = \mathbb{E}\big[G(X_T)\, \Gamma^0 \big], \\
&\Gamma^0 = -\mathrm{div}_\xi \left\{ \Xi\big(V_T^\xi\big) \big(V_T^\xi\big)^{-1} M_T^\xi \right\} \Big|_{\xi=0}.
\end{aligned}
$$

Proof From Lemma 12.6.7, we have

$$
\mathbb{E}\Big[G\big(X_T^\xi\big)\, \Xi\big(V_T^\xi\big) \big(V_T^\xi\big)^{-1} M_T^\xi \Big] = \mathbb{E}\big[G(X_T)\, \Xi(V_T)\, V_T^{-1} \big]. \tag{12.22}
$$

The assertion follows by taking the divergence in ξ in (12.22) and then evaluating it for $\xi = 0$.

Remark 12.6.9 We remark that

$$\begin{aligned}
&\operatorname{div}_\xi \left\{ \Xi\left(V_T^\xi\right)\left(V_T^\xi\right)^{-1} M_T^\xi \right\}\Big|_{\xi=0} \\
&= \left(\partial_\xi M_T^\xi \Big|_{\xi=0} \right) \Xi(V_T)\, V_T^{-1} + \operatorname{div}_\xi \left\{ \Xi\left(V_T^\xi\right)\left(V_T^\xi\right)^{-1} \right\}\Big|_{\xi=0}.
\end{aligned}$$

Using an argument similar to Lemma 12.5.6 for derivatives of flows, we can take the derivative at $\xi = 0$ of M_T^ξ, which gives

$$\partial_\xi M_T^\xi \big|_{\xi=0} = \int_{\mathbb{R}_0^m \times [0,T]} \partial_\xi k^\xi(z)\big|_{\xi=0} \widetilde{\mathcal{N}}(dz, ds).$$

Since the derivative of the determinant function can be computed explicitly, we have

$$\begin{aligned}
\partial_\xi k^\xi(z)\Big|_{\xi=0} &= \partial_\xi \left\{ \frac{f\left(H^\xi(z)\right)}{f(z)} \det \partial_z H^\xi(z) \right\}\Bigg|_{\xi=0} \\
&= \frac{\partial f(z)\, h(z)}{f(z)} + \partial_\xi \left\{ \det \partial_z H^\xi(z) \right\}\Bigg|_{\xi=0} \\
&= \frac{\partial f(z)\, h(z)}{f(z)} + (\operatorname{div} h)(z) \\
&= \frac{\operatorname{div}\{h(z)\, f(z)\}}{f(z)},
\end{aligned}$$

and

$$\operatorname{div}_\xi \left\{ \Xi\left(V_T^\xi\right)\left(V_T^\xi\right)^{-1} \right\}\Big|_{\xi=0} = \int_{\mathbb{R}_0^m \times [0,T]} \operatorname{tr}\left\{ D_z\left(\Xi(V_T)\, V_T^{-1}\right)\right\} \mathcal{N}(dz, dt)$$

from the definition of the operator D, we have

$$\begin{aligned}
\Gamma^0 &= -\int_{\mathbb{R}_0^m \times [0,T]} \frac{\operatorname{div}\{h(z)\, f(z)\}}{f(z)} \widetilde{\mathcal{N}}(dz, dt)\left(\Xi(V_T)\, V_T^{-1}\right) \\
&\quad - \int_{\mathbb{R}_0^m \times [0,T]} \operatorname{tr}\left\{ D_z\left(\Xi(V_T)\, V_T^{-1}\right)\right\} \mathcal{N}(dz, dt) \\
&= \tilde{\delta}\left(\Xi(V_T)\, V_T^{-1}\right).
\end{aligned}$$

Remark 12.6.10 Clearly a comparison of both methods is in order. In doing so, the reader will see that the conditions are almost the same except for some technical conditions that apply to each method.

12.6.3 Construction of Deformation Functions and Explicit Examples

We will build functions H^ξ, k^ξ so that they satisfy the conditions stated for each method. Recall that the required hypotheses related to the deformation functions were:

1. Norris approach: Hypothesis 12.3.2, and 12.3.3 or the condition in Remark 12.3.4, the ones in Exercises 12.4.18 and 12.5.1, or the one in Remark 12.5.2 and finally Hypothesis 12.5.3.
2. Bismut approach: Hypothesis 12.6.6.

On the other hand, we always assume that Hypothesis 12.3.1 is satisfied.

Example 12.6.11 (*Gamma processes*) Let Z be *the gamma process* with the parameters (a, b) introduced in Sect. 11.1. Now, we shall choose $H^\xi(z) = b\,z/(b-\xi)$, where $\xi \in \mathbb{R}$ such that $\xi < b$. Then, the measure change is the Esscher transform:

$$\left.\frac{d\mathbb{Q}_\xi}{d\mathbb{P}}\right|_{\mathscr{F}_T} = \frac{\exp(\xi\, Z_T)}{\mathbb{E}\big[\exp(\xi\, Z_T)\big]} = \left(\frac{b}{b-\xi}\right)^{-aT} \exp(\xi\, Z_T).$$

Moreover, all the conditions listed in Sect. 12.3 can be checked easily. In fact, the Hypotheses 12.2.1 and 12.3.1 are clear, because

$$\int_{\mathbb{R}_0} \nu(dz) \geq \int_0^1 \frac{a}{z}\, e^{-bz}\, dz = \infty,$$

$$\int_{\mathbb{R}_0} \big(|z|^2 \wedge 1\big)\, \nu(dz) = \int_0^1 a\, z\, \exp(-bz)\, dz + \int_1^\infty \frac{a}{z}\, e^{-bz}\, dz \leq a + \frac{a}{b}\, e^{-b},$$

$$\int_{|z|>1} |z|^p \nu(dz) = \int_1^\infty a\, z^{p-1} \exp(-bz)\, dz \leq \frac{a}{b^p}\, \Gamma(p)$$

for $p > 0$. Since $h(z) = \partial_\xi H^\xi(z)\big|_{\xi=0} = z/b$, we have $\operatorname{div}\{h(z)\, f(z)\} = -a \exp(-bz)$. So, we can check the Hypotheses 12.3.2 and 12.3.3. In fact, $\partial_z H^\xi(z)\big|_{\xi=0} = 1$, and

$$\int_{\mathbb{R}_0} \big|h(z)\big|\, \nu(dz) = \int_0^\infty \frac{a}{b}\, e^{-bz}\, dz = \frac{a}{b^2},$$

$$\int_{\mathbb{R}_0} \left|\frac{\operatorname{div}\big\{h(z)\, f(z)\big\}}{f(z)}\right| \nu(dz) = \int_0^{+\infty} a\, \exp(-bz)\, dz = \frac{a}{b}.$$

We proceed with Hypothesis 12.5.1. Let F be an open neighborhood in $\mathbb{R}$ including the origin with a radius smaller than b. The condition in Hypothesis 12.5.1 can be checked easily, because

$$\sup_{\xi\in F}\int_{|z|\le 1}\left|H^{\xi}(z)-z\right|^2\nu(dz)=\sup_{\xi\in F}\left(\frac{\xi}{b-\xi}\right)^2\int_0^1 a\,z\,\exp(-bz)\,dz\le C\,\frac{a}{b^2},$$

$$\sup_{\xi\in F}\int_{|z|>1}\left|H^{\xi}(z)\right|^p\nu(dz)=\sup_{\xi\in F}\left(\frac{b}{b-\xi}\right)^p\int_1^{\infty} a\,z^{p-1}\,e^{-bz}\,dz\le\frac{a}{b^p}\,\Gamma(p).$$

The condition in Hypothesis 12.5.3 is checked similarly.

Now in order to check the conditions on the Bismut method, we use the same deformation function which gives that $\ln k^{\xi}(z)=-zb\xi/(b-\xi)$. Then, all conditions in Hypothesis 12.6.6 can be verified easily.

Example 12.6.12 (*Variance gamma processes*) Let Z be *the variance gamma process* with the parameters $(a_p,\,a_n,\,b_p,\,b_n)$, where a_p, a_n, b_p and b_n are positive constants. The Lévy measure is

$$\nu(dz)=\left\{\frac{a_p}{z}\,\exp(-b_p z)\,\mathbf{1}_{(0,+\infty)}(z)+\frac{a_n}{|z|}\,\exp(b_n z)\,\mathbf{1}_{(-\infty,0)}(z)\right\}dz.$$

Similarly to Example 12.6.11, we can prove all the conditions in Sects. 12.2, 12.3, 12.5 and 12.6, by choosing

$$H^{\xi}(z)=\begin{cases} b_p\,z/(b_p-\xi) & \text{for } z>0,\\ b_n\,z/(b_n-\xi) & \text{for } z<0,\end{cases}$$

where $\xi\in\mathbb{R}$ such that $\xi<b_n\wedge b_p$.

Example 12.6.13 (*Tempered stable processes*) Let Z be *the tempered stable process* with the parameters $(a,\,b,\,\alpha)$, where a and b are positive constants, and $0<\alpha<2$, introduced in Sect. 11.4. The Lévy measure is

$$\nu(dz)=\frac{a}{z^{1+\alpha}}\,\exp(-bz)\,\mathbf{1}_{(0,+\infty)}(z)\,dz.$$

Let $\varphi\in C^1\big((0,+\infty)\,;\,\mathbb{R}\big)$ such that

$$|\varphi(z)|\le C\,(|z|^2\wedge 1),\quad |\varphi'(z)|\le C\,(|z|\wedge 1). \tag{12.23}$$

For example, the function $\varphi(z)=1-\exp(-z^2)$ satisfies the above conditions. Let $H^{\xi}(z)=z\,\exp\big(\xi\,\varphi(z)\big)$. Then, all the conditions listed in Sect. 12.3 can be checked easily. In fact, the Lévy measure $\nu(dz)$ satisfies the following conditions:

$$\int_{\mathbb{R}_0}\nu(dz)\ge\int_0^1\frac{a}{z^{1+\alpha}}\,e^{-bz}\,dz=\infty,$$

$$\int_{\mathbb{R}_0}\big(|z|^2\wedge 1\big)\,\nu(dz)=\int_0^1 a\,z^{1-\alpha}\,e^{-bz}\,dz+\int_1^{\infty} a\,z^{-1-\alpha}\,e^{-bz}\,dz\le\frac{a}{2-\alpha}+\frac{a}{b}\,e^{-b},$$

$$\int_{|z|>1} |z|^p \nu(dz) = \int_1^{+\infty} a\, z^{p-\alpha-1} \exp(-bz)\, dz \le \frac{a}{b^{p-\alpha}}\, \Gamma(p-\alpha)$$

for $p > \alpha$. Since $h(z) = \partial_\xi H^\xi(z)\big|_{\xi=0} = \varphi(z)\, z$, the conditions (12.23) on the function $\varphi(z)$ enable us to check Hypotheses 12.3.2 and 12.3.3 as follows: In fact, since $\partial_z H^\xi(z)\big|_{\xi=0} = 1$ and $\mathrm{div}\{h(z)\, f(z)\}/f(z) = z\,\varphi'(z) - \alpha\,\varphi(z) - b\,z\,\varphi(z)$, we have

$$\begin{aligned}
\int_{\mathbb{R}_0} |h(z)|\, \nu(dz) &= \int_0^\infty a\, \frac{\varphi(z)}{z^\alpha}\, e^{-bz}\, dz \\
&\le \int_0^\infty a\, \frac{C\,(z^2 \wedge 1)}{z^\alpha}\, e^{-bz}\, dz \le a\, C \left(\frac{1}{3-\alpha} + \frac{e^{-b}}{b}\right), \\
\int_0^\infty \left| \frac{\mathrm{div}\big\{h(z)\, f(z)\big\}}{f(z)} \right| \nu(dz) &= \int_{\mathbb{R}_0} \big|\varphi'(z)\, z - \alpha\, \varphi(z) - b\, \varphi(z)\, z\big|\, \frac{a}{z^{1+\alpha}}\, e^{-bz}\, dz \\
&\le C\, a\, (1+2\alpha) \left(\frac{1}{2-\alpha} + \frac{e^{-b}}{b}\right).
\end{aligned}$$

Let F be an open neighborhood in $\mathbb{R}$ around 0. Then, to verify Hypothesis 12.5.1, we have

$$\begin{aligned}
\sup_{\xi\in F} \int_{\mathbb{R}_0} \big|H^\xi(z) - z\big|^2 \nu(dz) &\le \sup_{\xi\in F} \big\{\exp(C|\xi|) + 1\big\}^2 \int_0^\infty a\, z^{1-\alpha}\, e^{-bz}\, dz \\
&\le \sup_{\xi\in F} \big\{\exp(C|\xi|) + 1\big\}^2\, \frac{a}{b^{2-\alpha}}\, \Gamma(2-\alpha), \\
\sup_{\xi\in F} \int_{|z|>1} \big|H^\xi(z)\big|^p\, \nu(dz) &\le \sup_{\xi\in F} e^{2C|\xi|} \int_0^\infty a\, z^{1-\alpha}\, e^{-bz}\, dz \\
&\le \sup_{\xi\in F} e^{2C|\xi|}\, \frac{a}{b^{2-\alpha}}\, \Gamma(2-\alpha).
\end{aligned}$$

Finally, the verification of Hypothesis 12.5.3 is trivial.

For the application of the Bismut method, note that

$$k^\xi(z) = e^{-\alpha\xi\varphi(z) - b(H^\xi(z) - z)} \big\{1 + \xi z \varphi'(z)\big\}, \quad \xi > 0.$$

We leave the rest of the verification as an exercise.

Example 12.6.14 (*CGMY processes, cf.* [17, 36]) Let Z be *the CGMY process*, that is, the Lévy process without a Gaussian component, with the parameters $\big(a_p,\, a_n\, b_p,\, b_n,\, \alpha_p,\, \alpha_n\big)$, where $a_p,\, a_n\, b_p,\, b_n$ are positive constants, and $0 < \alpha_p,\, \alpha_n < 2$. Its Lévy measure is

$$\nu(dz) = \left\{ \frac{a_p}{z^{1+\alpha_p}} \exp(-b_p z)\, \mathbf{1}_{(0,+\infty)}(z) + \frac{a_n}{|z|^{1+\alpha_n}} \exp(b_n z)\, \mathbf{1}_{(-\infty,0)}(z) \right\} dz.$$

Let $H^\xi(z) = z\ \exp\big(\xi\ \varphi(z)\big)$, where $\varphi \in C_1\big(\mathbb{R}_0\,;\ \mathbb{R}\big)$ such that $|\varphi(z)| \le C\,(|z|^2 \wedge 1)$ and $|\varphi'(z)| \le C\,(|z| \wedge 1)$, e.g. $\varphi(z) = 1 - \exp(-|z|^2)$ satisfies the conditions stated above. Then, we can prove all the conditions in Sects. 12.2, 12.3, 12.5 and 12.6, similarly to Example 12.6.13.

Exercise 12.6.15 Let Z be *the truncated stable process* with the parameters $(\kappa,\ \alpha)$, where κ is a positive constant, and $0 < \alpha < 2$. The Lévy measure is

$$\nu(dz) = \frac{\kappa}{z^{1+\alpha}}\,\mathbf{1}_{(0,1]}(z)\,dz.$$

Use again $H^\xi(z) = z\ \exp\big(\xi\ \varphi(z)\big)$ as a deformation. Find the appropriate conditions for φ in order to verify all conditions for the Norris method and Bismut method to be applicable.

Similarly, study the case of

$$\nu(dz) = \left\{\frac{\kappa_p}{z^{1+\alpha_p}}\,\mathbf{1}_{(0,1]}(z) + \frac{\kappa_n}{|z|^{1+\alpha_n}}\,\mathbf{1}_{[-1,0)}(z)\right\}dz,$$

where κ_p and κ_n be positive constants, and $0 < \alpha_p,\ \alpha_n < 2$.

For a hint, see Chap. 14.

Remark 12.6.16 Let Z be *the stable process* with the parameters $(\kappa,\ \alpha)$, where κ is the positive constants, and $0 < \alpha < 2$. The Lévy measure is

$$\nu(dz) = \frac{\kappa}{z^{1+\alpha}}\,\mathbf{1}_{(0,+\infty)}(z)\,dz.$$

Unfortunately, the Lévy measure $\nu(dz)$ does not satisfy Hypothesis 12.3.1 in Sect. 12.3, because the integral

$$\int_{|z|>1} |z|^p\ \nu(dz) = \int_1^{\infty} \kappa\, z^{p-\alpha-1}\,dz$$

is finite only for $0 < p < \alpha < 2$.

Similarly, the Lévy measure

$$\nu(dz) = \left\{\frac{\kappa_p}{z^{1+\alpha_p}}\,\mathbf{1}_{(0,+\infty)}(z) + \frac{\kappa_n}{|z|^{1+\alpha_n}}\,\mathbf{1}_{(-\infty,0)}(z)\right\}dz$$

of the pure-jump Lévy process Z does not satisfy the condition introduced in Sect. 12.3, where κ_p and κ_n are positive constants, and $0 < \alpha_p,\ \alpha_n < 2$. In fact, the integral

$$\int_{|z|>1} |z|^p\ \nu(dz) = \int_1^{+\infty} \kappa_p\, z^{p-\alpha_p-1}\,dz + \int_{-\infty}^{-1} \kappa_n\, |z|^{p-\alpha_n-1}\,dz$$

is finite only for $0 < p < \alpha_p \wedge \alpha_n < 2$.

The main problem is the existence of moments. Still, if one considers the fact that h appears in the definition of the dual operator δ and in DX as stated in Exercise 12.4.5, one sees that the Norris or Bismut method can still work with a proper choice of the function φ under the deformation function $H^{\xi}(z) = ze^{\xi\varphi(z)}$. Since this is a long subject, we do not discuss it here. This shows that the method has a versatility when choosing the function φ. In particular, it may also lead to local minimization arguments once the performance criteria is established.

Problem 12.6.17 Repeat the same steps taken through this chapter to build an IBP formula for an stable process. Some important points to consider are:

1. How to build the approximation Z^{ε} to a one-dimensional stable process Z so as to have all moment conditions for Z^{ε}, although this property will disappear in the limit.
2. Find a deformation function that may lead to an integration by parts formula. One may consider the deformation $H^{\xi}(z) = ze^{\xi\varphi(z)}$ with $\varphi(z) = O(|z|^p)$ as $|z|$ approaches zero with $p > \alpha$ and $\varphi(z) = O(|z|^p)$ as $|z|$ approaches infinity with $p < \alpha$.
3. Take the limit as in Theorem 12.4.19 and obtain the IBP formula. Find an appropriate hypothesis for the random variable Θ in order for the formula to be valid.
4. Define $(DZ)^{-1}$ and try to use a method like that in Sect. 11.5 in order to prove that its moments are finite.

Remark 12.6.18 One may consider a similar deformation on the time variables. We leave this for the interested student or researcher. See for example, [22].

Chapter 13
A Non-linear Example: The Boltzmann Equation

The purpose of this chapter is to show an application of the concepts of integration by parts introduced so far in an explicit example. With this motivation in mind, we have chosen as a model the Boltzmann equation. The Boltzmann equation is a non-linear equation related to the density of particles within the gas. We consider an imaginary situation where particles collide in a medium and we observe a section of it, say $\mathbb{R}^2$. These particles collide at a certain angle θ and velocity v, which generates a certain force within a gas. The Boltzmann equation main quantity of interest, $f_t(v)$, describes the density of particles traveling at speed v at time $t > 0$ supposing an initial distribution f_0. We assume that these densities are equal all over the section and therefore independent of the position within the section. The feature of interest to be proven here is that even if f_0 is a degenerate law in the sense that it may be concentrated at some points, the noise in the corresponding equation will imply that for any $t > 0$ f_t will be a well-defined function with some regularity. The presentation format in this section follows closely the one presented in [7] with some simplifications. This field of research is growing very quickly and therefore even at the present time the results presented here may be outdated. Still, our intention is to provide an explicit example of application of the method presented in the previous chapter. For the same reason, not all proofs are provided and some facts are relegated to exercises with some hints given in Chap. 14.

13.1 Two-Dimensional Homogeneous Boltzmann Equations

In this section, we provide the basics of the Boltzmann equation needed in order to proceed without going into too many (interesting) physical and mathematical details. In particular, our physical explanations below are just in order to give a very loose idea of the physical set-up without claiming exactness.

© Springer Nature Singapore Pte Ltd. 2019

A. Kohatsu-Higa and A. Takeuchi, *Jump SDEs and the Study of Their Densities*, Universitext, https://doi.org/10.1007/978-981-32-9741-8_13

Fix $T > 0$, $0 < \gamma \leq 1$ and $0 < \alpha < 1$ throughout the chapter. Define $b(\theta) := |\theta|^{-1-\alpha}$. In what follows, $f_t(v)$ denotes the density of particles with velocity v at time $t \geq 0$. Also let $\Pi := [-\pi/2, \pi/2]$ and $A(\theta) := (R_\theta - I)/2$, where I is the identity matrix and R_θ is the two-dimensional rotation (particle collision) matrix of angle θ so that:

$$R_\theta = \begin{pmatrix} \cos\theta & -\sin\theta \\ \sin\theta & \cos\theta \end{pmatrix}, \quad A(\theta) = \frac{1}{2}\begin{pmatrix} \cos\theta - 1 & -\sin\theta \\ \sin\theta & \cos\theta - 1 \end{pmatrix}.$$

Consider the two-dimensional spatially homogeneous gas modeled by the Boltzmann equation:

$$\frac{\partial f_t(v)}{\partial t} = \int_\Pi \int_{\mathbb{R}^2} \left\{ f_t(v')\, f_t(v'_*) - f_t(v)\, f_t(v_*) \right\} B(|v - v_*|, \theta)\, dv_*\, d\theta, \qquad (13.1)$$

where $B(|v - v_*|, \theta) = |v - v_*|^\gamma\, b(\theta)$ describes the cross-section physical structure in terms of the angle, θ and speed change, $|v - v_*|$, and

$$v' = v + A(\theta)\,(v - v_*), \quad v'_* = v_* - A(\theta)\,(v - v_*).$$

The family of finite measures $\left\{ f_t \,;\, 0 \leq t \leq T \right\}$ is called a weak solution of (13.1), if it satisfies

$$\frac{d}{dt}\int_{\mathbb{R}^2} \psi(v)\, f_t(dv) = \int_\Pi \int_{\mathbb{R}^2} \int_{\mathbb{R}^2} \left\{ \psi\big(v + A(\theta)\,(v - v_*)\big) - \psi(v) \right\} \\ \times B(|v - v_*|, \theta)\, f_t(dv)\, f_t(dv_*)\, d\theta$$

for any $t \in (0, T]$ and any globally Lipschitz continuous function $\psi : \mathbb{R}^2 \to \mathbb{R}$.

Loosely speaking, Eq. (13.1) describes the time evolution of a gas due to the collision of particles which is given by the integral operator on the right side of (13.1). In this right side, the difference $f_t(v')\, f_t(v'_*) - f_t(v)\, f_t(v_*)$ describes essentially the change of density due to the result of the collision of particles. Some of them are lost as they gain other speeds and others are added as they achieve a certain speed. The function B describes the interaction between the particles, such as the strength of the collision. For more details and the exact physical meaning of the various terms, we refer the reader to [1, 57].

From the mathematical viewpoint, notice that the equation described in (13.1) is non-linear in f_t. In fact, it is linear in this variable on the left-hand side of the equation and quadratic on the right-hand side of the equation. This issue provokes additional problems that did not appear in the case of the approximation of stochastic equations treated in Sect. 12.5 which may be considered as a linear equation as the coefficients of the equation were Lipschitz functions.

Moreover, if we assume the principles of conservation of mass, momentum and kinetic energy hold, then the following restrictions are imposed onto the solution of (13.1):

$$\int_{\mathbb{R}^2} f_t(dv) = \int_{\mathbb{R}^2} f_0(dv),$$
$$\int_{\mathbb{R}^2} v\, f_t(dv) = \int_{\mathbb{R}^2} v\, f_0(dv),$$
$$\int_{\mathbb{R}^2} |v|^2\, f_t(dv) = \int_{\mathbb{R}^2} |v|^2\, f_0(dv).$$

Without loss of generality, we shall suppose that $\int_{\mathbb{R}^2} f_0(dv) = 1$ and $\int_{\mathbb{R}^2} v\, f_0(dv) = 0$. Furthermore, we will suppose that f_0 is not a Dirac mass. In fact, we will suppose the slightly stronger condition $\int_{\mathbb{R}^2} |v|^2\, f_0(dv) > 0$. In fact, the conservation of momentum and kinetic energy is only heuristic in the above understanding of weak solutions. They are obtained by choosing the appropriate function ψ.

Exercise 13.1.1 Prove that in the case that there is an $x \in \mathbb{R}^2$ such that $f_0(\{x\}) = 1$ then the solution to (13.1) is the trivial solution. That is, one needs at least two different speeds in the gas in order to create a non-trivial solution.

The goal of the line of research we intend to introduce here is to prove that even in the case that f_0 does not have a smooth density, the collision strength characterized by B implies the existence of a somewhat smooth density for f_t, for any $t > 0, \alpha \in (0, 1)$ and $\gamma \in (0, 1]$. A number of these cases have an associated physical interpretation. We refer the reader again to [57] for further explanation and we just remark that there are the following cases in $\mathbb{R}^3$: for $\gamma = (s-5)/(s-1)$ and $\alpha = 2/(s-1)$,

- $s > 5$ is called the hard potential case,
- $s = 5$ is called the Maxwellian potential,
- $7/3 < s < 5$ is called the moderate soft potential case,
- $2 < s < 7/3$ is called the very soft potential case,
- $s = 2$ is called the Coulomb potential,

where s is the parameter that characterizes the force exerted between particles, that is, two particles separated by a distance r in $\mathbb{R}^3$ exert each other a force of the order r^{-s} (which is expressed by B in (13.1)).

As it seems to be a matter of opinion, technique, taste and goal as to which case is the hardest to treat mathematically, we refrain from commenting on this matter and proceed with the main matter of discussion.

Recall that according to our notation, vectors x are always column vectors and with matrix norm $\|A\| := \sup_{|x|=1} |Ax|$. For a sequence of fixed real numbers $\{x_i\}_{i\in\mathbb{N}}$, we write $x_{1\to k} = (x_1, \ldots, x_k) \in \mathbb{R}^k$ and $x^{(j)}_{1\to k} = (x_{1\to j-1}, a, x_{j+1\to k})\big|_{a=x_j}$ with a certain abuse of notation. In particular, this notation allows us to understand the vector $x_{1\to k}$ as a function of its j-th variable when all other components are left fixed. Also, we use the following convention: $\sum_{k=1}^{0} x_k := 0$.

In this chapter many constants will appear. They are generically positive and denoted by C. Sometimes we will write explicitly the dependence of these constants on parameters. As usual, the exact value may change from one line to the next.

In order to reduce long expressions within the proofs we will shorten the length of the notation. In particular, we will denote the norm for a random matrix X in $L^p(\Omega)$ indistinctly as $\|X\|_p \equiv \|X\|_{L^p(\Omega)}$.

As we will use various different probabilistic representations for the approximation of (13.1), each one of them needs an appropriate filtration. We leave their proper definition as a task to the reader.

13.2 Stochastic Representation

Let V_0 be an $\mathbb{R}^2$-valued random variable with law $f_0(dv)$. Consider the $\mathbb{R}^2$-valued process $V = \{V_t\,;\ 0 \le t \le T\}$ with marginal laws given by $\{f_t;\ 0 \le t \le T\}$ determined by the stochastic differential equation:

$$V_t = V_0 + \int_0^t \int_\Pi \int_{\mathbb{R}^2} \int_0^\infty A(\theta)\left(V_{s-} - v\right) \mathbf{1}_{[u,\infty)}\left(|V_{s-} - v|^\gamma\right) d\mathcal{N}, \tag{13.2}$$

where $d\mathcal{N} = \mathcal{N}(ds, d\theta, dv, du)$ is the Poisson random measure over $[0, T] \times \Pi \times \mathbb{R}^2 \times [0, \infty)$ with the intensity measure

$$d\widehat{\mathcal{N}} = \widehat{\mathcal{N}}(ds, d\theta, dv, du) := ds\, b(\theta)d\theta\ f_s(dv)\, du,$$

independent of V_0. This means that the space of jumps is a subset of $\mathbb{R}^4$.

Exercise 13.2.1 Show that the total mass of the compensator is infinite. That is, $\widehat{\mathcal{N}}([0, T] \times A \times \mathbb{R}^2 \times B) = \infty$ if either $A \subset \Pi$ includes a neighborhood of 0 or $B \subset [0, \infty)$ is a set of infinite measure. This implies that the construction of the above Poisson random measure has to be understood as the limit of Poisson random measures as done in e.g. Corollary 4.1.6 and Exercise 5.4.1 or 7.1.1.

Exercise 13.2.2 Prove that if V is a solution for (13.2) then $\mathbb{E}[V_t] = \mathbb{E}[V_0] = 0$ and $\mathbb{E}[|V_t|^2] = \mathbb{E}[|V_0|^2] > 0$ for all $t > 0$.

For a hint, see Chap. 14.

It is known that there exists a unique strong solution to the Eq. (13.2) such that the probability law of V_t is $f_t(dv)$ (cf. [7]-Proposition 2.1-(i), [55]-Theorems 4.1 and 4.2). In particular, we note that this is an infinite activity driving process, because $\int_0^\varepsilon b(\theta)d\theta = \infty$ for any $\varepsilon > 0$. The reason why the stochastic integral in (13.2) is well defined in the variable θ is due to the behavior of A. In fact, $A_{ij}(\theta)\, b(\theta) = O(\theta^{1-\alpha})\delta_{ij} + O(\theta^{-\alpha})$ as $\theta \to 0$ for $i,\ j \in \{1,\ 2\}$, $\alpha \in (0, 1)$. As for the variable u, in order to understand why the stochastic integral is well defined, one has to prove that the moments of the variable $|V_{s-} - v|^\gamma$ are bounded. In fact, even its exponential moments are bounded as stated in Theorem 13.2.3.

With all these preparatory results one can tackle the study of the existence of the density for the measures f_t for $t > 0$.

The sources of problems from the point of view of the Malliavin calculus for jump processes are:

1. The coefficient $A(\theta)\,(x-y)\,\mathbf{1}_{[u,\infty)}\big(|x-y|^{\gamma}\big)$ is a non-smooth function of $x \in \mathbb{R}^2$, $y \in \mathbb{R}^2$ and $u \in [0,\infty)$.
2. The compensator of the driving Poisson measure depends on f_t which is already the law associated with the solution process V_t.
3. $b(\theta)$ and its derivatives explode as $\theta \to 0$ at a rate of at least $|\theta|^{-(1+\alpha)}$.

The objective here is therefore to deal with each of these problems. In short, the solution for the first problem is to change the stochastic representation so as to obtain a smoother version, even if this demands the use of an approximation and Theorem 9.1.10. For the second, we will perform a conditional form of the Malliavin calculus where the component related to f_t in the compensator is conditioned, and therefore this will not be the noise which will generate the integration by parts formula. For the third problem, we will use the Norris approach with the corresponding analysis of what is the proper direction of differentiation (i.e. the choice of H^{ξ}) to be able to obtain the best result possible within this methodology.

The fact that we are using an approximation instead of the solution V_t itself will be later solved by using an alternative technique where, on the one hand, one measures the regularity of the approximating law and on the other one measures the rate of convergence of the approximation to the solution of (13.2). Finally, we will apply Theorem 9.1.10. This will lead to an estimate of the characteristic function of V_t which will finally lead to the main result. These techniques of integration by parts with respect to the jump size were briefly sketched in Chap. 10, Exercise 10.4.5.

We shall start with a moment estimate.

Theorem 13.2.3 (cf. [7]-Theorem 1.2, [25]-Corollary 2.3) *Suppose that there exists $\gamma < \delta < 2$ with*

$$\int_{\mathbb{R}^2} \exp\left(|v|^{\delta}\right) f_0(dv) < \infty.$$

Then, there exists a unique weak solution for (13.1) *such that for all $0 < \kappa < \delta$, it holds that*

$$\sup_{0\le t\le T} \int_{\mathbb{R}^2} \exp\left(|v|^{\kappa}\right) f_t(dv) < \infty.$$

Note that the above conclusion has its equivalent in terms of V as follows:

$$\sup_{0\le t\le T} \mathbb{E}\Big[\exp\left(|V_t|^{\kappa}\right)\Big] < \infty.$$

From now on, we assume that the hypothesis of the above theorem is satisfied, and δ is therefore fixed through the rest of the chapter, which depends only on γ.

Exercise 13.2.4 In this exercise we will try to discuss a much simpler model which may help us understand some part of the structure of (13.2). Consider

$$Z_t := Z_0 + \int_0^t \int_{\Pi} \int_0^{\infty} \theta\, \mathbf{1}_{[u,\infty)}(X) \mathcal{N}(ds, d\theta, du).$$

Here X is a positive random variable with finite non-zero expectation independent of the Poisson random measure $\mathcal{N}$ which has as a compensator $\widehat{\mathcal{N}}(ds, d\theta, du) := dsb(\theta)d\theta du$. Compute the characteristic function of Z_t and discuss the properties of the law of Z_t using the results in Chap. 9.

Furthermore, discuss a similar problem when one considers instead

$$Z'_t := Z_0 + \int_0^t \int_\Pi \int_0^\infty \theta \, \mathbf{1}_{[u,\infty)}(\theta) \mathcal{N}(ds, d\theta, du).$$

Also notice the similarities and differences of (13.2) with the setting in Exercise 4.2.4.

For a hint, see Chap. 14.

13.3 Approximation to the Process V

In this section, we describe the approximation procedure that is used in order to carry out the integration by parts (IBP). For this, we will use the parameter of approximation $\varepsilon > 0$. This parameter will be used for the approximation of the domain of the stable law as was done in Chap. 5. It will also be used to control the range of possible values for u. Now we proceed to define the functions used for the approximation.

Define

$$\phi_\varepsilon(u) := \int_{\mathbb{R}} \{(z \vee 2\varepsilon) \wedge \Gamma_\varepsilon\} \frac{\chi\big((u-z)/\varepsilon\big)}{\varepsilon} dz,$$

where $\chi : \mathbb{R} \to [0, \infty)$ is a smooth and even function supported in $(-1, 1)$ with $\int_{\mathbb{R}} \chi(z)\, dz = 1$,[1] and $\Gamma_\varepsilon = \big(\log(\varepsilon^{-1})\big)^{\eta_0}$ with $1 < \eta_0 < 1/\gamma$. We remark that $\Gamma_\varepsilon \to \infty$ as $\varepsilon \to 0$.

Exercise 13.3.1 Prove that the first derivative of the function ϕ_ε is uniformly bounded for $\varepsilon \in (0, e^{-4})$. Furthermore prove that ϕ_ε is an increasing function which satisfies the following properties:

$$\phi_\varepsilon(u) = \begin{cases} 2\varepsilon & (u \le \varepsilon), \\ u & (3\varepsilon \le u \le \Gamma_\varepsilon - 1), \\ \Gamma_\varepsilon & (u \ge \Gamma_\varepsilon + 1). \end{cases}$$

Moreover, check that the inequality $3\varepsilon < \Gamma_\varepsilon - 1$ is satisfied for $\varepsilon \in (0, e^{-4})$.

The next approximation function χ_ε is used for the variable θ. It is defined as follows: For $0 < \varepsilon < 1/2$, let $\chi_\varepsilon \in C_b^\infty(\Pi\,;\ [0, 1])$ be a function which satisfies that $\chi_\varepsilon(\theta) = 1$ for $\varepsilon^a \pi \le |\theta| \le \pi/2$, $a > 0$ and that $\chi_\varepsilon(\theta) = 0$ for $|\theta| \le \varepsilon^a\pi/2$. We also define $\Pi_\varepsilon := [-\varepsilon^a\pi/2,\ \varepsilon^a\pi/2]$, and $A_\varepsilon(\theta) := A(\theta)\,\chi_\varepsilon(\theta)$.

[1] One such function is the renormalization of the function $\exp\big\{-1/(x^2-1)\big\}$.

Exercise 13.3.2 Using a similar construction as in the statement before Exercise 13.3.1, we will use from now on the explicit function

$$\chi_\varepsilon(\theta) = C\varepsilon^{-a} \int_{\varepsilon^a \pi/2}^{|\theta|} \exp\left(-\varepsilon^{ap}|x - \varepsilon^a\pi/2|^{-p}\right) \exp\left(-\varepsilon^{ap}|x - \varepsilon^a\pi|^{-p}\right) dx$$

for $|\theta| \in (\varepsilon^a\pi/2, \varepsilon^a\pi)$ and $p \geq 1$. Find the value of C which depends on p so that the properties stated above are satisfied. Prove that $|\chi_\varepsilon^{(k)}(\theta)| \leq C\varepsilon^{-ak}$. Therefore choosing the value of p the rate of explosion of these derivatives may be somewhat controlled. To standardize the calculation we will assume from now on that $p = 1$.

The first approximation is defined as follows:

$$V_t^\varepsilon = V_0 + \int_0^t \int_\Pi \int_{\mathbb{R}^2} \int_0^{2\Gamma_\varepsilon^\gamma} A_\varepsilon(\theta)\,(V_{s-}^\varepsilon - v)\; \mathbf{1}_{[u,\infty)}\left(\phi_\varepsilon(|V_{s-}^\varepsilon - v|)^\gamma\right) d\mathcal{N}.$$

Note that as $0 \leq \phi_\varepsilon(x)^\gamma \leq \Gamma_\varepsilon^\gamma$ then $A_\varepsilon(\theta)\,(x - y)\; \mathbf{1}_{[u,\infty)}\left(\phi_\varepsilon(|x - y|)^\gamma\right) = 0$ if $u > \Gamma_\varepsilon^\gamma$ or $\theta \in \Pi_\varepsilon$. The choice of $2\Gamma_\varepsilon^\gamma$ may seem arbitrary here but, as we will see, it is important to choose a value strictly larger than $\Gamma_\varepsilon^\gamma$. For more details on this see the proof of Lemma 13.6.17 where this fact is used in the proof.

For fixed $\varepsilon > 0$, note that the above stochastic equation has a structure like that of a compound Poisson process (but it is not a compound Poisson process!).

Exercise 13.3.3 Consider the Poisson random measure defined as

$$\mathcal{N}^\varepsilon := \mathbf{1}_{\bar{\Pi}_\varepsilon}(|\theta|)\mathbf{1}_{[0,2\Gamma_\varepsilon^\gamma]}(u)\mathcal{N},$$

where $\bar{\Pi}_\varepsilon := \Pi \cap \Pi_\varepsilon^c = [-\pi/2, -\varepsilon^a\pi/2) \cup (\varepsilon^a\pi/2, \pi/2]$. Prove that this Poisson random measure corresponds to a time non-homogeneous Poisson process (see Sect. 5.5) with parameter $\lambda_\varepsilon(t) := \frac{2^{2+\alpha}}{\alpha\pi^\alpha}(\varepsilon^{-a\alpha} - 1)\,\Gamma_\varepsilon^\gamma \int_0^t f_s(\mathbb{R}^2)ds$. Clearly, this parameter depends on the law of V and it explodes as $\varepsilon \downarrow 0$. Furthermore prove that

$$\mathbb{P}(V_t^\varepsilon = V_0) \geq \sum_{j=0}^{\infty} \frac{e^{-\lambda_\varepsilon(t)}}{j!} \left\{\mathbb{E}\left[\int_0^t \lambda'_\varepsilon(s)\left(1 - \frac{\phi_\varepsilon(|V_s^\varepsilon - v|)^\gamma}{2\Gamma_\varepsilon^\gamma}\right) ds\right]\right\}^j.$$

Note that this result implies that the law of V_t^ε has at least a point mass, although the probability of this point mass will eventually disappear as $\varepsilon \downarrow 0$.[2] On the other hand, the analysis of V^ε is easier than its limit process.

The main goal of the calculations to follow is to prove that on a large part of the sample space, the law of V^ε admits a density. Clearly, this may not be enough as there are trivial counterexamples of sequences of random variables which admit a density but whose limit does not have a density (recall Exercises 9.1.5 and 9.1.11).

---

[2]A harder exercise/problem is to give an upper bound for $\mathbb{P}(V_t^\varepsilon = V_0)$. To answer this question, it requires all calculations to follow.

As stated in Theorem 9.1.10, this problem can be circumvented if the rate of convergence of characteristic functions of the approximating sequence can be appropriately controlled. Before discussing this, we need to have an estimate on the moments of the approximating process.

Using the same non-trivial arguments as in Proposition 2.1 in [7] one can prove the following properties for the above approximation:

Proposition 13.3.4 *Assume the same conditions as in Theorem 13.2.3. For ε sufficiently small and with κ as in Theorem 13.2.3, we have*

$$\mathbb{E}\left[\sup_{t\in[0,T]}\left(e^{|V_t^\varepsilon|^\kappa}+e^{|V_t|^\kappa}\right)\right]<\infty.$$

Furthermore for any $\tilde{\beta}\in(\alpha,1]$ and $\tilde{r}>0$ such that $\tilde{\beta}-\alpha>\tilde{r}$,

$$\sup_{t\in[0,T]}\mathbb{E}\left[\left|V_t-V_t^\varepsilon\right|^{\tilde{\beta}}\right]\leq C\varepsilon^{a(\tilde{\beta}-\alpha-\tilde{r})}.$$

Here a is the parameter used in the definition of Π_ε.

Therefore the above result shows that the approximation process V^ε converges to the process V in the above sense. In particular, the marginal laws converge. This is the first step towards the use of Theorem 9.1.10.

Exercise 13.3.5 Suppose that $\eta_0\gamma<1$. Prove that for any $\eta>1$ there exists $k\in(0,e^{-1})$ such that if $\varepsilon\in(0,k)$ then $\exp(C\Gamma_\varepsilon^\gamma)\leq C_{\eta_0,\eta,\gamma}\,\varepsilon^{-\eta}$. Similarly, prove that for any $\eta>1$ and $\eta_0\,\xi>1$ then $\exp(-C\Gamma_\varepsilon^\xi)\leq\varepsilon^\eta$ for $\varepsilon\leq e^{-(\frac{\eta}{C})^{\frac{1}{\eta_0\xi-1}}}$. Furthermore, use Proposition 13.3.4 to prove that for ε small enough and $p>0$, one has that for $0\leq a<b<T$,

$$\int_a^b\int\mathbb{E}[|V_s^\varepsilon-v|^p]f_s(dv)ds>0.$$

For a hint, see Chap. 14.

Exercise 13.3.6 Prove the following inequality for the difference between the characteristic functions of V_t^ε and V_t:

$$\left|\varphi_{V_t}(\zeta)-\varphi_{V_t^\varepsilon}(\zeta)\right|\leq\mathbb{E}\left[2\sin^2\left(\zeta\left(\frac{V_t^\varepsilon-V_t}{2}\right)\right)+\left|\sin\left(\zeta\left(V_t^\varepsilon-V_t\right)\right)\right|\right].$$

Use the inequality $|\sin(x)|\leq|x|^a\wedge 1$ for any $a\in[0,1]$ in order to prove the first hypothesis in Theorem 9.1.10.

When using Theorem 9.1.10, one may be worried about the fact that the random variable V_t^ε has a discrete mass. But, as stated in Exercise 13.3.3 this is a problem of no significant importance as the size of the probability mass is small. Explaining this is the purpose of the following exercise.

Exercise 13.3.7 Let X^n be a sequence of random variables which satisfies the properties stated in Theorem 9.1.10 except that the asymptotic regularity property is replaced by: For $u > 0$ and $k > 0$,

$$|\varphi_{X_n}(\zeta) - \varphi_{X_n}(0)| \leq \frac{Cn^u}{(1 + |\zeta|)^k},$$

where $\mathbb{P}(X_n = 0) = Cn^{-a}$ with $a > 0$. Prove a conclusion similar to Theorem 9.1.10 which will now depend on the value of a.

Now, in order to control the term $f_t(dv)$ we will use the following change of probabilistic representation which already appeared in Exercise 10.4.4 and Example 10.4.5. That is, from the Skorokhod representation theorem (cf. [7]-Sect. 3, [18]), we can find a measurable mapping $v_t : [0, 1] \to \mathbb{R}^2$ such that, for all $\psi_1 : \mathbb{R}^2 \to [0, \infty)$,

$$\int_{\mathbb{R}^2} \psi_1(v)\, f_t(dv) = \int_0^1 \psi_1\big(v_t(\rho)\big)\, d\rho. \tag{13.3}$$

This change of variable allows the reduction of the approximation process into another one, where the jump structure on the non-linear term, $f_t(v)$, is easier to deal with. But now the representation is not an almost sure equivalent representation of the approximation process V^ε but a weak representation in the sense that the laws of the solution processes will remain the same, as will be proven in Proposition 13.3.9.

Now we define the approximation process $\bar{V} = \{\bar{V}_t(= \bar{V}_t^\varepsilon);\ 0 \leq t \leq T\}$ to be the solution to the stochastic differential equation:

$$\begin{aligned}\bar{V}_t &= V_0 + \int_0^t \int_\Pi \int_0^1 \int_0^\infty A_\varepsilon(\theta)\,(\bar{V}_{s-} - v_s(\rho))\ \mathbf{1}_{[u,\infty)}\big(\phi_\varepsilon(|\bar{V}_{s-} - v_s(\rho)|)^\gamma\big)\, d\mathcal{N}^\varepsilon \qquad (13.4)\\ &= V_0 + \int_0^t \int_{\bar{\Pi}_\varepsilon} \int_0^1 \int_0^{2\Gamma_\varepsilon^\gamma} A_\varepsilon(\theta)\,(\bar{V}_{s-} - v_s(\rho))\ \mathbf{1}_{[u,\infty)}\big(\phi_\varepsilon(|\bar{V}_{s-} - v_s(\rho)|)^\gamma\big) d\mathcal{N}^\varepsilon.\end{aligned}$$

Here, $\mathcal{N}^\varepsilon(ds, d\theta, d\rho, du)$ denotes a Poisson random measure with compensator given by

$$\widehat{\mathcal{N}^\varepsilon}(ds, d\theta, d\rho, du) = ds b(\theta)\ \mathbf{1}_{\bar{\Pi}_\varepsilon}(\theta) d\theta d\rho\ \mathbf{1}_{[0, 2\Gamma_\varepsilon^\gamma]}(u) du.$$

Note that the interval $\bar{\Pi}_\varepsilon := \Pi \cap \Pi_\varepsilon^c = [-\pi/2, -\varepsilon^a\pi/2) \cup (\varepsilon^a\pi/2, \pi/2]$ does not contain a neighborhood of zero.

Now the above Poisson random measure has a compensator with finite total mass given for fixed $\varepsilon \in (0, e^{-4})$ (in order to assure that $\Gamma_\varepsilon \geq 1$) by

$$\lambda_\varepsilon T := \widehat{\mathcal{N}^\varepsilon}([0, T] \times \bar{\Pi}_\varepsilon \times [0, 1] \times [0, 2\Gamma_\varepsilon^\gamma]) < \infty.$$

Exercise 13.3.8 Prove that $\lambda_\varepsilon = \frac{2^{2+\alpha}}{\alpha\pi^\alpha}(\varepsilon^{-a\alpha} - 1)\Gamma_\varepsilon^\gamma$.

We also perform the same change of space on the Poisson random measure $\mathcal{N}$ which will now have jumps supported on $\Pi \times [0, 1] \times \mathbb{R}_+$.

In fact, the Eq. (13.4) can be solved using the explicit techniques explained in Chap. 3. That is, the solution can be explicitly written as

$$\bar{V}_t = V_0 + \sum_{j=1}^{N_t^\varepsilon} A_\varepsilon(\theta_j)\,(\bar{V}_{T_{j-1}} - v_{T_j}(\rho_j))\ \mathbf{1}_{[U_j,\infty)}\left(\phi_\varepsilon(|\bar{V}_{T_{j-1}} - v_{T_j}(\rho_j)|)^\gamma\right). \quad (13.5)$$

Here N^ε is the corresponding Poisson process which counts the jumps with a rate parameter $\lambda_\varepsilon T$ with corresponding jump times $T_j \equiv T_j^\varepsilon$, $j \in \mathbb{N}$ such that $T_{j+1} - T_j$ is a sequence of i.i.d exponentially distributed r.v.s with parameter λ_ε. Similarly, $(\theta_j, \rho_j, U_j) \equiv (\theta_j^\varepsilon, \rho_j^\varepsilon, U_j^\varepsilon)$ is an i.i.d. sequence of random vectors in $\bar{\Pi}_\varepsilon \times [0, 1] \times [0, 2\Gamma_\varepsilon^\gamma]$ which are independent between themselves and the Poisson process and which have as a density the renormalization of the function $b(\theta)\,\mathbf{1}_{\bar{\Pi}_\varepsilon\times[0,1]\times[0,2\Gamma_\varepsilon^\gamma]}(\theta, \rho, u)$.

One can now prove the following result which we leave as an exercise.

Proposition 13.3.9 *The law of the process* $\{V_t^\varepsilon; t \in [0, T]\}$ *is the same as the law of the process* $\{\bar{V}_t; t \in [0, T]\}$.

Our efforts now will be concentrated on obtaining an integration by parts formula for $\bar{V}$. We remark here that the Eq. (13.5) does not have the same structure as a random equation driven by a compound Poisson process due to the term v_{T_j} in Eq. (13.5).

Denote by $\{T_k(= T_k^\varepsilon)\,;\, k \in \mathbb{Z}_+\}$ the sequence of $(\mathscr{F}_t)$-stopping times which are defined by

$$T_0 := 0,$$
$$T_k := \inf\left\{t > T_{k-1}\,;\ \int_{T_{k-1}}^t \int_{\bar{\Pi}_\varepsilon} \int_0^1 \int_0^{2\Gamma_\varepsilon^\gamma} d\mathcal{N} \neq 0\right\} \wedge T \quad (k \in \mathbb{N}).$$

Let $N^\varepsilon = \{N_t^\varepsilon\,;\, 0 \le t \le T\}$ be a Poisson process with the rate λ_ε and $\{(\theta_k, \rho_k, U_k)\ (= (\theta_k^\varepsilon, \rho_k^\varepsilon, U_k^\varepsilon))\,;\, k \in \mathbb{N}\}$ a family of $\bar{\Pi}_\varepsilon \times [0, 1] \times [0, 2\Gamma_\varepsilon^\gamma]$-valued, independent and identically distributed random variables with the common law

$$\mathbb{P}\big[\theta_k \in d\theta,\ \rho_k \in d\rho,\ U_k \in du\big] = b(\theta)\,\mathbf{1}_{\bar{\Pi}_\varepsilon\times[0,1]\times[0,2\Gamma_\varepsilon^\gamma]}(\theta, \rho, u)d\theta\, d\rho\, du\big/\lambda_\varepsilon, \quad (13.6)$$

which are also independent of the process N^ε. Then, we have

$$\begin{aligned}\bar{V}_t &= V_0 + \int_0^t \int_{\bar{\Pi}_\varepsilon} \int_0^1 \int_0^{2\Gamma_\varepsilon^\gamma} A_\varepsilon(\theta)\,(\bar{V}_{s-} - v_s(\rho))\ \mathbf{1}_{[u,\infty)}\big(\phi_\varepsilon(|\bar{V}_{s-} - v_s(\rho)|)^\gamma\big)\, d\mathcal{N}\\ &= V_0 + \sum_{k=1}^{N_t^\varepsilon} A_\varepsilon(\theta_k)\,(\bar{V}_{T_{k-1}} - v_{T_k}(\rho_k))\ \mathbf{1}_{[U_k,\infty)}\big(\phi_\varepsilon(|\bar{V}_{T_{k-1}} - v_{T_k}(\rho_k)|)^\gamma\big). \quad (13.7)\end{aligned}$$

Note that in the above expression, we have rewritten the stochastic equation in the most simple notation associated with equations driven by compound Poisson processes. Then, for any Borel measurable function $F : \mathbb{R}^2 \to \mathbb{R}$, we see that for any $N \in \mathbb{N}$

$$
\begin{aligned}
&\mathbb{E}\left[F(\bar{V}_{T_N})\middle|\bar{V}_{T_{N-1}}, T_{N-1}, T_N\right] \\
&= \mathbb{E}\Big[F\Big(\bar{V}_{T_{N-1}} + A_\varepsilon(\theta_N)\left(\bar{V}_{T_{N-1}} - v_{T_N}(\rho_N)\right) \\
&\qquad \times \mathbf{1}_{[U_N,+\infty)}\left(\phi_\varepsilon\left(|\bar{V}_{T_{N-1}} - v_{T_N}(\rho_N)|\right)^\gamma\right)\Big)\Big|\bar{V}_{T_{N-1}}, T_{N-1}, T_N\Big] \\
&= \int_{\bar{\Pi}_\varepsilon}\int_0^1\int_0^{+\infty} F\Big(\bar{V}_{T_{N-1}} + A_\varepsilon(\theta)\left(\bar{V}_{T_{N-1}} - v_{T_N}(\rho)\right) \mathbf{1}_{[u,+\infty)}\left(\phi_\varepsilon\left(|\bar{V}_{T_{N-1}} - v_{T_N}(\rho)|\right)^\gamma\right)\Big) \\
&\qquad \times \left\{\mathbf{1}_{[0,\phi_\varepsilon(|\bar{V}_{T_{N-1}} - v_{T_N}(\rho)|)^\gamma]}(u) + \mathbf{1}_{[\phi_\varepsilon(|\bar{V}_{T_{N-1}} - v_{T_N}(\rho)|)^\gamma, 2\Gamma_\varepsilon^\gamma]}(u)\right\} \frac{b(\theta)}{\lambda_\varepsilon}\, d\theta\, d\rho\, du \\
&= \int_{\bar{\Pi}_\varepsilon}\int_0^1 F\Big(\bar{V}_{T_{N-1}} + A_\varepsilon(\theta)\left(\bar{V}_{T_{N-1}} - v_{T_N}(\rho)\right)\Big) \\
&\qquad \times \phi_\varepsilon\left(|\bar{V}_{T_{N-1}} - v_{T_N}(\rho)|\right)^\gamma b(\theta)\, \frac{d\theta\, d\rho}{\lambda_\varepsilon} \\
&\quad + F(\bar{V}_{T_{N-1}})\left\{1 - \frac{1}{2\Gamma_\varepsilon^\gamma}\int_0^1 \phi_\varepsilon\left(|\bar{V}_{T_{N-1}} - v_{T_N}(\rho)|\right)^\gamma d\rho\right\} \\
&= \int_\Pi\int_0^1 d\theta\, d\rho\, F\Big(\bar{V}_{T_{N-1}} + A_\varepsilon(\theta)\left(\bar{V}_{T_{N-1}} - v_{T_N}(\rho)\right)\Big) \\
&\qquad \times \Bigg\{\frac{\phi_\varepsilon\left(|\bar{V}_{T_{N-1}} - v_{T_N}(\rho)|\right)^\gamma b(\theta)}{\lambda_\varepsilon} \mathbf{1}_{\bar{\Pi}_\varepsilon}(\theta) \\
&\qquad + \left(1 - \frac{1}{2\Gamma_\varepsilon^\gamma}\int_0^1 \phi_\varepsilon\left(|\bar{V}_{T_{N-1}} - v_{T_N}(\rho)|\right)^\gamma d\rho\right) \varepsilon^{a\alpha}\, \tilde{\chi}\left(\frac{\theta}{\varepsilon^a}\right) b(\theta)\, \mathbf{1}_{\Pi_\varepsilon}(\theta)\Bigg\} \\
&=: \int_\Pi\int_0^1 F\Big(\bar{V}_{T_{N-1}} + A_\varepsilon(\theta)\left(\bar{V}_{T_{N-1}} - v_{T_N}(\rho)\right)\Big)\, q_\varepsilon(T_N, \bar{V}_{T_{N-1}}, \theta, \rho)\, d\theta\, d\rho
\end{aligned}
$$

by using, in that order, (13.7) in the first equality, (13.6) in the second equality, the explicit integral on u and the definition of λ_ε in the third equality, and $\chi_\varepsilon(\theta) = 0$ for $\theta \in \Pi_\varepsilon$ in the fourth equality, where $\tilde{\chi} \in C^\infty(\mathbb{R}\,;\, [0,1])$ is a function such that $\mathrm{Supp}[\tilde{\chi}] \subset \Pi$ and

$$\int_\Pi \tilde{\chi}(\theta)\, b(\theta) d\theta = 1. \tag{13.8}$$

In the last equality stated above, we have defined for $t \in [0, T]$ and $w \in \mathbb{R}^2$,

$$
\begin{aligned}
q_\varepsilon(t, w, \theta, \rho) &= \frac{\phi_\varepsilon\left(|w - v_t(\rho)|\right)^\gamma b(\theta)}{\lambda_\varepsilon} \mathbf{1}_{\bar{\Pi}_\varepsilon}(\theta) \\
&\quad + \left(1 - \frac{1}{2\Gamma_\varepsilon^\gamma}\int_0^1 \phi_\varepsilon\left(|w - v_t(r)|\right)^\gamma dr\right) \varepsilon^{a\alpha}\, \tilde{\chi}\left(\frac{\theta}{\varepsilon^a}\right) b(\theta)\, \mathbf{1}_{\Pi_\varepsilon}(\theta).
\end{aligned}
\tag{13.9}
$$

Then, it is easy to see that for fixed (t, w)

$$
\begin{aligned}
&\int_{\Pi}\int_0^1 q_\varepsilon(t, w, \theta, \rho)\, d\theta\, d\rho \\
&= \frac{1}{\lambda_\varepsilon}\int_{\bar{\Pi}_\varepsilon}\int_0^1 \phi_\varepsilon\big(|w - v_t(\rho)|\big)^\gamma\, b(\theta)\, d\rho\, d\theta \\
&\quad + \int_{\Pi_\varepsilon}\left(1 - \frac{1}{2\Gamma_\varepsilon^\gamma}\int_0^1 \phi_\varepsilon\big(|w - v_t(\rho)|\big)^\gamma\, d\rho\right)\varepsilon^{a\alpha}\,\tilde{\chi}\left(\frac{\theta}{\varepsilon^a}\right) b(\theta)\, d\theta \\
&= 1.
\end{aligned}
$$

That is, $q_\varepsilon(t, w, \theta, \rho)\, d\theta\, d\rho$ is a probability measure on $\Pi \times [0, 1]$, which represents the conditional density of $\bar{V}_{T_N}$ conditioned on $\bar{V}_{T_{N-1}}, T_{N-1}, T_N$.

Remark 13.3.10 This step is so important in the construction that some comments are in order.

(i) Note that this step is a change of probabilistic representation for the process $\bar{V}$. In fact, the randomness of the random variable U_N has been integrated into the representation for q_ε.
(ii) The above fact has allowed us to have some regularity in the problem. The indicator function is no longer part of the model and therefore the possibility of an IBP formula appears.
(iii) The newly introduced function $\tilde{\chi}$ serves as a parameter that needs to be chosen later but also it may help us to build an IBP formula with no boundary conditions due to the restriction that $\mathrm{Supp}[\tilde{\chi}] \subset \Pi$. Note that the integrability condition (13.8) is used in order to have that q_ε is a density function.
(iv) Note that the density function q_ε is not continuous at $\theta = \pm\varepsilon^a\,\pi/2$. Therefore we will need a localization procedure later on to avoid boundary effects when using the IBP formula. For this, recall Exercise 10.1.8.
(v) Note that in the construction of q_ε there has been an exchange of "noise" in the sense that the probability that $\mathbf{1}_{[U_N,\infty)}\big(\phi_\varepsilon\big(|\bar{V}_{T_{N-1}} - v_{T_N}(\rho_N)|\big)^\gamma\big) = 0$ has been exchanged by the probability that $|\theta_N| < \varepsilon^a\pi/2$. This is a crucial point of the method.

This change of probabilistic representation creates a change of dynamics. Following this dynamics, we can construct a new process $\tilde{V}$ which is going to be equivalent in the sense of probability law to the process $\bar{V}$. That is, construct a sequence of $\Pi \times [0, 1]$-valued random variables $\big\{(\tilde{\theta}_k, \tilde{\rho}_k)\,;\ k \in \mathbb{N}\big\}$, and the process $\tilde{V} = \big\{\tilde{V}_t (= \tilde{V}_t^\varepsilon)\,;\ 0 \le t \le T\big\}$ such that

$$
\begin{aligned}
&\tilde{V}_0 = V_0,\\
&\tilde{V}_t = \tilde{V}_{T_k} \quad \big(t \in [T_k, T_{k+1})\big),\\
&\tilde{V}_{T_{k+1}} = \tilde{V}_{T_k} + A_\varepsilon(\tilde{\theta}_{k+1})\left(\tilde{V}_{T_k} - v_{T_{k+1}}(\tilde{\rho}_{k+1})\right), \qquad (13.10)\\
&\mathbb{P}\left[\tilde{\theta}_{k+1} \in d\theta,\ \tilde{\rho}_{k+1} \in d\rho \middle| \tilde{V}_{T_k}, T_k, T_{k+1}\right] = q_\varepsilon\big(T_{k+1}, \tilde{V}_{T_k}, \theta, \rho\big)\, d\theta\, d\rho.
\end{aligned}
$$

We remark that in this new sequence of jumps $\tilde{\theta}_k$ can now take values close to zero. Still, due to the fact that $A_\varepsilon(\theta) = 0$, if $|\theta| < \varepsilon^a \pi/2$ then $\bar{V} = \tilde{V}$ in law. Therefore studying the density of $\bar{V}$ or $\tilde{V}$ are equivalent and the latter will be the model for which we will apply the integration by parts formula. In particular, all moments estimates for $\bar{V}$ are applicable to $\tilde{V}$ without any further mention.

In the following exercise, we give the explicit construction of the r.v.s to be used from now on.

Exercise 13.3.11 Define an r.v. $\tilde{\rho}_{k+1}$ which has a mixed law. First, it takes values in the interval $(0, 1]$ with density given by

$$
\frac{\phi_\varepsilon\big(|w - v_t(\rho)|\big)^\gamma}{2\Gamma_\varepsilon^\gamma}.
$$

Second, $\tilde{\rho}_{k+1}$ takes the value zero with probability

$$
1 - \frac{1}{2\Gamma_\varepsilon^\gamma} \int_0^1 \phi_\varepsilon\big(|w - v_t(r)|\big)^\gamma\, dr.
$$

Conditioned on the value of $\tilde{\rho}_{k+1} \in (0, 1]$, the r.v. $\tilde{\theta}_{k+1}$ is defined with the density $\frac{2\Gamma_\varepsilon^\gamma}{\lambda_\varepsilon} b(\theta) \mathbf{1}_{\bar{\Pi}_\varepsilon}(\theta)$. On the other hand, if $\tilde{\rho}_{k+1} = 0$ then the density for $\tilde{\theta}_{k+1}$ is given by

$$
\varepsilon^{a\alpha}\, \tilde{\chi}\left(\frac{\theta}{\varepsilon^a}\right) b(\theta)\, \mathbf{1}_{\Pi_\varepsilon}(\theta).
$$

Prove that this definition gives an r.v. $(\tilde{\theta}_{k+1}, \tilde{\rho}_{k+1})$ with the law $q_\varepsilon(t, w, \cdot, \cdot)$. It is important to note that in the above construction there is dependence between the value of $\tilde{\rho}_{k+1}$ and the value of $\tilde{\theta}_{k+1}$, although in the case that $\tilde{\theta}_{k+1} = 0$ there is no effect on the functional $F\Big(\bar{V}_{T_{N-1}} + A_\varepsilon(\theta)\left(\bar{V}_{T_{N-1}} - v_{T_N}(\rho)\right)\Big)$.

Define by induction the functions $\mathscr{H}_k : \mathbb{R}^2 \times \big([0, \infty) \times \Pi \times [0, 1]\big)^k \to \mathbb{R}^2$ ($k \in \mathbb{Z}_+$) by

$$\begin{aligned}
&\mathscr{H}_0(v) = v,\\
&\mathscr{H}_1\big(v, (t_1, \theta_1, \rho_1)\big) = \mathscr{H}_0(v) + A_\varepsilon(\theta_1)\big(\mathscr{H}_0(v) - v_{t_1}(\rho_1)\big),\\
&\mathscr{H}_{k+1}\big(v, (t, \theta, \rho)_{1\to k+1}\big) = \mathscr{H}_k\big(v, (t, \theta, \rho)_{1\to k}\big)\\
&\qquad\qquad + A_\varepsilon(\theta_{k+1})\Big(\mathscr{H}_k\big(v, (t, \theta, \rho)_{1\to k}\big) - v_{t_{k+1}}(\rho_{k+1})\Big)
\end{aligned}$$

for $k \in \mathbb{N}$, where we recall the notation $(t, \theta, \rho)_{1\to k} = \big((t_1, \theta_1, \rho_1), \ldots, (t_k, \theta_k, \rho_k)\big)$. Then, one proves by induction that

$$\tilde{V}_{T_k} = \mathscr{H}_k\big(V_0, (T, \tilde{\theta}, \tilde{\rho})_{1\to k}\big),$$

where $(T, \tilde{\theta}, \tilde{\rho})_{1\to k} \Big(= (T_{1\to k}, \tilde{\theta}_{1\to k}, \tilde{\rho}_{1\to k})\Big) = \big((T_1, \tilde{\theta}_1, \tilde{\rho}_1), \ldots, (T_k, \tilde{\theta}_k, \tilde{\rho}_k)\big)$.

For any Borel measurable bounded function $F : \mathbb{R}^2 \to \mathbb{R}$, we have

$$\begin{aligned}
\mathbb{E}\Big[F(\tilde{V}_t)\Big] &= \sum_{N=0}^{\infty} \frac{\left(t\,\lambda_\varepsilon\right)^N}{N!} e^{-t\lambda_\varepsilon}\, \mathbb{E}\Big[F\big(\mathscr{H}_N\big(V_0, (T, \tilde{\theta}, \tilde{\rho})_{1\to N}\big)\big)\Big]\\
&= \sum_{N=0}^{\infty} \frac{\left(t\,\lambda_\varepsilon\right)^N}{N!} e^{-t\lambda_\varepsilon}\, \mathbb{E}\Big[\mathbb{E}\big[F\big(\mathscr{H}_N\big(V_0, (T, \tilde{\theta}, \tilde{\rho})_{1\to N}\big)\big)\big|V_0, T_{1\to N}\big]\Big]\\
&= \sum_{N=0}^{\infty} \frac{\left(t\,\lambda_\varepsilon\right)^N}{N!} e^{-t\lambda_\varepsilon} \int_{\Pi^N} d\theta_{1\to N} \int_{[0,1]^N} d\rho_{1\to N}\\
&\qquad \times \mathbb{E}\Big[F\big(\mathscr{H}_N\big(V_0, (T, \theta, \rho)_{1\to N}\big)\big)\, g_N\big(V_0, (T, \theta, \rho)_{1\to N}\big)\Big],
\end{aligned}$$

where $d\theta_{1\to N} \equiv d\theta_1 \ldots d\theta_N$ and similarly for $d\rho_{1\to N}$. Furthermore,

$$\begin{aligned}
&g_N\big(V_0, (T, \theta, \rho)_{1\to N}\big)\\
&= q_\varepsilon\Big(T_1, \mathscr{H}_0(V_0), \theta_1, \rho_1\Big) \prod_{k=2}^{N} q_\varepsilon\Big(T_k, \mathscr{H}_{k-1}\big(V_0, (T, \theta, \rho)_{1\to k-1}\big), \theta_k, \rho_k\Big).
\end{aligned}$$

13.4 IBP Formula

Obtaining the IBP formula is a long process as explained in Sect. 12.5.

In the first step, we will study the derivative process of $\tilde{V}$ and its inverse similarly to how they were studied in Lemma 12.5.6 for the SDE case. We define the process $\tilde{Y} = \big\{\tilde{Y}_t(= \tilde{Y}_t^\varepsilon)\,;\, 0 \le t \le T\big\}$ as the $\mathbb{R}^2 \otimes \mathbb{R}^2$-valued process given by

$$\tilde{Y}_t = I + \sum_{k=1}^{N_t^\varepsilon} A_\varepsilon(\tilde{\theta}_k)\, \tilde{Y}_{T_{k-1}},$$

which can be solved explicitly as

$$\tilde{Y}_t = \prod_{k=1}^{N_t^\varepsilon} \big(I + A_\varepsilon(\tilde{\theta}_k)\big).$$

Write $\tilde{Z}_t = \tilde{Y}_t^{-1}$. Then, we have:

Proposition 13.4.1 *It holds that* $\sup_{0<s\le t\le T} \|\tilde{Y}_t\, \tilde{Z}_s\| \le 1$ *a.s.*

Exercise 13.4.2 Prove Proposition 13.4.1. For a hint, just compute as explicitly as possible $\|I + A(\theta)\|$.

For a hint, see Chap. 14.

Exercise 13.4.3 Prove that the product of the matrices $I + A_\varepsilon(\tilde{\theta}_k)$ and $I + A_\varepsilon(\tilde{\theta}_j)$ for $k \neq j$ commute. Therefore the order in which the product in the explicit expression of $\tilde{Y}$ is written is irrelevant.

Proposition 13.4.4 *For all* $t \in [0, T]$, *the* $\mathbb{R}^2 \otimes \mathbb{R}^2$*-valued random matrix* $\tilde{Y}_t$ *is invertible a.s. Furthermore,* $\tilde{Z}_t$ *satisfies the following estimate for any* $p > 1$*:*

$$\mathbb{E}\left[\sup_{t\in[0,T]} \|\tilde{Z}_t\|^p\right] \le \exp\left[C_{p,T}\,\Gamma_\varepsilon^\gamma\right] \tag{13.11}$$

Proof Since $\mathbb{E}[\,N_t^\varepsilon\,] = t\,\lambda_\varepsilon < \infty$, we have

$$\begin{aligned}\Delta := \det\big(I + A_\varepsilon(\theta)\big) &= \left(1 + \chi_\varepsilon(\theta)\,\frac{\cos\theta - 1}{2}\right)^2 + \left(\chi_\varepsilon(\theta)\,\frac{\sin\theta}{2}\right)^2 \\ &= \frac{1-\cos\theta}{2}\,\big(\chi_\varepsilon(\theta) - 1\big)^2 + \frac{1+\cos\theta}{2} \ge \frac{1}{2}\end{aligned}$$

for $\theta \in \Pi$. Then, since we have

$$\det \tilde{Y}_t = \prod_{k=1}^{N_t^\varepsilon} \det\big(I + A_\varepsilon(\tilde{\theta}_k)\big) \ge 2^{-N_t^\varepsilon},$$

$\tilde{Y}$ is invertible a.s.

On the other hand, for $\theta \in \Pi$ we have

$$
\begin{aligned}
&\left\|\left(I+A_\varepsilon(\theta)\right)^{-1}\right\|^2\\
&=\sup_{(x,y)^*\in B_1(0)}\left|\frac{1}{\Delta}\begin{pmatrix}1+\chi_\varepsilon(\theta)\dfrac{\cos\theta-1}{2} & \chi_\varepsilon(\theta)\dfrac{\sin\theta}{2}\\ -\chi_\varepsilon(\theta)\dfrac{\sin\theta}{2} & 1+\chi_\varepsilon(\theta)\dfrac{\cos\theta-1}{2}\end{pmatrix}(x,y)^*\right|^2\\
&=\frac{1}{\Delta}\leq\frac{2}{1+\cos\theta}\leq 1+\theta^2\leq e^{\theta^2}.
\end{aligned}
$$

Exercise 13.4.5 Prove that $\frac{2}{1+\cos\theta}\leq 1+\theta^2$ for $\theta\in\Pi$.

Here note that due to Remark 13.3.10-(v), $\tilde{N}_t^\varepsilon\leq N_t^\varepsilon$ corresponds to the number of jumps which have $|\tilde{\theta}_k|>\varepsilon^a\pi/2$. The difference here is that $\tilde{N}^\varepsilon$ is not a Poisson process. From here, we obtain

$$
\left\|\tilde{Z}_t\right\|^2\leq\prod_{k=1}^{N_t^\varepsilon}\left\|\left(I+A_\varepsilon(\tilde{\theta}_k)\right)^{-1}\right\|^2\leq\exp\left[\sum_{k=1}^{\tilde{N}_T^\varepsilon}\left(\tilde{\theta}_k\right)^2\right].
$$

Moreover, since

$$
\sum_{k=1}^{\tilde{N}_t^\varepsilon}(\tilde{\theta}_k)^2\overset{\mathscr{L}}{=}\int_0^t\int_{\bar{\Pi}_\varepsilon}\int_0^1\int_0^{2\Gamma_\varepsilon^\gamma}|\theta|^2\,\mathbf{1}_{[u,\infty)}\left(\phi_\varepsilon\left(|\bar{V}_{s-}-v_s(\rho)|\right)^\gamma\right)d\mathcal{N},\tag{13.12}
$$

it holds that, for $p>0$,

$$
\begin{aligned}
&\mathbb{E}\left[\exp\left(p\sum_{k=1}^{\tilde{N}_T^\varepsilon}(\tilde{\theta}_k)^2\right)\right]\\
&=\mathbb{E}\left[\exp\left(p\int_0^T\int_{\bar{\Pi}_\varepsilon}\int_0^1\int_0^{2\Gamma_\varepsilon^\gamma}|\theta|^2\,\mathbf{1}_{[u,\infty)}\left(\phi_\varepsilon\left(|\bar{V}_{s-}-v_s(\rho)|\right)^\gamma\right)d\mathcal{N}\right)\right]\\
&\leq\mathbb{E}\left[\exp\left(p\int_0^T\int_{\bar{\Pi}_\varepsilon}\int_0^1\int_0^{2\Gamma_\varepsilon^\gamma}|\theta|^2\,d\mathcal{N}\right)\right]\\
&=\exp\left(2\Gamma_\varepsilon^\gamma\,T\int_{\bar{\Pi}_\varepsilon}\left(e^{p|\theta|^2}-1\right)b(\theta)\,d\theta\right)\\
&\leq\exp\left(C_{p,T}\,\Gamma_\varepsilon^\gamma\right).
\end{aligned}\tag{13.13}
$$

Note that in the above we have used the Laplace transform associated with a compound Poisson process (see Theorem 3.1.2). Therefore, we can get

$$\mathbb{E}\left[\sup_{t\in[0,T]}\|\tilde{Z}_t\|^p\right] \le \mathbb{E}\left[\exp\left(p\sum_{k=1}^{\tilde{N}_T^\varepsilon}\tilde{\theta}_k^2\right)\right] \le \exp\left(C_{p,T}\,\Gamma_\varepsilon^\gamma\right).$$

The proof is complete.

Exercise 13.4.6 Note that the bound obtained in (13.11) explodes as $\varepsilon \downarrow 0$ as explained in Exercise 13.3.5. This is due to the second inequality in (13.13). Also this estimate is bad enough so that in many estimates to follow one does not strive for the optimal bounds due to the bound in (13.11).

On the other hand, use (13.12) and Proposition 13.3.4 to prove that for $p = 1$,

$$\sup_{\varepsilon\in(0,e^{-4})} \mathbb{E}\left[\left(\sum_{k=1}^{\tilde{N}_t^\varepsilon}\tilde{\theta}_k^2\right)^p\right] < \infty.$$

Discuss a strategy to prove the above statement for the case $p = 2$.

Now we introduce the rest of the ingredients in order to differentiate the process $\tilde{V}$. Here, it may be good for the reader to recall the set-up in Sect. 12.3. For each $k \in \mathbb{N}$ and $\xi \in \mathbb{R}^2$ with $|\xi| \le 1$, define[3]

$$\ell_k(\theta,\rho)\left(\equiv \ell_k\big(\theta,\rho\,;\,\tilde{V}_{T_{k-1}},\,T_{k-1},\,T_k\big)\right) = A_\varepsilon{}'(\theta)\left(\tilde{V}_{T_{k-1}} - v_{T_k}(\rho)\right)\varphi(\theta)^2,$$

where $\varphi \in C_c^\infty\left(\mathbb{R}_0\,;\,\mathbb{R}_+\right)$ is an even function and it satisfies that for $n = 0, \ldots, r$ and $\beta > 0$

$$\varphi^{(n)}(\theta) = O(|\theta|^{\beta-n}) \ \ (|\theta| \to 0), \quad |\varphi(\theta)| = o(1) \ \ (|\theta| \to \pi/2). \tag{13.14}$$

Exercise 13.4.7 Prove the existence of such a function φ. The value of β and r will be determined by the application. For example, prove that

$$\sup_{\varepsilon\in(0,e^{-4})} \mathbb{E}\left[\sum_{k=1}^{N_t^\varepsilon}\left(\varphi(\tilde{\theta}_k)\,\varphi'(\tilde{\theta}_k)\right)^2\right] < \infty$$

if $4\beta - 2 - \alpha > 0$.

For a hint, see Chap. 14.

We remark that as requested in Hypothesis 12.3.2, we have $H_k^0(\theta,\rho) = \theta$. Let

$$h_k(\theta,\rho) = (\partial_\xi)_0 H_k^\xi(\theta,\rho) = \theta\,\ell_k(\theta,\rho)^*,$$

---

[3] Although it will not be used directly in this chapter, the relation with the previous chapter is given through the deformation $H_k^\xi(\theta,\rho)\left(\equiv H_k^\xi\big(\theta,\rho\,;\,\tilde{V}_{T_{k-1}},\,T_{k-1},\,T_k\big)\right) = \theta\,\exp\left(\xi\cdot\ell_k(\theta,\rho)\right)$.

where $\left(\partial_\xi\right)_0$ denotes the partial derivative at $\xi = 0$. Let $j \le k$, and write $\tilde{\theta}^{(j)}_{1\to k} = (\tilde{\theta}_{1\to j-1}, \theta_j, \tilde{\theta}_{j+1\to k})$, $\theta_j \in \Pi$. Now we define

$$\tilde{V}^{(j)}_{T_k} \equiv \tilde{V}^{(j)}_{T_k}(\theta_j) = \mathscr{H}_k\big(V_0, (T, \tilde{\theta}^{(j)}, \tilde{\rho})_{1\to k}\big)$$
$$= \begin{cases} \tilde{V}_{T_{k-1}} + A_\varepsilon(\theta_k)\big(\tilde{V}_{T_{k-1}} - v_{T_k}(\tilde{\rho}_k)\big) & (j = k), \\ \tilde{V}^{(j)}_{T_{k-1}} + A_\varepsilon(\tilde{\theta}_k)\left(\tilde{V}^{(j)}_{T_{k-1}} - v_{T_k}(\tilde{\rho}_k)\right) & (1 \le j \le k-1). \end{cases}$$

That is, $\tilde{V}_{T_k} = \tilde{V}^{(j)}_{T_k}(\tilde{\theta}_j)$. With this definition, we can compute the derivative which gives

$$\partial_{\theta_j} \tilde{V}^{(j)}_{T_k} = \begin{cases} \tilde{Y}_{T_k} \tilde{Z}_{T_j} A_\varepsilon{}'(\theta_j)\left(\tilde{V}_{T_{j-1}} - v_{T_j}(\tilde{\rho}_j)\right) & (1 \le j \le k-1), \\ A_\varepsilon{}'(\theta_k)\left(\tilde{V}_{T_{k-1}} - v_{T_k}(\tilde{\rho}_k)\right) & (j = k), \\ 0 & (j > k). \end{cases} \tag{13.15}$$

Exercise 13.4.8 Prove the following properties:

(i) Prove by induction the equality (13.15) and $\partial_{\theta_j} \tilde{V}^{(j)}_{T_k} \varphi^2(\theta_j) = \tilde{Y}_{T_k} \tilde{Z}_{T_j} \ell_j(\theta_j, \tilde{\rho}_j)$.

(ii) Note that contrary to the behavior of A_ε, $A^{(k)}_\varepsilon$, $k \in \mathbb{N}$ does not converge to $A^{(k)}$ uniformly in θ as $\varepsilon \downarrow 0$ in general. On the other hand, note that using the explicit function χ_ε in Exercise 13.3.2, one obtains that $A_\varepsilon{}'(\theta)$ is uniformly bounded as $\varepsilon \downarrow 0$. In general, prove that $A^{(k)}_\varepsilon(\theta)$, $k \in \mathbb{N}$ is bounded by $C\varepsilon^{-ak}$.

For a hint, see Chap. 14.

Definition 13.4.9 Assume that V_t is a random variable which depends on the jumps $\tilde{\theta}_j$, $j \le N^\varepsilon_t$ which correspond to the density q_ε in the sense that it is adapted to the filtration associated with the present setting, $\mathscr{F}_t \equiv \mathscr{F}^\varepsilon_t = \sigma(N^\varepsilon_s; s \le t) \vee \sigma(Z_1, Z_2) \vee \sigma(\tilde{\theta}_k, \tilde{\rho}_k; k \le N^\varepsilon_t)$. Here $\left(Z_1, Z_2\right)^*$ is a two-dimensional normal vector with mean vector 0 and covariance matrix I_2 such that they are independent of all the random variables introduced before. These normal vectors will appear later in (13.16).

We define the stochastic derivative of any such random variable V_t with respect to the j-jump as

$$D_j V_t = (\partial_{\theta_j})_{\tilde{\theta}_j} V_t.$$

Here we assume that V_t can be written as a differentiable function of $\tilde{\theta}_{1\to k}$ on $N^\varepsilon_t = k$ for each $k \in \mathbb{N}$.

We also remark that according to this definition $D_j \tilde{\theta}_k = 0$ for $j \ne k$, although the law of $\tilde{\theta}_k$ depends on $\tilde{V}_{T_{k-1}}$.

Exercise 13.4.10 Prove the following assertions about the stochastic derivative:

(i) Note that (13.15) gives $D_j \tilde{V}_t$. From this expression, use Propositions 13.4.1, 13.4.4 and 13.3.4 to obtain an upper bound for the L^p-norm of $D_j \tilde{V}_t$ on the set $N^\varepsilon_t \ge j$.

(ii) Compute $D_{j_1} D_{j_2} \tilde{V}_t = D_{j_2} D_{j_1} \tilde{V}_t$ and $D_j^2 \tilde{V}_t$.
(iii) State and prove the fact that the stochastic derivative satisfies the usual properties of a differential operator such as the chain rule, product rule, etc.
(iv) Prove that for $N_t^\varepsilon \geq k$ and $r \in \mathbb{N}$, $D_k^r \tilde{Y}_t = \tilde{Y}_t \tilde{Z}_{T_k} A_\varepsilon^{(r)}(\tilde{\theta}_k) \tilde{Y}_{T_{k-1}}$. Find explicit formulas for $D_{j_1} ... D_{j_k} \tilde{Y}_t$ and $D_{j_1} ... D_{j_k} \tilde{Z}_t$ when $N_t^\varepsilon \geq j_k > j_{k-1} > ... > j_1$. Then try to consider the case $N_t^\varepsilon \geq j_k \geq j_{k-1} \geq ... \geq j_1$.
(v) Prove that for $N_t^\varepsilon \geq j_k > j_{k-1} > ... > j_1 \geq 1$,

$$\|D_{j_1} ... D_{j_k} \tilde{V}_t\| \leq C \left|\tilde{V}_{T_{j_1}-} - v_{T_{j_1}}(\tilde{\rho}_{j_1})\right|.$$

(vi) Prove that for $0 < s < t \leq T$ and $N_t^\varepsilon \geq j_k > j_{k-1} > ... > j_1 > N_s^\varepsilon \geq 1$,

$$\|D_{j_1} ... D_{j_k} \tilde{Y}_t \tilde{Z}_s\| \leq C.$$

Here C is a constant which is independent of ε, s, t and $j_1, ..., j_k$. In the case that $j_1 \leq N_s^\varepsilon$ then the above derivative equals zero. We remark that this bound is not necessarily true in the case that $N_t^\varepsilon \geq j_k \geq j_{k-1} \geq ... \geq j_1$.
(vii) Prove that for $N_t^\varepsilon \geq j_k > j_{k-1} > ... > j_1 > N_s^\varepsilon \geq 1$ and $l_1, ..., l_k \in \mathbb{N}$,

$$\|D_{j_1}^{l_1} ... D_{j_k}^{l_k} \tilde{Y}_t \tilde{Z}_s\| \leq C \, \varepsilon^{-a(l_1+...+l_k)}.$$

Here C is a constant which is independent of ε, s, t, $j_1, ..., j_k$ and $l_1, ..., l_k$.
(viii) Use (vi) and (vii) to prove that

$$\|D_{j_1}^{l_1} ... D_{j_k}^{l_k} \tilde{V}_t\| \leq C\varepsilon^{-a(l_1+...+l_k)} \left|\tilde{V}_{T_{j_1}-} - v_{T_{j_1}}(\tilde{\rho}_{j_1})\right|.$$

For a hint, see Chap. 14.

As explained in Exercise 13.3.3, the random processes V^ϵ or similarly $\bar{V}$ or $\tilde{V}$ do not have densities because there is a non-zero probability that there will be no jump or that all jumps in the variable θ fall in value where $|\theta| < \varepsilon^a \pi/2$. As that probability is small, we will define a further approximation so that there will always be a smooth density for the random variable under consideration.

We shall prepare the process $X^\varepsilon = \{X_t^\varepsilon ; \, t \in [0, T]\}$ defined as

$$X_t^\varepsilon = \sqrt{u_\varepsilon(t)} \begin{pmatrix} Z_1 \\ Z_2 \end{pmatrix}, \tag{13.16}$$

where[4] $0 < \alpha < 1$, $u_\varepsilon(t) = t\,(\varepsilon^a \pi)^{4\beta+2-\alpha}$, and $(Z_1, Z_2)^*$ is a two-dimensional normal vector with mean vector 0 and covariance matrix I_2 such that they are independent of all the random variables introduced before. Here, $I_2 \in \mathbb{R}^2 \otimes \mathbb{R}^2$ is the identity matrix.

[4] The reason for this choice of u_ε appears in the study of the finiteness of the inverse moments of the determinant of the Malliavin covariance matrix in (13.27).

Since $\bar{V}_t\big(= \bar{V}_t^\varepsilon\big) = V_0 = \tilde{V}_t$ on the event $\{T_1 > t\}$, no regularization phenomena will occur. In order to avoid such degenerate situation, instead of considering the density of $\tilde{V}_t$ we will consider the density of $X_t^\varepsilon + \tilde{V}_t$.

Furthermore, in order to obtain a stable IBP formula, we will apply the idea introduced in Theorem 10.2.1.

Define for $z = \big(z_1, z_2\big)^* \in \mathbb{R}^2$ with $|z| \leq 1$, the following positive definite matrices:

$$
\begin{aligned}
\tilde{C}_{1,t} &:= (\partial_z)_0\big(X_t^\varepsilon + u_\varepsilon(t)\, z\big) = u_\varepsilon(t)\, I_2,\\
\tilde{C}_{2,t} &:= \sum_{k=1}^{N_t^\varepsilon} D_k \tilde{V}_t\, \varphi(\tilde{\theta}_k)^4 \left(D_k \tilde{V}_t\right)^*\\
&= \sum_{k=1}^{N_t^\varepsilon} \tilde{Y}_t\, \tilde{Z}_{T_k}\, \ell_k(\tilde{\theta}_k, \tilde{\rho}_k) \left(\tilde{Y}_t\, \tilde{Z}_{T_k}\, \ell_k(\tilde{\theta}_k, \tilde{\rho}_k)\right)^*. \qquad (13.17)
\end{aligned}
$$

Now, we define the Malliavin covariance matrix as $\tilde{C}_t := \tilde{C}_{1,t} + \tilde{C}_{2,t}$, which is always invertible for $\varepsilon > 0$. The random matrix $\tilde{C}_t$ can be regarded as the Malliavin covariance matrix for $X_t^\varepsilon + \tilde{V}_t$. In particular, as explained in Exercises 13.4.7 and 13.4.8, the above sum may not converge as $\varepsilon \downarrow 0$ unless one requires that[5] $\varphi(\theta)^4 = O(|\theta|)$ or $4\beta > \alpha$. On the other hand, note that the addition of the vector X^ε has been used here in order to ensure the existence of the inverse of the matrix $\tilde{C}$.

The final ingredient for the IBP formula is an additional smooth random variable in order to deduce an IBP formula for $\mathbb{E}\big[\partial F(X_t^\varepsilon + \tilde{V}_t)\, \tilde{G}_t\big]$. Recall that this was explained just before Sect. 10.4 and it has appeared as the random variable Θ in Theorem 12.4.19. Let $\tilde{G}_t$ be a smooth random process in the sense that under $N_t^\varepsilon = k$ then there exists a bounded smooth random function with bounded derivatives $G_k(\omega, \cdot) : \Pi^k \to \mathbb{R}$ which is independent of all $\tilde{\theta}_j$, $j > k$, but may depend on all the other first k-jumps such as $\tilde{\rho}_j$, $j \leq k$, X_t^ε and denote $\tilde{G}_t = G_k(\tilde{\theta}_{1\to k})$.

Now we have all the ingredients to start computing and finding bounds for the IBP formula. For this, let $F \in C_c^1\big(\mathbb{R}^2\, ;\, \mathbb{R}\big)$. We divide the calculation into parts:

$$
\begin{aligned}
\mathbb{E}\big[\nabla F(X_t^\varepsilon + \tilde{V}_t)\, \tilde{G}_t\big] &= \mathbb{E}\big[\nabla F(X_t^\varepsilon + \tilde{V}_t)\, \tilde{C}_{1,t}\, \tilde{C}_t^{-1}\, \tilde{G}_t\big]\\
&\quad + \mathbb{E}\big[\nabla F(X_t^\varepsilon + \tilde{V}_t)\, \tilde{C}_{2,t}\, \tilde{C}_t^{-1}\, \tilde{G}_t\big]\\
&=: L_1 + L_2.
\end{aligned}
$$

Consider the set $\{N_t^\varepsilon = N\}$, $N \in \mathbb{N}$ and recall that $\tilde{\theta}_{1\to N}^{(k)} = (\tilde{\theta}_{1\to k-1}, \theta_k, \tilde{\theta}_{k+1\to N})$ and similarly define $\tilde{\rho}_{1\to N}^{(k)} = (\tilde{\rho}_{1\to k-1}, \rho_k, \tilde{\rho}_{k+1\to N})$ and

$$
\tilde{Y}_{T_N}^{(k)} = \prod_{1 \leq j \leq N,\ j \neq k} \big(I + A_\varepsilon(\tilde{\theta}_j)\big) \big(I + A_\varepsilon(\theta_k)\big).
$$

[5]The reason for the power 4 will be clear later in Exercise 13.7.3-(iii).

Exercise 13.4.11 Prove that $I + A_\varepsilon(\theta)$ commutes with the matrix $I + A_\varepsilon(\theta_1)$ for any $\theta, \theta_1 \in \Pi$.

Denote by $\tilde{C}_t^{(k)}$ (or $\tilde{G}_t^{(k)}$) the random matrix where the random variable $\tilde{\theta}_k$ has been replaced by the real variable θ_k in $\tilde{C}_t$ ($\tilde{G}_t$, respectively), as in the definition of $\tilde{V}^{(k)}$. Moreover, we shall use the notation $\tilde{V}_t^{(1\to N)}$, $\tilde{C}_t^{(1\to N)}$ and $\tilde{G}_t^{(1\to N)}$ in the sense that the components $\{\tilde{\theta}_k\,;\ 1 \le k \le N\}$ in the random variables $\tilde{V}_t$, $\tilde{C}_t$ and $\tilde{G}_t$ are replaced by the real variables $\{\theta_k\,;\ 1 \le k \le N\}$, in the set $\{N_t^\varepsilon = N\}$ for $N \in \mathbb{N}$.

Since

$$\tilde{V}_t = \mathscr{H}_{N_t^\varepsilon}\big(V_0, (T, \tilde{\theta}, \tilde{\rho})_{1\to N_t^\varepsilon}\big), \quad \tilde{V}_t^{(1\to N_t^\varepsilon)} = \mathscr{H}_{N_t^\varepsilon}\big(V_0, (T, \theta, \tilde{\rho})_{1\to N_t^\varepsilon}\big)$$

the IBP formula in the usual sense implies that

$$\begin{aligned}
L_2 = &\sum_{N=0}^{\infty} \frac{(t\,\lambda_\varepsilon)^N}{N!}\, e^{-t\,\lambda_\varepsilon} \int_{\Pi^N} d\theta_{1\to N}\, \mathbb{E}\Big[\nabla F\big(X_t^\varepsilon + \mathscr{H}_N\big(V_0, (T, \theta, \tilde{\rho})_{1\to N}\big)\big)\, \tilde{Y}_t^{(1\to N)} \\
&\times \sum_{k=1}^{N} \tilde{Z}_{T_k}^{(1\to N)}\, \ell_k(\theta_k, \tilde{\rho}_k)\, \varphi(\theta_k)^2 \left(\partial_{\theta_k} \tilde{V}_t^{(1\to N)}\right)^* \big(\tilde{C}_t^{(1\to N)}\big)^{-1} \tilde{G}_t^{(1\to N)} \\
&\times g_N\big(V_0, (T, \theta, \tilde{\rho})_{1\to N}\big)\Big] \\
= &\sum_{N=0}^{\infty} \frac{(t\,\lambda_\varepsilon)^N}{N!}\, e^{-t\,\lambda_\varepsilon} \sum_{k=1}^{N} \int_{\Pi} d\theta_k\, \mathbb{E}\Big[\partial_{\theta_k}\Big\{F\big(X_t^\varepsilon + \mathscr{H}_N\big(V_0, (T, \tilde{\theta}^{(k)}, \tilde{\rho})_{1\to N}\big)\big)\Big\} \\
&\times \varphi(\theta_k)^4 \left(\partial_{\theta_k} \tilde{V}_t^{(k)}\right)^* \big(\tilde{C}_t^{(k)}\big)^{-1} \tilde{G}_t^{(k)}\, g_N\big(V_0, (T, \tilde{\theta}^{(k)}, \tilde{\rho})_{1\to N}\big)\Big] \\
= &\sum_{N=0}^{\infty} \frac{(t\,\lambda_\varepsilon)^N}{N!}\, e^{-t\,\lambda_\varepsilon} \sum_{k=1}^{N} \mathbb{E}\Big[\Big[F\big(X_t^\varepsilon + \mathscr{H}_N\big(V_0, (T, \tilde{\theta}^{(k)}, \tilde{\rho})_{1\to N}\big)\big) \\
&\times \varphi(\theta_k)^4 \big(\partial_{\theta_k} \tilde{V}_t^{(k)}\big)^* \big(\tilde{C}_t^{(k)}\big)^{-1} \tilde{G}_t^{(k)}\, g_N\big(V_0, (T, \tilde{\theta}^{(k)}, \tilde{\rho})_{1\to N}\big)\Big]_{\theta_k=0}^{\theta_k=\pi/2}\Big] \\
+ &\sum_{N=0}^{\infty} \frac{(t\,\lambda_\varepsilon)^N}{N!}\, e^{-t\,\lambda_\varepsilon} \sum_{k=1}^{N} \mathbb{E}\Big[\Big[F\big(X_t^\varepsilon + \mathscr{H}_N\big(V_0, (T, \tilde{\theta}^{(k)}, \tilde{\rho})_{1\to N}\big)\big) \\
&\times \varphi(\theta_k)^4 \big(\partial_{\theta_k} \tilde{V}_t^{(k)}\big)^* \big(\tilde{C}_t^{(k)}\big)^{-1} \tilde{G}_t^{(k)}\, g_N\big(V_0, (T, \tilde{\theta}^{(k)}, \tilde{\rho})_{1\to N}\big)\Big]_{\theta_k=-\pi/2}^{\theta_k=0}\Big] \\
- &\sum_{N=0}^{\infty} \frac{(t\,\lambda_\varepsilon)^N}{N!}\, e^{-t\,\lambda_\varepsilon} \sum_{k=1}^{N} \int_{\Pi} d\theta_k\, \mathbb{E}\Big[F\left(X_t^\varepsilon + \mathscr{H}_N\big(V_0, (T, \tilde{\theta}^{(k)}, \tilde{\rho})_{1\to N}\big)\right) \\
&\times \frac{\partial_{\theta_k}\Big\{\varphi(\theta_k)^4 \big(\partial_{\theta_k} \tilde{V}_t^{(k)}\big)^* \big(\tilde{C}_t^{(k)}\big)^{-1} \tilde{G}_t^{(k)}\, g_N\big(V_0, (T, \tilde{\theta}^{(k)}, \tilde{\rho})_{1\to N}\big)\Big\}}{g_N\big(V_0, (T, \tilde{\theta}^{(k)}, \tilde{\rho})_{1\to N}\big)}\Big].
\end{aligned}$$

On the other hand, in a similar fashion we have for L_1 that using the IBP formula with respect to Gaussian laws and the fact that $F \in C_c^1(\mathbb{R}^2; \mathbb{R})$.

$$
\begin{aligned}
L_1 &\equiv \mathbb{E}\left[\nabla F\left(X_t^\varepsilon + \tilde{V}_t\right) u_\varepsilon(t)\, I_2\, \tilde{C}_t^{-1}\, \tilde{G}_t\right] \\
&= \mathbb{E}\left[\int_{\mathbb{R}^2} \nabla F\left(\sqrt{u_\varepsilon(t)}\,(x, y)^* + \tilde{V}_t\right) \frac{u_\varepsilon(t)\, I_2}{2\pi} \exp\left(-\frac{x^2+y^2}{2}\right) dx\, dy\, \tilde{C}_t^{-1}\, \tilde{G}_t\right] \\
&= \mathbb{E}\left[\int_{\mathbb{R}^2} F\left(\sqrt{u_\varepsilon(t)}\,(x, y)^* + \tilde{V}_t\right) \frac{(x, y)}{2\pi} \exp\left(-\frac{x^2+y^2}{2}\right) dx\, dy\, \sqrt{u_\varepsilon(t)}\, \tilde{C}_t^{-1}\, \tilde{G}_t\right] \\
&= \mathbb{E}\left[F\left(X_t^\varepsilon + \tilde{V}_t\right)(Z_1, Z_2)\sqrt{u_\varepsilon(t)}\, \tilde{C}_t^{-1}\, \tilde{G}_t\right].
\end{aligned}
$$

Exercise 13.4.12 Prove that the boundary terms do not give any extra terms. That is,

$$
\begin{aligned}
&\Big[F\big(X_t^\varepsilon + \mathscr{H}_N\big(V_0, (T_{1\to N}, \tilde{\theta}_{1\to N}^{(k)}, \tilde{\rho}_{1\to N}^{(k)})\big)\big) \\
&\qquad \times \varphi(\theta_k)^4 \big(\partial_{\theta_k} \tilde{V}_t^{(k)}\big)^* \big(\tilde{C}_t^{(k)}\big)^{-1} \tilde{G}_t^{(k)}\, q_\varepsilon\big(T_k, \tilde{V}_{T_{k-1}}, \theta_k, \rho_k\big)\Big]_{\theta_k=0}^{\theta_k=\pi/2} \\
&+ \Big[F\big(X_t^\varepsilon + \mathscr{H}_N\big(V_0, (T_{1\to N}, \tilde{\theta}_{1\to N}^{(k)}, \tilde{\rho}_{1\to N}^{(k)})\big)\big) \\
&\qquad \times \varphi(\theta_k)^4 \big(\partial_{\theta_k} \tilde{V}_t^{(k)}\big)^* \big(\tilde{C}_t^{(k)}\big)^{-1} \tilde{G}_t^{(k)}\, q_\varepsilon\big(T_k, \tilde{V}_{T_{k-1}}, \theta_k, \rho_k\big)\Big]_{\theta_k=-\pi/2}^{\theta_k=0} = 0.
\end{aligned}
$$

In order to prove this recall that, as explained in Remark 13.3.10, the density q is discontinuous at $\theta = \pm\varepsilon^a\pi/2$, therefore the above expression has to actually be evaluated at the intervals $[-\pi/2, -\varepsilon^a\pi/2]$, $[-\varepsilon^a\pi/2, 0]$, $[0, \varepsilon^a\pi/2]$ and $[\varepsilon^a\pi/2, \pi/2]$. In each boundary, terms cancel because of the properties of the localization functions φ, χ_ε (recall the explicit calculation of $\partial_{\theta_k}\tilde{V}_t^{(k)}$ in (13.15)) and the invariance of $\mathscr{H}_N\big(V_0, (T_{1\to N}, \tilde{\theta}_{1\to N}^{(k)}, \tilde{\rho}_{1\to N}^{(k)})\big)$ when $|\theta_k| \le \varepsilon^a\pi/2$.

It is instructive to compare this fact with the handling of the term $\mathscr{R}_\varepsilon$ in Lemma 12.4.1.

From all the above arguments we obtain the following result.

Theorem 13.4.13 (First-order IBP formula) *For $F \in C_b^1(\mathbb{R}^2; \mathbb{R})$, it holds that*

$$
\mathbb{E}\left[\nabla F\left(X_t^\varepsilon + \tilde{V}_t\right) \tilde{G}_t\right] = \mathbb{E}\left[F\left(X_t^\varepsilon + \tilde{V}_t\right) \tilde{\Gamma}_t\left(X^\varepsilon + \tilde{V}, \tilde{G}\right)\right], \tag{13.18}
$$

where the random variable $\tilde{\Gamma}_t\left(X^\varepsilon + \tilde{V}, \tilde{G}\right)$ is assumed to be integrable and defined by

$$\tilde{\Gamma}_t\big(X^\varepsilon + \tilde{V}, \tilde{G}\big) = \big(Z_1, Z_2\big)\, \tilde{C}_t^{-1}\, \tilde{G}_t \sqrt{u_\varepsilon(t)} \\ - \sum_{k=1}^{N_t^\varepsilon} \frac{D_k\Big\{\varphi(\tilde{\theta}_k)^4 \big(D_k \tilde{V}_t\big)^* \tilde{C}_t^{-1}\, \tilde{G}_t\, g_{N_t^\varepsilon}\big(V_0, (T, \tilde{\theta}, \tilde{\rho})_{1\to N_t^\varepsilon}\big)\Big\}}{g_{N_t^\varepsilon}\big(V_0, (T, \tilde{\theta}, \tilde{\rho})_{1\to N_t^\varepsilon}\big)}.$$

Remark 13.4.14 In Theorem 13.4.13 the notation $\tilde{\Gamma}_t\big(X^\varepsilon + \tilde{V}, \tilde{G}\big)$ is not being used in the sense of function but to denote the dependence of the random variable on the random vector $\big(X^\varepsilon + \tilde{V}, \tilde{G}\big)$. Sometimes in order to shorten the length of the equation we shall let $\tilde{\Gamma}_t \equiv \tilde{\Gamma}_t\big(X^\varepsilon + \tilde{V}, \tilde{G}\big)$. We call this random variable the Malliavin weight in what follows.

13.5 Negative-Order Integrability of Malliavin Weight

In Theorem 13.4.13 of the previous section, we have provided a first IBP formula assuming that the random variable $\tilde{\Gamma}_t\big(X^\varepsilon + \tilde{V}, \tilde{G}\big)$ is integrable. In order to prove that this property is satisfied and find upper estimates one needs to break the analysis into various parts. The first one considers the estimates for $\det(\tilde{C}_t)$ as this is an important component of $\tilde{C}_t^{-1}$.

First of all, we shall introduce:

Exercise 13.5.1 For any $p > 1$ and all non-negative definite, symmetric matrix $A \in \mathbb{R}^d \otimes \mathbb{R}^d$, it holds that

$$d^{d(2p-1)/2} \frac{\Gamma(p)^d}{\big(\det A\big)^p} \le \int_{\mathbb{R}^d} |x|^{d(2p-1)} \exp\big[-x \cdot Ax\big]\, dx. \tag{13.19}$$

Prove also the following property for a non-negative definite symmetric matrix A:

$$\det(A)^{-1} \le \lambda_1(A)^{-d} = \sup\{|Ax|; |x| = 1\}^{-d}.$$

Here $\lambda_1(A)$ denotes the smallest eigenvalue of A.

For a hint, see Chap. 14.

Using this result, we have the following upper estimate on the negative-order moment of the determinant of $\tilde{C}_t$.

Lemma 13.5.2 *For any $\beta \ge 0$, $p > 1$ and $0 < t \le T$, it holds that*

$$\mathbb{E}\big[(\det \tilde{C}_t)^{-p}\big] \le C_{p,\alpha,\beta,t} \exp\big[C_{p,T}\, \Gamma_\varepsilon^\gamma\big].$$

Proof From Proposition 13.4.4, we have

$$\mathbb{E}\big[(\det \tilde{Y}_t)^{-4p}\big] \le \mathbb{E}\big[\|\tilde{Z}_t\|^{8p}\big] \le \mathbb{E}\Big[\sup_{0\le t\le T} \|\tilde{Z}_t\|^{8p}\Big] \le \exp\big[C_{p,T}\, \Gamma_\varepsilon^\gamma\big].$$

Therefore as $\tilde{C}_{1,t}$ and $\tilde{C}_{2,t}$ are non-negative definite then

$$\begin{aligned}\mathbb{E}\big[(\det \tilde{C}_t)^{-p}\big] &\le \mathbb{E}\big[(\det \tilde{Y}_t)^{-4p}\big]^{1/2}\, \mathbb{E}\Big[\big(\det(\tilde{Z}_t\, \tilde{C}_t\, \tilde{Z}_t^*)\big)^{-2p}\Big]^{1/2} \\ &\le \exp\left[\frac{C_{p,T}\, \Gamma_\varepsilon^\gamma}{2}\right] \mathbb{E}\big[(\det(\tilde{Z}_t\, \tilde{C}_t\, \tilde{Z}_t^*))^{-2p}\big]^{1/2}. \end{aligned} \tag{13.20}$$

Moreover, since $\|\tilde{Y}_t\| \le 1$ from Proposition 13.4.1, we have

$$u_\varepsilon(t)\, |\tilde{Z}_t^*\, \xi|^2 \ge u_\varepsilon(t)\, \|\tilde{Y}_t^*\|^2\, |\tilde{Z}_t^*\, \xi|^2 \ge u_\varepsilon(t)\, |\xi|^2. \tag{13.21}$$

Thus, in order to estimate the second term in the product on the right-hand side of (13.20), we use Exercise 13.5.1 in order to obtain

$$\begin{aligned} I_t^\varepsilon &:= \mathbb{E}\left[\det\left(u_\varepsilon(t)\, \tilde{Z}_t\, \tilde{Z}_t^* + \sum_{k=1}^{N_t^\varepsilon} \tilde{Z}_{T_k}\, [\ell_k\, \ell_k^*](\tilde{\theta}_k, \tilde{\rho}_k)\, \tilde{Z}_{T_k}^*\right)^{-p}\right] \\ &\le C_p \int_{\mathbb{R}^2} |\xi|^{4p-2}\, \mathbb{E}\left[\exp\left\{-u_\varepsilon(t)\, |\tilde{Z}_t^* \xi|^2 - \sum_{k=1}^{N_t^\varepsilon} \big(\xi \cdot \tilde{Z}_{T_k}\, \ell_k(\tilde{\theta}_k, \tilde{\rho}_k)\big)^2\right\}\right] d\xi \\ &\le C_p \int_{\mathbb{R}^2} |\xi|^{4p-2}\, \mathbb{E}\left[\exp\left\{-u_\varepsilon(t)\, |\xi|^2 - \sum_{k=1}^{N_t^\varepsilon} \big(\xi \cdot \tilde{Z}_{T_k}\, \ell_k(\tilde{\theta}_k, \tilde{\rho}_k)\big)^2\right\}\right] d\xi. \end{aligned} \tag{13.22}$$

Hence, we have only to consider the upper estimate of

$$I_{t,\xi}^\varepsilon := \mathbb{E}\left[\exp\left\{-\sum_{k=1}^{N_t^\varepsilon} \big(\xi \cdot \tilde{Z}_{T_k}\, \ell_k(\tilde{\theta}_k, \tilde{\rho}_k)\big)^2\right\}\right].$$

Let $0 \ne X \in \mathbb{R}^2$ and $0 \ne \xi \in \mathbb{R}^2$. We remark that $\det(I + A(\theta)) = (1 + \cos\theta)/2$ and

$$\begin{aligned} &\big(I + A(\theta)\big)^{-1} A'(\theta) \\ &= \frac{1}{\det(I + A(\theta))} \begin{pmatrix} \frac{\cos\theta + 1}{2} & \frac{\sin\theta}{2} \\ -\frac{\sin\theta}{2} & \frac{\cos\theta + 1}{2} \end{pmatrix} \begin{pmatrix} -\frac{\sin\theta}{2} & -\frac{\cos\theta}{2} \\ \frac{\cos\theta}{2} & -\frac{\sin\theta}{2} \end{pmatrix} \\ &= \frac{1}{2} \begin{pmatrix} -\frac{\sin\theta}{1 + \cos\theta} & -1 \\ 1 & -\frac{\sin\theta}{1 + \cos\theta} \end{pmatrix}. \end{aligned}$$

$$= \frac{1}{2}\left(-\frac{\sin\theta}{1+\cos\theta}\, I + R(\pi/2)\right),$$

and that

$$\begin{aligned}
&\{\xi \cdot (I + A(\theta))^{-1} A'(\theta)\, X\}^2 \\
&= \frac{1}{4}\left\{-\frac{\sin\theta}{1+\cos\theta}\, \xi \cdot X + \xi \cdot R(\pi/2)\, X\right\}^2 \\
&= \frac{1}{4}\left\{\left(\frac{\sin\theta}{1+\cos\theta}\right)^2 (\xi \cdot X)^2 + (\xi \cdot R(\pi/2)\, X)^2 - \frac{2\sin\theta}{1+\cos\theta}(\xi \cdot X)\ (\xi \cdot R(\pi/2)\, X)\right\}.
\end{aligned}$$

As $(\xi \cdot X)^2 + (\xi \cdot R(\pi/2)X)^2 = |\xi|^2|X|^2$ then we have that either

$$\left(\frac{\xi}{|\xi|} \cdot \frac{X}{|X|}\right)^2 \geq \frac{1}{2} \quad \text{or} \quad \left(\frac{\xi}{|\xi|} \cdot R(\pi/2)\, \frac{X}{|X|}\right)^2 \geq \frac{1}{2}.$$

Further, define $\mathscr{A}(\xi, X)$ as the subset of $\theta \in \Pi$ such that the following inequality is satisfied:

$$\left(\frac{\xi}{|\xi|} \cdot \frac{X}{|X|}\right)\left(\frac{\xi}{|\xi|} \cdot R(\pi/2)\, \frac{X}{|X|}\right) \sin\theta \leq 0.$$

Note that $\mathscr{A}(\xi, X)$ is either $[-\frac{\pi}{2}, 0]$ or $[0, \frac{\pi}{2}]$ depending on the sign of $(\xi \cdot X)(\xi \cdot R(\pi/2)X)$. Then, we have that for $\theta \in \mathscr{A}(\xi, X)$:

$$\begin{aligned}
\left\{\xi \cdot \left(I + A(\theta)\right)^{-1} A'(\theta)\, X\right\}^2 &= \frac{1}{4}\left(-\frac{\sin\theta}{1+\cos\theta}\, \xi \cdot X + \xi \cdot R(\pi/2)\, X\right)^2 \\
&\geq \frac{|X|^2\, |\xi|^2}{8} \min\left(\frac{\sin^2\theta}{\left(1+\cos\theta\right)^2},\ 1\right) \\
&\geq \frac{|X|^2\, |\xi|^2}{32} \sin^2\theta \\
&\geq \frac{|X|^2\, |\xi|^2\, |\theta|^2}{128},
\end{aligned} \tag{13.23}$$

because $\theta \in \Pi, 0 \leq \cos\theta \leq 1, \sin^2\theta/4 = (1 - \cos 2\theta)/8 \leq 1/4 \leq 1$, and $|\sin\theta| \geq |\theta|/2$.

Exercise 13.5.3 Prove that $|\sin\theta| \geq |\theta|/2$ for $\theta \in \Pi$.

On the other hand, since $\chi_\varepsilon(\theta) = 1$ for $\theta \in \bar{\Pi}_{\bar{\epsilon}}$ with $\bar{\epsilon} := 2^{1/a}\varepsilon$, and Proposition 13.4.1 yields that $|\xi|^2 \leq |\tilde{Z}^*_{T_{k-1}}\xi|^2$, we see that for $\bar{\mathscr{A}} := \bar{\Pi}_{\bar{\epsilon}} \cap \mathscr{A}(\xi \cdot \tilde{Z}_{T_{k-1}}, \tilde{V}_{T_{k-1}} - v_{T_k}(\tilde{\rho}_k))$,

$$
\begin{aligned}
&\sum_{k=1}^{N_t^\varepsilon}\left(\xi\cdot\tilde{Z}_{T_k}\,\ell_k(\tilde{\theta}_k,\tilde{\rho}_k)\right)^2\\
&=\sum_{k=1}^{N_t^\varepsilon}\left(\xi\cdot\tilde{Z}_{T_{k-1}}\left(I+A_\varepsilon(\tilde{\theta}_k)\right)^{-1}A_\varepsilon{}'(\tilde{\theta}_k)\left(\tilde{V}_{T_{k-1}}-v_{T_k}(\tilde{\rho}_k)\right)\varphi(\tilde{\theta}_k)^2\right)^2\\
&\geq\sum_{k=1}^{N_t^\varepsilon}\left(\xi\cdot\tilde{Z}_{T_{k-1}}\left(I+A(\tilde{\theta}_k)\right)^{-1}A'(\tilde{\theta}_k)\left(\tilde{V}_{T_{k-1}}-v_{T_k}(\tilde{\rho}_k)\right)\varphi(\tilde{\theta}_k)^2\right)^2\mathbf{1}_{\bar{\Pi}_{2\varepsilon}}(\tilde{\theta}_k)\\
&\geq\sum_{k=1}^{N_t^\varepsilon}\frac{|\xi\cdot\tilde{Z}_{T_{k-1}}|^2}{128}\left|\tilde{V}_{T_{k-1}}-v_{T_k}(\tilde{\rho}_k)\right|^2\left|\varphi(\tilde{\theta}_k)\right|^4|\tilde{\theta}_k|^2\,\mathbf{1}_{\bar{\mathscr{A}}}(\tilde{\theta}_k)\\
&\geq\sum_{k=1}^{N_t^\varepsilon}\frac{|\xi|^2 r_0^2}{128}\,|\varphi(\tilde{\theta}_k)|^4|\tilde{\theta}_k|^2\,\mathbf{1}_{\bar{\mathscr{A}}}(\tilde{\theta}_k)\,\mathbf{1}_{[r_0,\infty)}\left(|\tilde{V}_{T_{k-1}}-v_{T_k}(\tilde{\rho}_k)|\right)\\
&=\frac{|\xi|^2 r_0^2}{128}\int_0^t\int_{\bar{\Pi}_\varepsilon}\int_0^1\int_0^{+\infty}\mathbf{1}_{\bar{\mathscr{A}}}(\theta)|\varphi(\theta)|^4|\theta|^2\\
&\quad\times\mathbf{1}_{[u,\infty)}\left(\phi_\varepsilon(|\bar{V}_{s-}-v_s(\rho)|)^\gamma\right)\mathbf{1}_{[r_0,\infty)}\left(\left|\bar{V}_{s-}-v_s(\rho)\right|\right)d\mathscr{N}\\
&\geq\frac{|\xi|^2 r_0^2}{128}\int_0^t\int_{\bar{\Pi}_\varepsilon}\int_0^1\int_0^{r_0^\gamma}\mathbf{1}_{\bar{\mathscr{A}}}(\theta)|\varphi(\theta)|^4|\theta|^2\,\mathbf{1}_{[r_0,\infty)}\left(|\bar{V}_{s-}-v_s(\rho)|\right)d\mathscr{N}.
\end{aligned}
\tag{13.24}
$$

Here we have used (13.23) in the third inequality, and (13.21) in the fourth inequality; the sixth inequality holds because of $\phi_\varepsilon\left(|\bar{V}_{s-}-v_s(\rho)|\right)\geq r_0$ on $|\bar{V}_{s-}-v_s(\rho)|>r_0$ for sufficiently small $\varepsilon>0$.

The constant $r_0>0$ is chosen according to the following probability concentration result in Lemma 4.9 of [7].

Lemma 13.5.4 *There exists $r_0>0$ and $q_0>0$ such that for any $w\in\mathbb{R}^2$ and any $t\in[0,T]$, we have*

$$
f_t(\{v;\,|v-w|\geq r_0\})\geq q_0.
$$

This result is based on the fact that the variance $\int|v|^2 f_t(dv)=c>0$ and the weak continuity of f_t.

Define

$$
L_t:=\int_0^t\int_{\bar{\Pi}_\varepsilon}\int_0^1\int_0^{r_0^\gamma}\mathbf{1}_{\bar{\mathscr{A}}}(\theta)|\varphi(\theta)|^4|\theta|^2\,\mathbf{1}_{[r_0,\infty)}\left(|\bar{V}_{s-}-v_s(\rho)|\right)d\mathscr{N}.
$$

Let $x\geq 0$ be a constant. Then, we have:

Exercise 13.5.5 Use Lemma 13.5.4 in order to prove that for any $\beta>0$ we have

$$\mathbb{E}\left[e^{-xL_t}\right] \leq \exp\left\{-t\, r_0^{\gamma}\, q_0 \int_{\bar{\Pi}'_{\bar{\epsilon}}} \left(1-e^{-cx|\theta|^{4\beta+2}}\right) b(\theta)d\theta\right\}. \tag{13.25}$$

Here $\bar{\Pi}'_{\bar{\epsilon}} = \bar{\Pi}_{\bar{\epsilon}} \cap [-(1-\varepsilon')\pi/2, (1-\varepsilon')\pi/2]$ for some $0 < \varepsilon' < 1$ which can be chosen as small as required independently of the rest of the problem variables. Here it is good to recall the techniques introduced in Sect. 11.5 and in particular Exercise 11.5.1.

For a hint, see Chap. 14.

In particular, taking $x = |\xi|^2 r_0^2/128$ in Exercise 13.5.5 yields that

$$\mathbb{E}\left[\exp\left\{-\frac{|\xi|^2 r_0^2}{128} L_t\right\}\right] \leq \exp\left\{-t\, r_0^{\gamma}\, q_0 \int_{\bar{\Pi}'_{\bar{\epsilon}}} \left(1-e^{-\frac{c|\xi|^2 r_0^2}{128}|\theta|^{4\beta+2}}\right) b(\theta)d\theta\right\}.$$

Now we apply the change of variables $|\theta|^{4\beta+2} = u$ so that we obtain for $\bar{\Pi}^{\beta}_{\bar{\epsilon}} = \left[(\varepsilon^a\pi)^{4\beta+2}, ((1-\varepsilon')\pi/2)^{4\beta+2}\right]$

$$\begin{aligned}
&\mathbb{E}\left[\exp\left\{-\frac{|\xi|^2 r_0^2}{128} L_t\right\}\right]\\
&\leq \exp\left\{-t\, r_0^{\gamma}\, q_0 \int_{\bar{\Pi}^{\beta}_{\bar{\epsilon}}} \left(1-e^{-\frac{c|\xi|^2 r_0^2}{128}u}\right) \frac{du}{u^{1+\frac{\alpha}{4\beta+2}}}\right\} \qquad (13.26)\\
&\leq \exp\left\{-t\, r_0^{\gamma}\, q_0 \left(1-e^{-\frac{cr_0^2}{128}}\right) \int_{(\varepsilon^a\pi)^{4\beta+2}\leq u\leq((1-\varepsilon')\pi/2)^{4\beta+2},\ |\xi|^2 u\geq 1} \frac{du}{u^{1+\frac{\alpha}{4\beta+2}}}\right\}\\
&= \exp\left\{-\frac{2t\, r_0^{\gamma}\, q_0 \left(1-e^{-\frac{cr_0^2}{128}}\right)}{\alpha}\left((\varepsilon^a\pi)^{-\alpha}\wedge|\xi|^{\frac{\alpha}{2\beta+1}} - \left(\frac{2}{\pi(1-\varepsilon')}\right)^{\alpha}\right)\right\}.
\end{aligned}$$

$$(13.27)$$

Now, recall that $u_\varepsilon(t) = t\,(\varepsilon^a\pi)^{4\beta+2-\alpha}$. For $|\xi| \geq (\varepsilon^a\pi)^{-(2\beta+1)}$, it is easy to see that if $4\beta+2-\alpha \geq 0$ then

$$u_\varepsilon(t)\,|\xi|^2 = t\,(\varepsilon^a\pi)^{4\beta+2-\alpha}\,|\xi|^2 \geq t\,|\xi|^{\frac{\alpha}{2\beta+1}}.$$

The reason for the above choice of $u_\varepsilon(t)$ is clear in the following estimate for (13.22). In fact, using (13.24) and (13.27) one obtains that

$$\begin{aligned}
I_t^{\varepsilon} &\leq C_p \int_{\mathbb{R}^2} |\xi|^{4p-2} e^{-u_\varepsilon(t)|\xi|^2}\\
&\times \exp\left\{-\frac{2t\, r_0^{\gamma}\, q_0 \left(1-e^{-\frac{cr_0^2}{128}}\right)}{\alpha}\left\{(\varepsilon^a\pi)^{-\alpha}\wedge|\xi|^{\frac{\alpha}{2\beta+1}} - \left(\frac{2}{\pi(1-\varepsilon')}\right)^{\alpha}\right\}\right\} d\xi
\end{aligned}$$

$$
\begin{aligned}
&= C_p \int_{|\xi|\le(\varepsilon^a\pi)^{-2\beta-1}} |\xi|^{4p-2} e^{-u_\varepsilon(t)|\xi|^2} \\
&\quad \times \exp\left\{ -\frac{2t\, r_0^\gamma\, q_0 \left(1-e^{-\frac{cr_0^2}{128}}\right)}{\alpha} \left\{ |\xi|^{\frac{\alpha}{2\beta+1}} - \left(\frac{2}{\pi(1-\varepsilon')}\right)^\alpha \right\} \right\} d\xi \\
&\quad + C_p \int_{|\xi|\ge(\varepsilon^a\pi)^{-2\beta-1}} |\xi|^{4p-2} e^{-u_\varepsilon(t)|\xi|^2} \\
&\quad \times \exp\left\{ -\frac{2t\, r_0^\gamma\, q_0 \left(1-e^{-\frac{cr_0^2}{128}}\right)}{\alpha} \left\{ (\varepsilon^a\pi)^{-\alpha} - \left(\frac{2}{\pi(1-\varepsilon')}\right)^\alpha \right\} \right\} d\xi \\
&\le C_p \int_{\mathbb{R}^2} |\xi|^{4p-2} \exp\left\{ -\frac{2t\, r_0^\gamma\, q_0 \left(1-e^{-\frac{cr_0^2}{128}}\right)}{\alpha} \left\{ |\xi|^{\frac{\alpha}{2\beta+1}} - \left(\frac{2}{\pi(1-\varepsilon')}\right)^\alpha \right\} \right\} d\xi \\
&\quad + C_p \int_{\mathbb{R}^2} |\xi|^{4p-2} e^{-t|\xi|^{\frac{\alpha}{2\beta+1}}} \\
&\quad \times \exp\left\{ -\frac{2t\, r_0^\gamma\, q_0 \left(1-e^{-\frac{cr_0^2}{128}}\right)}{\alpha} \left\{ (\varepsilon^a\pi)^{-\alpha} - \left(\frac{2}{\pi(1-\varepsilon')}\right)^\alpha \right\} \right\} d\xi \\
&\le C_p \exp\left\{ \frac{2\, r_0^\gamma\, q_0}{\alpha} \left(1-e^{-\frac{cr_0^2}{128}}\right) \left(\frac{2}{\pi(1-\varepsilon')}\right)^\alpha t \right\} \\
&\quad \times \left[1 + \exp\left\{ -\frac{2\, r_0^\gamma\, q_0}{\alpha} \left(1-e^{-\frac{cr_0^2}{128}}\right) (\varepsilon^a\pi)^{-\alpha} t \right\} \right] \\
&\quad \times \int_{\mathbb{R}^2} |\xi|^{4p-2} \exp\left[-\left\{ \left(\frac{2\, r_0^\gamma\, q_0}{\alpha} \left(1-e^{-\frac{cr_0^2}{128}}\right) \right) \wedge 1 \right\} t |\xi|^{\frac{\alpha}{2\beta+1}} \right] d\xi \\
&= C_p \exp\left\{ \frac{2\, r_0^\gamma\, q_0}{\alpha} \left(1-e^{-\frac{cr_0^2}{128}}\right) \left(\frac{2}{\pi(1-\varepsilon')}\right)^\alpha t \right\} \\
&\quad \times \left[1 + \exp\left\{ -\frac{2\, r_0^\gamma\, q_0}{\alpha} \left(1-e^{-\frac{cr_0^2}{128}}\right) (\varepsilon^a\pi)^{-\alpha} t \right\} \right] \\
&\quad \times \frac{2\pi\,(2\beta+1)}{\alpha} \left[t \left\{ \left(\frac{2\, r_0^\gamma\, q_0}{\alpha} \left(1-e^{-\frac{cr_0^2}{128}}\right) \right) \wedge 1 \right\} \right]^{-\frac{4p(2\beta+1)}{\alpha}} \Gamma\left(\frac{4p\,(2\beta+1)}{\alpha} \right) \\
&\le \tilde{C}_{4,p,\alpha,\beta,t}.
\end{aligned}
$$

The constant $\tilde{C}_{4,p,\alpha,\beta,t}$ may be defined by just replacing

$$
1 + \exp\left\{ -\frac{2\, r_0^\gamma\, q_0}{\alpha} \left(1-e^{-\frac{cr_0^2}{128}}\right) (\varepsilon^a\pi)^{-\alpha} t \right\}
$$

by 2 so that the constant is independent of ε.

Exercise 13.5.6 Obtain the last equality by evaluating the integral using cylindrical coordinates.

Hence, the conclusion of Lemma 13.5.2 follows from (13.20) and

$$\mathbb{E}\Big[\Big\{\det\Big(u_\varepsilon(t)\,\tilde{Z}_t\,\tilde{Z}_t^* + \sum_{k=1}^{N_t^\varepsilon} \tilde{Z}_{T_k}\,[\ell_k\,\ell_k^*](\tilde{\theta}_k,\tilde{\rho}_k)\,\tilde{Z}_{T_k}^*\Big)\Big\}^{-2p}\Big]^{\frac{1}{2}} \le \tilde{C}^{\frac{1}{2}}_{4,2p,\alpha,\beta,t}.$$

Remark 13.5.7 The following remarks are made in order to understand two important points in the proof.

1. We note that the proof of the finiteness of the determinant of the Malliavin covariance matrix in Lemma 13.5.2 uses not only the stable noise but also the Gaussian noise X^ε in order to obtain the result.
2. We also remark that one important step is the use of the change of variables $u = \theta^\beta$. Let $\delta > 0$. This gives for a positive constant c

$$\int_\delta^\pi \Big(1 - e^{-cx\theta^\beta}\Big)\frac{d\theta}{\theta^{1+\alpha}} = \beta^{-1}\int_{\delta^\beta}^{\pi^\beta} \big(1 - e^{-cxu}\big)\frac{du}{u^{1+\frac{\alpha}{\beta}}}.$$

   That is, the fact that β may be large does not change the stable property of the underlying jump measure, although it reduces its index from α to $\frac{\alpha}{\beta}$.

13.6 Upper Estimate of Malliavin Weight

The estimate obtained in Lemma 13.5.2 is crucial in the estimates for the characteristic function. We now go on to a more routine but important calculation in order to bound all the rest of the terms that appear in the weight $\tilde{\Gamma}_t(X^\varepsilon + \tilde{V}, \tilde{G})$ of Theorem 13.4.13. In fact, the goal of this section is to prove Proposition 13.6.18. This will be done in parts.

This section is technical and it is therefore important to find the right notation for norms.

Definition 13.6.1 We will extend the notion of norms of random matrices to stochastic processes as follows. Let W be any jointly measurable càdlàg matrix-valued stochastic process on the time interval $[0, t]$ such that it only changes values with each jump of the process N^ε. Then we define (extend) the norm for $p, q \ge 1$:

$$\|W\|_{p,q} := \mathbb{E}\left[\left(\sum_{k=1}^{N_t^\varepsilon} |W_{T_k}|^q\right)^{p/q}\right]^{1/p}.$$

Clearly, this norm depends on ε and $t > 0$ (which are supposed to be fixed) and no confusion should arise with $\|X\|_p \equiv \|X\|_{L^p(\Omega)}$ as X is a random matrix in the latter. In particular, one should be careful with $\|W\|_p$ which corresponds to the norm of a stochastic process and $\|W_t\|_{p,1} = \|W_t\|_p \equiv \|W_t\|_{L^p(\Omega)}$ which corresponds to the norm of a random matrix.

We will also use the notation (φ_t, DW_t) for the process for which its value at T_k is $(\varphi(\tilde{\theta}_k), D_k W_t)$ considered when $N_t^\varepsilon \geq k$. This notation will always be used with these norms.

Exercise 13.6.2 Prove that the above definition satisfies the properties of a norm. Define the Banach space that this norm generates.

Remark 13.6.3 In this section, it is important to know that the parameters β and r in φ and the function χ_ε can be chosen so that $\sup_{\theta\in\Pi} \left|A_\varepsilon^{(k)}(\theta)\varphi^{(l)}(\theta)\right|$ is uniformly bounded in ε for a limited number of values of $k, l \in \mathbb{N}^*$. This fact follows from the property that $\sup_{\theta\in\Pi} \left|A_\varepsilon^{(k)}(\theta)\theta^{k-1}\right| \leq C$, which is related to Exercise 13.4.8-(ii).

Lemma 13.6.4 *For any $p > 1$, $j \in \mathbb{N}$, $m \in \mathbb{N}^*$ and $\beta j > \alpha$ it holds that*

$$I_t(j, m, p) := \left\| \sum_{k=1}^{N_t^\varepsilon} |\varphi(\tilde{\theta}_k)|^j |v_{T_{k-}}(\tilde{\rho}_k)|^m \right\|_p \leq C_{p,j,m,t,\beta}\, \Gamma_\varepsilon^\gamma .$$

Similarly for $(2\beta - 1)j > \alpha$, one has

$$\left\| \sum_{k=1}^{N_t^\varepsilon} \left|\varphi(\tilde{\theta}_k)\, \varphi'(\tilde{\theta}_k)\right|^j |v_{T_{k-}}(\tilde{\rho}_k)|^m \right\|_p \leq C_{p,j,m,t,\beta}\, \Gamma_\varepsilon^\gamma .$$

Proof We only prove the first assertion, leaving the second as an exercise. Recalling the arguments in (13.12), we see that

$$I_t(j, m, p) \leq \mathbb{E}\left[\left(\int_0^t \int_{\bar{\Pi}_\varepsilon} \int_0^1 \int_0^{2\Gamma_\varepsilon^\gamma} |\varphi(\theta)|^j |v_s(\rho)|^m \, d\mathcal{N} \right)^p \right].$$

Then the Burkholder-type inequality in Proposition 7.1.2 leads us to see that

$$\begin{aligned} I_t(j, m, p) &\leq C_p \left(J_t(2j, 2m, p/2) + J_t(j, m, p) \right), \\ J_t(j, m, p) &:= \mathbb{E}\left[\left(\int_0^t \int_{\bar{\Pi}_\varepsilon} \int_0^1 \int_0^{2\Gamma_\varepsilon^\gamma} |\varphi(\theta)|^j |v_s(\rho)|^m \, d\widehat{\mathcal{N}} \right)^p \right]. \end{aligned}$$

An explicit calculation of the integral involved in the above expression gives

$$J_t(j, m, p) \leq C_{j,p,t} \Gamma_\varepsilon^{p\gamma} \left(\int_{\bar{\Pi}_\varepsilon} |\varphi(\theta)|^j b(\theta) d\theta \sup_{s \in [0,T]} \int |v|^m f_s(dv) \right)^p .$$

Therefore the condition $\beta j - \alpha > 0$ is required in order to obtain the finiteness of the above integral in θ. The last supremum is bounded due to Theorem 13.2.3. The condition $2\beta j > \alpha$ appears when one considers the finiteness of $J_t(2j, 2m, p/2)$.

Lemma 13.6.5 *Let $p > 1$ and W be any jointly measurable càdlàg real-valued stochastic process such that $\|W\|_{2p} < \infty$ and suppose that $2\beta j > \alpha$. Then it holds that*

$$\left\| \sum_{k=1}^{N_t^\varepsilon} W_{T_k} |\varphi(\tilde{\theta}_k)|^j |v_{T_k-}(\tilde{\rho}_k)|^m \right\|_p \leq \|W\|_{2p} C_{p,j,m,t,\beta} \Gamma_\varepsilon^\gamma .$$

Similarly, for $2(2\beta - 1)j > \alpha$, one has

$$\left\| \sum_{k=1}^{N_t^\varepsilon} W_{T_k} \left| \varphi(\tilde{\theta}_k) \varphi'(\tilde{\theta}_k) \right|^j |v_{T_k-}(\tilde{\rho}_k)|^m \right\|_p \leq \|W\|_{2p} C_{p,j,m,t,\beta} \Gamma_\varepsilon^\gamma .$$

In the particular case that $\| \sup_s |W_s| \|_{2p} < \infty$, one has the above inequalities for $\beta j > \alpha$ and $(2\beta - 1)j > \alpha$ and the norm $\|W\|_{2p}$ is replaced by $\| \sup_s |W_s| \|_{2p} < \infty$.

The proof of the above result, which follows from Hölder's inequality and Lemma 13.6.4, is easy and left as an exercise.

Lemma 13.6.6 *For any $2\beta > \alpha$ and $p > 1$, it holds that*

$$\left\| (Z_1, Z_2) \tilde{C}_t^{-1} \tilde{G}_t \right\|_p \leq C_{p,\alpha,\beta,T} \exp\left[C_{p,T} \Gamma_\varepsilon^\gamma \right] \left\| \tilde{G}_t \right\|_{2p}.$$

Proof This is a direct consequence of an easy application of the Cauchy–Schwarz inequality. In fact, $(Z_1, Z_2)^*$ is a two-dimensional normal vector with the mean vector 0 and the covariance matrix I_2. Furthermore using Lemma 13.5.2, one obtains the other results once one realizes that

$$\|\tilde{C}_t\|_{4p} = \left\| \frac{\check{C}_t}{\det \tilde{C}_t} \right\|_{4p} .$$

Here $\check{C}_t$ is the cofactor matrix of $\tilde{C}_t$. As this matrix can be explicitly written so as to fit the requirements of Lemma 13.6.5 for $j = 1$ and $m = 0, 1$, the result follows. The needed estimates follow from Theorem 13.2.3, Propositions 13.4.1 and 13.4.4.

Another direct application of Lemma 13.6.5 gives the following result.

Lemma 13.6.7 *Assume that $2q\beta > \alpha$. Then*

$$\|\varphi_t\, D\tilde{V}_t\|_{p,q} \leq C_{p,T,\alpha}\, \Gamma_\varepsilon^\gamma .$$

Proof Using Exercise 13.4.10-(v), one obtains that

$$\left\| \sum_{k=1}^{N_t^\varepsilon} \left| \varphi(\tilde{\theta}_k)\, D_k \tilde{V}_t \right|^q \right\|_p \leq C_{p,q,T} \left\| \sum_{k=1}^{N_t^\varepsilon} \varphi(\tilde{\theta}_k)^q \{ |\tilde{V}_{T_{k-1}}| + |v_{T_k}(\tilde{\rho}_k)| \}^q \right\|_p .$$

Therefore the result follows from Lemmas 13.6.4 and 13.6.5, and Propositions 13.3.4 and 13.4.1.

Now we start considering all the terms that appear in the second term of $\tilde{\Gamma}_t(X^\varepsilon + \tilde{V}, \tilde{G})$ one by one.

Lemma 13.6.8 *Assume that* $4\beta - 1 > \alpha$. *Then*

$$\left\| D\varphi_t^4 \left\{ D\tilde{V}_t \right\}^* \tilde{C}_t^{-1} \tilde{G}_t \right\|_{p,1} \leq C_{p,T,\alpha}\, \Gamma_\varepsilon^\gamma\, \|\tilde{C}_t^{-1}\|_{4p}\, \|\tilde{G}_t\|_{2p}.$$

Proof The assertion follows from the appropriate use of Hölder's inequality. In fact, one obtains that

$$\left\| \sum_{k=1}^{N_t^\varepsilon} \varphi^3(\tilde{\theta}_k)\, \varphi'(\tilde{\theta}_k) \left\{ D_k \tilde{V}_t \right\}^* \tilde{C}_t^{-1} \tilde{G}_t \right\|_p \leq C_{p,T,\alpha} \left\| \varphi_t^3\, \varphi_t'\, D\tilde{V}_t \right\|_{4p} \|\tilde{C}_t^{-1}\|_{4p}\, \|\tilde{G}_t\|_{2p}.$$

Therefore the result follows from Lemma 13.6.7.

Lemma 13.6.9 *Suppose that* $4\beta > \alpha$. *Then*

$$\left\| \varphi_t^4 \left(D^2 \tilde{V}_t \right)^* \tilde{C}_t^{-1}\, \tilde{G}_t \right\|_{p,1} \leq C_{p,T,\alpha}\, \Gamma_\varepsilon^\gamma\, \|\tilde{C}_t^{-1}\|_{4p}\, \|\tilde{G}_t\|_{2p}.$$

Proof Since

$$\partial_{\theta_k}^2 \tilde{V}_t^{(k)} = \tilde{Y}_t\, \tilde{Z}_{T_k}\, A_\varepsilon''(\theta_k) \left(\tilde{V}_{T_{k-1}} - v_{T_k}(\tilde{\rho}_k) \right),$$

we see from Remark 13.6.3 and Lemma 13.6.5 that

$$\begin{aligned} &\left\| \sum_{k=1}^{N_t^\varepsilon} \varphi(\tilde{\theta}_k)^4 \left\{ (\partial_{\theta_k}^2)_{\tilde{\theta}_k} \tilde{V}_t^{(k)} \right\}^* \right\|_{4p} \\ &= \left\| \sum_{k=1}^{N_t^\varepsilon} \varphi(\tilde{\theta}_k)^4 \left\{ \tilde{Y}_t\, \tilde{Z}_{T_k}\, A_\varepsilon''(\tilde{\theta}_k) \left(\tilde{V}_{T_{k-1}} - v_{T_k}(\tilde{\rho}_k) \right) \right\}^* \right\|_{4p} \end{aligned}$$

$$\leq \left\| \sum_{k=1}^{N_t^\varepsilon} \varphi(\tilde{\theta}_k)^4 \left\{ A_\varepsilon''(\tilde{\theta}_k)\left(\tilde{V}_{T_{k-1}} - v_{T_k}(\tilde{\rho}_k)\right)\right\}^* \right\|_{4p}$$
$$\leq C_{p,T,\alpha}\, \Gamma_\varepsilon^\gamma$$

from Theorem 13.2.3, Propositions 13.3.4, 13.4.1 and Lemma 13.6.4.

The proofs of the following lemmas are obtained in a similar fashion, so in each proof we will only note the particularity of each case, leaving the detailed proofs for the reader.

Lemma 13.6.10 *Suppose that* $4\beta > \alpha$. *Then*

$$\left\| \varphi_t^4 \left(D\tilde{V}_t\right)^* D\tilde{C}_t^{-1}\, \tilde{G}_t \right\|_{p,1} \leq C_{p,T,\alpha}\, \Gamma_\varepsilon^{2\gamma}\, \|\tilde{C}_t^{-1}\|_{4p}^2\, \|\tilde{G}_t\|_{2p}.$$

Proof Using the product rule for derivatives and (13.15), we have

$$(\partial_{\theta_k})_{\tilde{\theta}_k}(\tilde{C}_t^{(k)})^{-1} = -\tilde{C}_t^{-1}\,(\partial_{\theta_k})_{\tilde{\theta}_k}\tilde{C}_t^{(k)}\,\tilde{C}_t^{-1},$$
$$(\partial_{\theta_j})_{\tilde{\theta}_j}\tilde{V}_t^{(j,k)} = \tilde{Y}_t^{(k)}\,\tilde{Z}_{T_j}^{(k)}\,A_\varepsilon'(\tilde{\theta}_j)\left(\tilde{V}_{T_{j-1}}^{(k)} - v_{T_j}(\tilde{\rho}_j)\right),$$
$$(\partial_{\theta_k})_{\tilde{\theta}_k}\left((\partial_{\theta_j})_{\tilde{\theta}_j}\tilde{V}_t^{(j,k)}\varphi(\tilde{\theta}_j)^2\right)$$
$$= \begin{cases} (\partial_{\theta_k})_{\tilde{\theta}_k}\left(\tilde{Y}_t^{(k)}\,\tilde{Z}_{T_j}^{(k)}\right) A_\varepsilon'(\tilde{\theta}_j)\,\varphi(\tilde{\theta}_j)^2 \left(\tilde{V}_{T_{j-1}} - v_{T_j}(\tilde{\rho}_j)\right) & (j \leq k-1), \\ \tilde{Y}_t\,\tilde{Z}_{T_k}\,(\partial_{\theta_k})_{\tilde{\theta}_k}\left\{A_\varepsilon'(\theta_k)\,\varphi(\theta_k)^2\right\}\left(\tilde{V}_{T_{k-1}} - v_{T_k}(\tilde{\rho}_k)\right) & (j = k), \\ \tilde{Y}_t\,\tilde{Z}_{T_j}\,A_\varepsilon'(\tilde{\theta}_j)\,\varphi(\tilde{\theta}_j)^2\,(\partial_{\theta_k})_{\tilde{\theta}_k}\tilde{V}_{T_{j-1}}^{(k)} & (j \geq k+1). \end{cases}$$

Here, we have used the extended notation $\tilde{V}_t = \tilde{V}_t^{(j,k)}(\tilde{\theta}_j, \tilde{\theta}_k)$. Therefore using the explicit expression for $\tilde{C}_t$ in (13.17), the above equalities and using the notation simplification $(\partial_{\theta_j})_{\tilde{\theta}_j}\tilde{V}_t^{(j)}\varphi(\tilde{\theta}_j)^2 = \tilde{Y}_t\tilde{Z}_{T_j}\ell_j(\tilde{\theta}_j, \tilde{\rho}_j)$ and dividing the derivatives according to the adaptedness of the terms we obtain

$$\sum_{k=1}^{N_t^\varepsilon} \varphi(\tilde{\theta}_k)^4\left((\partial_{\theta_k})_{\tilde{\theta}_k}\tilde{V}_t^{(k)}\right)^*(\partial_{\theta_k})_{\tilde{\theta}_k}(\tilde{C}_t^{(k)})^{-1}\tilde{G}_t$$
$$= -\sum_{k=1}^{N_t^\varepsilon} \varphi(\tilde{\theta}_k)^4\left(\tilde{Y}_t\,\tilde{Z}_{T_k}\,A_\varepsilon'(\tilde{\theta}_k)\left(\tilde{V}_{T_{k-1}} - v_{T_k}(\tilde{\rho}_k)\right)\right)^*\tilde{C}_t^{-1}$$
$$\times \left\{ \sum_{j=1}^{k-1} (\partial_{\theta_k})_{\tilde{\theta}_k}\left(\tilde{Y}_t^{(k)}\,\tilde{Z}_{T_j}^{(k)}\right)\ell_j(\tilde{\theta}_j, \tilde{\rho}_j)\left\{(\partial_{\theta_j})_{\tilde{\theta}_j}\tilde{V}_t^{(j)}\varphi(\tilde{\theta}_j)^2\right\}^* \right.$$
$$+ \sum_{j=1}^{k-1} (\partial_{\theta_j})_{\tilde{\theta}_j}\tilde{V}_t^{(j)}\varphi(\tilde{\theta}_j)^2\left\{(\partial_{\theta_k})_{\tilde{\theta}_k}\left(\tilde{Y}_t^{(k)}\,\tilde{Z}_{T_j}^{(k)}\right)\ell_j(\tilde{\theta}_j, \tilde{\rho}_j)\right\}^*$$
$$+ \tilde{Y}_t\,\tilde{Z}_{T_k}\,(\partial_{\theta_k})_{\tilde{\theta}_k}\left\{A_\varepsilon'(\theta_k)\,\varphi(\theta_k)^2\right\}\left(\tilde{V}_{T_{k-1}} - v_{T_k}(\tilde{\rho}_k)\right)\left\{(\partial_{\theta_k})_{\tilde{\theta}_k}\tilde{V}_t^{(k)}\varphi(\tilde{\theta}_k)^2\right\}^*$$

$$+ (\partial_{\theta_k})_{\tilde{\theta}_k} \tilde{V}_t^{(k)} \varphi(\tilde{\theta}_k)^2 \left\{ \tilde{Y}_t \tilde{Z}_{T_k} (\partial_{\theta_k})_{\tilde{\theta}_k} \left\{ A_\varepsilon{}'(\theta_k) \varphi(\theta_k)^2 \right\} \left(\tilde{V}_{T_{k-1}} - v_{T_k}(\tilde{\rho}_k) \right) \right\}^*$$

$$+ \sum_{j=k+1}^{N_t^\varepsilon} \tilde{Y}_t \tilde{Z}_{T_j} A_\varepsilon{}'(\tilde{\theta}_j) \varphi(\tilde{\theta}_j)^2 \left(\partial_{\theta_k} \right)_{\tilde{\theta}_k} \tilde{V}_{T_{j-1}}^{(k)} \left\{ (\partial_{\theta_j})_{\tilde{\theta}_j} \tilde{V}_t^{(j)} \varphi(\tilde{\theta}_j)^2 \right\}^*$$

$$+ \sum_{j=k+1}^{N_t^\varepsilon} (\partial_{\theta_j})_{\tilde{\theta}_j} \tilde{V}_t^{(j)} \varphi(\tilde{\theta}_j)^2 \left\{ \tilde{Y}_t \tilde{Z}_{T_j} A_\varepsilon{}'(\tilde{\theta}_j) \varphi(\tilde{\theta}_j)^2 \left(\partial_{\theta_k} \right)_{\tilde{\theta}_k} \tilde{V}_{T_{j-1}}^{(k)} \right\}^* \Bigg\} \tilde{C}_t^{-1} \tilde{G}_t.$$

We remark that, for $j < k$, using Exercise 13.4.3, we obtain

$$(\partial_{\theta_k})_{\tilde{\theta}_k} \left(\tilde{Y}_t^{(k)} \tilde{Z}_{T_j}^{(k)} \right) = \tilde{Y}_t \tilde{Z}_{T_j} \left(I + A_\varepsilon(\tilde{\theta}_k) \right)^{-1} A_\varepsilon{}'(\tilde{\theta}_k).$$

Therefore the result follows as before after long calculations, as in the proof of Lemma 13.6.9.

So far no particular localization process $\tilde{G}_t$ has been defined. Recall that in Theorems 12.5.9 and 12.6.8 a localization function Ξ was used in order to control the degeneration of the Malliavin covariance matrix. We now define a similar localization for this case. We shall use a non-decreasing smooth function $\Phi_\varepsilon : [0, \infty) \to [0, 1]$ such that the derivatives of all orders of Φ_ε are bounded uniformly for[6] $\varepsilon \in (0, e^{-4})$ and that

$$\Phi_\varepsilon(x) = \begin{cases} 0 & (x \le \Gamma_\varepsilon - 1), \\ 1 & (x \ge \Gamma_\varepsilon). \end{cases}$$

Moreover, we shall introduce a smooth function $\Psi : [0, +\infty) \to [0, 1]$ such that

$$\Psi(x) = \begin{cases} 0 & (x \ge 3/4)\,, \\ 1 & (x \le 1/4)\,. \end{cases}$$

Write

$$\tilde{\Sigma}_N = \Phi_\varepsilon(|V_0|) + \sum_{k=1}^{N} \Phi_\varepsilon\left(|\tilde{V}_{T_k}|\right), \quad \tilde{G}_N = \Psi\left(\tilde{\Sigma}_N\right).$$

Then, we see that

$$\mathbf{1}_{\{\sup_{s\in[0,t]} |\tilde{V}_s| \le \Gamma_\varepsilon - 1\}} \le \tilde{G}_t := \tilde{G}_{N_t^\varepsilon} \le \mathbf{1}_{\{\sup_{s\in[0,t]} |\tilde{V}_s| \le \Gamma_\varepsilon\}}. \tag{13.28}$$

Exercise 13.6.11 Prove the above inequality.

Remark 13.6.12 By the definition of the function Ψ, we have

[6]Recall that this restriction on ε appears due to Exercise 13.3.1.

$$\sup_{1\le k\le l}\left|\Psi^{(k)}(x)\right| \le C_l\,\mathbf{1}_{(1/4\le x\le 3/4)}.$$

Moreover, by the definition of the function Φ_ε, we have

$$\tilde{\Sigma}_{N_t^\varepsilon} \in \left[\frac{1}{4},\frac{3}{4}\right] \Longrightarrow \sup_{s\in[0,t]}|\tilde{V}_s| \in \left[\Gamma_\varepsilon-1,\Gamma_\varepsilon\right].$$

We remark that the function $\mathbb{R}^2 \ni x \longmapsto \Phi_\varepsilon(|x|) \in [0,1]$ has uniformly bounded derivatives in ε, because $\Phi_\varepsilon(|x|) = 0$ for $|x| \le \Gamma_\varepsilon - 1$.

Lemma 13.6.13 *For any $\kappa \in (0,\delta)$ as in Theorem 13.2.3, we have for $2q\beta > \alpha$*

$$\|\varphi_t\, D\tilde{G}_t\|_{p,q} \le C_{p,T}\,\lambda_\varepsilon\,\Gamma_\varepsilon^\gamma\,\exp\left(-C_{p,T}\Gamma_\varepsilon^\kappa\right).$$

Proof From the definition of Ψ, we have

$$\left|\Psi'(\tilde{\Sigma}_t^{(k)})\right| \le C\,\mathbf{1}_{(1/4\le\tilde{\Sigma}_t^{(k)}\le 3/4)}.$$

On the other hand, the derivative of $\tilde{\Sigma}_{N_t^\varepsilon}$ will be non-zero only if

$$\sup_{0\le s\le t}\left|\tilde{V}_s^{(k)}\right| \in [\Gamma_\varepsilon-1,\Gamma_\varepsilon]$$

under $1/4 \le \tilde{\Sigma}_t^{(k)} \le 3/4$. Define $\overline{\Phi}_\varepsilon(x) := \Phi_\varepsilon(|x|)$ for $x \in \mathbb{R}^2$. We remark that on $N_t^\varepsilon \ge k$

$$\begin{aligned}
\partial_{\theta_k}\tilde{\Sigma}_t^{(k)} &= \partial_{\theta_k}\left(\overline{\Phi}_\varepsilon(V_0) + \sum_{j=1}^{N_t^\varepsilon}\overline{\Phi}_\varepsilon(\tilde{V}_{T_j}^{(k)})\right)\\
&= \sum_{j=k}^{N_t^\varepsilon}\overline{\Phi}_\varepsilon'(\tilde{V}_{T_j}^{(k)})\,\partial_{\theta_k}\tilde{V}_{T_j}^{(k)}\\
&= \sum_{j=k}^{N_t^\varepsilon}\overline{\Phi}_\varepsilon'(\tilde{V}_{T_j}^{(k)})\left\{\tilde{Y}_{T_j}\,\tilde{Z}_{T_k}\,A_\varepsilon{}'(\theta_k)\left(\tilde{V}_{T_{k-1}} - v_{T_k}(\tilde{\rho}_k)\right)\right\}.
\end{aligned}$$

Hence, we see that as $\overline{\Phi}_\varepsilon$ has uniformly bounded derivatives then using the same technique as in the proof of Lemma 13.6.8, we have

$$D_k\tilde{G}_t = \Psi'(\tilde{\Sigma}_t)\sum_{j=k}^{N_t^\varepsilon}\overline{\Phi}_\varepsilon'(\tilde{V}_{T_j})\left\{\tilde{Y}_{T_j}\,\tilde{Z}_{T_k}\,A_\varepsilon{}'(\tilde{\theta}_k)\left(\tilde{V}_{T_{k-1}} - v_{T_k}(\tilde{\rho}_k)\right)\right\}.$$

Therefore

$$\sum_{k=1}^{N_t^\varepsilon} \left|\varphi(\tilde{\theta}_k) D_k \tilde{G}_t\right|^q$$
$$\leq C(N_t^\varepsilon)^q \sum_{k=1}^{N_t^\varepsilon} \left|\varphi(\tilde{\theta}_k)\left(\tilde{V}_{T_{k-1}} - v_{T_k}(\tilde{\rho}_k)\right)\right|^q \mathbf{1}_{(\sup_{0\leq s\leq t}|\tilde{V}_s|\in[\Gamma_\varepsilon - 1, \Gamma_\varepsilon])}.$$

The final result follows by using exponential Chebyshev's inequality. That is,

$$\mathbb{P}\left[\sup_{0\leq s\leq t}|\tilde{V}_s| \geq \Gamma_\varepsilon - 1\right] \leq \exp(-C(\Gamma_\varepsilon - 1)^\kappa)\,\mathbb{E}\left[\sup_{0\leq s\leq t}\exp(C|\tilde{V}_s|^\kappa)\right].$$

The above expectation is finite due to the equality in law between $\tilde{V}$ and $\bar{V}$ and Proposition 13.3.4.

Lemma 13.6.14 *Assume that $\beta > \alpha$. Then*

$$\left\|\varphi_t^4\left(D\tilde{V}_t\right)^* \tilde{C}_t^{-1} D\tilde{G}_t\right\|_{p,1} \leq C_{p,T,\alpha}\,\lambda_\varepsilon \Gamma_\varepsilon^{C_p\gamma}\,\exp\left[-C_p\,\Gamma_\varepsilon^\kappa\right]\left\|\tilde{C}_t^{-1}\right\|_{2p}.$$

Proof Using Hölder's inequality one obtains that

$$\left\|\sum_{k=1}^{N_t^\varepsilon}\varphi(\tilde{\theta}_k)^4\left(D_k\tilde{V}_t\right)^*\tilde{C}_t^{-1} D_k\tilde{G}_t\right\|_p \leq \|\varphi_t^3 D\tilde{V}_t\|_{4p,2}\|\tilde{C}_t^{-1}\|_{2p}\|\varphi_t D\tilde{G}_t\|_{4p,2}.$$

Therefore the proof follows the same pattern as the proof of previous Lemmas 13.6.8 and 13.6.9, and uses Lemmas 13.6.7 and 13.6.13.

Remark 13.6.15 Note that the following bound is also valid:

$$\left\|\sum_{k=1}^{N_t^\varepsilon}\varphi(\tilde{\theta}_k)^4\left(D_k\tilde{V}_t\right)^*\tilde{C}_t^{-1} D_k\tilde{G}_t\right\|_p \leq C_{p,T,\alpha}\,\Gamma_\varepsilon^{C_p\gamma}\,\exp\left[C_{p,T}\,\Gamma_\varepsilon^\gamma\right]\|\varphi_t D\tilde{G}_t\|_{4p,2}.$$

Recall that the goal of this long section is to obtain an estimate for moments of the weight $\tilde{\Gamma}_t(X^\varepsilon + \tilde{V}, \tilde{G})$ of Theorem 13.4.13. Now we start to study its main term which is the logarithmic derivative. First, note that

$$\log g_{N_t^\varepsilon}\left(V_0, (T, \tilde{\theta}, \tilde{\rho})_{1\to N_t^\varepsilon}\right) = \log q_\varepsilon(T_1, \mathscr{H}_0(V_0), \tilde{\theta}_1, \tilde{\rho}_1)$$
$$+ \sum_{j=2}^{N_t^\varepsilon}\log q_\varepsilon\left(T_j, \mathscr{H}_{j-1}(V_0, (T, \tilde{\theta}, \tilde{\rho})_{1\to j-1}), \tilde{\theta}_j, \tilde{\rho}_j\right),$$

we can compute that

$$D_{N_t^\varepsilon}\Big\{ \log g_{N_t^\varepsilon}\big(V_0, (T, \tilde\theta, \tilde\rho)_{1\to N_t^\varepsilon}\big)\Big\}$$
$$= D_{N_t^\varepsilon}\Big\{ \log q_\varepsilon\big(T_{N_t^\varepsilon}, \mathscr{H}_{N_t^\varepsilon-1}(V_0, (T, \tilde\theta, \tilde\rho)_{1\to N_t^\varepsilon-1}), \tilde\theta_{N_t^\varepsilon}, \tilde\rho_{N_t^\varepsilon}\big)\Big\},$$

$$D_k\Big\{ \log g_{N_t^\varepsilon}\big(V_0, (T, \tilde\theta, \tilde\rho)_{1\to N_t^\varepsilon}\big)\Big\}$$
$$= D_k\Big\{ \log q_\varepsilon\big(T_k, \mathscr{H}_{k-1}(V_0, (T, \tilde\theta, \tilde\rho)_{1\to k-1}), \tilde\theta_k, \tilde\rho_k\big)$$
$$+ \sum_{j=k+1}^{N_t^\varepsilon} \log q_\varepsilon\big(T_j, \mathscr{H}_{j-1}(V_0, (T, \tilde\theta, \tilde\rho)_{1\to j-1}), \tilde\theta_j, \tilde\rho_j\big)\Big\}$$

for $1 \le k \le N_t^\varepsilon - 1$. These derivatives will generate a number of terms in the weight $\tilde\Gamma_t\big(X^\varepsilon + \tilde V, \tilde G\big)$ of Theorem 13.4.13. We analyze them in various parts.

Lemma 13.6.16 *Assume that* $4\beta - 1 > \alpha$. *Then*

$$\left\| \sum_{k=1}^{N_t^\varepsilon} \varphi(\tilde\theta_k)^4 \left(D_k \tilde V_t\right)^* \tilde C_t^{-1}\, \tilde G_t\, \frac{D_k q_\varepsilon\big(T_k, \tilde V_{T_{k-1}}, \tilde\theta_k, \tilde\rho_k\big)}{q_\varepsilon\big(T_k, \tilde V_{T_{k-1}}, \tilde\theta_k, \tilde\rho_k\big)} \right\|_p$$
$$\le C_{p,T,\alpha}\, \Gamma_\varepsilon^\gamma \left\|\tilde C_t^{-1}\right\|_{4p} \left\|\tilde G_t\right\|_{2p}.$$

Proof We remark that $\chi_\varepsilon(\theta) = 0$ for $|\theta| \le \varepsilon^a \pi/2$ implies that $A_\varepsilon(\theta) = 0$ and that $\tilde\chi(\theta) = 0$ for $|\theta| \ge \pi/2$. Moreover, we remark that

$$q_\varepsilon\big(T_k, \tilde V_{T_{k-1}}, \theta_k, \tilde\rho_k\big) = \frac{1}{\lambda_\varepsilon}\, \phi_\varepsilon\big(|\tilde V_{T_{k-1}} - v_{T_k}(\tilde\rho_k)|\big)^\gamma\, b(\theta_k)$$

for $|\theta_k| \ge \varepsilon^a\pi/2$. Then, we see that using (13.9) and Proposition 13.4.1 we obtain

$$\left\| \sum_{k=1}^{N_t^\varepsilon} \varphi(\tilde\theta_k)^4 \left((\partial_{\theta_k})_{\tilde\theta_k} \tilde V_t^{(k)}\right)^* \tilde C_t^{-1}\, \tilde G_t\, \frac{(\partial_{\theta_k})_{\tilde\theta_k} q_\varepsilon\big(T_k, \tilde V_{T_{k-1}}, \theta_k, \tilde\rho_k\big)}{q_\varepsilon\big(T_k, \tilde V_{T_{k-1}}, \tilde\theta_k, \tilde\rho_k\big)} \right\|_p$$
$$= \Bigg\| \sum_{k=1}^{N_t^\varepsilon} \varphi(\tilde\theta_k)^4 \left\{\tilde Y_t\, \tilde Z_{T_k}\, A_\varepsilon{}'(\tilde\theta_k) \left(\tilde V_{T_{k-1}} - v_{T_k}(\tilde\rho_k)\right)\right\}^*$$
$$\times\, \tilde C_t^{-1}\, \tilde G_t\, (\partial_{\theta_k})_{\tilde\theta_k} \ln q_\varepsilon\big(T_k, \tilde V_{T_{k-1}}, \theta_k, \tilde\rho_k\big)\Bigg\|_p$$
$$= \Bigg\| -\sum_{k=1}^{N_t^\varepsilon} \varphi(\tilde\theta_k)^4 \left\{\tilde Y_t\, \tilde Z_{T_k}\, A_\varepsilon{}'(\tilde\theta_k) \left(\tilde V_{T_{k-1}} - v_{T_k}(\tilde\rho_k)\right)\right\}^*$$
$$\times\, \tilde C_t^{-1}\, \tilde G_t\, \frac{1+\alpha}{\tilde\theta_k}\, \mathbf{1}_{(|\tilde\theta_k| \ge \varepsilon^a\pi/2)}\Bigg\|_p$$

$$
\begin{aligned}
&\leq (1+\alpha) \left\| \sum_{k=1}^{N_t^\varepsilon} \frac{\varphi(\tilde{\theta}_k)^4}{\tilde{\theta}_k} \left\{ \tilde{Y}_t\, \tilde{Z}_{T_k}\, A_\varepsilon{}'(\tilde{\theta}_k) \left(\tilde{V}_{T_{k-1}} - v_{T_k}(\tilde{\rho}_k) \right) \right\}^* \right\|_{4p} \|\tilde{C}_t^{-1}\|_{4p} \|\tilde{G}_t\|_{2p} \\
&\leq (1+\alpha) \left\| \sum_{k=1}^{N_t^\varepsilon} \frac{\varphi(\tilde{\theta}_k)^4}{\tilde{\theta}_k} \|A_\varepsilon{}'(\tilde{\theta}_k)\| \, |\tilde{V}_{T_{k-1}} - v_{T_k}(\tilde{\rho}_k)| \right\|_{4p} \|\tilde{C}_t^{-1}\|_{4p} \|\tilde{G}_t\|_{2p} \\
&\leq C_{\kappa,p,\alpha,T}\, \Gamma_\varepsilon^\gamma \, \|\tilde{C}_t^{-1}\|_{4p} \|\tilde{G}_t\|_{2p}
\end{aligned}
$$

from Proposition 13.4.1, Theorem 13.2.3, and Lemma 13.6.5 . Note that in the second equality, we have used that $A_\varepsilon'(\theta) = 0$ for $|\theta| \leq \varepsilon^a \pi/2$. The proof is complete.

Lemma 13.6.17 *Suppose that* $4\beta > \alpha$. *Then, it holds that*

$$
\begin{aligned}
&\left\| \sum_{k=1}^{N_t^\varepsilon} \varphi(\tilde{\theta}_k)^4 \left(D_k \tilde{V}_t\right)^* \tilde{C}_t^{-1}\, \tilde{G}_t \sum_{j=k+1}^{N_t^\varepsilon} \frac{D_k q_\varepsilon\big(T_j, \tilde{V}_{T_{j-1}}, \tilde{\theta}_j, \tilde{\rho}_j\big)}{q_\varepsilon\big(T_j, \tilde{V}_{T_{j-1}}, \tilde{\theta}_j, \tilde{\rho}_j\big)} \right\|_p \\
&\leq C_{\kappa,p,\alpha,T,\gamma}\, \varepsilon^{-a\alpha-1}\, \Gamma_\varepsilon^{2\gamma} \, \|\tilde{C}_t^{-1}\|_{4p} \|\tilde{G}_t\|_{2p}.
\end{aligned}
$$

Proof Using the definition of q_ε given in (13.9), we have for $k+1 \leq j \leq N_t^\varepsilon$,

$$
\begin{aligned}
&D_k \left(\ln q_\varepsilon(T_j, \tilde{V}_{T_{j-1}}, \tilde{\theta}_j, \tilde{\rho}_j) \right) \\
&= D_k \left(\ln \left\{ \frac{1}{\lambda_\varepsilon} \phi_\varepsilon\big(|\tilde{V}_{T_{j-1}} - v_{T_j}(\tilde{\rho}_j)|\big)^\gamma b(\tilde{\theta}_j) \right\} \right) \mathbf{1}_{\bar{\Pi}_\varepsilon}(\tilde{\theta}_j) \\
&\quad + D_k \left(\ln \left\{ 1 - \frac{1}{2\Gamma_\varepsilon^\gamma} \int_0^1 \phi_\varepsilon\big(|\tilde{V}_{T_{j-1}} - v_{T_j}(r)|\big)^\gamma dr \right\} \varepsilon^{a\alpha}\, \tilde{\chi}\left(\frac{\tilde{\theta}_j}{\varepsilon^a}\right) b(\tilde{\theta}_j) \right) \mathbf{1}_{\Pi_\varepsilon}(\tilde{\theta}_j) \\
&= D_k \left(\gamma \ln \phi_\varepsilon\big(|\tilde{V}_{T_{j-1}} - v_{T_j}(\tilde{\rho}_j)|\big) \right) \mathbf{1}_{\bar{\Pi}_\varepsilon}(\tilde{\theta}_j) \\
&\quad + D_k \left(\ln \left\{ 1 - \frac{1}{2\Gamma_\varepsilon^\gamma} \int_0^1 \phi_\varepsilon\big(|\tilde{V}_{T_{j-1}} - v_{T_j}(r)|\big)^\gamma dr \right\} \right) \mathbf{1}_{\Pi_\varepsilon}(\tilde{\theta}_j) \\
&= \gamma\, (\ln \phi_\varepsilon)'\big(|\tilde{V}_{T_{j-1}} - v_{T_j}(\tilde{\rho}_j)|\big) \frac{\tilde{V}_{T_{j-1}} - v_{T_j}(\tilde{\rho}_j)}{|\tilde{V}_{T_{j-1}} - v_{T_j}(\tilde{\rho}_j)|} \cdot D_k \big\{ \tilde{V}_{T_{j-1}} - v_{T_j}(\tilde{\rho}_j) \big\} \mathbf{1}_{\bar{\Pi}_\varepsilon}(\tilde{\theta}_j) \\
&\quad - \frac{1}{2\Gamma_\varepsilon^\gamma - \int_0^1 \phi_\varepsilon\big(|\tilde{V}_{T_{j-1}} - v_{T_j}(r)|\big)^\gamma dr} \\
&\quad \times \gamma \int_0^1 \phi_\varepsilon\big(|\tilde{V}_{T_{j-1}} - v_{T_j}(r)|\big)^\gamma (\ln \phi_\varepsilon)'\big(|\tilde{V}_{T_{j-1}} - v_{T_j}(r)|\big) \\
&\quad \times \frac{\tilde{V}_{T_{j-1}} - v_{T_j}(r)}{|\tilde{V}_{T_{j-1}} - v_{T_j}(r)|} \cdot D_k \big\{ \tilde{V}_{T_{j-1}} - v_{T_j}(r) \big\} dr\, \mathbf{1}_{\Pi_\varepsilon}(\tilde{\theta}_j) \\
&=: K_{1,j}^\varepsilon + K_{2,j}^\varepsilon.
\end{aligned}
$$

We remark here that $2\Gamma_\varepsilon^\gamma - \int_0^1 \phi_\varepsilon\big(|\tilde{V}_{T_{j-1}} - v_{T_j}(r)|\big)^\gamma dr \geq \Gamma_\varepsilon^\gamma$, which means that the term in the denominator of $K_{2,j}^\varepsilon$ does not vanish.

Now in order to estimate each term in the above expressions, we first note that from the definition of $\tilde{V}$ in (13.10), we have

$$D_k\big\{\tilde{V}_{T_{j-1}} - v_{T_j}(\tilde{\rho}_j)\big\} = \begin{cases} A_\varepsilon'(\tilde{\theta}_k)\,\big\{\tilde{V}_{T_{k-1}} - v_{T_k}(\tilde{\rho}_k)\big\}, & (j = k+1), \\ Y_{T_{j-1}} Z_{T_k}^{-1} A_\varepsilon'(\tilde{\theta}_k)\,\big\{\tilde{V}_{T_{k-1}} - v_{T_k}(\tilde{\rho}_k)\big\}, & (k+2 \leq j \leq N_t^\varepsilon). \end{cases}$$

Since $\big\|I + A_\varepsilon(\tilde{\theta}_k)\big\| \leq 1$, $\big\|A_\varepsilon'(\tilde{\theta}_k)\big\| \leq C$, $\varepsilon \leq \phi_\varepsilon \leq \Gamma_\varepsilon$ and Proposition 13.4.1, we have

$$\begin{aligned}
&\left\| \sum_{k=1}^{N_t^\varepsilon} \varphi(\tilde{\theta}_k)^4 \left(D_k \tilde{V}_t\right)^* \tilde{C}_t^{-1} \tilde{G}_t \sum_{j=k+1}^{N_t^\varepsilon} \frac{D_k q_\varepsilon\big(T_j, \tilde{V}_{T_{j-1}}, \tilde{\theta}_j, \tilde{\rho}_j\big)}{q_\varepsilon\big(T_j, \tilde{V}_{T_{j-1}}, \tilde{\theta}_j, \tilde{\rho}_j\big)} \right\|_p \\
&\leq \left\{ \left\| \sum_{k=1}^{N_t^\varepsilon} \varphi(\tilde{\theta}_k)^4 \left|D_k \tilde{V}_t\right| \sum_{j=k+1}^{N_t^\varepsilon} K_{1,j}^\varepsilon \right\|_{4p} + \left\| \sum_{k=1}^{N_t^\varepsilon} \varphi(\tilde{\theta}_k)^4 \left|D_k \tilde{V}_t\right| \sum_{j=k+1}^{N_t^\varepsilon} K_{2,j}^\varepsilon \right\|_{4p} \right\} \\
&\quad \times \left\|\tilde{C}_t^{-1}\right\|_{4p} \left\|\tilde{G}_t\right\|_{2p} \\
&\leq \frac{\gamma}{2\varepsilon} \left\| \sum_{k=1}^{N_t^\varepsilon} \varphi(\tilde{\theta}_k)^4 \left|D_k \tilde{V}_t\right| \sum_{j=k+1}^{N_t^\varepsilon} \left|D_k\{\tilde{V}_{T_{j-1}} - v_{T_j}(\tilde{\rho}_j)\}\right| \right\|_{4p} \left\|\tilde{C}_t^{-1}\right\|_{4p} \left\|\tilde{G}_t\right\|_{2p} \\
&\quad + \frac{2}{\varepsilon} \left\| \sum_{k=1}^{N_t^\varepsilon} \varphi(\tilde{\theta}_k)^4 \left|D_k \tilde{V}_t\right| \sum_{j=k+1}^{N_t^\varepsilon} D_k \tilde{V}_{T_{j-1}} \right\|_{4p} \left\|\tilde{C}_t^{-1}\right\|_{4p} \left\|\tilde{G}_t\right\|_{2p} \\
&\leq C \left(\frac{\gamma}{2\varepsilon} + \frac{2}{\varepsilon}\right) \left\| N_t^\varepsilon \sum_{k=1}^{N_t^\varepsilon} \varphi(\tilde{\theta}_k)^4 \left|D_k \tilde{V}_t\right| \left|\tilde{V}_{T_{k-1}} - v_{T_k}(\tilde{\rho}_k)\right| \right\|_{4p} \left\|\tilde{C}_t^{-1}\right\|_{4p} \left\|\tilde{G}_t\right\|_{2p} \\
&\leq C_{K,p,\alpha,T,\gamma}\, \varepsilon^{-a\alpha-1}\, \Gamma_\varepsilon^{2\gamma} \left\|\tilde{C}_t^{-1}\right\|_{4p} \left\|\tilde{G}_t\right\|_{2p}
\end{aligned}$$

by Theorems 13.2.3, 13.3.4 and Proposition 13.4.4. Here, we have used the upper estimate of

$$\left|\big(\log \phi_\varepsilon\big)'\big(|\tilde{V}_{T_{j-1}} - v_{T_j}(\tilde{\rho}_j)|\big)\right| \leq \frac{1}{2\varepsilon}$$

in the third inequality and the techniques used in previous lemmas (such as Lemma 13.6.7), while the Poisson random variable N_t^ε with the parameter $\lambda_\varepsilon t$ has the following property:

$$\left\|N_t^\varepsilon\right\|_q \leq C_{q,\alpha}\, \varepsilon^{-a\alpha}\, \Gamma_\varepsilon^\gamma.$$

Proposition 13.6.18 *Assume that* $4\beta - 1 > \alpha$ *and* $0 < t \leq T$. *Then it holds that*

$$\left\| \tilde{\Gamma}_t\left(X^\varepsilon + \tilde{V}, \tilde{G}\right) \right\|_p$$
$$\leq C_{p,T,\alpha} \exp\left[C_{p,T}\, \Gamma_\varepsilon^\gamma\right] \left(\varepsilon^{-a\alpha - 1}\, \Gamma_\varepsilon^{2\gamma} \left\| \tilde{G}_t \right\|_{2p} + \varepsilon^{-a\alpha}\, \Gamma_\varepsilon^{C_p \gamma} \exp\left[- C_p\, \Gamma_\varepsilon^\kappa \right] \right).$$

Proof The proof is a direct consequence of Lemmas 13.5.2, 13.6.6, 13.6.8–13.6.10, 13.6.14, 13.6.16 and 13.6.17. We remark that the condition $0 < \varepsilon < e^{-4}$ ensures that $\Gamma_\varepsilon > 1$.

13.7 Higher-Order IBP Formula

We can now discuss the higher-order IBP formula, inductively. This is a technical section which therefore involves a lot of careful calculations which we do not completely give here for the sake of space and clarity.

For $p_1,\ p_2 \in \mathbb{Z}_+$ and an $\mathbb{R}$-valued process $\tilde{K} = \left\{\tilde{K}_t\, ;\ 0 \leq t \leq T\right\}$ with nice properties, define the operator $\tilde{\mathscr{L}}_{p_1, p_2, t}$ by

$$\begin{aligned}
\tilde{\mathscr{L}}_{0,0,t}(\tilde{K}) &= \tilde{K}_t,\\
\tilde{\mathscr{L}}_{1,0,t}(\tilde{K}) &= \tilde{\Gamma}_t\left(X^\varepsilon + \tilde{V}, \tilde{K}\right) (1, 0)^*,\\
\tilde{\mathscr{L}}_{0,1,t}(\tilde{K}) &= \tilde{\Gamma}_t\left(X^\varepsilon + \tilde{V}, \tilde{K}\right) (0, 1)^*,\\
\tilde{\mathscr{L}}_{p_1+1, p_2, t}(\tilde{K}) &= \tilde{\mathscr{L}}_{1,0,t}\left(\tilde{\mathscr{L}}_{p_1, p_2, t}(\tilde{K})\right),\\
\tilde{\mathscr{L}}_{p_1, p_2+1, t}(\tilde{K}) &= \tilde{\mathscr{L}}_{0,1,t}\left(\tilde{\mathscr{L}}_{p_1, p_2, t}(\tilde{K})\right).
\end{aligned}$$

Denote by $\partial_i F$ $(i = 1, 2)$ the partial derivative in the i-th component in the order they appear in the function F, and write $\partial = \left(\partial_1, \partial_2\right)$.

Theorem 13.7.1 (Higher-order IBP formula) *For $F \in C_c^\infty\left(\mathbb{R}^2\, ;\ \mathbb{R}\right)$, it holds that*

$$\mathbb{E}\left[\left(\partial_1^{p_1} \partial_2^{p_2} F\right)\left(X_t^\varepsilon + \tilde{V}_t\right) \tilde{G}_t\right] = \mathbb{E}\left[F\left(X_t^\varepsilon + \tilde{V}_t\right) \tilde{\mathscr{L}}_{p_1, p_2, t}(\tilde{G})\right]. \tag{13.29}$$

As before (see Theorem 13.4.13) we assume that $\tilde{\mathscr{L}}_{p_1, p_2, t}(\tilde{G})$ is integrable.

The proof follows from the following exercise:

Exercise 13.7.2 Prove

$$\mathbb{E}\left[\left(\partial_1^{p_1} \partial_2^{p_2} F\right)\left(X_t^\varepsilon + \tilde{V}_t\right) \tilde{G}_t\right] = \mathbb{E}\left[F\left(X_t^\varepsilon + \tilde{V}_t\right) \tilde{\mathscr{L}}_{p_1, p_2, t}(\tilde{G})\right]. \tag{13.30}$$

For a hint, see Chap. 14.

Rather than proceeding with these long but instructive calculations, we propose the following generalization which will help.

Exercise 13.7.3 In this exercise, we define the Sobolev norms associated with the derivative we have defined here. Under the setting of Definition 13.4.9, define the following norm for $m \in \mathbb{N}^*$ and $p \geq q \geq 1$:

$$\|V_t\|_{p,q,m}^p = \mathbb{E}\left[\sum_{l=0}^{m}\left(\sum_{j_1\leq\ldots\leq j_l\leq N_t^\varepsilon}\prod_{i=1}^{l}\varphi(\tilde{\theta}_{j_i})^q\|D_{j_1}\ldots D_{j_l}V_t\|^q\right)^{p/q}\right].$$

Provide a sufficient hypothesis in order to prove the following properties:

(i) Prove that $\|\cdot\|_{p,q,m}$ is a norm and characterize its corresponding Banach space. In the particular case that $m = 0$, prove that $\|V_t\|_{p,q,0} = \|V_t\|_{p,q}$ as defined in Definition 13.6.1. Note that one can also extend the definition of the norm for processes as follows:

$$\|V\|_{p,q,m}^p = \mathbb{E}\left[\sum_{l=0}^{m}\left(\sum_{k=1}^{N_t^\varepsilon}\sum_{j_1\leq\ldots\leq j_l\leq k}\prod_{i=1}^{l}\varphi(\tilde{\theta}_{j_i})^q\|D_{j_1}\ldots D_{j_l}V_{T_k}\|^q\right)^{p/q}\right].$$

State and prove the coincidence between the above norm and the one defined in Definition 13.6.1 in the particular case $m = 0$.

(ii) For appropriate random variables X, Y prove that

$$\|D(XY)\|_{p,q} \leq \|Y\|_{2p}\|X\|_{2p,q,1} + \|X\|_{2p}\|Y\|_{2p,q,1}.$$

Extend the above inequality for appropriate random processes X and Y.

(iii) For appropriate (m, p, q) and smooth random variables (X, Y), there exists (m, p_1, q_1, p_2, q_2) such that

$$\|XY\|_{p,q,m} \leq C\|X\|_{p_1,q_1,m}\|Y\|_{p_2,q_2,m}.$$

Extend the above result to stochastic processes. In the above inequality and the inequalities to follow using the norms $\|\cdot\|_{p,q,m}$, we may have that the parameters β and r in the function φ used on the left side may be different from the those on the right. Still, as their use will be limited to a finite number of times this will not pose any restrictions later.

(iv) Prove that for any appropriate $(m, p, q, j) \in \mathbb{N}^4$, there exists (p_1, q_1) such that

$$\left\|\sum_{k=1}^{N_t^\varepsilon}\varphi(\tilde{\theta}_k)D_k^j\tilde{V}_t\right\|_{p,q,m} \leq C\|\tilde{V}_t\|_{p_1,q_1,m+j},$$

(v) For any smooth random variable $\tilde{G}_t$ which is $\mathscr{F}_t$-adapted and appropriate $(m, p, q) \in \mathbb{N}^3$, there exists (q_i, s_i), $i = 1, 2$ and (p_i, r_i), $i = 1, \ldots, 4$ such that

$$\|\tilde{\Gamma}_t(X^\varepsilon + \tilde{V}, \tilde{G})\|_{p,q,m} \le C\sqrt{u_\varepsilon(t)}\|\tilde{C}_t^{-1}\|_{s_1,q_1,m}\|\tilde{G}_t\|_{s_2,q_2,m}$$
$$+ C\|\tilde{V}_t\|_{p_1,r_1,m+2}\,\|\tilde{C}_t^{-1}\|_{p_2,r_2,m+1}\|\tilde{G}_t\|_{p_3,r_3,m+1}\|(\varphi\chi_{\varepsilon/2^{1/a}}(1+|\theta|^{-1}))_t\|_{p_4,r_4,m}.$$

Note that in the last term above, there is a slight abuse of notation. The process being considered there is the one which at each time T_k the process has a jump of size $\varphi\chi_{\varepsilon/2^{1/a}}(\tilde{\theta}_k)(1+|\tilde{\theta}_k|^{-1})$. This will be always well defined if $r = m$ and $\beta - m > 1 + \alpha$.

(vi) Use Exercise 13.4.10 to prove that for appropriate large values of β and r which depend on m, one has $\|\tilde{V}_t\|_{p,q,m} \le C\Gamma_\varepsilon^{C_{p,q,m}\gamma}$. Similarly, prove $\|\tilde{C}_t\|_{p,q,m} \le C\Gamma_\varepsilon^{C_{p,q,m}\gamma}$.

(vii) Prove that for the localization function provided in (13.28), one has

$$\|D\tilde{G}_t\|_{p,q,m} \le C_{p,q,m}\lambda_\varepsilon^{C_{p,q,m}}\Gamma_\varepsilon^{C_{p,q,m}\gamma}\exp\left(-C_{p,T}\Gamma_\varepsilon^\kappa\right).$$

Therefore if we choose $0 < \gamma < \kappa < \delta < 2$ and $\eta_0\kappa > 1$ then one has that due to Exercise 13.3.5 and Proposition 13.3.4 that the above is smaller than $C\varepsilon^\eta$ for any $\eta > 1$ once ε is chosen small enough. A similar conclusion follows for $\|\varphi_t D\tilde{G}_t\|_{p,q,m}$. Therefore from the above it follows that $\sup_\varepsilon \|\tilde{G}_t\|_{p,q,m} \le C_{p,q,m}$.

(viii) State and prove a similar result for $\sup_\varepsilon \|(\varphi\chi_{\varepsilon/2^{1/a}}(1+|\theta|^{-1}))_t\|_{p,q,m}$.

(ix) Prove the following estimate for appropriate m, p, q, p_1, p_2, q_1 and a positive constant C:

$$\|\tilde{C}_t^{-1}\|_{p,q,m} \le C\Gamma_\varepsilon^{C_{p,q,m}\gamma}\exp\left[C_{p,q,m}\,\Gamma_\varepsilon^\gamma\right].$$

(x) Recall that $u_\varepsilon(t) = t\,(\varepsilon^a\pi)^{4\beta+2-\alpha}$ and conclude from all the above that if β and r are chosen appropriately large with $\tilde{G}_t$ given by (13.28) then

$$\|\tilde{\Gamma}_t(X^\varepsilon + \tilde{V}, \tilde{G})\|_{p,q,m} \le C_{p,q,m}\,(1+\varepsilon^{-m(a\alpha+1)}\,\Gamma_\varepsilon^{C_{p,q,m}\gamma}\exp\left[C_{p,q,m}\,\Gamma_\varepsilon^\gamma\right]).$$

For a hint, see Chap. 14.

Following the above exercise and using induction one obtains the following result.

Theorem 13.7.4 (Upper estimate of the higher-order Malliavin weight) *For* $0 < t \le T$, $u_\varepsilon(t) = t\,(\varepsilon^a\pi)^{4\beta+2-\alpha}$, $\Gamma_\varepsilon = (\log(\varepsilon^{-1}))^{\eta_0}$ *with* $\eta_0\gamma < 1$ *and* $0 < \gamma < \kappa < \delta < 2$, $\kappa\eta_0 > 1$ *and it holds that if* β *and* r *are chosen appropriately large with* $\tilde{G}_t$ *given by* (13.28), *one has that there exists a small constant* C_{p_1,p_2} *such that for* $0 < \varepsilon \le C_{p_1,p_2}$ *one has*

$$\left\|\tilde{\mathscr{L}}_{p_1,p_2,t}(\tilde{G})\right\|_p \le C_{p_1,p_2,p}(1+\varepsilon^{-(p_1+p_2)(a\alpha+1)}\,\Gamma_\varepsilon^{C_{p_1,p_2,p}\gamma}\exp\left[C_{p_1,p_2,p}\,\Gamma_\varepsilon^\gamma\right]).$$

Rather than carrying out the proof it is a good exercise to try to understand where each term comes from. In fact, if we compare the above estimate with the one in Proposition 13.6.18, we see that one of the terms is considered above just as a constant and we only considered the largest term of the two terms appearing in Proposition 13.6.18.

Now because we are considering a higher-order IBP formula, this will involve higher-order derivatives. In the above result, they are expressed in the terms $\varepsilon^{-(p_1+p_2)(a\alpha+1)}\,\Gamma_\varepsilon^{C_{p_1,p_2,p}\gamma}$. We see that the appearance of the power $p_1 + p_2$ is due to the higher derivatives that imply a larger number of terms which are of the order of $(N_t^\varepsilon)^{p_1+p_2}$. Similarly, higher-order derivatives of ϕ_ε imply a larger power of ϕ_ε appearing in the denominators of the derivatives which lead to $\varepsilon^{-(p_1+p_2)}$. Note that higher moments of $\tilde{C}_t^{-1}$ will always lead to the same estimate $\exp\left[C_{p_1,p_2,p}\,\Gamma_\varepsilon^\gamma\right]$ with a different constant $C_{p_1,p_2,p}$.

13.8 Decay Order of the Characteristic Function of V_t

In this section, rather than using the full result obtained so far we will conclude with a brief argument in order to show the existence of the density of V_t. The method to be used in order to prove this will be Theorem 9.1.10 as explained some lines before Theorem 13.2.3.

Let $\zeta \in \mathbb{R}^2$ with $|\zeta| \geq 1$, and $0 < t \leq T$. Then, since $\bar{V}_t$ and $\tilde{V}_t$ are equivalent in law, we have

$$\begin{aligned}
\left|\mathbb{E}\left[\exp\left(i\,\zeta^*\,V_t\right)\right]\right| \leq{}& \left|\mathbb{E}\left[\exp\left(i\,\zeta^*\,V_t\right)\right] - \mathbb{E}\left[\exp\left(i\,\zeta^*\bar{V}_t\right)\right]\right| \\
&+ \left|\mathbb{E}\left[\exp\left(i\,\zeta^*\,\tilde{V}_t\right) - \exp\left\{i\,\zeta^*\left(X_t^\varepsilon + \tilde{V}_t\right)\right\}\right]\right| \\
&+ \left|\mathbb{E}\left[\exp\left\{i\,\zeta\left(X_t^\varepsilon + \tilde{V}_t\right)\right\}(1 - \tilde{G}_t)\right]\right| \\
&+ \left|\mathbb{E}\left[\exp\left\{i\,\zeta^*\left(X_t^\varepsilon + \tilde{V}_t\right)\right\}\tilde{G}_t\right]\right| \\
=:{}& I_1 + I_2 + I_3 + I_4.
\end{aligned}$$

We now estimate each term. The mean value theorem gives that

$$\begin{aligned}
\left|\exp\left(i\,\zeta^*\,x\right) - \exp\left(i\,\zeta^*\,y\right)\right| &= \left|i\,\zeta^*\,(x-y)\exp\left\{i\,\zeta^*\left(x + \theta_1(y-x)\right)\right\}\right| \\
&\leq \min\left(|\zeta||x-y|, 2\right) \\
&\leq 2^{1-\tilde{\beta}}\,|\zeta|^{\tilde{\beta}}\,|x-y|^{\tilde{\beta}},
\end{aligned}$$

where θ_1 is a $[0, 1]$-valued random variable and $0 < \tilde{\beta} \leq 1$. Therefore using this result, the fact that $\bar{V} \stackrel{\mathscr{L}}{=} V^\varepsilon$ and Proposition 13.3.4, we have

$$\begin{aligned}
I_1 &\equiv \left|\mathbb{E}\left[\exp\left(i\,\zeta^*\,V_t\right)\right] - \mathbb{E}\left[\exp\left(i\,\zeta^*\,V_t^\varepsilon\right)\right]\right| \\
&\le 2^{1-\tilde{\beta}}\,|\zeta|^{\tilde{\beta}}\,\mathbb{E}\left[|V_t - V_t^\varepsilon|^{\tilde{\beta}}\right] \\
&\le 2^{1-\tilde{\beta}}\,C_{\tilde{\beta},\eta}\,|\zeta|^{\tilde{\beta}}\,\varepsilon^{a(\tilde{\beta}-\alpha-\tilde{r})}.
\end{aligned}$$

In particular, recall that $\tilde{\beta} \in (\alpha, 1]$ and $\Gamma_\varepsilon = (\log(\varepsilon^{-1}))^{\eta_0}$ with $\eta_0\gamma < 1$ for any $\tilde{r} > 0$ such that $\tilde{\beta} - \alpha > \tilde{r}$.

To bound the term I_2, note that $\mathbb{E}\left[|X_t^\varepsilon|\right] \le \sqrt{2\,u_\varepsilon(t)}$, therefore from the Cauchy–Schwarz inequality, we can get, via the mean value theorem, that

$$\begin{aligned}
I_2 &\equiv \left|\mathbb{E}\left[\exp\left(i\,\zeta^*\,\tilde{V}_t\right) - \exp\left\{i\,\zeta^*\left(X_t^\varepsilon + \tilde{V}_t\right)\right\}\right]\right| \\
&= \left|\mathbb{E}\left[i\,\zeta^*\,X_t^\varepsilon\,\exp\left\{i\,\zeta^*\left(\tilde{V}_t + \theta_3\,X_t^\varepsilon\right)\right\}\right]\right| \\
&\le |\zeta|\,\mathbb{E}\left[|X_t^\varepsilon|\right] \\
&\le \sqrt{2\,u_\varepsilon(t)}\,|\zeta| \\
&\le \sqrt{2T}\,|\zeta|\,(\varepsilon^a\pi)^{(4\beta+2-\alpha)/2},
\end{aligned}$$

where θ_3 is a $[0,1]$-valued random variable.

Since

$$\mathbf{1}_{(\sup_{s\in[0,T]}|\tilde{V}_s|\ge\Gamma_\varepsilon)} \le 1 - \tilde{G}_t \le \mathbf{1}_{(\sup_{s\in[0,t]}|\tilde{V}_s|\ge\Gamma_\varepsilon-1)},$$

we see that using again Proposition 13.3.4 and exponential Chebyshev's inequality,

$$\begin{aligned}
I_3 &\equiv \left|\mathbb{E}\left[\exp\left\{i\,\zeta^*\left(X_t^\varepsilon + \tilde{V}_t\right)\right\}(1 - \tilde{G}_t)\right]\right| \\
&\le \mathbb{P}\left[\sup_{s\in[0,t]} |\tilde{V}_s| \ge \Gamma_\varepsilon - 1\right] \\
&\le \exp\left[-(\Gamma_\varepsilon - 1)^\kappa\right]\mathbb{E}\left[\sup_{t\in[0,T]} \exp\left(|\tilde{V}_t|^\kappa\right)\right] \\
&\le C_\kappa\,\exp\left[-(\Gamma_\varepsilon - 1)^\kappa\right] \\
&\le C_\kappa\,\exp\left(-\frac{\Gamma_\varepsilon^\kappa}{2}\right) \\
&\le C_\kappa\,\varepsilon^{\tilde{p}}
\end{aligned}$$

for any $\alpha < \kappa < \delta$ and $\tilde{p} > 0$.

Exercise 13.8.1 Use the arguments in Exercise 13.6.11 to prove that

$$\mathbf{1}_{(\sup_{s\in[0,T]}|\tilde{V}_s|\ge\Gamma_\varepsilon)} \le 1 - \tilde{G}_t \le \mathbf{1}_{(\sup_{s\in[0,t]}|\tilde{V}_s|\ge\Gamma_\varepsilon-1)}.$$

The upper bound estimate for I_4 follows; since

$$\exp\left\{i\,\zeta^*\left(X_t^\varepsilon+\tilde V_t\right)\right\}=(-1)^p\,|\zeta|^{-2p}\left(\partial_1^2+\partial_2^2\right)^p\,\exp\left\{i\,\zeta^*\left(X_t^\varepsilon+\tilde V_t\right)\right\}$$

for any $p\in\mathbb{N}$, we have

$$\begin{aligned}
I_4&\equiv\left|\mathbb{E}\left[\exp\left\{i\,\zeta^*\left(X_t^\varepsilon+\tilde V_t\right)\right\}\tilde G_t\right]\right|\\
&=|\zeta|^{-2p}\left|\mathbb{E}\left[\left(\partial_1^2+\partial_2^2\right)^p\,\exp\left\{i\,\zeta^*\left(X_t^\varepsilon+\tilde V_t\right)\right\}\tilde G_t\right]\right|\\
&\le|\zeta|^{-2p}\sum_{k=0}^{p}\binom{p}{k}\left|\mathbb{E}\left[\partial_1^{2k}\partial_2^{2p-2k}\left(\exp\left\{i\,\zeta^*\left(X_t^\varepsilon+\tilde V_t\right)\right\}\right)\tilde G_t\right]\right|\\
&\le C_p\,|\zeta|^{-2p}\left\{1+\varepsilon^{-2p(a\alpha+1)}\,\Gamma_\varepsilon^{C_p\gamma}\,\exp\left[C_p\,\Gamma_\varepsilon^\gamma\right]\right\}.
\end{aligned}$$

In the fourth inequality of the above estimate of I_4, we have used Theorems 13.7.1 and 13.7.4 with the parameter conditions stated in the latter.

Putting all these results together and using Exercise 13.3.5, we obtain the following estimate on the characteristic function of V_t.

Theorem 13.8.2 *Let $0<t\le T$ and assume the parameter conditions in Theorem 13.7.4 and $\tilde\beta\in(\alpha,1]$ with $\tilde\beta-\alpha>\tilde r>0$. Then, it holds that for any $p\in\mathbb{N}$ and $\zeta\in\mathbb{R}^2$ with $|\zeta|\ge1$*

$$\begin{aligned}
\left|\mathbb{E}\left[\exp\left(i\,\zeta^*\,V_t\right)\right]\right|\le C_{\tilde\beta}\,|\zeta|^{\tilde\beta}\,\varepsilon^{a(\tilde\beta-\alpha-\tilde r)}+\sqrt{2T}\,|\zeta|\,(\varepsilon^a\pi)^{(4\beta+2-\alpha)/2}&\\
+C_p|\zeta|^{-2p}\left(1+\varepsilon^{-2p(a\alpha+1)}\,\Gamma_\varepsilon^{C_p\gamma}\,\exp\left[C_p\,\Gamma_\varepsilon^\gamma\right]\right).&
\end{aligned}\tag{13.31}$$

In particular, choose $\varepsilon=|\zeta|^{-b}$, where the positive constant b will be given later. Write

$$\sigma:=\left\{ab(\tilde\beta-\alpha-\tilde r)-\tilde\beta\right\}\wedge\left\{ab\left(2\beta+1-\frac{\alpha}{2}\right)-1\right\}\wedge\left\{2p-2pb(a\alpha+1)-b\eta\right\}.\tag{13.32}$$

Then, since $\left|\mathbb{E}\left[\exp\left(i\,\zeta^*\,V_t\right)\right]\right|\le1$ for $\zeta\in\mathbb{R}^2$ is trivial, we see from (13.31) that

$$\left|\mathbb{E}\left[\exp\left(i\,\zeta^*\,V_t\right)\right]\right|\le\left(C_{\tilde\beta,T,p,\eta_0,\eta,\gamma}\vee1\right)\left(1+|\zeta|\right)^{-\sigma}$$

for $\zeta\in\mathbb{R}^2$.

Remark 13.8.3 Some final remarks are in order to understand the range of the results obtained.

1. Note that each term in the estimate in (13.31) has a meaning. In fact, the first term corresponds to the approximation error of V_t by V_t^ε. The second corresponds to the small Gaussian noise term X_t^ε and the last term corresponds to the IBP with respect to the approximation process $\tilde V_t$.

2. If the reader compares the above methodology with the one provided in [7], one finds that although technically similar, all differences in the results appear due to the choice of the function φ. A straightforward comparison is complicated due to the number of parameters in the problem.
3. Now we will try to explain again the meaning and the restrictions on each parameter in the above result. First, recall that $\eta_0\gamma < 1$ is used in order to apply Exercise 13.3.5.
4. The restriction $\tilde{\beta} \in (\alpha, 1]$ with $\tilde{\beta} - \alpha > \tilde{r} > 0$ appears due to the use of Proposition 13.3.4. Similarly, the restriction $0 < \gamma < \kappa < \delta < 2$ appears in Theorem 13.2.3 and Proposition 13.3.4. We leave to the reader to verify that this set of restrictions is non-empty.
5. Note that $\beta \equiv \beta(p)$ in (13.14), is chosen so that the application of p-th order IBP formula is possible. Note that we can choose $\beta(p)$ as large as desired without changing any of the estimates (except for constants).
6. In order to obtain that the density of V_t exists and has certain regularities, let us consider the following situation: Let $L \geq 0$, $0 < \gamma < 1$ and $0 < \alpha < 1/(L+2)$. We shall choose the constants β, $\eta > 1$ (in Exercise 13.3.5), $\tilde{\beta}$, p (the number of IBP formulas), $\tilde{r}$, a (parameter of χ_ε) and b (parameter which defines ε as a function of ζ in the above theorem) so that the following inequalities are satisfied:

$$\frac{\alpha\,(L+1)}{1-\alpha} < \tilde{\beta} \leq 1, \quad \beta > \frac{\alpha(2L+3)-2}{4},$$

$$p > \frac{L\,(\tilde{\beta}-\alpha)}{2\left((1-\alpha)\,\tilde{\beta}-\alpha\,(L+1)\right)} \vee \frac{L\,(4\beta+2-\alpha)}{2\left(4\beta+2-\alpha(2L+3)\right)} \vee \frac{L}{2} \vee 1, \tag{13.33}$$

$$0 < \tilde{r} < \tilde{\beta} - \alpha - \frac{2p\alpha\,(\tilde{\beta}+L)}{2p-L},$$

$$a > \frac{(\tilde{\beta}+L)\,(2p+\eta)}{(2p-L)\,(\tilde{\beta}-\alpha-\tilde{r}) - 2p\alpha\,(\tilde{\beta}+L)} \vee \frac{2(L+1)\,(2p+\eta)}{(2p-L)\,(4\beta+2-\alpha)-4p\alpha(L+1)},$$

$$\frac{\tilde{\beta}+L}{a\,(\tilde{\beta}-\alpha-\tilde{r})} \vee \frac{2(L+1)}{a\,(4\beta+2-\alpha)} < b < \frac{2p-L}{2p\,(a\alpha+1)+\eta}. \tag{13.34}$$

Then, we can get that $\sigma > L$ via easy but tedious computations. In fact, the above inequalities are obtained from bottom to top. First, one obtains the conditions on b that ensure that $\sigma > L$. Then in order for the inequalities for b to be satisfied we need that both lower bounds be smaller that the upper bound in (13.34) from which the inequalities for a follow. Similarly, one continues to the top. In the case of inequalities involving $\tilde{r}$ one optimizes in order to find that no restriction on p appears in (13.33).

(i) If we choose $L = 0$, then our setting is in the case of $\sigma > 0$, which implies that, for any $t \in (0, T]$, $f_t \in H^r(\mathbb{R}^2)$ for all $r < \frac{\sigma-1}{2}$. Here, $H^r(\mathbb{R}^2)$ is the set of probability measures g on $\mathbb{R}^2$ such that $\|g\|_{H^r(\mathbb{R}^2)} < \infty$, where

$$\|g\|^2_{H^r(\mathbb{R}^2)} = \int_{\mathbb{R}^2} \left(1 + |\zeta|^2\right)^r \left|\hat{g}(\zeta)\right|^2 d\zeta, \quad \hat{g}(\zeta) = \int_{\mathbb{R}^2} e^{i\,\zeta\,x}\, g(dx).$$

One may also check the relation with Corollary 9.1.3.

(ii) If we choose $L = 1$, then $\sigma > 1$, which implies that, for any $t \in (0, T]$, f_t admits an $\mathbb{L}^2(\mathbb{R}^2)$-density with respect to the Lebesgue measure on $\mathbb{R}^2$.

(iii) If we choose $L = 2$, then $\sigma > 2$, which implies that, for any $t \in (0, T]$, f_t has a continuous and bounded density with respect to the Lebesgue measure on $\mathbb{R}^2$ (see Theorem 9.1.2). In any case, we see that in order to obtain more regularity we need α to be closer to zero.

7. Note that higher-order IBP formula are needed in order to carry out the above arguments. Then, the function $\varphi(\theta)$ has to approach to zero as $\theta \to 0$ faster (in polynomial order). On the other hand, as seen in Remark 13.5.7, this fact does not create a problem in the existence of moments of the determinant of the Malliavin covariance matrix. This fact was crucial in order to obtain the proof. The above result it is not optimal and by far there are a number of related issues to be tackled such as its three-dimensional extension, the regularity properties of the density and its estimates.

Chapter 14
Further Hints for the Exercises

This chapter collects hints to solve various exercises. These solutions are not complete by any means, the given arguments can only be considered at best to be heuristic and need to be completed by the reader. These are a level above the hints given in each corresponding exercise which are given because we believe that some of the exercises may be difficult or even that some misunderstanding may occur.

Hint (Problem 2.1.41) *For example, one simple improvement of the result is*

$$\lim_{x\to+\infty}\frac{\log\mathbb{P}(N_t=x)}{x\log(x)}=-1.$$

Using appropriate decreasing properties of $\mathbb{P}(N_t = x)$ *as a function of* x, *one can also obtain results for* $\mathbb{P}(N_t > x)$.

Hint (Exercise 3.1.4) *For example, one has* $Y_1 := \mathbf{1}_{\tau_1^{(1)}<\tau_2^{(2)}} - \mathbf{1}_{\tau_1^{(1)}>\tau_2^{(2)}}$. *As* $N^{(1)} + N^{(2)}$ *is a Poisson process, it has its corresponding jump times which we denote by* τ_i, $i \in \mathbb{N}$. *Now we can define in general* $Y_j := a\mathbf{1}_{\tau_j=\tau_k^{(1)},\text{ for some } k\in\mathbb{N}} + b\mathbf{1}_{\tau_j=\tau_k^{(2)},\text{ for some } k\in\mathbb{N}}$. *Another possibility is* $Y_j := a\mathbf{1}_{\Delta Z_{\tau_j}=a} + b\mathbf{1}_{\Delta Z_{\tau_j}=b}$ *for* $Z := aN^{(1)} + bN^{(2)}$. *Then to find the distribution of* Y_1 *just note that* $\mathbb{P}(Y_1 = a) = \mathbb{P}(\tau_1^{(1)} < \tau_1^{(2)})$.

In fact, compute $\mathbb{P}(N_{\tau+t}^{(i)} - N_\tau^{(i)} = k)$ *for* $i = 1, 2$ *by dividing this computation in cases according to* $\tau = \tau_j^{(1)}$ *for some* j *or* $\tau = \tau_j^{(2)}$ *for some* j.

Hint (Exercise 3.1.9) *There are two ways of solving the problem. The first and shorter is to compute the characteristic function. The second, which is longer but more informative, is to give an explicit construction in the spirit of Theorem 2.1.31.*

Hint (Exercise 3.1.11) *We discuss only the latter exercises. In order to prove that* N^i *are independent is useful to compute the characteristic function. In fact, the*

© Springer Nature Singapore Pte Ltd. 2019
A. Kohatsu-Higa and A. Takeuchi, *Jump SDEs and the Study of Their Densities*, Universitext, https://doi.org/10.1007/978-981-32-9741-8_14

sum $\sum_{j=1}^{2}\theta_j N_t^j$ *corresponds to a compound Poisson process with jumps of size* $\sum_{j=1}^{2}\theta_j \mathbf{1}_{Y_i\in A_j}$. *From here, one can compute the characteristic function*

$$\mathbb{E}\left[\exp\left(i\sum_{j=1}^{2}\theta_j N_t^j\right)\right]$$

and deduce the necessary properties.

Hint (Exercise 3.4.9) *For each point* $t\in[0,T)$, f *is right-continuous. Therefore for fixed* $\varepsilon/2$, *there exists a* $\delta(t)>0$ *such that* $0<s-t<\delta(t)$ *implies that* $|f(s)-f(t)|<\varepsilon/2$. *As the interval* $[0,T]$ *is compact, one has that the interval is covered by a finite number of intervals* $(t_i,t_i+\delta(i))$, $i=1,\cdots,n$ *for some* $n\in\mathbb{N}$. *Use this to determine* $\delta:=\min_{i=1,\dots,n}\delta_i$ *which satisfies the properties.*

For the second problem, prove that there can only be a finite number of points where the discontinuity is greater than $\varepsilon\in\mathbb{Q}$. *From here the second problem can be solved.*

Hint (Exercise 3.4.28) *Just a short note to say that the relation with Exercise 2.1.42 is obtained by taking h and indicator function for the number of jumps.*

Hint (Exercise 3.5.7) *This exercise may not be considered totally fair to the reader as it does not state clearly what we mean by generalizing the Itô formula. Here is one option. Suppose that* $\int_{\mathbb{R}\times[0,T]}|g(z,s)|\widehat{\mathcal{N}}(dz,ds)<\infty$, *a.s. Then one can proceed by considering the function* $(g\vee(-n))\wedge n$ *instead of* g *and then try to take limits with respect to* n *in the Itô formula. For this, one will need to suppose some continuity condition on* h.

Hint (Exercise 3.5.9) *Clearly if* $f\equiv 1$ *the law of* N^1N^2 *is not the law of a Poisson process. Therefore the first process is not necessarily a compound Poisson process. The required construction as iterated integral will require one first to determine a compound Poisson process which depends on the value of the jump size in one coordinate. This is easier to understand with discrete distributions. Suppose that* Y^i *has a discrete distribution concentrated on* $\{a_j^i;\ j\in\mathbb{N}\}$. *Then one needs to define the process* $N_t(a):=\sum_{i=1}^{N_t}\mathbf{1}\{X_i^2=a\}$. *Prove that this process is a Poisson process. Use it for the iterated representation. In the literature this procedure is usually called "thinning".*

Hint (Problem 3.6.12) *If we condition on the first jump of* N *and the value of the jump* Y_1 *we obtain the following equation:*

$$\Psi(u)=\int_0^\infty dr\lambda e^{-\lambda r}\left(\int_0^{u+cr}\Psi(u+cr-z)f(z)dz+\int_{u+cr}^\infty f(z)dz\right).$$

Then one performs the change of variables $u+cr=\theta$ *and differentiates with respect to* u. *This gives the first result.*

Another way of doing this is to use Itô's formula to rewrite $\mathbb{E}[\Psi(S_t)] - \Psi(u)$. *Then one has to prove that this difference is small (i.e.* $o(t)$*; this requires some long calculations, try it). To obtain the quoted equation we need to consider*

$$\begin{aligned}
&\mathbb{E}[\Psi(S_t)] - \Psi(u)\\
&= (\Psi(u+ct) - \Psi(u))e^{-\lambda t}\\
&\quad + \int_0^t ds\lambda e^{-\lambda s}\Bigg(\int_0^{u+cs} (\mathbb{E}[\Psi(u+cs-y+\sum_{i=0}^{N_{t-s}} Y_i)] - \Psi(u))f(y)dy\\
&\qquad\qquad + (1-\Psi(u))\bar{F}(u+cs)\Bigg).
\end{aligned}$$

If one considers the first-order terms in the above equation one obtains the result.

To solve this linear equation one may use Laplace transform techniques.

Hint (Problem 3.6.13) *First, we prove that* Ψ *is a continuous function. In fact* $\Psi(u) = \mathbb{P}(\sup_{s>0} Z_s > u)$*, where* Z *is the associated compound Poisson process. Conditioning on the first jump we have*

$$\Psi(u) = \int_0^\infty dr\lambda e^{-\lambda r}\int_{-\infty}^u \Psi(z)f(u-z)dz = \int_{-\infty}^u \Psi(z)f(u-z)dz.$$

Therefore one obtains first that Ψ *is continuous. By a similar argument one obtains also the differentiability of* Psi*. This argument is essentially the same for the case* $c \neq 0$.

Hint (Exercise 4.1.12) *Just use the fact that there is a finite number of jumps larger than one and the fact that* $|\Delta Z_s|^p \leq |\Delta Z_s|$ *for* $p \geq 1$ *if* $|\Delta Z_s| \leq 1$.

Hint (Exercise 4.1.21) *In fact, first prove that the Laplace transform of a gamma law with parameters* (a, b) *is essentially given by*

$$\frac{b^a}{\Gamma(a)}\int_0^\infty e^{-\theta x}x^{a-1}e^{-bx}dx = \left(\frac{b}{b+\theta}\right)^a.$$

From Corollary 4.1.13 deduce that the Laplace transform of the gamma process is given by

$$\exp\left(-t\int_0^\infty (e^{-\theta x}-1)\frac{e^{-\lambda x}}{x}dx\right).$$

The exponent above can be expanded for $|\theta| < a$ *as* $-t\sum_{p=1}^\infty(\frac{-\theta}{a})^p\frac{1}{p}$. *Finally, find the values of* a *and* b *to match the two Laplace transforms. One may try to do the same using characteristic functions.*[1] *Although the final result is the same the solution requires techniques from complex analysis (like the calculation of integrals using residues and the Cauchy theorem.)*

A different methodology which is more model specific goes as follows: Use the mean value theorem to obtain that

$$\int_0^\infty (e^{-\theta x}-1)\frac{e^{-\lambda x}}{x}dx = -\int_0^\infty \int_0^\theta e^{-vx}dve^{-\lambda x}dx.$$

Now just use Fubini's theorem.

Hint (Exercise 4.1.22) *Given the approximations used in the definition of Z, one can use Proposition 3.3.7 in order to show that the characteristic function of W is given by*

$$\mathbb{E}[e^{i\theta W_t}] = \exp[ct \int_{\mathbb{R}_+} (e^{i\theta x^r}-1)\frac{1}{x^{1+\alpha}}dx].$$

From here, using the change of variables $z = x^r$ in the above integral one obtains that the Lévy measure of W is given by $\frac{1}{z^{1+\frac{\alpha}{r}}}\mathbf{1}_{\{z>0\}}$. The other properties of Lévy processes follow by using the approximation procedure.

Hint (Exercise 4.1.23) *In order to take limits of the characteristic functions one needs to prove some that one can take the limits inside the integral. For this one needs to use that for $|x| \le 1$ one has that*

$$\frac{|e^{i\theta x}-1|}{|x|^{1+\alpha}} \le \frac{|\theta|}{|x|^{\alpha}}.$$

Similarly for large values of x is enough to use that

$$\frac{|e^{i\theta x}-1|}{|x|^{1+\alpha}} \le \frac{2}{|x|^{1+\alpha}}.$$

Therefore the limit can be taken inside the integrals and the integral is finite. For 2., try to use the symmetry properties of the cosine and sine functions.

For 3., we just need to remark that for a particular choice of $\theta = \kappa t^{-1/\alpha}$ one obtains the characteristic function of $V := t^{-1/\alpha}Z_t$. Prove that this is independent of t.

For 4., just prove that the rate of decrease of the characteristic function is faster than any polynomial. That is, $\limsup_{|\theta|\to\infty} \frac{\varphi_{Z_t}(\theta)}{|\theta|^p} = 0$ for any $p > 0$.

Hint (Exercise 4.1.25) *For 3., once 2. is proven one only needs to prove that the following asymptotic result is satisfied for an infinite number of i of big enough value:*

---

[1] As you may have noticed the fact that the infinite Taylor series of the considered functions are equal is not yet enough to say that the functions are equal. There are well-known counterexamples. You still need to use that the considered functions are analytic.

$$\mathcal{N}^{(\varepsilon_i,\varepsilon_{i-1})}((0,1]\times[0,t]) \approx 2^{i\alpha}\frac{Ct}{\alpha}2^{-\alpha}(1-2^{-\alpha}), \text{ a.s.}$$

Adding these terms one obtains the result.

Hint (Exercise 4.1.26) *According to Corollary 4.1.13 we have the characteristic function of Z_t, therefore the mean and the variance can be computed. For the martingale property use the fact that the increments of Z are independent.*

Hint (Exercise 4.1.27) *To prove the blow-up of moments just consider the case of one jump and then compute the respective moment. For finiteness, one needs to consider all cases for all numbers of jumps using the independence of the jumps.*

Hint (Exercise 4.1.32) *The example for f may be $f(x) = \frac{e^{-\lambda|x|}}{|x|^{1+\alpha}}\mathbf{1}_{\{x\neq 0\}}$, where $\lambda > 0$ and $\alpha \in [0, 1)$.*

Hint (Exercise 4.2.6) *Just follow the same proof using $(k-1)^p|X_t^{(k)} - X_t^{(k-1)}|$ instead of $|X_t^{(k)} - X_t^{(k-1)}|$. Then the fact that the factorial converges to zero faster than any polynomial can be used in order to obtain the same bounds. Then one considers $k^p|X_t^{(l)} - X_t^{(k-1)}|$. Take $l \to \infty$ and then take the limits with respect to k. Stirling's approximation should tell you that the convergence is actually much faster than a polynomial.*

Hint (Exercise 4.2.7) *One way to understand why it is important to be careful with how the statement to be proved is to try to find explicit bounds for $\mathbb{E}[|X_t^{(k)}|]$ for $k = 1$ and $k = 2$. Depending on how you do it you may end with a mess of expressions or a neatly written expression which will tell you what to do in general.*

The final answer is to consider by induction a bound for $\mathbb{E}[|X_t^{(k)}|]$ using Theorem 4.1.38 or other similar results in this section (can you guess these other results?).

For the sake of the argument and to decide how to do the induction, we suppose momentarily that all expressions to follow are well defined. Then using triangular inequality, the Lipschitz property of a and Theorem 4.1.7,

$$\begin{aligned}\mathbb{E}[|X_t^{(k)}|] &\leq |x| + \int_0^t \mathbb{E}[|a(X_s^{(k-1)})|]ds + \mathbb{E}[\int_{\mathbb{R}\times[0,t]} |X_{s-}^{(k-1)}b(z)|\mathcal{N}(dz, ds)] \\ &\leq |x|(1+Ct) + |a(0)|t + C\int_0^t \mathbb{E}[|X_s^{(k-1)}|]ds \\ &\quad + c\int_{\mathbb{R}} |b(z)|f(z)dz\mathbb{E}[\int_0^t |X_{s-}^{(k-1)}|ds].\end{aligned}$$

Consider that $t \in [0, T]$ and use Gronwall's inequality. The resulting formula has to be iterated down to $k = 1$ or $k = 0$ (depending on how you have done the first part of the hint) and this will give you the right hypothesis for the induction procedure.

Hint (Exercise 4.2.15) *Define*

$$Y_t = \int_{\mathbb{R}\times[0,t]} b(z,s)\mathcal{N}(dz,ds).$$

Note that $\mathbb{E}[Y_t]$ has already been computed in Theorem 4.1.38. Applying Itô's formula and expectations, we have

$$\mathbb{E}[Y_t^2] - \mathbb{E}[Y_0^2] = \mathbb{E}[\int_{\mathbb{R}\times[0,t]} b(z,s)(b(z,s) + 2Y_{s-})\widehat{\mathcal{N}}(dz,ds)].$$

The result follows from here.

Hint (Problem 4.2.16) *To expand the solution, one just needs to use the Itô formula in Theorem 4.2.12 repeatedly. In fact,*

$$X_t = x + a(x)\int_{\mathbb{R}\times[0,t]} z\mathcal{N}(dz,ds) + \int_{\mathbb{R}\times[0,t]} (a(X_{s-}) - a(x))z\mathcal{N}(dz,ds).$$

Now

$$a(X_{s-}) - a(x) = \int_{\mathbb{R}\times[0,s)} \{a(X_{u-} + a(X_{u-})z) - a(X_{u-})\}\mathcal{N}(dz,du).$$

Next, we apply a mean value theorem to obtain

$$a(X_{u-} + a(X_{u-})z) - a(X_{u-}) = \int_0^1 a'(X_{u-} + \alpha a(X_{u-})z)d\alpha a(X_{u-})z.$$

One can do a Taylor expansion of this term. The main term will be $a'a(x)z$. This will generate an integral of the type

$$a'a(x)\int_{\mathbb{R}\times[0,t]}\int_{\mathbb{R}\times[0,s)} z_1\mathcal{N}(dz_1,du)z_2\mathcal{N}(dz_2,ds)$$

Then continue doing this expansion.

Hint (Problem 4.2.17) *In the case of the expectation of X, obtaining a Taylor expansion is much simpler because the expectation simplifies the problem structure. In fact,*

$$\mathbb{E}[X_t] = x + a(x)t\int_{\mathbb{R}} zf(z)dz + \mathbb{E}\left[\int_{\mathbb{R}\times[0,t]} (a(X_{s-}) - a(x))zf(z)dzds\right].$$

The term which was analyzed in the hint of Exercise 4.2.16 has now as expectation

$$a'a(x)\frac{t^2}{2}\int_{\mathbb{R}}\int_{\mathbb{R}} z_1 f(z_1)z_2 f(z_2)dz_1dz_2,$$

where the last integral becomes $\frac{1}{2}\left(\int_{\mathbb{R}} zf(z)dz\right)^2$. *A further interesting question is to analyze what will be the expansion for* $\mathbb{E}[g(X_t)]$, *where* $g \in C_b^\infty(\mathbb{R})$.

Hint (Exercise 5.1.9) *The topology is determined by the corresponding characteristic functions and given by the convergence of* $\int(|x|^2 \wedge 1)|f_\varepsilon(x) - f(x)|dx$.

Hint (Exercise 5.1.17) *In fact, note that the calculation in this case is not so different to the calculation in Exercise 4.1.23 due to the symmetry of* f. *This fact is essential when dealing with the parallel to Exercise 4.1.23.2.*

Hint (Exercise 5.1.19) *In fact, take* $a_0 = 1$ *and let us suppose that we compensate the point process so as to obtain a martingale which converges in* L^2. *Then the characteristic function for the constructed process at time* 1 *can be written as*

$$\exp\left(-\sum_{i=0}^{\infty} 2\sin^2\left(\frac{\theta a_i}{2}\right)\frac{(a_i - a_{i-1})}{a_i^{1+\alpha}}\right).$$

The above sum converges because $\sin^2\left(\frac{\theta a_i}{2}\right) \leq \frac{\theta^2 a_i^2}{4}$ *and* $\sum_{i=1}^{\infty} \frac{(a_i - a_{i-1})}{a_i^{-1+\alpha}} \leq \int_0^1 \frac{dx}{x^{-1+\alpha}} < \infty$ *for* $\alpha \in (1, 2)$.

In order to prove that the density is smooth one needs to prove that the rate of decrease of the above characteristic function is faster than any polynomial.

In order to do this, one can use the following trick, which is similar to what happens with integrals. Consider the sequence $u_i := \theta a_i$. *Then we have that the exponent in the characteristic function is*

$$\theta^\alpha \sum_{i=0}^{\infty} 2\sin^2\left(\frac{u_i}{2}\right)\frac{(u_i - u_{i-1})}{u_i^{1+\alpha}}.$$

Now one needs to prove that the sum is non-zero. This will be the case as u_i *decreases to zero and its values change from* 0 *to* θ.

Hint (Exercise 5.1.20) *This exercise is based on a very well-known result by Asmussen and Rosiński [3]. One has to compute the variance of each of the jump sizes associated with* $\bar{Z}^{\varepsilon_n,+}$. *We first rewrite it in the more convenient form of*

$$\bar{Z}_t^{(\varepsilon',\,\varepsilon,+)} := \sum_{i=1}^{N_t^{\varepsilon_n}} (Y_i^{(\varepsilon_n,1)} - \tilde{\mu}^{(\varepsilon_n)}),$$

$$N_t^{\varepsilon_n} := \int_{\mathbb{R}\times[0,t]} x\mathcal{N}^{(\varepsilon_n,1)}(dx, ds).$$

Therefore as $N_t^{\varepsilon_n} \to \infty$ *as* $n \to \infty$, *we can apply the central limit theorem with care (note that the number of terms is random). Other elements that appear in the proof is the study of the asymptotic behavior of the moments associated with the jump sizes. In fact,*

$$\lambda_{\varepsilon_n,1} = \int_{\varepsilon_n}^{1} \frac{1}{x^{1+\alpha}} \approx \varepsilon_n^{-\alpha}.$$

The moments of order p of $Y_i^{(\varepsilon_n,1)}$ *are of order* $\varepsilon_n^{p-2\alpha}$. *With these estimates one can apply a version of the central limit theorem. Note that the random variables* $Y_i^{(\varepsilon_n,1)}$ *also depend on n, therefore the type of central limit theorem is not the most basic version for i.i.d. r.v.s. In particular, note that the mean of the considered r.v.s is zero, therefore the mean of Y is zero. The asymptotic behavior of the variance should give the answer for the variance of Y.*

Hint (Exercise 5.1.21) *We only give a hint for part 2. According to Exercise 5.1.20 we can assume that the small jumps can be approached by a Gaussian random variable of mean zero. Therefore the probability of being negative is close to* $1/2$ *For the jumps bigger than 1, one may consider the probability that no jump bigger than one occurs which is also strictly positive for any* $t > 0$.

Hint (Exercise 5.1.22) *First note that the statement given in words should be*

$$\mathbb{P}(\Delta \bar{Z}^+_{T_i} \textit{ for some } i \in \mathbb{N}) = 0.$$

In order to prove this, note that $\{T_i; i \in \mathbb{N}\}$ *is independent of* $\bar{Z}^+$. *Then use appropriate extensions of Exercise 3.1.11.*

Hint (Exercise 5.1.23) *This exercise is not as easy as it seems at the beginning. In order to obtain the characteristic function use calculus of residues from complex analysis. In order to compute the integrals related to the expression in Theorem 4.1.7 one has to divide this integral in two domains as usual and find the corresponding expansions in each domain.*

Hint (Exercise 5.2.4) *(Hint for Sect. 5.2) We will retake the proof at the equation*

$$\begin{aligned} h(Y_t^n) &= h(Y_0^n) + \int_{[0,\infty)\times[0,t]} \left\{h(Y_{s-}^n + z) - h(Y_{s-}^n) - h'(Y_{s-}^n)z\mathbf{1}_{|z|\le 1}\right\} \mathcal{N}^{(\varepsilon_n)}(dz, ds) \\ &\quad + \int_{[0,1]\times[0,t]} h'(Y_s^n) z \widetilde{\mathcal{N}}^{(\varepsilon_n)}(dz, ds) \\ &=: I_t^n + J_t^n. \end{aligned}$$

Clearly the left-hand side converges uniformly in time as $n \to \infty$. *We need to prove that the two integrals,* I^n *and* J^n *on the right-hand side converge also. For this, we will prove that they are Cauchy sequences in appropriate spaces. Before doing this we will need to separate the integrals according to the size of the jumps: On the one hand, the jumps bigger than one and on the other the jumps smaller than one in absolute value. Here, we give the proof assuming that there are no jumps bigger than one and leave the case for jumps bigger than one for the reader.*

Note also the following estimate using the Taylor expansion formula with integral remainder

$$|h(z) - h(z_0) - h'(z_0)(z - z_0)| = \left| \int_0^1 h''(z_0 + r(z - z_0))(1 - r)dt(z - z_0)^2 \right|$$
$$\leq \frac{\|h''\|_\infty}{2}(z - z_0)^2.$$

This estimate gives that for $n \geq m$

$$|I_t^n - I_t^m| \leq \frac{\|h''\|_\infty}{2} \int_{[0,1]\times[0,t]} |z|^2 \mathcal{N}^{(\varepsilon_n,\varepsilon_m)}(dz, ds)$$
$$+ \left| \int_{[0,1]\times[0,t]} g_{n,m}(z, s) z^2 \mathcal{N}^{(\varepsilon_m)}(dz, ds) \right|.$$

Here

$$g_{n,m}(z, s) := \int_0^1 (h''(z + rY_{s-}^n) - h''(z + rY_{s-}^m))(1 - r)dt.$$

Now one needs to use analytical tools to prove that as

$$\lim_{n,m\to\infty} \sup_{s\in[0,t]} |Y_{s-}^n - Y_{s-}^m| = 0,$$

$g_{n,m}(z, s)$ also converges uniformly. For J^n a similar argument is used but instead of using convergence of the measures $z^2\mathcal{N}^{(\varepsilon_m)}(dz, ds)$ one needs to use the martingale estimates and the martingale convergence theorem that already appeared in the construction of Z. That is, we upper bound the difference $J^n - J^m$ as follows:

$$\left| \int_{[0,1]\times[0,t]} h'(Y_s^n) z \widetilde{\mathcal{N}}^{(\varepsilon_n,\varepsilon_m)}(dz, ds) \right| + \left| \int_{[0,1]\times[0,t]} (h'(Y_s^n) - h'(Y_{s-}^m)) z \widetilde{\mathcal{N}}^{(\varepsilon_m)}(dz, ds) \right|.$$

In each term, one needs to compute the L^2-norms, prove that they converge to zero and then apply the same arguments as in proof of Theorem 5.1.5.

Hint (Exercise 5.3.5) *For simplicity, let us first assume that g is positive. Now, let g_n be a sequence of simple functions of the type:*

$$g_n(z, s) = \sum_{i,j=1}^{n} a_{i,j} \mathbf{1}_{A_{i,j}}(z, s).$$

Here $a_{i,j} \geq 0$ and $A_{i,j}$ is a sequence of disjoint measurable sets such that $g_n \uparrow g$ (a.e.). One further modifies the sequence using $g_n' = g_n \mathbf{1}_{|z|>\varepsilon_n}$, where ε_n is a sequence of positive real numbers decreasing to zero. The definition of $\int_{[0,1]\times[0,t]} g_n'(z, s)\widetilde{\mathcal{N}}(dz, ds) = \int_{[0,1]\times[0,t]} g_n'(z, s)\widetilde{\mathcal{N}}^{(\varepsilon_n)}(dz, ds)$ is clear as this is the definition for stochastic integrals with respect to compound Poisson processes. Next, one proves the following $L^2(\Omega)$ estimate using Theorem 3.4.31:

$$\mathbb{E}\left[\left|\int_{[0,1]\times[0,t]} g_n'(z,s) - g_m'(z,s)\widetilde{\mathcal{N}}(dz,ds)\right|^2\right]$$
$$= \int_{[0,1]\times[0,t]} \left|g_n'(z,s) - g_m'(z,s)\right|^2 \widehat{\mathcal{N}}(dz,ds).$$

Therefore the stochastic integrals converge to what we call the stochastic integral of g. Furthermore the $L^2(\Omega)$ norm of g is given by

$$\mathbb{E}\left[\left|\int_{[0,1]\times[0,t]} g(z,s)\widetilde{\mathcal{N}}(dz,ds)\right|^2\right] = \int_{[0,1]\times[0,t]} |g(z,s)|^2 \widehat{\mathcal{N}}(dz,ds).$$

Using a similar argument one can prove that any other sequence used will give the same limit, therefore proving the uniqueness of definition.

Hint (Exercise 5.3.20) *First construct a function h_n such that it satisfies:*

- $h_n(x) = x$ *for* $|x| \leq n$.
- $|h_n'(x)| \leq 2|x|$ *for* $x \in \mathbb{R}$. *Furthermore* $|h_n''(x)| \leq 2$ *for* $x \in \mathbb{R} - \{n, -n\}$.

Apply Itô's formula to $h_n(X_t - Y_t)$ and apply expectations (you will need to use the appropriate theorems in order to take expectations in each stochastic integral) in order to obtain

$$\begin{aligned}
&\mathbb{E}[h_n(X_t - Y_t)]\\
&= \int_{[-1,1]^c\times[0,t]} \mathbb{E}[h_n(X_{s-} - Y_{s-} + (a(X_{s-}) - a(Y_{s-}))z) - h_n(X_{s-} - Y_{s-})]\widehat{\mathcal{N}}(dz,ds)\\
&+ \int_{[-1,1]\times[0,t]} \mathbb{E}[h_n(X_{s-} - Y_{s-} + (a(X_{s-}) - a(Y_{s-}))z) - h_n(X_{s-} - Y_{s-})\\
&\qquad - h_n'(X_{s-} - Y_{s-})(a(X_{s-}) - a(Y_{s-}))z]\widehat{\mathcal{N}}(dz,ds).
\end{aligned}$$

Now use the Lipschitz properties of h_n and a in order to prove that the integrals above are bounded by

$$\mathbb{E}[h_n(X_t - Y_t)] \leq C \int_{\mathbb{R}\times[0,t]} \mathbb{E}[|X_{s-} - Y_{s-}|^2]|z|^2 \widehat{\mathcal{N}}(dz,ds).$$

Now we need to take limits with respect to n using the appropriate limit theorems. This will give you an inequality for which you can apply Gronwall's inequality.

For existence, define as in previous chapters, the approximation sequence:

$$\begin{aligned}
X_t^n = x &+ \int_{[-1,1]\times[0,t]} a(X_{s-}^n)z\widetilde{\mathcal{N}}^{(\varepsilon_n)}(dz,ds)\\
&+ \int_{[-1,1]^c\times[0,t]} a(X_{s-}^n)z\mathcal{N}^{(\varepsilon_n)}(dz,ds).
\end{aligned}$$

Prove that the uniform limit of X^n exists. For this, one proves that X^n is a Cauchy sequence using the sup norm in compact time intervals. In order to do this one uses Doob's inequality for the martingale parts and classical inequalities for the bounded variation parts. The limit will then be a càdlàg process.

To prove that $\mathbb{E}[X_t^2]$ one may use the same ideas as in Exercise 5.3.21.

Hint (Exercise 5.3.21) *First, one proves that $\mathbb{E}[X_t] = 0$.*

In order to compute $\mathbb{E}[X_t^2]$ we apply Itô's formula to obtain

$$\mathbb{E}[X_t^2] = x^2 + \int_{[-1,1]\times[0,t]} \mathbb{E}[X_{s-}^2] z^2 f(z) dz ds.$$

Note that some effort has to be made in order to prove that the martingale term in the Itô formula has expectation zero. Now one needs to argue that $\mathbb{E}[X_{s-}^2] = \mathbb{E}[X_s^2]$ for all $s \geq 0$.

Finally, one just needs to solve a linear equation.

Hint (Exercise 5.3.22) *For this exercise it is useful to recall the solution of Exercise 4.1.27. A similar calculation needs to be done here.*

Hint (Exercise 5.4.1) *In this exercise, clearly question 3 only makes sense if $\lambda(A) < \infty$. Therefore we assume this condition. We realize that $\mathcal{N}^n(A)$ is a compound Poisson process with Bernoulli jump increments. In fact, for $\lambda_n = \lambda(A_n)$ we have*

$$\begin{aligned}
\mathbb{P}(\mathcal{N}^n(A) = r) &= e^{-\lambda_n t} \sum_{j=r}^{\infty} \frac{t^j}{r!(j-r)!} \lambda(A \cap A_n)^r \lambda(A \cap A_n^c)^{j-r} \\
&= \frac{e^{-\lambda_n t} t^r}{r!} \lambda(A \cap A_n)^r e^{t\lambda(A\cap A_n^c)}.
\end{aligned}$$

Therefore $\mathcal{N}^n(A)$ is a Poisson random variable with parameter $\lambda(A \cap A_n)$. For the second statement, this is an issue that has appeared repeatedly through the text. See for example, Corollary 3.1.3, Proposition 3.1.8 or Corollary 3.3.9. The rest of the exercise consists in computing characteristic functions.

Hint (Exercise 5.4.2) *Note that for $Z_t := \mathcal{N}([0,t] \times \mathbb{R})$, we have*

$$\mathbb{P}(T_1 > t) = \mathbb{P}(Z_t = 0) = e^{-\lambda([0,\infty)\times\mathbb{R})}.$$

In the case that $\lambda([0,\infty) \times \mathbb{R}) = \infty$ then one has that $\mathbb{P}(T_1 > t) = 0$. That is, a jump occurs immediately after time zero.

Hint (Exercise 5.5.18)

1. *First show that λ satisfies*

$$\lambda_t = \lambda_0 + \int_0^t e^{-(t-s)} dN_s. \tag{14.1}$$

Finally, compute its differential.

2. *Define*

$$\gamma_t = \lambda_0 + \int_0^t h(t-s)\, d\tilde{N}_s.$$

Note that the iteration of the formula (14.1) *will lead to an application of the (stochastic) Fubini-type theorem. One will need to use formulas such as*

$$\int_0^t h(t-s) \left(\int_0^s h(s-u)\, \lambda_u\, du \right) ds = \int_0^t h^{2*}(t-s)\, \lambda_s\, ds,$$
$$\int_0^t \Psi(t-s) \left(\int_0^s h(s-u)\, d\tilde{N}_u \right) ds = \int_0^t (\Psi * h)(t-s)\, d\tilde{N}_s.$$

Then it holds that

$$\begin{aligned} \lambda_t &= \gamma_t + \int_0^t h(t-s)\, \lambda_s\, ds \\ &= \gamma_t + \int_0^t h(t-s)\, \gamma_s\, ds + \int_0^t h^{2*}(t-s)\, \lambda_s\, ds. \end{aligned}$$

Iterating such a procedure leads us to the conclusion.

3. *We have only to take the expectation of (5.11). Note that*

$$\|\Psi\|_1 = \sum_{n\geq 1} \|h^{n*}\|_1 = \frac{\|h\|_1}{1 - \|h\|_1}.$$

For more information about these processes, one may refer to [16, 23, 28, 29, 32].

Hint (Exercise 5.6.4) *First, note that using the change of variables $-az = b/u$ one obtains an integral of characteristics similar to the original one with the exception that one obtains an extra term $-b/(az^2)$ in the integrand. On the other hand, if in the integrand, we had the extra term $a - \frac{b}{u^2}$ the integral can be computed explicitly. This gives the result for 1.*

For 2., one needs to complete the squares in the exponent of the exponential and then use the transformation $u = s^{-1/2}$ in order to use the result of 1.

Using Fubini, we just compute

$$\int_0^\infty \frac{e^{-\frac{x^2}{2s}}}{\sqrt{2\pi s}} \nu(ds).$$

In order to do this it is better to compute the Laplace transform of 2. In fact, note that the differentiation of the above expression with respect to x is related to the function for which the Laplace transform is computed in 2.

In fact the Laplace transform in 2. is linked with the following density function:

$$f(s) = \frac{\sqrt{x}}{\sqrt{2}} \frac{e^{-\frac{x^2}{2s}}}{\sqrt{2\pi s^3}}, \quad s > 0.$$

Here $x > 0$ is the parameter of the process. This is called the Lévy distribution.

Hint (Exercise 6.2.2) *For a non-absolutely continuous measure it is enough to consider a covariance matrix which is a combination of an invertible matrix on a strict subspace and zero otherwise. The simplest example of such a case is*

$$A = \begin{pmatrix} 1 & 0 \\ 0 & 0 \end{pmatrix}.$$

One the other hand, the generated distribution measure is continuous because on the first component it is absolutely continuous.

Hint (Exercise 7.1.5) *A solution is a cádlág stochastic process X adapted to the filtration generated by the Poisson random measure such that the stochastic integrals in (7.3) are well defined and the equation is satisfied with the property that $\mathbb{E}\left[\sup_{t \le T_n^\sigma} |X_t|^2\right] < \infty$ for all $n \in \mathbb{N}$. Let Y be another solution such that $\mathbb{E}\left[\sup_{t \le T_n^\sigma} |Y_t|^2\right] < \infty$ for all $n \in \mathbb{N}$. Then because Y solves the Eq. (7.5) on the time interval $[0, \sigma_1)$ then by uniqueness of solutions one has that $X_t = Y_t$ for all $t \in [0, \sigma_1)$ a.s. It is not difficult now to prove that the same property is satisfied in $[0, \sigma_1]$. Now an iteration procedure can be used.*

Hint (Exercise 7.2.6) *In fact, we have that*

$$\mathbb{E}\left[|X_{t\wedge\sigma_1}(x) - X_{t\wedge\sigma_1}(y)|^p\right] \le C\,|x-y|^p \exp\left[C\,T\right].$$

Then similarly, we have

$$\mathbb{E}\left[|X_{\sigma_1}(x) - X_{\sigma_1}(y)|^p\right] \le 2^{p-1} C(1 + \int_{[-1,1]^c} |z|^p \nu(dz))\,|x-y|^p \exp\left[C\,T\right].$$

In general, note that $\mathbb{P}(\sigma_n > T > \sigma_{n-1}) = \frac{e^{\lambda T}(\lambda T)^n}{n!}$ where $\lambda = \nu([-1,1]^c) < \infty$. and therefore

$$\begin{aligned} &\mathbb{E}\left[|X_{\sigma_1}(x) - X_{\sigma_1}(y)|^p\right] \\ &\le C(1 + \int_{[-1,1]^c} |z|^p \nu(dz))\,|x-y|^p \exp\left[C\,T\right] \sum_{n=1}^{\infty} \frac{e^{\lambda T}(\lambda T)^n (2n)^{p-1}}{n!}. \end{aligned}$$

Hint (Exercise 9.0.10) *This exercise is interesting as in probability theory, one may think that when building approximations by convolution for the Dirac delta function, the time variable plays the role of the parameter converging to zero. In this case it* *is the time variable which becomes the variable for the convolution and the space variable is used for the approximation. In fact, define*

$$g(s) := \frac{1}{s^{3/2}} e^{-\frac{1}{s}}.$$

Then let $C := \int_0^\infty g(s)ds < \infty$*. Note that the finiteness follows because the behavior of* g *around infinity is of the order* $s^{-3/2}$ *and at zero it is a continuous function. Now, we build the kernel function* $g(\frac{2s}{z^2})\frac{2}{z^2}$ *for* $z \neq 0$*. This is an approximation to the Dirac delta in the same way that the Gaussian density was an approximation in Lemma 9.0.7. Now it is just a matter of applying the technique used to solve Exercise 9.0.7. This exercise is related to Exercise 5.6.4.*

Hint (Exercise 9.1.1) *In the one-dimensional case there are two ways of finding the characteristic function*

$$\varphi(\theta) = 2\int_0^\infty \frac{1}{\sqrt{2\pi a}} e^{-\frac{z^2}{2a}} \cos(\theta z)dz.$$

Prove that $\varphi'(\theta) = -\theta\varphi(\theta)$ *is satisfied and solve the ordinary differential equation with appropriate initial conditions.*

Hint (Problem 9.1.5) *Note that the density function can be bounded as follows:*

$$|f_X(w)| \leq \frac{1}{(2\pi)^d} \int \frac{1}{1 + |\theta|^2} d\theta.$$

From here the result follows. The extension of the corollary is as follows: If $\varphi(\theta)$ *satisfies* $|\varphi(\theta)| \leq \frac{C}{1+|\theta|^{\lambda+1+\varepsilon}}$ *for some* $\varepsilon > 0$ *then the density function belongs to* C_b^λ*.*

For other bounds, they can be achieved with some additional tricks such as integration by parts formulas. Note that this exercise was proposed for real-valued random variables. The multi-dimensional case also follows with the proper changes.

For the last item just note that one can assume without loss of generality that $x_0 = 0$ *then the characteristic function will be* $\varphi(\theta) = \varepsilon + (1-\varepsilon)e^{-\frac{\theta^2}{2}}$*. Therefore as long as* $\varepsilon > 0$ *the characteristic function will not converge to zero as* $\theta \uparrow \infty$*. This is a feature of discrete distributions.*

Hint (Exercise 9.1.6) *In fact, we use complex analysis in order to bound the density which we already know it exists. By the inversion formula, we have that this density is given by*

$$f_X(x) = \frac{1}{2\pi} \int e^{-i\theta x} \varphi(\theta) d\theta.$$

Suppose without loss of generality that $x > 0$*. Now we will change the domain of integration using the Cauchy integral theorem. The domain of integration is* *a square composed of* $A_1 := [-R, R]$, $A_2 := \{R - i\eta; \eta \in [0, r]\}$, $A_3 := \{-R - i\eta; \eta \in [0, r]\}$ *and* $A_4 := \{\theta - ir; \theta \in [-R, R]\}$ *for* $r > 0$. *One proves that due to the hypotheses the integrals over* A_2 *and* A_3 *converge to zero as* $R \to \infty$ *for fixed* $r > 0$.

Therefore it suffices to compute the integral over A_4. *For this, one uses the hypotheses to obtain*

$$\left| \int_{-R}^{R} e^{-i\theta x - rx} \varphi(\theta - ir) d\theta \right| \le e^{-rx} \int_{B_1 \cup B_2} |\varphi(\theta - ir)| d\theta.$$

Here $B_1 := [-a, a]$ *and* $B_2 := [-R, R] \backslash B_1$. *The argument now finishes by using the hypotheses.*

In many applications calculations of this type work well. One just needs to compute upper bounds like the functions F_1 *and* F_2 *stated in the hypotheses.*

Hint (Problem 9.1.7) *In fact, using the hint, we have that as*

$$\begin{aligned}
\mathbb{E}\left[\left((W^0_{Y^0_t})^2 + (W^1_{Y^1_s})^2\right)^{-1}\right] &= \int_0^\infty \mathbb{E}\left[\exp\left(-\lambda (W^0_{Y^0_t})^2\right)\right] \mathbb{E}\left[\exp\left(-\lambda (W^1_{Y^1_s})^2\right)\right] d\lambda \\
&= \int_0^\infty \mathbb{E}[\exp(-Y^0_t \lambda |Z|^2)] \mathbb{E}[\exp(-Y^1_s \lambda |Z|^2)] d\lambda \\
&= \int_0^\infty \mathbb{E}[(1 + 2\lambda Y^0_t)^{-1}] \mathbb{E}[(1 + 2\lambda Y^1_s)^{-1}] d\lambda \\
&\le C + \int_C^\infty \lambda^{-2} \mathbb{E}[Y_t^{-1}] \mathbb{E}[Y_s^{-1}] d\lambda.
\end{aligned}$$

In the above, we have used the explicit expression for the Laplace transform of the square of a standard normal, and the expectation of the inverse of a gamma random variable with rate parameter a *and shape parameter* t. *Finally, one can explicitly compute*

$$\mathbb{E}[Y_t^{-1}] = \int_0^\infty x^{t-2} a^t \Gamma(t)^{-1} e^{-ax} dx < \infty.$$

Hint (Problem 10.1.8) *The IBP requires a boundary term. In fact, for* $p(x) := \frac{2}{\sqrt{2\pi}} e^{-\frac{x^2}{2}}$

$$\int_0^\infty G'(x) p(x) dx = -G(0) \frac{2}{\sqrt{2\pi}} + \int_0^\infty G(x) x p(x) dx.$$

The discrete part in the formula is designed in order to obtain the boundary terms. In fact, just note that

$$\sum_{n=1}^{\infty}(a_{n+1}-a_n)=a_\infty-a_1.$$

Defining $a_n := n(n-1)\mathbb{E}\left[\mathbf{1}(\frac{1}{n} < X < \frac{1}{n-1})G(X)\right]$ *one obtains the result.*

Hint (Problem 10.1.10) *We only give the hint for part 2 of the exercise as for part 1 it is similar. In fact, the integral is*

$$\int G'(\varphi(x))\exp(-\frac{x^2}{2})dx = \alpha^{-1}\int_0^\infty G'(x)\exp(-\frac{x^2}{2\alpha^2})dx$$
$$+\beta^{-1}\int_0^\infty G'(x)\exp(-\frac{x^2}{2\beta^2})dx.$$

This gives the weight $H = -\alpha^{-1}\varphi(W_1)\mathbf{1}_{\varphi(W_1)>0} - \beta^{-1}\varphi(W_1)\mathbf{1}_{\varphi(W_1)<0}$. *Note that the case* $\alpha = -\beta = 1$ *corresponds to part 1.*

For part 3, one has to take into account when doing the IBP that integrals have to be finite. For this reason, the additional condition on G has been added.

Hint (Problem 10.1.11) *In fact, it is enough to write the conditional density of the random variable Z with respect to U and then perform the IBP. This gives*

$$\mathbb{E}[G'(Z)] = \mathbb{E}\left[\int_0^\infty G'\left(\frac{A(U)}{w^{\frac{1-\alpha}{\alpha}}}\right)e^{-w}dw\right]$$
$$= \mathbb{E}\left[\int_0^\infty G\left(\frac{A(U)}{w^{\frac{1-\alpha}{\alpha}}}\right)\left(-w^{1/\alpha}+\frac{w^{\frac{1-\alpha}{\alpha}}}{\alpha}\right)A(U)^{-1}e^{-w}dw\right]$$
$$A(U) := \frac{\sin(\alpha U)\cos((1-\alpha)U)^{\frac{1-\alpha}{\alpha}}}{\cos^{1/\alpha}(U)}.$$

In the above calculation we have to assume that G satisfies the following condition: $G(aw^{1-\frac{1}{\alpha}})w^{1/\alpha}e^{-w}|_{w=0}^{w=\infty} = 0$ *and that the final expression is integrable with finite expectation.*

Hint (Exercise 10.1.22) *Here as in Theorem 10.1.16, we only show the unstable formula. One may start using an IBP formula in each direction of* Z_k, *therefore using only partial derivatives. That is,*

$$\mathbb{E}[G'(X_{k+1})/X_k] = \mathbb{E}[\int G'(X_k + \langle A(X_k), \mathbf{a}\rangle)\Delta^{-1}g((\sqrt{\Delta})^{-1}\mathbf{a})da].$$

Then one may perform the IBP on any of the two integrals $da = da_1 da_2$.

One may also use the divergence theorem in order to obtain a full two-dimensional IBP formula. In fact, define the function

$$F(\mathbf{a}) = \frac{1}{2}G(X_k + \langle A(X_k), \mathbf{a}\rangle)\Delta^{-1}g((\sqrt{\Delta})^{-1}\mathbf{a})(A_1(X_k)^{-1}, A_2(X_k)^{-1}).$$

Use this function in the divergence theorem in order to obtain the IBP formula. That is,

$$\int\int \nabla \cdot F dxdy = 0.$$

An interesting option which will reduce the restrictions on A is the choice

$$F(\mathbf{a}) = G(X_k + \langle A(X_k), \mathbf{a}\rangle)\|A(X_k)\|^{-2}\Delta^{-1}g((\sqrt{\Delta})^{-1}\mathbf{a})(A_1(X_k), A_2(X_k)).$$

An important remark here is that this formula may also work even if g were a well-defined function on a domain D with boundary. This is the case for Exercise 10.1.23.

Hint (Exercise 10.1.24) *In fact, note that if G denotes the distribution function associated with g and Φ the distribution function associated with a standard Gaussian law, we have that $F(u) := \Phi^{-1}(G(u))$. This idea is commonly used in simulation methods.*

Hint (Exercise 10.4.7) *Try to find a function $h(x)$ which is smooth such that it is constant except for the interval $(-\varepsilon, \varepsilon)$, where it should go from the value zero to one. In order to achieve this, one may use the function $h_1(x) = C\exp\left(-\frac{c}{(x+\varepsilon)^2(x-\varepsilon)^2}\right)$.*

Hint (Exercise 10.5.1) *In fact, note that for a simple Poisson process with parameter λ, we have*

$$\begin{aligned}\mathbb{E}[i\theta e^{i\theta N_t}] &= i\theta\exp(-\lambda t(1-e^{i\theta})),\\ \mathbb{E}[e^{i\theta N_t}N_t] &= \lambda t e^{i\theta}\exp(-\lambda t(1-e^{i\theta})).\end{aligned}$$

Do the same for a compound Poisson process and verify that the equality does not follow in general.

Hint (Exercise 10.5.5) *One just needs to repeat the calculation used in the proof of Theorem 10.5.2.*

Hint (Exercise 10.5.8) *Clearly the required condition for integrability is $\int |h(z)|\,\nu(dz) < \infty$. If one uses the fact that $\int h(z)\nu(dz) = 0$ for density functions g_i such that g_i' are integrable then we will have that we will only require the integrability of $\int_0^t \int h(z)\widetilde{\mathcal{N}}(dz, ds)$. This integrability will be satisfied if $\int |h(z)|^2\nu(dz) < \infty$.*

Hint (Problem 10.5.18) *In fact, in this case, one has for an even function w, $\int (e^{i\theta z} - 1)w(z)\nu(dz) = -4\sum_{j=0}^{\infty}\sin^2(\frac{\theta a_j}{2})\frac{a_j - a_{j+1}}{|a_j|^{1+\alpha}}w(a_j)$. Note that the Burkholder–Davis–Gundy inequalities also apply to this case.*

Hint (Exercise 10.5.22) *Using $a(k) := k - \lambda t$, one finds that $b(k) := k(k-1) - 2k\lambda t + (\lambda t)^2$. Therefore*

$$H := (\lambda t)^{-2}\left(N_t(N_{t-1}) - 2N_t\lambda t + (\lambda t)^2\right).$$

This is related to the Charlier polynomial of order 2 and as expected is also related to the iterated stochastic integral $\int_0^t \tilde{N}_{s-} d\tilde{N}_s = \frac{(\lambda t)^2}{2} H$.

Hint (Exercise 10.5.23) *Apply this result for a particular type of process w, which is a stochastic integral, in order to obtain the result of Exercise 10.5.22. As suggested in Exercise 10.5.15, one defines* $u(t-s,x) := \mathbb{E}[e^{i\theta N_t} \mid N_s = x]$. *Then applying the Itô formula one has the result, noting that* $\partial_s u(t-s,x) = -\lambda(e^{i\theta}-1)u(t-s,x)$. *Finally, take* $w(s,z) := N_{s-}$ *in order to obtain the result in Exercise 10.5.22.*

Hint (Problem 11.1.4) *Since the marginal* Z_t *has the density (11.2), the divergence formula yields that*

$$\begin{aligned}
\mathbb{E}\left[G(S_t)\, S_t\, \rho\, Z_t\right] &= \int_0^\infty G'\left(S_0\, e^{\rho y + ct}\right) S_0\, e^{\rho y + ct}\, \rho\, y\, p_t^Z(y)\, dy \\
&= \int_0^\infty \partial_y \left\{G\left(S_0\, e^{\rho y + ct}\right)\right\} \frac{b^{at}}{\Gamma(at)}\, y^{at}\, e^{-by}\, dy \\
&= \left[G\left(S_0\, e^{\rho y + ct}\right) \frac{b^{at}}{\Gamma(at)}\, y^{at}\, e^{-by}\right]_0^\infty \\
&\quad - \int_0^\infty G\left(S_0\, e^{\rho y + ct}\right) \frac{b^{at}}{\Gamma(at)}\, \partial_y \left(y^{at}\, e^{-by}\right) dy \\
&= \int_0^\infty G\left(S_0\, e^{\rho y + ct}\right) \left(-at + by\right) p_t^Z(y)\, dy \\
&= \mathbb{E}\left[G(S_t)\left(-at + b\, Z_t\right)\right].
\end{aligned}$$

For 2., using the density function the deduction is easier. For example, perform an IBP on

$$\mathbb{E}\left[G'(S_t)\right] = \int_0^\infty G'(S_0 \exp(\rho y + ct)) \frac{b^{at}}{\Gamma(at)}\, y^{at-1}\, e^{-by} dy.$$

In the case that G is not differentiable just use instead

$$\partial_\rho \mathbb{E}\left[G(S_t)\right] = \partial_\rho \int_0^\infty G(S_0 \exp(\rho y + ct)) \frac{b^{at}}{\Gamma(at)}\, y^{at-1}\, e^{-by} dy.$$

Then perform the change of variables $\rho y = u$ *before carrying out the exchange between derivatives and integrals.*

Hint (Problem 11.2.3) *Since* $\tilde{Z}_t$ *has the same law of* Z_t, *we have*

$$
\begin{aligned}
\mathbb{E}\big[G'(Z_t)\,Z_t\big] &= \mathbb{E}\big[G'(\tilde{Z}_t)\,\tilde{Z}_t\big] \\
&= \int_0^\infty \int_0^\infty \mathbb{E}\big[G'(\tilde{Z}_t^{y,z})\,\tilde{Z}_t^{y,z}\big]\,e^{-y}\,e^{-z}\,dy\,dz \\
&= \int_0^\infty \int_0^\infty \mathbb{E}\Big[G'(\tilde{Z}_t^{y,z})\,(\tilde{Z}_{-,t}^{y} + \tilde{Z}_{+,t}^{z})\Big]\,e^{-y}\,e^{-z}\,dy\,dz \\
&= \frac{1}{\delta}\int_0^\infty \int_0^\infty \bigg\{\mathbb{E}\Big[\frac{\partial}{\partial y}\Big(G(\tilde{Z}_t^{y,z}) - G(0)\Big)\,y\Big] \\
&\qquad + \mathbb{E}\Big[\frac{\partial}{\partial z}\Big(G(\tilde{Z}_t^{y,z}) - G(0)\Big)\,z\Big]\bigg\}\,e^{-y}\,e^{-z}\,dy\,dz \\
&=: I_1 + I_2.
\end{aligned}
$$

Since $G \in C_b^1(\mathbb{R})$ *and* $\big(y\,e^{-y}\big)' = (1-y)\,e^{-y}$, *the IBP in the usual sense tells us that*

$$
\begin{aligned}
I_1 &= -\frac{1}{\delta}\int_0^\infty \int_0^\infty \mathbb{E}\Big[\big\{G(\tilde{Z}_t^{y,z}) - G(0)\big\}\,\big(y\,e^{-y}\big)'\Big]\,e^{-z}\,dy\,dz \\
&= -\frac{1}{\delta}\int_0^\infty \int_0^\infty \mathbb{E}\Big[\big\{G(\tilde{Z}_t^{y,z}) - G(0)\big\}\,\big(1-y\big)\Big]\,e^{-y}\,e^{-z}\,dy\,dz \\
&= -\frac{1}{\delta}\,\mathbb{E}\Big[\big\{G(\tilde{Z}_t) - G(0)\big\}\,\big(1 - E_-\big)\Big].
\end{aligned}
$$

Similarly, we can get

$$
I_2 = -\frac{1}{\delta}\,\mathbb{E}\Big[\big\{G(\tilde{Z}_t) - G(0)\big\}\,\big(1 - E_+\big)\Big].
$$

Hence, we can obtain the formula (11.15).

Hint (Problem 11.2.5) *Since*

$$
G'(\tilde{Z}_t^y)\,(\tilde{Z}_t^y - \gamma\,t) = \frac{1}{\delta}\,\frac{\partial}{\partial y}\big\{G(\tilde{Z}_t^y) - G(\gamma\,t)\big\}\,y,
$$

the IBP in the usual sense leads us to see that

$$
\begin{aligned}
\mathbb{E}\big[G'(Z_t)\,(Z_t - \gamma\,t)\big] &= \mathbb{E}\big[G'(\tilde{Z}_t^E)\,(\tilde{Z}_t^E - \gamma\,t)\big] \\
&= \int_0^\infty \mathbb{E}\big[G'(\tilde{Z}_t^y)\,(\tilde{Z}_t^y - \gamma\,t)\big]\,e^{-y}\,dy \\
&= \frac{1}{\delta}\int_0^\infty \mathbb{E}\Big[\frac{\partial}{\partial y}\big\{G(\tilde{Z}_t^y) - G(\gamma\,t)\big\}\Big]\,y\,e^{-y}\,dy \\
&= -\frac{1}{\delta}\int_0^\infty \mathbb{E}\Big[\big\{G(\tilde{Z}_t^y) - G(\gamma\,t)\big\}\Big]\,\big(y\,e^{-y}\big)'\,dy
\end{aligned}
$$

$$= -\frac{1}{\delta} \int_0^\infty \mathbb{E}\Big[\big\{G(\tilde{Z}_t^y) - G(\gamma\, t)\big\}\Big]\, (1-y)\, e^{-y}\, dy$$
$$= \mathbb{E}\left[\big\{G(\tilde{Z}_t^E) - G(\gamma\, t)\big\}\, \frac{E-1}{\delta}\right].$$

Hint (Exercise 11.4.2) *Let* $\varphi(z) = 1 - \exp\big(-z^{\alpha+1}\big)$. *Then, we can easily check the conditions mentioned above. In fact, it is clear that* $|\varphi(z)| \le |z|^{1+\alpha} \wedge 1$, *and that*

$$\left|\frac{\varphi(\varepsilon)}{\varepsilon^\alpha}\right| \le \sum_{k=1}^\infty \frac{\varepsilon^k}{k!}\, \varepsilon^{\alpha(k-1)} \le e^\varepsilon - 1 \to 0 \quad (\varepsilon \searrow 0).$$

Hint (Exercise 11.5.1) *The condition on* a, b *and* t *is that* $t > 1/a$ *and* $1 \le p < at$. *In fact,*

(i) Since the density function of Z_t *is*

$$p_{Z_t}(z) = \frac{b^{at}}{\Gamma(at)}\, z^{at-1}\, e^{-bz}\, \mathbb{I}_{(0,\infty)}(z),$$

we have

$$\mathbb{E}[Z_t^{-p}] = \frac{\Gamma(at-p)}{\Gamma(at)}\, b^p,$$

which is bounded for any $1 \le p < at$.

(ii) In fact, using the method in Exercise 11.1.1, one obtains that

$$\mathbb{E}[e^{-\theta Z_t}] = \big(1 + \theta b^{-1}\big)^{at}$$

for $\theta > -b$, *as explained in (11.1). Now we just need to prove the integrability of* $\int_0^\infty \frac{\theta^{p-1}}{(1+\theta b^{-1})^{at}}\, d\theta$. *This quantity is finite if* $at > p$. *Note that one of the important conclusions of this calculation is that the behavior of the Laplace transform* $\mathbb{E}[e^{-\theta Z_t}]$ *at* $\theta \to \infty$ *determines the behavior of the density of* Z_t *close to zero. That is,*

$$\mathbb{E}\big[Z_t^{-p}\big] = \frac{1}{\Gamma(p)} \int_0^\infty \theta^{p-1}\, \mathbb{E}[\exp(-\theta\, Z_t)]\, d\theta$$
$$= \frac{b^{at}}{\Gamma(p)} \int_0^\infty \frac{\theta^{p-1}}{(\theta+b)^{at}}\, d\theta,$$

which is bounded for any $1 \le p < at$.

Hint (Exercise 11.5.2) *Since*

$$\mathbb{E}\big[\exp(-\theta\, Z_t)\big] = \exp(-C_\alpha\, t\, |\theta|^\alpha),$$

where C_α is a positive constant depending on α, we see that

$$\begin{aligned}\mathbb{E}\big[Z_t^{-p}\big] &= \frac{1}{\Gamma(p)}\int_0^\infty \theta^{p-1}\,\mathbb{E}[\exp(-\theta\,Z_t)]\,d\theta \\ &= \frac{1}{\Gamma(p)}\int_0^\infty \theta^{p-1}\,\exp(-C_\alpha\,t\,|\theta|^\alpha)\,d\theta,\end{aligned}$$

which is bounded for any $p \geq 1$. Note that this result compared with the result for Exercise 11.5.1 reveals that inverse moments for stable processes are generally better behaved in comparison with gamma processes.

Hint (Exercise 11.5.3) *The Itô formula leads us to see that*

$$\mathbb{E}\big[\exp(-\theta\,\Gamma_t)\big] = 1 + \left\{\int_0^\infty \big(\exp(-\theta\,z\,\varphi(z)) - 1\big)\,\nu(dz)\right\}\int_0^t \mathbb{E}\big[\exp(-\theta\,\Gamma_s)\big]\,ds,$$

which implies that

$$\mathbb{E}\big[\exp(-\theta\,\Gamma_t)\big] = \exp\left\{-t\int_0^\infty \big(1 - \exp(-\theta\,z\,\varphi(z))\big)\,\nu(dz)\right\}.$$

Since $\varphi(z) = c\,z$ for $|z| \leq 1$, we have

$$\begin{aligned}\int_0^\infty \{1 - \exp(-\theta\,z\,\varphi(z))\}\,\nu(dz) &= \int_0^1 \big(1 - \exp(-\theta\,z\,\varphi(z))\big)\,\frac{\kappa}{z^{1+\alpha}}\,dz \\ &\geq \int_0^1 (1 - e^{-c})\,\theta\,z\,\varphi(z)\,\frac{\kappa}{z^{1+\alpha}}\,dz \\ &= \frac{\kappa\,(1 - e^{-c})\,c}{2-\alpha}\,\theta\end{aligned}$$

for $\theta \geq 1$. Hence, we see that

$$\mathbb{E}\big[\exp(-\theta\,\Gamma_t)\big] \leq \exp\left(-\frac{t\,\kappa\,(1 - e^{-c})\,c}{2-\alpha}\,\theta\right)$$

for $\theta \geq 1$, and we can get

$$\begin{aligned}\mathbb{E}[\Gamma_t^{-p}] &= \frac{1}{\Gamma(p)}\int_0^\infty \theta^{p-1}\,\mathbb{E}\big[\exp(-\theta\,\Gamma_t)\big]\,d\theta \\ &\leq \frac{1}{\Gamma(p)}\int_0^1 \theta^{p-1}d\theta + \frac{1}{\Gamma(p)}\int_1^\infty \theta^{p-1}\,\exp\left(-\frac{t\,\kappa\,(1 - e^{-c})\,c}{2-\alpha}\,\theta\right)\,d\theta,\end{aligned}$$

which is bounded for any $p \geq 1$.

Hint (Exercise 11.5.4) *As seen in the previous exercise, we can obtain that*

$$\mathbb{E}\big[\exp(-\theta\,\Gamma_t)\big] = \exp\left\{-t\int_0^\infty \big(1-\exp(-\theta\, z\,\varphi(z))\big)\,\nu(dz)\right\}.$$

Since

$$\begin{aligned}\int_0^\infty \big(1-\exp(-\theta\, z\,\varphi(z))\big)\,\nu(dz) &= \int_0^\infty \big(1-\exp(-\theta\, z\,\varphi(z))\big)\,\frac{a}{z^{1+\alpha}}\,e^{-bz}\,dz\\ &\geq \int_0^1 \big(1-\exp(-\theta\, z\,\varphi(z))\big)\,\frac{a}{z^{1+\alpha}}\,e^{-bz}\,dz\\ &\geq e^{-b}\int_0^1 \big(1-\exp(-\theta\, z\,\varphi(z))\big)\,\frac{a}{z^{1+\alpha}}\,dz,\end{aligned}$$

the computation in the previous exercise for the truncated stable process helps us to see that

$$\mathbb{E}\big[\exp(-\theta\,\Gamma_t)\big] \leq \exp\left(-\frac{t\,a\,e^{-b}\,(1-e^{-c})\,c}{2-\alpha}\,\theta\right)$$

for $\theta \geq 1$, *because of* $\varphi(z) = c\,z$ *for* $|z| \leq 1$. *Then, it holds that*

$$\begin{aligned}\mathbb{E}[\Gamma_t^{-p}] &= \frac{1}{\Gamma(p)}\int_0^\infty \theta^{p-1}\,\mathbb{E}[\exp(-\theta\,\Gamma_t)]\,d\theta\\ &\leq \frac{1}{\Gamma(p)}\int_0^1 \theta^{p-1}d\theta + \frac{1}{\Gamma(p)}\int_1^\infty \theta^{p-1}\exp\left(-\frac{t\,a\,e^{-b}\,(1-e^{-c})\,c}{2-\alpha}\,\theta\right)d\theta,\end{aligned}$$

which is bounded for any $p \geq 1$.

Hint (Exercise 11.5.7) *We will not discuss this in full detail. Instead we will just give a way to obtain the result in general. First we remark as in Exercises 11.5.3 through 11.5.4 we have that*

$$\mathbb{E}\big[\exp(-\theta\,\Gamma_t)\big] = \exp\left\{-t\int_0^\infty \big(1-\exp(-\theta\, z\,\varphi(z))\big)\,\nu(dz)\right\}.$$

If $\varphi(z) > 0$ *is set up such that* $\varphi(z) = O(|z|^p)$ *for* $p \geq 1$ *then the exponent* $\int_0^\infty \big(1-\exp(-\theta\, z\,\varphi(z))\big)\,\nu(dz)$ *will be large enough so that the decrease of the Laplace transform is fast enough. Note that in the case that* ν *has also support on the negative values of the real line, the sign of* φ *has to be changed.*

Hint (Exercise 12.2.2) *In fact, when* $\int f(z)dz < \infty$, *the driving process is the equivalent of a Poisson random measure associated with a compound Poisson process where the jump size has a density constructed from the renormalization of* f. *Therefore, there is always the probability of having no jump, which gives a point mass in the law of* X_t. *The point mass will be given as a solution to an ordinary differential equation. On the other hand, if* $a(x, z)$ *is a function that has a differentiable inverse in* z *then the conditional density of* X_t *conditioned to a positive number of jumps will have a density.*

Hint (Exercise 12.4.5) *The first part is clear, because*

$$\sum_{j=1}^{N^\varepsilon} D_j X^\varepsilon = \int_{\bar{B}_1^c \times \mathbb{R}_+} D_z X^\varepsilon \, \mathcal{N}(dz, dt).$$

Note that the integrand in this case is not adapted to the underlying filtration.

For the second part, let $\tau_j^\varepsilon \le t < \tau_{j+1}^\varepsilon$. Since the process X^ε satisfies the equation in Exercise 12.2.5, one has

$$\begin{aligned}
\partial_{z_j} X_t^{\varepsilon,(j)} &= \partial_{z_j} a\big(X_{\tau_j^\varepsilon -}^\varepsilon, z_j\big) + \int_{\tau_j^\varepsilon}^t \nabla b\big(X_u^{\varepsilon,(j)}\big)\, \partial_{z_j} X_u^{\varepsilon,(j)}\, du \\
&\quad + \int_{(\bar{B}_1 \cap \bar{B}_\varepsilon^c)\times(\tau_j^\varepsilon, t]} \nabla a\big(X_{u-}^{\varepsilon,(j)}, z\big)\, \partial_{z_j} X_{u-}^{\varepsilon,(j)}\, \widetilde{\mathcal{N}}^{(\varepsilon)}(dz, du) \\
&\quad + \int_{\bar{B}_1^c \times (\tau_j^\varepsilon, t]} \nabla a\big(X_{u-}^{\varepsilon,(j)}, z\big)\, \partial_{z_j} X_{u-}^{\varepsilon,(j)}\, \mathcal{N}(dz, du),
\end{aligned}$$

which enables us to get the equation for $D_{z_j} X^\varepsilon = \big\{D_{z_j} X_t^\varepsilon = \partial_{z_j} X_t^{\varepsilon,(j)}\, h(z_j)\,;\ t \ge 0\big\}$.

Hint (Exercise 12.4.8) *In fact, we will only consider the uniform integrability. Clearly,*

$$\left| \int_{U_\varepsilon \times [0,T]} \frac{\operatorname{div}\{h(z)\, f(z)\}}{f(z)} \mathcal{N}(dz, ds) \right| \le \int_{U_\varepsilon \times [0,T]} \left| \frac{\operatorname{div}\{h(z)\, f(z)\}}{f(z)} \right| \mathcal{N}(dz, ds).$$

This last sequence of random variables is increasing in ε and it converges in $L^1(\Omega)$ due to Hypothesis 12.3.3.

Hint (Exercise 12.4.14) *In fact, recall Exercise 12.4.2, and in order to obtain uniform boundedness it is enough to consider an appropriate localization function and apply it to the approximating sequence.*

Hint (Exercise 12.4.22) *In fact, the formula is a direct application of the following argument using the chain rule and the duality formula:*

$$\begin{aligned}
DG(X) &= G'(X) DX \\
DG(X)\,(DX)^{-1}\, \Theta &= G'(X)\Theta.
\end{aligned}$$

Taking expectations in the above formula we obtain an IBP if we assume that:

1. *$G \in C_c^1$, $X \in \mathbb{D}_{2,1}(\mathbb{R})$.*
2. *$(DX)^{-1}$ is a well-defined r.v.*
3. *$\tilde{\delta}((DX)^{-1}\, \Theta)$ exists and is integrable. Note that this condition is hiding a complex structure which we studied in Theorems 12.4.19 and 12.4.25.*

Hint (Exercise 12.6.2) *In fact, using classical stopping time techniques one can reduce the statement to proving that the process X is a martingale. That is, define*

$$\tau_n := \inf\left\{t;\ \int_{\bar{B}_1\times[0,T]} |g(t,z)|^2\, \widehat{\mathscr{N}}(dz,dt) > n \right.$$
$$\left. \text{or} \int_{\bar{B}_1^c\times[0,T]} \left(\exp\left[g(t,z)\right]-1\right) \widehat{\mathscr{N}}(dz,dt) > n\right\}.$$

Then instead of proving that X is a local martingale, one proves that the process $X_{t\wedge\tau_n}$, $t \geq 0$ is a martingale. To prove this fact it is enough to apply the Itô formula to $X_{t\wedge\tau_n}$.

Hint (Exercise 12.6.3) *In fact, we use an alternative form in order to compute $\mathbb{Q}$-conditional expectations with respect to $\mathscr{F}_s$. For this, let f_s be $\mathscr{F}_s$-measurable and compute the conditional expectation as follows:*

$$\mathbb{E}_{\mathbb{Q}}[Yf_s] = \mathbb{E}[Yf_s\alpha_T] = \mathbb{E}[f_s\mathbb{E}[Y\alpha_T|\mathscr{F}_s]] = \mathbb{E}_{\mathbb{Q}}[\alpha_T^{-1} f_s\mathbb{E}[Y\alpha_T|\mathscr{F}_s]].$$

From here the formula follows. The formula can be simplified if we define $\alpha_s = \mathbb{E}[\alpha_T|\mathscr{F}_s]$.

Hint (Exercise 12.6.5) *By using the Itô formula, we can compute*

$$\mathbb{E}_{\mathbb{Q}}\left[e^{i\theta Z_t}\right] = \exp\left[i\theta \int_{\bar{B}_1\times[0,t]} \left(\exp[g(s,z)]-1\right) \widehat{\mathscr{N}}(dz,ds)\right.$$
$$\left. + \int_{\mathbb{R}_0^m\times[0,t]} \left(e^{i\theta z} - 1 - i\theta z\,\mathbf{1}_{\bar{B}_1}(z)\right) \exp[g(s,z)]\, \widehat{\mathscr{N}}(dz,ds)\right].$$

Hint (Exercise 12.6.15) *It is easy to check that*

$$\int_{\mathbb{R}_0} \nu(dz) = \infty, \quad \int_{\mathbb{R}_0} (|z|^2\wedge 1)\,\nu(dz) < \infty, \quad \int_{\bar{B}_1^c} |z|^p\,\nu(dz) < \infty$$

for any $p \geq 2$. As before, $\varphi \in C^1\big((0,+\infty)\,;\,\mathbb{R}\big)$ such that

$$|\varphi(z)| \leq C\,(|z|^2\wedge 1), \quad 0 < |\varphi'(z)| \leq C\,(|z|\wedge 1). \tag{14.2}$$

Since $h(z) = \partial_\xi H^\xi(z)\big|_{\xi=0} = \varphi(z)\,z$, the condition (14.2) *enables us to check that*

$$\int_{\mathbb{R}_0} |h(z)|\,\nu(dz) = \kappa\int_0^t \tfrac{|\varphi(z)|}{z^\alpha}\,dz \leq \tfrac{C\kappa}{2-\alpha},$$
$$\int_{\mathbb{R}_0} \left|\frac{\operatorname{div}\{h(z)\,f(z)\}}{f(z)}\right| \nu(dz) = \int_0^1 \left|\varphi'(z)\,z - \alpha\,\varphi(z)\right| \tfrac{\kappa}{z^{1+\alpha}}\,dz \leq \tfrac{C\kappa\,(1+\alpha)}{2-\alpha}.$$

Let F be a open neighborhood in $\mathbb{R}$ around 0. *Then, we have*

$$\sup_{\xi\in F}\int_{\mathbb{R}_0}\left|H^{\xi}(z)-z\right|^2\nu(dz)\leq\sup_{\xi\in F}\left|\exp\left(C\,\xi\right)-1\right|^2\int_0^1\kappa\,z^{1-\alpha}\,dz$$
$$=\sup_{\xi\in F}\left|\exp\left(C\,\xi\right)-1\right|^2\frac{\kappa}{2-\alpha},$$
$$\sup_{\xi\in F}\int_{\bar{B}_1^c}\left|H^{\xi}(z)\right|^2\nu(dz)=0,$$

because of $0<\alpha<2$ *and the condition* (14.2).

As for the Bismut method, note that

$$k^{\xi}(z)=\left(1+\xi\,z\,\varphi'(z)\right)\,\exp\left(-\alpha\,\xi\,\varphi(z)\right).$$

Then, all conditions in Hypothesis 12.6.6 can be verified easily.

Hint (Exercise 13.2.2) *In fact, just apply expectations (you need to ensure that this is allowed) to obtain that*

$$\mathbb{E}[V_t]=\mathbb{E}[V_0]+\int_0^t\int_{\Pi}\int_{\mathbb{R}^2}\frac{\cos(\theta)-1}{2}\mathbb{E}[(V_{s-}-v)|V_{s-}-v|^{\gamma}]\,ds b(\theta)d\theta f_s(dv).$$

As the law of V_s *is* f_s*, the result follows by symmetry in the space integrals. In the case of* $|V_t|^2$ *the calculation is similar but a bit longer and also uses the symmetry in* θ.

Hint (Exercise 13.2.4) *Computing the characteristic function by an approximation procedure which has already been used for the stable laws, one has (see e.g. Exercise 4.1.23 and Sect. 9.2) for* $r\in\mathbb{R}$

$$\mathbb{E}[e^{irZ_t}]=\mathbb{E}\left[e^{irZ_0}\right]\exp\left(t\mathbb{E}[X]\int_{\Pi}\left(e^{ir\theta}-1\right)b(\theta)d\theta\right).$$

Now the analysis can be completed as in Exercise 4.1.23-(ii). From here follows the regularity of the law of Z_t*. In the case of* Z' *one has*

$$\mathbb{E}[e^{irZ'_t}]=\mathbb{E}\left[e^{irZ_0}\right]\exp\left(t\int_{\Pi}\left(e^{ir\theta}-1\right)b(\theta)\theta_+d\theta\right).$$

Hint (Exercise 13.3.5) *In fact, the inequality* $\exp(C\Gamma_{\varepsilon}^{\gamma})\leq C_{\eta_0,\eta,\gamma}\,\varepsilon^{-\eta}$ *reduces to* $Cx^{\gamma\eta_0}\leq\log(C_{\eta_0,\eta,\gamma})+\eta x$ *for* $x=\log(\varepsilon^{-1})>0$*. If one defines* $k=(\frac{\eta}{C\gamma\eta_0})^{\frac{1}{\gamma\eta_0-1}}$*, then* $\log(C_{\eta_0,\eta,\gamma})=C(1-\gamma\eta_0)k^{\gamma\eta_0}$.

For the second assertion, suppose without loss of generality (here we are using the weak continuity of the processes involved) that $\int\mathbb{E}[|V_s^{\varepsilon}-v|^p]f_s(dv)=0$ *for* $s\in[a,b]$*. Then* $V_s^{\varepsilon}=v$ *a.s. for* v-f_s *a.e., which is clearly a contradiction.*

Hint (Exercise 13.4.2) *Since*

$$I + A(\theta) = \begin{pmatrix} 1 + \dfrac{\cos\theta - 1}{2} & -\dfrac{\sin\theta}{2} \\ \dfrac{\sin\theta}{2} & 1 + \dfrac{\cos\theta - 1}{2} \end{pmatrix},$$

we have

$$\begin{aligned} \|I + A(\theta)\|^2 &= \sup_{(x,y)^* \in \mathbb{S}^1} \left|(I + A(\theta))\,(x, y)^*\right|^2 \\ &= \left(1 + \frac{\cos\theta - 1}{2}\right)^2 + \left(\frac{\sin\theta}{2}\right)^2 \\ &= \frac{1 + \cos\theta}{2} \\ &\le 1. \end{aligned}$$

A careful modification of the above argument also gives that $\|I + A_\varepsilon(\theta)\| \le 1$. *See also the solution to Exercise 13.4.8.*

Hint (Exercise 13.4.7) *It is enough to consider* $\varphi(\theta) := |\theta|^\beta \exp\left(-(|\theta| - \frac{\pi}{2})^{-2}\right)$. *For the second part recall Exercise 13.4.6.*

Hint (Exercise 13.4.8) *Since*

$$I + A_\varepsilon(\theta) = \begin{pmatrix} 1 + \chi_\varepsilon(\theta)\,\dfrac{\cos\theta - 1}{2} & -\chi_\varepsilon(\theta)\,\dfrac{\sin\theta}{2} \\ \chi_\varepsilon(\theta)\,\dfrac{\sin\theta}{2} & 1 + \chi_\varepsilon(\theta)\,\dfrac{\cos\theta - 1}{2} \end{pmatrix},$$

and $0 \le \chi_\varepsilon(\theta) \le 1$, *we have*

$$\begin{aligned} \|I + A_\varepsilon(\theta)\|^2 &= \sup_{(x,y)^* \in \mathbb{S}^1} \left|(I + A_\varepsilon(\theta))\,(x, y)^*\right|^2 \\ &= \left(1 + \chi_\varepsilon(\theta)\frac{\cos\theta - 1}{2}\right)^2 + \chi_\varepsilon(\theta)^2 \left(\frac{\sin\theta}{2}\right)^2 \\ &= \frac{1 - \cos\theta}{2}\left(\chi_\varepsilon(\theta) - 1\right)^2 + \frac{1 + \cos\theta}{2} \\ &\le \frac{1 - \cos\theta}{2} + \frac{1 + \cos\theta}{2} = 1. \end{aligned}$$

Hence, we can get

$$\left\|\tilde{Y}_t\right\| \le \prod_{k=1}^{N_t^\varepsilon} \left\|I + A_\varepsilon(\tilde{\theta}_k)\right\| \le 1.$$

In order to prove point 3., we can check inductively that

$$\partial_{\theta_k} \tilde{V}_{T_k}^{(k)} = A_\varepsilon{}'(\theta_k)\big(\tilde{V}_{T_{k-1}} - v_{T_k}(\tilde{\rho}_k)\big),$$

and that

$$\begin{aligned}
\partial_{\theta_j} \tilde{V}_{T_k}^{(j)} &= \big(I + A_\varepsilon(\tilde{\theta}_k)\big)\,\partial_{\theta_j} \tilde{V}_{T_{k-1}}^{(j)} \\
&= \big(I + A_\varepsilon(\tilde{\theta}_k)\big)\big(I + A_\varepsilon(\tilde{\theta}_{k-1})\big)\,\partial_{\theta_j} \tilde{V}_{T_{k-2}}^{(j)} \\
&= \cdots \\
&= \prod_{i=j+1}^{k} \big(I + A_\varepsilon(\tilde{\theta}_i)\big)\,\partial_{\theta_j} \tilde{V}_{T_j}^{(j)} \\
&= \tilde{Y}_{T_k}\,\tilde{Z}_{T_j}\,A_\varepsilon{}'(\theta_j)\big(\tilde{V}_{T_{j-1}} - v_{T_j}(\tilde{\rho}_j)\big)
\end{aligned}$$

for $1 \le j \le k-1$. *Since* $\tilde{V}_{T_k}^{(j)} = \tilde{V}_{T_k}$ *for* $j > k$, *it is trivial to see that* $\partial_{\theta_j} \tilde{V}_{T_k}^{(j)} = 0$. *In order to obtain the estimates is important to use Proposition 13.4.1.*

Hint (Exercise 13.4.10) *(v) An easy way to see how to compute this derivative is to first compute* $D_{j_1}\tilde{V}_t$. *Then one has to observe that for* $j \ge j_1$, *one has* $D_j(\tilde{V}_{T_{j_1}} - v_{T_{j_1}}(\tilde{\rho}(j_1))) = 0$. *(vi) The reason why the high-order derivatives are not bounded in the general case* $N_t^\varepsilon \ge j_k \ge j_{k-1} \ge \ldots \ge j_1$ *is due to the high-order derivatives of* A_ε *which then leads to (iv) and Exercise 13.4.8. This issue leads to (vii).*

Hint (Exercise 13.6.11) *Suppose that* $|\tilde{V}_s| \le \Gamma_\varepsilon - 1$ *for any* $s \in [0, t]$. *Since*

$$\mathbf{1}_{\{\sup_{s\in[0,t]} |\tilde{V}_s| \le \Gamma_\varepsilon - 1\}} = 1, \quad \tilde{G}_{N_t^\varepsilon} = \Psi(0) = 1,$$

we can get the left-side inequality. On the other hand, suppose that there exists $s \in [0, t]$ *satisfying with* $|\tilde{V}_s| \ge \Gamma_\varepsilon - 1$. *Then, we see that*

$$\mathbf{1}_{\{\sup_{s\in[0,t]} |\tilde{V}_s| \le \Gamma_\varepsilon - 1\}} = 0, \quad \tilde{G}_{N_t^\varepsilon} \ge 0,$$

so the left-side inequality holds. Similarly, we can prove easily the right-side inequality.

Hint (Exercise 13.5.1) *(cf. [9]-Lemma 7.29) We have only to study the case where* A *is a diagonal matrix, because a symmetric matrix in* $\mathbb{R}^d \otimes \mathbb{R}^d$ *can be diagonalized via an orthogonal matrix. Denote by* $\lambda_1, \ldots, \lambda_d$ *the non-negative eigenvalues of the matrix* A.

At the beginning, we shall consider the case where there exists $k \in \{1, \ldots, d\}$ *such that* $\lambda_k = 0$. *Then, one trivially has that*

$$\det A = 0, \quad \int_{\mathbb{R}^d} |x|^{d(2p-1)} \exp\big[-x\cdot Ax\big]\,dx = \infty,$$

which proves the assertion in this case.

Now, we shall consider the case where all eigenvalues λ_k ($k \in \{1, \ldots, d\}$) are strictly positive. The Jensen inequality enables us to see that

$$\prod_{k=1}^{d} |x_k| \le \left(\frac{|x_1| + \cdots + |x_d|}{d} \right)^d \le d^{-d/2} |x|^d .$$

Then, we have

$$\begin{aligned}
\frac{\Gamma(p)^d}{(\det A)^p} &= \prod_{k=1}^{d} \frac{\Gamma(p)}{\lambda_k^p} = \prod_{k=1}^{d} \int_{\mathbb{R}} |x_k|^{2p-1} \exp\big[-\lambda_k x_k^2 \big] dx_k \\
&= \int_{\mathbb{R}^d} \Big(\prod_{k=1}^{d} |x_k| \Big)^{2p-1} \exp\big[-x \cdot Ax \big] dx \\
&\le d^{-d(2p-1)/2} \int_{\mathbb{R}^d} |x|^{(2p-1)d} \exp\big[-x \cdot Ax \big] dx .
\end{aligned}$$

Hint (Exercise 13.5.5) *First, note that Itô formula leads us to*

$$\begin{aligned}
e^{-xL_t} = 1 - \int_0^t \int_{\bar{\Pi}_{\bar{\varepsilon}}} \int_0^1 \int_0^{r_0^\gamma} & \mathbf{1}_{\bar{\mathscr{A}}}(\theta) e^{-xL_{s-}} \big(1 - e^{-x|\varphi(\theta)|^4|\theta|^2}\big) \\
& \times \mathbf{1}_{[r_0, +\infty)}\big(|\bar{V}_{s-} - v_s(\rho)|\big)\, d\mathscr{N} .
\end{aligned}$$

Take expectations in the above expression, then using the fact that $\bar{\mathscr{A}}$ is either $[\varepsilon^a \pi, \frac{\pi}{2}]$ or $[-\frac{\pi}{2}, -\varepsilon^a \pi]$ and the symmetry of the integrand with respect to θ, we obtain

$$\begin{aligned}
\mathbb{E}\big[e^{-xL_t}\big] &= 1 - r_0^\gamma \int_0^t \int_{\bar{\Pi}_{\bar{\varepsilon}}} \int_0^1 \mathbb{E}\Big[e^{-xL_s} \big(1 - e^{-x|\varphi(\theta)|^4|\theta|^2}\big) \mathbf{1}_{\bar{\mathscr{A}}}(\theta) \\
&\qquad \times \mathbf{1}_{[r_0, +\infty)}\big(|\bar{V}_{s-} - v_s(\rho)|\big) \Big] ds\, b(\theta) d\theta\, d\rho \\
&= 1 - r_0^\gamma \int_0^t \int_{\bar{\Pi}_{\bar{\varepsilon}}} \int_{\mathbb{R}^2} \mathbb{E}\Big[e^{-xL_s} \big(1 - e^{-x|\varphi(\theta)|^4|\theta|^2}\big) \\
&\qquad \times \mathbf{1}_{[r_0, +\infty)}\big(|\bar{V}_{s-} - v|\big) \Big] ds\, b(\theta) d\theta\, f_s(dv) \\
&\le 1 - r_0^\gamma q_0 \int_0^t \int_{\bar{\Pi}_{\bar{\varepsilon}}} \big(1 - e^{-x|\varphi(\theta)|^4|\theta|^2}\big) \mathbb{E}\big[e^{-xL_s}\big] ds\, b(\theta) d\theta \\
&\le 1 - r_0^\gamma q_0 \int_0^t \int_{\bar{\Pi}'_{\bar{\varepsilon}}} \big(1 - e^{-cx|\theta|^{4\beta+2}}\big) \mathbb{E}\big[e^{-xL_s}\big] ds\, b(\theta) d\theta ,
\end{aligned}$$

where $q_0 > 0$ is the constant given in Lemma 13.5.4. Note that in the last inequality we have also used the fact that given the conditions on the function φ in (13.14) this function satisfies $|\varphi(\theta)| \ge c|\theta|^\beta$ for $\theta \in \bar{\Pi}'_{\bar{\varepsilon}}$ for some $\varepsilon' > 0$. Then, we see that

$$\mathbb{E}\left[e^{-xL_t}\right] \leq \exp\left\{-tr_0^{\gamma}\, q_0 \int_{\bar{\Pi}'_{\bar{\varepsilon}}} \left(1 - e^{-cx|\theta|^{4\beta+2}}\right) b(\theta)d\theta\right\}.$$

Hint (Exercise 13.7.2) *Iterative application of Theorem 13.4.13 enables us to see that*

$$\begin{aligned}
\mathbb{E}\left[(\partial_1^{p_1}\partial_2^{p_2}F)(X_t^{\varepsilon}+\tilde{V}_t)\,\tilde{G}_t\right] &= \mathbb{E}\left[\left(\partial(\partial_1^{p_1-1}\partial_2^{p_2}F)\right)(X_t^{\varepsilon}+\tilde{V}_t)\,(1,0)^*\,\tilde{G}_t\right] \\
&= \mathbb{E}\left[(\partial_1^{p_1-1}\partial_2^{p_2}F)(X_t^{\varepsilon}+\tilde{V}_t)\,\tilde{\mathscr{L}}_{1,0,t}(\tilde{G})\right] \\
&= \cdots \\
&= \mathbb{E}\left[(\partial_2^{p_2}F)(X_t^{\varepsilon}+\tilde{V}_t)\,\tilde{\mathscr{L}}_{p_1,0,t}(\tilde{G})\right] \\
&= \mathbb{E}\left[(\partial_2^{p_2-1}F)(X_t^{\varepsilon}+\tilde{V}_t)\,\tilde{\mathscr{L}}_{p_1,1,t}(\tilde{G})\right] \\
&= \cdots \\
&= \mathbb{E}\left[F(X_t^{\varepsilon}+\tilde{V}_t)\,\tilde{\mathscr{L}}_{p_1,p_2,t}(\tilde{G})\right].
\end{aligned}$$

Hint (Exercise 13.7.3) *(ii) We will give the idea in the case of random variables X and Y adapted to $\mathscr{F}_t$. In fact, by the product rule one has $D(XY) = XDY + XDY$, then by triangle (Minkowski) inequality one obtains*

$$\|D(XY)\|_{p,q} \leq \mathbb{E}\left[|X|^p\left(\sum_{j=1}^{N_t^{\varepsilon}}|D_jY|^q\right)^{p/q}\right]^{1/p} + \mathbb{E}\left[|Y|^p\left(\sum_{j=1}^{N_t^{\varepsilon}}|D_jX|^q\right)^{p/q}\right]^{1/p}.$$

Therefore the result follows from Hölder's inequality.

(iii)

$$D_{j_1}\ldots D_{j_l}(XY) = \sum_{n=0}^{l}\sum_{r} D_{r_1}\ldots D_{r_n}X\,D_{r_{n+1}}\ldots D_{r_l}Y.$$

Here the second sum is taken over all recombinations $r = (r_1, \ldots, r_l)$ of $(j_1, \ldots, j_l)$. Therefore by Hölder's inequality, we have

$$\begin{aligned}
\prod_{i=1}^{l}\varphi(\tilde{\theta}_{j_i})^q\left\|D_{j_1}\ldots D_{j_l}(XY)\right\|^q \leq &\{(l+1)!\}^{q-1}\sum_{n=0}^{l}\sum_{r}\prod_{i=1}^{n}\varphi(\tilde{\theta}_{r_i})^q\left\|D_{r_1}\ldots D_{r_n}X\right\|^q \\
&\times\prod_{i=n+1}^{l}\varphi(\tilde{\theta}_{r_i})^q\left\|D_{r_{n+1}}\ldots D_{r_l}Y\right\|^q.
\end{aligned}$$

In a similar fashion one obtains that

$$\left(\sum_{j_1\le\ldots\le j_l\le N_t^\varepsilon}\prod_{i=1}^{l}\varphi(\tilde{\theta}_{j_i})^q\|D_{j_1}\ldots D_{j_l}(XY)\|^q\right)^{p/q}$$

$$\le C_m\left(\sum_{l=0}^{m}\left(\sum_{j_1\le\ldots\le j_l\le N_t^\varepsilon}\prod_{i=1}^{l}\varphi(\tilde{\theta}_{j_i})^q\|D_{j_1}\ldots D_{j_l}X\|^q\right)^k\right)^{p/(qk)}$$

$$\times\left(\sum_{l=0}^{m}\left(\sum_{j_1\le\ldots\le j_l\le N_t^\varepsilon}\prod_{i=1}^{l}\varphi(\tilde{\theta}_{j_i})^q\|D_{j_1}\ldots D_{j_l}Y\|^q\right)^{k'}\right)^{p/(qk')}.$$

Therefore the result follows from summing over l on the left-hand side and using Hölder's inequality with the appropriate choice of k and k'.

(v) What is important is to understand that in the proof of (iii) it was important that the terms being considered on the left-hand side are always included on the right. The same idea solves this exercise.

(vii) If one uses Exercise 13.4.10 together with Exercise 13.4.8-(ii), one sees that if one takes $\beta = m+1$ and $r = m$ will lead to the estimation of sums of the type $\sum_{j=1}\left|\tilde{\theta}_j\right|^p$ where $p \ge 1$. Therefore the bounds will follow from Lemma 13.6.5. This technique applies to many of the other estimates.

(x) In order to prove this estimate, one has first to use (iii) in order to realize that it is enough to bound $\|\check{C}_t\|_{p_1,q_1,m}$ and $\|\det(\tilde{C}_t)^{-1}\|_{p_2,q_2,m}$. But the latter is bounded by $C\|\det(\tilde{C}_t)^{-1}\|_{p_3}$ and a polynomial function of $\|\tilde{C}_t\|_{p_4,q_4,m}$. From here the estimate follows using (vii) and Lemma 13.5.2.

Appendix A
Afterword

In this book we have tried to present a guide for students who would like to study a particular aspect of jump processes which arise through stochastic equations. In particular, the second part of the book gives hints at how to obtain integration by parts (IBP) formulas which will lead to the study of densities of random variables that arise in situations where jump processes are involved.

We have not tried to obtain general results as we believe that the value for the student is in the method rather than a lengthy detailed proof of a specific result which the reader can find in the references.

Although the book is not thorough in the many topics that abound the literature of jump processes, we hope that this effort will open the door to young students.

We also reinforce the statement that the second part of the book is merely an overview and neither a detailed nor systematic introduction to the subject. For this, we recommend the reader to look at the books in the references.

There is also a large amount of recent literature hinting at the possibility of using approximations in clever ways in order to obtain the regularity properties studied in the second part of the book. For three examples of this, we cite [4, 5, 37]. On the specific subject of the Boltzmann equation, we suggest the reader to look at [24] for an alternative method using Lemma 9.1.14.

© Springer Nature Singapore Pte Ltd. 2019

A. Kohatsu-Higa and A. Takeuchi, *Jump SDEs and the Study of Their Densities*,
Universitext, https://doi.org/10.1007/978-981-32-9741-8

References

1. Alexandre, R.: A review of Boltzmann equation with singular kernels. Kinet. Relat. Models **2**(4), 551–646 (2009). https://doi.org/10.3934/krm.2009.2.551
2. Applebaum, D.: Lévy Processes and Stochastic Calculus. Cambridge Studies in Advanced Mathematics, vol. 116, 2nd edn. Cambridge University Press, Cambridge (2009). https://doi.org/10.1017/CBO9780511809781
3. Asmussen, S., Rosiński, J.: Approximations of small jumps of Lévy processes with a view towards simulation. J. Appl. Probab. **38**(2), 482–493 (2001)
4. Bally, V., Caramellino, L.: Integration by parts formulas, Malliavin calculus, and regularity of probability laws. In: Stochastic Integration by Parts and Functional Itô Calculus. Advanced Courses in Mathematics - CRM Barcelona, pp. 1–114. Birkhäuser/Springer, Berlin (2016)
5. Bally, V., Caramellino, L.: Convergence and regularity of probability laws by using an interpolation method. Ann. Probab. **45**(2), 1110–1159 (2017). https://doi.org/10.1214/15-AOP1082
6. Bally, V., Clément, E.: Integration by parts formula and applications to equations with jumps. Probab. Theory Relat. Fields **151**(3–4), 613–657 (2011). https://doi.org/10.1007/s00440-010-0310-y
7. Bally, V., Fournier, N.: Regularization properties of the 2D homogeneous Boltzmann equation without cutoff. Probab. Theory Relat. Fields **151**(3–4), 659–704 (2011). https://doi.org/10.1007/s00440-010-0311-x
8. Bertoin, J.: Lévy Processes, Cambridge Tracts in Mathematics, vol. 121. Cambridge University Press, Cambridge (1996)
9. Bichteler, K., Gravereaux, J.B., Jacod, J.: Malliavin Calculus for Processes with Jumps. Stochastics Monographs, vol. 2. Gordon and Breach Science Publishers, New York (1987)
10. Billingsley, P.: Probability and Measure, 3rd edn. Wiley, New York (1995)
11. Billingsley, P.: Convergence of Probability Measures, 2nd edn. Wiley Series in Probability and Statistics: Probability and Statistics. Wiley (A Wiley-Interscience Publication), New York (1999). https://doi.org/10.1002/9780470316962
12. Bismut, J.M.: Calcul des variations stochastique et processus de sauts. Z. Wahrsch. Verw. Gebiete **63**(2), 147–235 (1983). https://doi.org/10.1007/BF00538963
13. Bouleau, N., Denis, L.: Energy image density property and the lent particle method for Poisson measures. J. Funct. Anal. **257**(4), 1144–1174 (2009). https://doi.org/10.1016/j.jfa.2009.03.004
14. Bouleau, N., Denis, L.: Application of the lent particle method to Poisson-driven SDEs. Probab. Theory Relat. Fields **151**(3–4), 403–433 (2011). https://doi.org/10.1007/s00440-010-0303-x
15. Bouleau, N., Denis, L.: Dirichlet forms Methods for Poisson Point Measures and Lévy Processes. Probability Theory and Stochastic Modelling, vol. 76. Springer, Cham (2015). https://doi.org/10.1007/978-3-319-25820-1. With emphasis on the creation-annihilation techniques

© Springer Nature Singapore Pte Ltd. 2019

A. Kohatsu-Higa and A. Takeuchi, *Jump SDEs and the Study of Their Densities*, Universitext, https://doi.org/10.1007/978-981-32-9741-8

16. Brémaud, P., Massoulié, L.: Stability of nonlinear Hawkes processes. Ann. Probab. **24**(3), 1563–1588 (1996). https://doi.org/10.1214/aop/1065725193
17. Carr, P., Geman, H., Madan, B.D., Yor, M.: The fine structure of asset returns: an empirical investigation. J. Business **75**, 303–325 (2002)
18. Cortissoz, J.: On the Skorokhod representation theorem. Proc. Am. Math. Soc. **135**(12), 3995–4007 (electronic) (2007). https://doi.org/10.1090/S0002-9939-07-08922-8
19. Davydov, Y.A., Lifshits, M.A., Smorodina, N.V.: Local Properties of Distributions of Stochastic Functionals. In: Nazaĭkinskiĭ, V.E., Shishkova, M.A. (eds.) Mathematical Monographs (Translated from the 1995 Russian original), vol. 173. American Mathematical Society, Providence, RI (1998)
20. Debussche, A., Fournier, N.: Existence of densities for stable-like driven SDEs with Hölder continuous coefficients. J. Funct. Anal. **264**(8), 1757–1778 (2013). https://doi.org/10.1016/j.jfa.2013.01.009
21. Debussche, A., Romito, M.: Existence of densities for the 3d Navier-Stokes equations driven by gaussian noise. Probab. Theory Relat. Fields **158**(3–4), 575–596 (2014). https://doi.org/10.1007/s00440-013-0490-3
22. Denis, L., Nguyen, T.M.: Malliavin calculus for markov chains using perturbations of time. Stochastics **88**(6), 813–840 (2016). https://doi.org/10.1080/17442508.2016.1148150
23. Embrechts, P., Liniger, T., Lin, L.: Multivariate Hawkes processes: an application to financial data. J. Appl. Probab. **48A**, 367–378 (2011). https://doi.org/10.1239/jap/1318940477
24. Fournier, N.: Finiteness of entropy for the homogeneous Boltzmann equation with measure initial condition. Ann. Appl. Probab. **25**(2), 860–897 (2015). https://doi.org/10.1214/14-AAP1012
25. Fournier, N., Mouhot, C.: On the well-posedness of the spatially homogeneous Boltzmann equation with a moderate angular singularity. Comm. Math. Phys. **289**(3), 803–824 (2009). https://doi.org/10.1007/s00220-009-0807-3
26. Friedman, A.: Stochastic Differential Equations and Applications. Dover Publications, Inc., Mineola, NY (2006). Two volumes bound as one, Reprint of the 1975 and 1976 original published in two volumes
27. Fujiwara, T., Kunita, H.: Stochastic differential equations of jump type and Lévy processes in diffeomorphisms group. J. Math. Kyoto Univ. **25**(1), 71–106 (1985)
28. Hawkes, A.G.: Spectra of some self-exciting and mutually exciting point processes. Biometrika **58**(1), 83–90 (1971)
29. Hawkes, A.G., Oakes, D.: A cluster process representation of a self-exciting process. J. Appl. Probab. **11**(3), 493–503 (1974)
30. Ikeda, N., Watanabe, S.: Stochastic Differential Equations and Diffusion Processes. North-Holland Mathematical Library, vol. 24, 2nd edn. North-Holland Publishing Co., Amsterdam; Kodansha, Ltd., Tokyo (1989)
31. Ishikawa, Y.: Stochastic Calculus of Variations for Jump Processes. De Gruyter Studies in Mathematics, vol. 54. De Gruyter, Berlin (2013). https://doi.org/10.1515/9783110282009
32. Jaisson, T., Rosenbaum, M.: Limit theorems for nearly unstable Hawkes processes. Ann. Appl. Probab. **25**(2), 600–631 (2015). https://doi.org/10.1214/14-AAP1005
33. Kahane, J., Salem, R.: Ensembles parfaits et séries trigonométriques. Actualités scientifiques et industrielles, Hermann (1963)
34. Karatzas, I., Shreve, S.E.: Brownian Motion and Stochastic Calculus. Graduate Texts in Mathematics, vol. 113, 2nd edn. Springer, New York (1991). https://doi.org/10.1007/978-1-4612-0949-2
35. Kawai, R., Kohatsu-Higa, A.: Computation of Greeks and multidimensional density estimation for asset price models with time-changed Brownian motion. Appl. Math. Finance **17**(4), 301–321 (2010). https://doi.org/10.1080/13504860903336429
36. Kawai, R., Takeuchi, A.: Computation of Greeks for asset price dynamics driven by stable and tempered stable processes. Quant. Finance **13**(8), 1303–1316 (2013). https://doi.org/10.1080/14697688.2011.589403

37. Knopova, V., Kulik, A.: Parametrix construction of the transition probability density of the solution to an sde driven by α-stable noise. Ann. Inst. H. Poincaré Probab. Statist. **54**(1), 100–140 (2018)
38. Komatsu, T., Takeuchi, A.: On the smoothness of PDF of solutions to SDE of jump type. Int. J. Differ. Equ. Appl. **2**(2), 141–197 (2001)
39. Kunita, H.: Representation of martingales with jumps and applications to mathematical finance. In: Stochastic Analysis and Related Topics in Kyoto. Advanced Studies in Pure Mathematics, vol. 41, pp. 209–232. Mathematical Society of Japan, Tokyo (2004)
40. Kunita, H., Watanabe, S.: On square integrable martingales. Nagoya Math. J. **30**, 209–245 (1967)
41. Kyprianou, A.E.: Introductory lectures on fluctuations of Lévy processes with applications. Universitext. Springer, Berlin (2006)
42. Léandre, R.: Applications of the Malliavin calculus of Bismut type without probability. WSEAS Trans. Math. **5**(11), 1205–1210 (2006)
43. Malliavin, P.: Stochastic Analysis. Grundlehren der Mathematischen Wissenschaften [Fundamental Principles of Mathematical Sciences], vol. 313. Springer, Berlin (1997). https://doi.org/10.1007/978-3-642-15074-6
44. Malliavin, P., Thalmaier, A.: Stochastic Calculus of Variations in Mathematical Finance. Springer Finance. Springer, Berlin (2006)
45. Norris, J.R.: Integration by Parts for Jump Processes. In: J. Azéma, M. Yor, P.A. Meyer (eds.) Séminaire de Probabilités XXII, Lecture Notes in Mathematics, vol. 1321, pp. 271–315. Springer, Berlin/Heidelberg (1988). https://doi.org/10.1007/bfb0084144
46. Nualart, D.: The Malliavin Calculus and Related Topics. Probability and Its Applications, 2nd edn. Springer, Berlin (2006)
47. Picard, J.: On the existence of smooth densities for jump processes. Probab. Theory Relat. Fields **105**(4), 481–511 (1996). https://doi.org/10.1007/BF01191910
48. Protter, P.E.: Stochastic Integration and Differential Equations. Stochastic Modelling and Applied Probability, vol. 21. Springer, Berlin (2005). https://doi.org/10.1007/978-3-662-10061-5. 2nd edn. Version 2.1, Corrected third printing
49. Revuz, D., Yor, M.: Continuous Martingales and Brownian Motion. Grundlehren der Mathematischen Wissenschaften [Fundamental Principles of Mathematical Sciences], vol. 293, 3rd edn. Springer, Berlin (1999). https://doi.org/10.1007/978-3-662-06400-9
50. Rudin, W.: Real and Complex Analysis, 3rd edn. McGraw-Hill Book Co., New York (1987)
51. Sato, K.i.: Lévy Processes and Infinitely Divisible Distributions. Cambridge Studies in Advanced Mathematics, vol. 68. Cambridge University Press, Cambridge (1999). Translated from the 1990 Japanese original, Revised by the author
52. Steutel, F.W., van Harn, K.: Infinite Divisibility of Probability Distributions on the Real Line. Monographs and Textbooks in Pure and Applied Mathematics, vol. 259. Marcel Dekker Inc, New York (2004)
53. Stroock, D.W., Varadhan, S.R.S.: Multidimensional Diffusion Processes. Classics in Mathematics. Springer, Berlin (2006). Reprint of the 1997 edition
54. Takeuchi, A.: Bismut-Elworthy-Li-type formulae for stochastic differential equations with jumps. J. Theoret. Probab. **23**(2), 576–604 (2010). https://doi.org/10.1007/s10959-010-0280-0
55. Tanaka, H.: Probabilistic treatment of the Boltzmann equation of Maxwellian molecules. Z. Wahrsch. Verw. Gebiete **46**(1), 67–105 (1978/79). https://doi.org/10.1007/BF00535689
56. Triebel, H.: Theory of function spaces. No. v. 1 in Monographs in mathematics. Birkhäuser (1983)
57. Villani, C.: A review of mathematical topics in collisional kinetic theory. In: Handbook of mathematical fluid dynamics, vol. I, pp. 71–305. North-Holland, Amsterdam (2002). https://doi.org/10.1016/S1874-5792(02)80004-0
58. Wagner, W.: Unbiased Monte Carlo estimators for functionals of weak solutions of stochastic differential equations. Stoch. Stoch. Rep. **28**(1), 1–20 (1989). https://doi.org/10.1080/17442508908833581

59. Weron, R.: On the Chambers-Mallows-Stuck method for simulating skewed stable random variables. Statist. Probab. Lett. **28**(2), 165–171 (1996)
60. Williams, D.: Probability with Martingales. Cambridge Mathematical Textbooks. Cambridge University Press, Cambridge (1991). https://doi.org/10.1017/CBO9780511813658
61. Zolotarev, V.: One-Dimensional Stable Distributions. Translations of Mathematical Monographs. American Mathematical Society, Providence (1986)

Index

© Springer Nature Singapore Pte Ltd. 2019
A. Kohatsu-Higa and A. Takeuchi, *Jump SDEs and the Study of Their Densities*, Universitext, https://doi.org/10.1007/978-981-32-9741-8

Printed in the United States
By Bookmasters